Y0-BZB-295

DATE DUE

DISCARD

DEMCO 38-296

Alternative Water Sources and Wastewater Management

About the Author

E. W. "Bob" Boulware is a registered professional engineer and President of Design-Aire Engineering, a Consulting Engineering firm located in Indianapolis, Indiana. He is a Life Member of the American Society of Heating Refrigeration and Air Conditioning Engineers (ASHRAE.org), is past president of the Central Indiana Chapter of American Society of Plumbing Engineers (ASPE.org) and serves on the National Technical Standards Committee. Mr. Boulware is principal author of the ARCSA/ASPE Rainwater Catchment Design and Installations Standard which is being considered as an ANSI National Standard. He is past national president of the American Rainwater Catchment Systems Association (ARCSA.org) and an Accredited Rainwater Systems Design Professional. He is a member of the International Association of Plumbing and Mechanical Officials' (IAPMO.org) Green Technology Committee, where he contributed to the development of the rainwater catchment and graywater systems standards included in the 2012 Green Plumbing Supplement and the Uniform Plumbing Code. Mr. Boulware is also on the International Rain Harvesting Alliance (IRHA-h20.org) advisory committee in Geneva, Switzerland.

Alternative Water Sources and Wastewater Management

E. W. Bob Boulware

New York   Chicago   San Francisco
Lisbon   London   Madrid   Mexico City
Milan   New Delhi   San Juan
Seoul   Singapore   Sydney   Toronto

The McGraw·Hill Companies

Cataloging-in-Publication Data is on file with the Library of Congress.

McGraw-Hill books are available at special quantity discounts to use as premiums and sales promotions, or for use in corporate training programs. To contact a representative please e-mail us at bulksales@mcgraw-hill.com.

Alternative Water Sources and Wastewater Management

1 2 3 4 5 6 7 8 9 0   DOC/DOC   1 9 8 7 6 5 4 3 2

ISBN 978-0-07-171951-3
MHID 0-07-171951-2

The pages within this book were printed on acid-free paper.

Sponsoring Editor	Copy Editor	Composition
Larry S. Hager	Naomi R. Beesen	Cenveo Publisher Services
Acquisitions Coordinator	**Proofreader**	
Bridget Thoreson	Meera Abraham	**Art Director, Cover**
		Jeff Weeks
Editorial Supervisor	**Indexer**	
David E. Fogarty	Robert Swanson	
Project Manager	**Production Supervisor**	
Harsimran K. Tikka, Cenveo Publisher Services	Richard C. Ruzycka	

Contents

Preface xv

1 **Introduction** 1
   The Water Cycle 3
   How Much Water Is on Earth? 5
        Liquid Fresh Water 5
        Water in Lakes and Rivers 6
   Sources of Water 7
   Sanitation and Waste Management 9
   Sanitation and Water Pollution 9
   Sanitation and the Future 10
   Alternative Waste Management 12
   Conclusion 14
   Notes and References 14

2 **Water from Springs** 15
   What Is a Spring? 15
   Depression Springs 15
   Gravity Springs 15
   Spring Water Protection 16
   Benefits of Spring Development Structures 18
   Basic Design Features 18
   Types of Spring Box Construction 19
   Designing the Water Distribution System 21
   Tank Construction 22
   Preparing the Site 22
   Constructing the Spring Box 25
   Installation 27
   Seep Collection System 28
   Horizontal Well (Horizontal Bore Hole) 28
   Maintenance 30
   Conclusion 31
   Notes and References 32

3 **Air Conditioning Condensate Recovery** 33
   Introduction 33
   Estimation of Condensate Volume 34
        Calculation Procedure 34
   Application Examples 38

Conclusion 40
Notes and References 41

4   **Harvesting Water from Dew** **43**
What Is Dew and How Does It Occur? 43
Historical Applications 44
Typical Dew Collection Installation 46
Dew Collection Research 47
Roof Mounted Systems 48
Ground Installed Dew Harvesting 51
Other Applications 54
Conclusions and Recommendations 54
Notes and References 55

5   **Water Gathered from Clouds** **57**
What Is Fog? 57
Suitability 58
Site Selection 59
Satellite Imaging 60
Technical Description of a Fog Collector 61
Wind and Site 63
Water Storage 64
        Step 1: Estimating Water Demand 65
        Step 2: Determining the Maximum Water
            That Can Be Collected 65
        Step 3: Calculating the Size of the Tank 65
Water Purification 66
Operation and Maintenance 66
Advantages of a Fog Harvesting System 67
Disadvantages of a Fog Harvesting
    System 68
The Need for Community Support 69
Costs 69
Extent of Use 70
Conclusion 71
Notes and References 73

6   **Glacier Water Reclaim** **75**
Introduction 75
Artificial Glaciers 76
Glacier Grafting 77
Glacier Impoundment 81
Construction 82
Conclusion 86
Notes and References 88

7   **Harvesting the Rain** **89**
    Introduction   89
    Estimating Demand   90
        Step 1: Estimating Rainwater Supply   93
        Step 2: Sizing the Collection System   93
        Step 3: Sizing the Storage   94
        Step 4: Sizing the Water Storage
            Tank (Cistern)   94
        Step 5: System Adjustment   95
    Acceptable Piping Schematics for Potable and
        Non-Potable Water   95
        Underground Exterior Cistern for Potable
            Application   95
        Non-Potable Water   95
        Freezing Environments   99
    Installation Details to Consider   99
        Tank Overflow   99
        Water Pickup   99
    Maintaining Water Quality   101
        Roof Washer   102
    Conclusion   104
    Notes and References   105
    Appendix Average Rainfall Data   106

8   **Solar Water Distillation** **119**
    Introduction   119
    How a Simple Solar Still Operates   119
    Would a Solar Still Suit Your Needs?   123
    Energy Requirements for Water Distillation and
        Performance of Solar Stills   124
    Design Objectives for an Efficient Solar Still   127
    Which Solar Still?   127
    Material Requirements of Basin Stills   130
    Basic Components   130
        The Basin   130
        Support Structures   133
        Glazing Cover   134
        Distillate Trough   137
    Ancillary Components   138
        Insulation   138
        Sealants   138
        Piping   139
        Still Operation   140
        Storage Reservoir   140

Conclusion   141
Notes and References   143
Appendix   143

9 **Graywater Systems** **149**
Introduction   149
Discussion of Graywater Systems   149
Graywater System Types and Description   151
Design Procedure   152
    Step 1: Establishing Applications and
        Sources for Graywater   154
    Step 2: Quantifying the Amount of
        Graywater That Can Be Used   154
The Simplified Approach   154
The Detailed Approach   155
    Estimating Graywater Yield Metric
        Analysis (See Appendix for English
        Units Conversion)   155
    Estimating Graywater Demand
        (Interior Building Applications)   156
Graywater Demand (External Application, e.g.,
    Garden Watering)   157
Storage Capacity   158
Combined Rainwater/Graywater Systems   158
Collection   159
Storage Tanks and Cisterns   159
Aboveground Tanks and Cisterns   160
Belowground Tanks   160
Treatment   160
Materials and Fittings   161
Backup Water Supply and Backflow Prevention . . .   161
Pumping   161
Overflow, Bypass, and Drainage   162
System Controls   162
Piping and System Installation   163
    Piping   163
    Cistern Installation   163
Water Quality   164
Maintenance   164
Risk Assessment   167
Conclusion   168
Notes and References   168

10 **Managing Water Quality** **169**
Sand Filters   170

Filter Operation 171
How It Works 172
Advantages 172
Disinfection Methods 173
    Chlorine Tablets 175
    Iodine 176
    Nano Silver 176
    Ultraviolet Light 179
    Ozone 179
Testing for Water Quality 180
Conclusion 181
Notes and References 181

11 **Artificial Groundwater Recharge** **183**
Introduction 183
Historical Applications 186
Factors That Affect Infiltration 187
    Precipitation 187
    Soil Characteristics 187
    Soil Saturation 188
    Land Cover 188
    Slope of the Land 188
    Evapo-Transpiration 188
Subsurface Water 189
How Infiltration Replenishes
   Aquifers Naturally 189
How Groundwater Moves within the Aquifer 189
Effects of Water Withdrawal 190
Groundwater Flow 192
Artificial Recharge Techniques and Designs 193
Classification of Recharging Methods 193
    Sources and Mechanisms of Recharge 193
    Direct Recharge 194
    Surface Spreading Techniques 195
    Wells 198
    Indirect Recharge 201
Recharge for Domestic Supplies 203
Cost 204
Maintenance 204
Conclusion 204
Notes and References 205

12 **Aquatic Plants as a Waste Management System** **207**
Introduction 207
Conventional versus Aquatic-Plant Treatment 207

Plants Used for Wastewater Treatment 208
Water Hyacinth Use 209
Design Parameters for a Water Hyacinth
  Wastewater Treatment Plant 212
    Wastewater Characteristics 212
    Location, Climate, and Temperature 212
    Site Requirements 213
    Land Requirements 213
    Pre-Treatment 213
    Post-Treatment 214
    Pond Size, Number, and Configuration 214
    Organic and Surface Loading Rates 214
    Hydraulic Residence (Detention) Time 215
    Hydraulic Loading Rates 215
Duckweed Wastewater Treatment 216
    Mineral Concentrations 217
    Water Depth 218
    Wastewater Treatment 219
    Example 220
    Recovery of Biomass 220
Natural Treatment for Sewage Treatment
  Facility 222
    Introduction 222
    Benefits of Using a Constructed Wetland ... 224
    Discharge Water Quality Requirements 224
    Climatic Conditions Are a Challenge
      in Alaska 225
    Constructed Wetland Design Overview 225
    Plant Selection 226
    Site Preparation for the Constructed
      Wetland 227
    Plants and the Planting Process 227
    Project Timeline 228
    Water Quality Results for the First
      Operating Season 229
    Problem Areas 229
    Lessons Learned 230
Conclusion 231
Notes and References 231

13 Biological Filter and Constructed
    Wetland Systems 233
    Introduction 233
    Discussion 233
    How This Happens 234

Filter Types 234
    Trickling Filter 234
    Trickling Filter Design 235
    Trickling Filter Performance 238
    Trickling Filter Advantages and Limitations 238
    Conclusion 239
Constructed Wetlands 239
    Plant Rock Filters 240
    Rock Filter Design Criteria 242
    Design Example #1 244
    Rock Filter Conclusion 246
Evapo-Transpiration Filter 246
    Evapo-Transpiration Design Criteria 249
    Evapo-Transpiration Design Parameters . . . 250
    Precipitation 251
    Evapo-Transpiration Rate 252
    Evapo-Transpiration Bed Design 253
    Drainage 254
    ET and ETI System Advantages
      and Limitations 255
    Limitations 255
    Conclusion 256
Recirculating Vertical Flow Constructed Wetlands . . . 257
    Who Should Consider RVF Constructed
      Wetlands? 257
    Design Criteria 257
    Construction 257
    Operation 260
    Maintenance Requirements 260
    Expected Performance 261
    Summary 261
Conclusion 261
Notes and References 264

14 Blackwater Recycling **267**
Introduction 267
Blackwater Recycling Systems 269
System Design 271
Establishing Water Balance 272
Fit-for-Purpose Water 275
Benefits of Blackwater Recycling 276
Limitations to Blackwater Recycling 277
Other Uses of Blackwater 278
Conclusion 280
Notes and References 280

15 Septic System Design **281**
Introduction 281
How Does a Septic System Work? 281
    Septic Tank 281
    Drainfield (Also Called a Leachfield) 283
    Soil 284
Operation and Maintenance 285
Anaerobic Baffled Reactor 285
Adequacy 286
Health Aspects/Acceptance 287
How to Run a Percolation Test 288
    Part 1: Site Preparation 288
    Part 2: Testing Procedure 289
    Part 3: What Can Go Wrong? 290
    Part 4: Test Results 290
    Part 5: Calculations and Sizing the
       Trench(es) 290
Conclusion 293
Notes and References 293
General Instructions 296
Soils Description 296
Percolation Test 296

16 Latrines and Privies **299**
Introduction 299
Latrines and Privies 299
Recommendations 299
Ventilated Improved Pit Latrine 300
Pour-Flush Latrine 302
Privy with Septic Tank 305
Composting and the Use of the End Product
    as Fertilizer 307
Utilization of Undiluted Urine 307
Maintenance 309
Conclusion 309
Notes and References 311

17 Composting Toilets **313**
Introduction 313
The Composting Process 315
Factors That Affect the Composting Process 316
    Environmental Factors 316
    Aeration 316
    Moisture 317
    Temperature 318

Carbon-to-Nitrogen (C:N) Ratio 319
Time 319
Process Control 321
Composting Toilets 321
Residential Composting Toilets 326
Advantages 330
Disadvantages 330
Operation and Maintenance 331
Design Standards 331
Costs 332
Conclusion 332
Notes and References 333

18 Net Zero Water **335**
Introduction 335
Discussion 335
A Net Zero Example 336
Conclusion 338
Notes and References 339

A1 Pipe Loss Data for Polyvinyl Pipe and Fittings . . . **341**

A2 Metric Conversion Factors **345**

A3 Equivalents of Pressure and Head **347**

Index **349**

Preface

Bonnie, my Welsh corgi, likes to herd turtles in our back yard. She sees a need, it seems important to her, and there is no one else doing it. My motivation to write this book is much the same. Changing weather patterns, coupled with increasing stress on our planet from a growing world population, are creating a need for alternative sources of water. There is also a need for better, less polluting waste disposal methods in order to preserve the water we now have available. This book is a summary of the work done by many individuals, far smarter than I, who saw a need and worked to address this need because no one else was doing it. It is my hope that this book will give their ideas wider visibility so they can be more broadly applied.

Individuals who should be recognized in this effort are as follows:

- Jack Shultz, an engineer from Santa Cruz, California, was active in green projects before they were popular. He was designing and selling solar energy systems in the late 1960s to mid-1970s. He also paddled down the Amazon River and sailed to Florida in a dugout canoe in his youth.

- Billy Kniffen, of the Texas A&M Agri-Life Department, has been tirelessly promoting rainwater catchment and xeriscape gardening in his native Texas and beyond. He also shared vital knowledge gained from living in his off-grid home.

- British architectural designer Matthew Parkes visited Africa while earning his master's at Oxford's Brookes University. There, he learned about the Namibian desert beetle, which uses its shell to collect moisture from the fog created when cool air off the Atlantic meets the warm landmass of the African coast.

- Dr. Daniel Beysens, who works with OPUR in Bordeaux, France, developed a coating that emulates the dew-gathering selective surface of the Namibian beetle. This can be applied to a thin metal foil to gather moisture from the air.

- Professor Girja Sharan, an engineer from India, saw dew collecting on his greenhouse and developed the concept into a water collection system so schools in India can have water during the dry season.

- Chewang Norphel, a retired government civil engineer with the Indian Department of Rural Development, saw how climate change affected the nearby glacier, causing a declining water supply. He developed artificial glaciers as a high altitude water conservation technique to preserve the water supply to a village in the Himalayan cold desert region of India.

And there are many other creators of alternative water sources and waste management solutions who saw a need that no one else was addressing.

E. W. Bob Boulware

CHAPTER 1

Introduction

Access to clean water can be considered one of the basic needs and rights of a human being. The health of people and a dignified life are based on access to clean water and proper sanitation. It can be said that together these two tenants are key to the health and economic prosperity of a civilized society.

According to WHO and UNICEF, 82 percent of the world's population has access to an adequate water supply (Fig. 1.1). The amount of proper sanitation has increased from 49 percent in 1990 to 58 percent in 2002. Still, approximately one out of six (1.1 billion people) are lacking access to an adequate water supply, and two out of five (2.6 billion people) are without access to proper sanitation services (Fig. 1.2). Most of these people live in areas of Asia, where as much as half of the population lack proper sanitation services, and in areas of Africa, two out of five do not have access to an adequate water supply. The situation is especially alarming in rural areas, where half of the people do not have access to proper sanitation and water supply services. In bigger cities, the problem is intense population growth and concentration on population centers. This will burden existent services in the decades to come.[1, 2]

In most parts of the world, freshwater is transported from another region or location by boat, train, truck, or pipeline. The source of this water can be from a spring, well, lake, or river. As demand for clean water increases and sources become more limited due to expanding pollution, alternative sources of water and waste management need to be found.

Freshwater supplies are unlikely to keep up with global demand by 2040, which will cause increasing political instability, hobbled economic growth, and endangered world food markets, according to a U.S. intelligence assessment.[3] This book gathers together technologies and ideas from around the world to illustrate alternative ideas for providing clean drinking water when local utility water or waste management systems are not available.

1

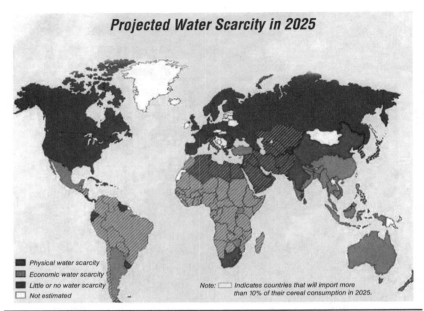

FIGURE 1.1   Water stress: 1.2 billion people—or almost 1 out of 5 people in the world—are without access to safe drinking water, and half of the world's population lacks adequate water purification systems. (See also Color Plates.)

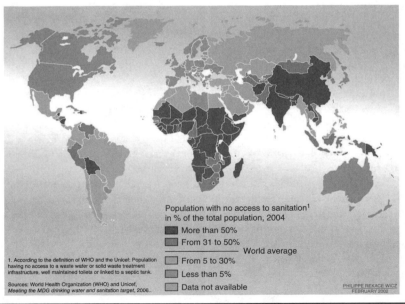

FIGURE 1.2   Sanitation stress: Compounding the shortage of water is that 40 percent of the world's population, 2.4 million people, does not have access to proper sanitation, which further decreases the availability of safe drinking water.

The Water Cycle

The water cycle describes the existence and movement of water on, in, and above the earth. Earth's water is always in movement and is always changing states, from liquid to vapor to ice and back again. The water cycle has been working for billions of years, and all life on earth depends on it continuing to work. In spite of water shortages that are being experienced worldwide, it is important to note that there is still the same amount of water on the planet as there was when dinosaurs roamed the earth.

All water is rainwater. Conventional wisdom in the Western world is to intercept the water after it has picked up debris and chemicals as surface runoff flowing across the ground, and then picked up more chemicals from our rivers and streams. Then it is filtered to take out the debris, and more chemicals are added to make it potable. Chapters 2 through 10 of this book explore other options of obtaining fresh water used successfully elsewhere in the world.

Of the water in the world water supply, 97 percent is salt water. Of the 3 percent that is freshwater, most is locked up in the ice caps, deep subsurface deposits, and in the atmosphere. Actual available water that is readily obtainable is less than 1 percent (Fig. 1.4). And with increasing pollution, this 1 percent is a diminishing asset. Sometimes there is too much water (floods), and sometimes there is too little (drought). Given the apparent problematic nature of water, it seems advisable to take a new look at our water use and where our water needs are to come from.

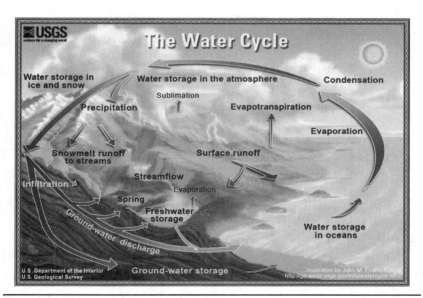

FIGURE 1.3   Water cycle. (See also Color Plates.)

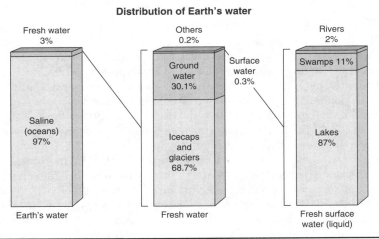

FIGURE 1.4   Distribution of earth's water.

It is instructive to note that, with the exception of food service buildings, very little water consumed in commercial buildings needs to be potable water (Fig. 1.5). For residences, potable water consumption is approximately 15 to 20 percent of the water consumed (Fig. 1.6). Clearly, opportunities exist for more efficient use and reuse of water provided to our facilities.[4]

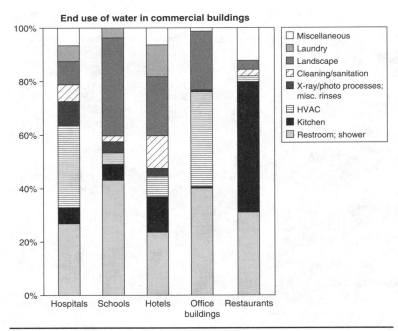

FIGURE 1.5   Water use profile for various buildings.

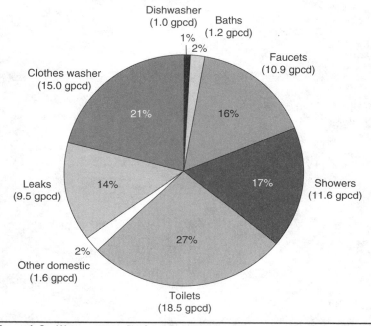

**Average indoor water use in a
non-conserving North American single-family home**

Dishwasher
(1.0 gpcd)   Baths
1%   (1.2 gpcd)
              2%

Faucets
(10.9 gpcd)

Clothes washer
(15.0 gpcd)

21%   16%

Leaks   14%   17%   Showers
(9.5 gpcd)   (11.6 gpcd)

27%

2%

Other domestic
(1.6 gpcd)

Toilets
(18.5 gpcd)

FIGURE 1.6   Water use profile for residential facilities.

Benjamin Franklin said, "You only know the value of water when the well runs dry." The world's well *is* running dry on many parts of the planet. Determining what alternatives and opportunities exist if you find your particular well running dry is the purpose of this book.

How Much Water Is on Earth?

The spheres in Fig. 1.7 represent all of Earth's water, Earth's liquid fresh water, and water in lakes and rivers. The largest sphere represents all of Earth's water, and its diameter is about 860 miles (the distance from Salt Lake City, Utah, to Topeka, Kansas). It would have a volume of about 332,500,000 cubic miles (mi^3) (1,386,000,000 cubic kilometers (km^3)). The sphere includes all the water in the oceans, ice caps, lakes, and rivers, as well as groundwater, atmospheric water, and even the water in you, your dog, and your tomato plant.

Liquid Fresh Water

How much of the total water is fresh water, which people and many other life forms need to survive? The blue sphere over Kentucky

Figure 1.7   How much water is on earth? (See also Color Plates.)

represents the world's liquid fresh water (groundwater, lakes, swamp water, and rivers). The volume comes to about 2,551,100 mi^3 (10,633,450 km^3), of which 99 percent is groundwater, much of which is not accessible to humans. The diameter of this sphere is about 169.5 miles (272.8 kilometers).

Water in Lakes and Rivers

And that "tiny" bubble over Atlanta, Georgia? That one represents fresh water in all the lakes and rivers on the planet, and most of the water people and life of earth need every day comes from these surface-water sources. The volume of this sphere is about 22,339 mi^3 (93,113 km^3). The diameter of this sphere is about 34.9 miles (56.2 kilometers). Yes, Lake Michigan looks way bigger than this sphere, but you have to try to imagine a bubble almost 35 miles high—whereas the average depth of Lake Michigan is less than 300 feet (91 meters).[3]

Sources of Water

Water sources are manifold. Many communities get their water from reservoirs. In 500 B.C.E., the Greeks supplemented local city wells with water supplied from the mountains as far as 10 miles away. In later times, the Romans built aqueducts that were many miles long— there were more than 200 that were still standing in the year 2001. Cities and other communities often provide for their water supply by allocating an open area that is pristine and protected as a watershed. The water was usually of high quality and free from chemical and microbial contamination.

The energy associated in producing clean water suitable for drinking is often overlooked. This can vary from between 1.6 kW/1000 gallons (3800 liters) for well water to 12 kW/1000 gallons for desalinization of seawater (Fig. 1.8). Conversely, it is illuminating to see the amount of water consumed in generating electric power (Fig. 1.9). Clearly, if we continue the existing way of managing our water, the overall result will be that the quality of our waters will further deteriorate and the availability will decline.

In certain parts of the world, water is supplied to communities from groundwater sources from deep wells, often many thousands of feet down. Water from these sources is usually free of chemical and microbial contamination. Groundwater is the main source of drinking water for almost half of the population in the United States. It is relatively inexpensive but is limited in volume and sometimes not readily replaceable if demand exceeds supply.[5] The Ogallala Aquifer,

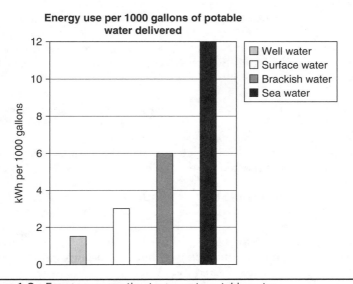

FIGURE 1.8   Energy consumption to generate potable water.

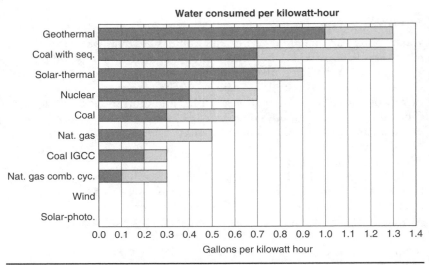

FIGURE 1.9   Amount of water consumed to generate electric power.

which is the sole source of water in the central western part of the United States, is said to have first been tapped in 1911 by a hand-dug well. Now an estimated 12 billion cubic meters of water is removed from the aquifer each year. Because it would take 6000 years to replace the water in the Ogallala Aquifer, it is expected that 6 percent of the aquifer will dry up every 25 years, leaving the residents and farmers of the Great Plains unsure about how long their water supply will last.[6]

While well water is usually free of solids and bacteria, as well as other chemical pollutants, it has often become contaminated by disposal of liquid waste, mining operations, and agricultural runoff. By providing protection to the source, either through buffers from the reservoirs or by protecting the well head for the deep wells, water is available without much treatment.

However, with increasing population and the increased use of water, there are less uncontaminated water supplies available. It is now commonplace that water must be treated prior to consumption. Disinfection is an important step in the water treatment process to destroy pathogenic bacteria and other harmful agents. Most water is treated with chlorine, as it is a very effective and economical method of treatment. An important advantage to using chlorine is that it has residual properties and continues to provide germ-killing potential as the water travels from the distribution point to the end users. There are concerns, however, about the formation of disinfection by-products from the reaction of the chlorine with humic substances in the water. These by-products are referred to as trihalomethanes, or THMs. The most common THM is chloroform, which is a carcinogen. In addition, some of the bacteria and viruses we want to treat are becoming resistant to traditional means of disinfection.

So it is apparent that the customary water supplies are becoming more and more inadequate, from both a quality and quantity point of view. To supplement the traditional sources of water, the following alternative ideas, which have been used successfully, will be presented for your consideration and evaluation:

- Springs
- Air conditioning condensate recovery
- Dew harvesting
- Fog harvesting
- Glacier water harvesting
- Rainwater catchment
- Solar water still
- Gray water systems

Sanitation and Waste Management

Sanitation includes the appropriate disposal of human and industrial wastes and the protection of the water sources. Waterborne agents are the cause of many diseases in the United States and elsewhere in the world.[7] These diseases may be caused by bacteria, viruses, and protozoans. Bacterial diseases include typhoid, shigellosis, and cholera. Viral agents cause diseases such as polio and hepatitis. Parasites include the protozoa *Entamoeba histolytica* and *Giardia lambdia*, which cause amebiasis and giardiasis, respectively. For the last decade, the primary agents in waterborne disease outbreaks in the United States have been the protozoal parasite *Giardia* and the bacteria *Shigella*. Another common agent is *Cryptosporidium*. Sanitation and water supply are interrelated issues that until now have been treated as separate issues. Looking forward, management of the entire water cycle will be needed.

Sanitation and Water Pollution

Sanitation is directly related to water quality and water pollution. The basic concept of collecting domestic liquid waste in waterborne sewer systems, treating the wastewater in centralized treatment plants, and discharging the effluent to surface water bodies became the accepted approach to sanitation in the last century. Although conventional sewer systems have significantly improved the public health situation for the communities that can afford to install and operate them, continued use adds pollution to our water supplies and reduces the purity of water needed for potable application later. This is becoming a worldwide problem, not only for the developing world with its inadequate sanitation, but also for the developed world with its aging infrastructures that cannot meet the needs of an increasing population.

The conventional sewer system was developed at a time, in regions, and under environmental conditions that made it an appropriate solution for removing liquid wastes from cities. Today, with increased populations, changes in consumer habits, and increasing pressure on freshwater and other resources, this human waste disposal system is no longer able to meet the pressing global needs. Thus, newer ideas need to be developed to address the issue of water quality conservation along with better means of waste disposal that do not pollute our water sources.

As serious as the disadvantages of water-based sanitation systems are, a far more fundamental problem is that they allow the nutrients in the wastewater to become a liability to the health of humanity rather than an asset. This lack of nutrient recovery leads to a linear flow of nutrients from agriculture, via humans, to recipient water bodies. The valuable nutrients and trace elements contained in human excrement are very rarely rechanneled back into agriculture in conventional systems. Even when sewage sludge is used in agriculture, only a very small fraction of the nutrients contained in the excrement are reintroduced into the living soil layer. Most are either destroyed in the treatment process (e.g., by nitrogen elimination) or enter the water cycle, where they pollute the environment and cause the eutrophication of lakes and rivers.

Not returning the nutrients to the soil has led to an increasing demand for chemical fertilizers in response to the problem of decreasing soil fertility. To produce the required chemical fertilizers, large amounts of energy are needed, and finite mineral resources, such as phosphorous, must be exploited. Farmers around the world yearly require 135 million tons of mineral fertilizer for their crops, while at the same time conventional sanitation dumps 50 million tons of fertilizer equivalents into our water bodies.

Sanitation and the Future

Our conventional drinking and wastewater systems are largely linear, end-of-pipe systems where drinking water is misused to transport waste into the water cycle, causing environmental damage and hygienic hazards and contributing to the water crisis. Figure 1.10 schematically illustrates the main limitations of conventional wastewater management systems, while Fig. 1.11 shows a more enlightened approach to water and waste management that preserves potable water for the vital processes that require it, while constructively managing the waste process to preserve the quality of the water resource.

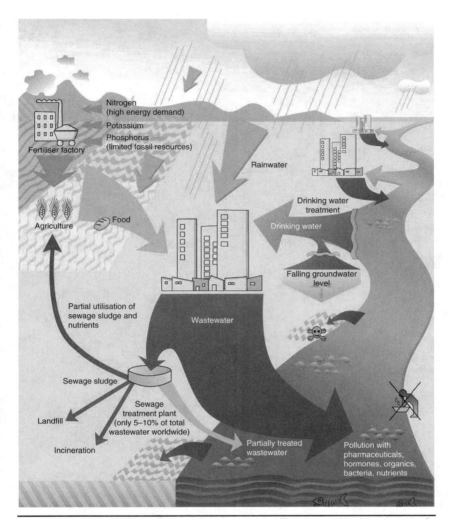

FIGURE 1.10 Existing water and wastewater stream: The conventional treatment of water requires chemicals and energy to generate potable water, then it adds contamination back into the water resource after use.

As a population increases, there is a greater demand for water. There is also increased pollution, which further depletes the available water. Because of water shortages and wastewater treatment plant overflows into waterways, the availability of clean water is decreasing. The new technologies that require chemically treated, high-pressure water to produce natural gas promise to further increase the stress on our water supplies. As the availability of clean, potable water resources becomes more limited, finding alternative sources for water and improved ways to manage our waste will become all-important.

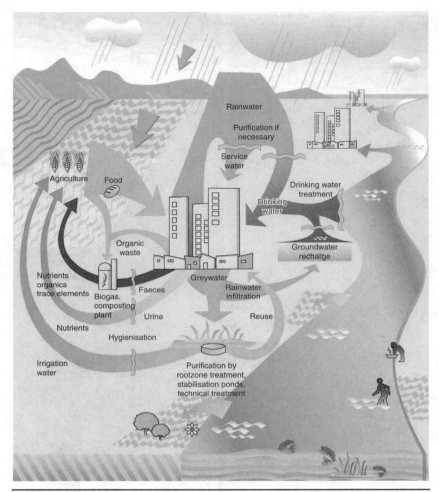

Figure 1.11 Optimal water and wastewater stream: A more enlightened water and waste management uses the waste constructively rather than contributing to the pollution of the water resource.

Alternative Waste Management

World Health Organization (WHO) defines *sanitation* as a group of methods to collect human excreta and urine, along with other community waste, in a hygienic way to maintain community health. Western plumbing systems are commonly water-based systems that use drinking-quality water to carry waste away to be addressed elsewhere. "Elsewhere" is presumed to be a septic field or a sewage treatment plant. Older cities have sewer systems that combine waste and storm water in the same pipe. When rainstorms occur that overwhelm the retention capacity of the sewage treatment plant, the combined storm water and sewage overflow into the local river or stream. This waterway might be someone's water source downstream. So an argument can be made that the typical western sewer system doesn't actually treat the problem as much as move the problem downstream to become someone else's problem.

Rural and suburban communities are confronted with problems that are unique to their size and population density, and are often unable to superimpose solutions typically applicable to larger urban areas. A good example of such a problem can be found in wastewater services.

In the past, priorities for water pollution control focused on the cities, since waste generation from these areas was most evident. In such high-density areas, the traditional sanitary engineering approach was to construct a network of sewers to convey wastewater to a central location for treatment and disposal to surface waters. Since a large number of users existed per unit length of sewer line, the costs of construction and operation could be divided among many people, thus keeping the financial burden on each user relatively low.

Within the past several decades, migration of the population from cities to suburban and rural areas has been significant. With this shift came the problems of providing utility services to the residents. Unfortunately, in many cases, solutions to wastewater problems in urban areas have been applied to rural communities. With the advent of federal programs that provide grants for construction of wastewater facilities, sewers and centralized treatment plants were constructed in these low-density rural settings. In many cases, the cost of operating and maintaining such facilities imposes severe economic burdens on the communities.

Although wastewater treatment and disposal systems serving single homes have been used for many years, they have often been considered an inadequate or temporary solution until sewers could be constructed. However, research has demonstrated that such systems, if constructed and maintained properly, can provide a reliable and efficient means of wastewater treatment and disposal at relatively low cost.

The following chapters provide technical information on alternative on-site wastewater treatment and disposal systems. They do not contain standards nor do they contain rules or regulations pertaining to on-site systems. The methods proposed herein aim to manage excreta and chemical waste in ways that serve to protect the water supply, and by so doing, reduce the causes of disease.[8] They also serve to address waste management at the source in an environmentally efficient manner without polluting the watershed.

The systems discussed include the following:

- Aquatic plants
- Constructed wetlands
- Biological filters
- On-site blackwater systems
- Septic fields
- Latrines and privies
- Composting toilets

Conclusion

The objective of this book is to illustrate alternative sources of water and means of waste disposal such that water quality will be preserved and the nutrients in our wastewater will be constructively reused. The presentation of innovative concepts being researched and developed from international sources is the topic being presented herein. The intended audience for this information includes those involved in the design, construction, operation, maintenance, and regulation of on-site wastewater systems and those that are concerned about preserving a vital resource for future generations.

Notes and References

1. *Global Water Supply and Sanitation Assessment 2000 Report*, WHO, UNICEF, 2000 (accessed May 2, 2005). *www.who.int/water_sanitation_health/monitoring/jmp2000.pdf*
2. *Meeting the Millennium Development Goals Drinking Water and Sanitation Target. A Mid-Term Assessment of Progress*, WHO, UNICEF (accessed June 15, 2005). *www.who.int/water_sanitation_health/monitoring/jmp04.pdf*
3. Quinn, A., "U.S. Intelligence Sees Global Water Conflict Risks Rising," Reuters, March 22, 2012.
4. Credit: Howard Perlman, USGS; globe illustration by Jack Cook, Woods Hole Oceanographic Institution (©); Adam Nieman. Data source: Igor Shiklomanov's chapter "World fresh water resources" in Peter H. Gleick (editor), 1993, Water in Crisis: A Guide to the World's Fresh Water Resources (Oxford University Press, New York).
5. The United States leads the world in water consumption at 150 gallons per person per day. Looking forward, this is not a sustainable number, given the projected water shortages for a significant part of the country. Therefore, the United States, along with the other parts of the world, should have a strong interest in better management of the available freshwater supply.
6. Prater, A. M., *Depletion of the Ogallala Aquifer*, April 23, 2010, www.helium.com/items/1812428-ogallala-aquifer-depletion.
7. *Gale Encyclopedia of Public Health: Sanitation*.
8. *Technology for Water Supply and Sanitation in Developing Countries*, WHO, Geneva, 1987.
9. *Injury and Illness Prevention Program, Appendix 1, Glossary of Terms*, San Diego State University (accessed April 17, 2005).

CHAPTER 2

Water from Springs

What Is a Spring?

A spring is a place where groundwater naturally seeps or gushes from the earth's surface, typically moving downhill through soils or through cracks and fissures in the bedrock, until the ground's surface intersects the water table. Springs fall under two general categories: depression springs and gravity or artesian springs.

- Depression springs occur when the land's surface dips below the level of the water table.

- Gravity springs, which include contact springs, fracture or tubular springs, and artesian springs.

Depression Springs

Yield from depression springs is highly variable, depending on the level of the water table. In areas that experience a pronounced dry season, depression springs may not be a suitable source of drinking water if the water table drops below the level of the depression, causing the spring to become seasonally dry.

Gravity Springs

Gravity contact springs occur when an impervious layer beneath the earth's surface restricts surface water infiltration. Water is channeled along the impervious layer until it eventually comes in contact with the earth's surface. This type of spring typically has a very high yield and makes a good source of drinking water (Fig. 2.1).

Artesian springs occur when water under pressure is trapped between two impervious layers. Because the water in these springs is under pressure, flow is generally greater than that of gravity springs. Artesian fissure springs are similar to fracture and tubular springs, in that water reaches the surface through cracks and fissures in rocks. These springs make excellent community water sources because of their relatively high flow rates and single discharge points.

FIGURE 2.1 A newly completed spring box in the Dominican Republic suitable for public access and livestock watering. (See also Color Plates; *Will Hart Photo*)

Another type of artesian spring that can be developed as a high-quality water source is the artesian flow spring. These occur when water confined between two impervious layers emerges at a lower elevation. Artesian flow springs often occur on hillsides, making protection a fairly easy process (Fig 2.2).

Spring Water Protection

Fracture and tubular springs are formed when water is forced upward through cracks and fissures in rocks. Because the discharge is often concentrated at one point, protecting the source from surface contamination is more easily done than springs with multiple or widely dispersed outlets (Fig. 2.2). Before reaching the surface, spring water is generally considered high quality, depending on the composition of the surrounding soils and bedrock. However, groundwater can become easily contaminated as it exits the ground's surface. Contamination sources include livestock, wildlife, crop fields, forestry activities, septic systems, and fuel tanks located upslope from the spring outlet.

Therefore, spring water sources, if intended for domestic use, need to be protected at the source, or eye. Protection can be as simple as providing fencing around the spring outlet, but proper application of a spring development structure also serves to maintain water quality.

Color Plates

Introduction

The spheres in Plate 1 represent all of Earth's water, Earth's liquid fresh water, and water in lakes and rivers.

- The largest sphere represents all of Earth's water, and its diameter is about 860 miles (the distance from Salt Lake City, Utah, to Topeka, Kansas). It would have a volume of about 332,500,000 cubic miles (mi^3) [1,386,000,000 cubic kilometers (km^3)]. The sphere includes all the water in the oceans, ice caps, lakes, and rivers, as well as groundwater, atmospheric water, and even the water in you, your dog, and your tomato plant.

- The blue sphere over Kentucky represents the world's liquid fresh water (groundwater, lakes, swamp water, and rivers). The volume comes to about 2,551,100 mi^3 (10,633,450 km^3), of which 99 percent is groundwater, much of which is not accessible to humans. The diameter of this sphere is about 169.5 miles (272.8 km).

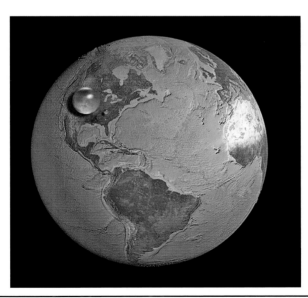

PLATE 1 How much water is on the Earth? [*Courtesy of Howard Perlman, USGS; globe illustration by Jack Cook, Woods Hole Oceanographic Institution ©; Adam Nieman. Source: Igor Shiklomanov, "World fresh water resources,"* in Peter H. Gleick (Ed.), Water in Crisis: A Guide to the World's Fresh Water Resources, *Oxford University Press, New York, 1993.*]

- And that "tiny" bubble over Atlanta, Georgia represents fresh water in all the lakes and rivers on the planet. The volume of this sphere is about 22,339 mi³ (93,113 km³). The diameter of this sphere is about 34.9 miles (56.2 km). Yes, Lake Michigan looks way bigger than this sphere, but you have to try to imagine a bubble almost 35 miles high—whereas the average depth of Lake Michigan is less than 300 ft (91 m).

Although the same amount of water is available that was available during the dinosaur reign on earth, the United Nations estimates that about 80% of the world's population (5.6 billion in 2011) live in areas with threats to water security (Fig. 1.1). Water security is a shared threat to humans and nature, and it is pandemic. Existing human water-management strategies increasingly affect wildlife such as migrating fish. Regions with intensive agriculture and dense population have an increasing threat of water shortage that can lead to food shortages. Finding alternative sources and encouraging better management of the supply now at hand is increasingly becoming a pressing necessity.

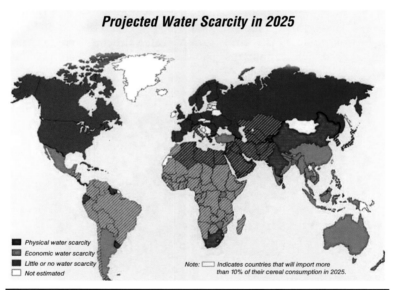

FIGURE 1.1 Water stress: 1.2 billion people—or almost 1 out of 5 people in the world—are without access to safe drinking water, and half of the world's population lacks adequate water purification systems. (*Source: International Water Management Institute*)

Water Cycle

The water cycle has been an essential part of the earth's living process since the earth began (Fig. 1.3). The difference now is that the added pollution caused when the rain passes through the atmosphere and runs across the land diminishes the supply of clean water, resulting in increasingly more energy applied and chemicals added to remove the pollutants such that it can become suitable for consumption.

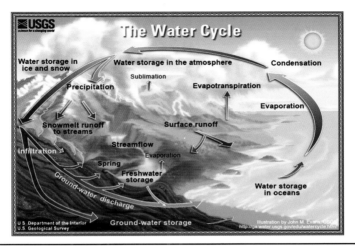

Figure 1.3 Water cycle.

Springs

Harnessing groundwater by developing a spring was one of the earliest means of managing water once man moved away from the rivers and lakes. A spring or seep is water that reaches the surface from some underground supply, appearing as small water holes or wet spots on hillsides or along river banks (Fig. 2.1). The flow of water from springs and seeps may come from small openings in porous ground or from joints or fissures in solid rock.

The two basic categories of springs: gravity and artesian, are described in Chapter 2 with insight as to how to optimize their production for constructive use.

Figure 2.1 A newly completed spring box in the Dominican Republic suitable for public access and livestock watering. (*Will Hart Photo*)

Air Conditioning Condensate Recovery

The psychrometric chart was developed by Dr. Willis Carrier in 1911 as a means to illustrate the capacity of air to hold moisture as a function of temperature (Fig. 3 1). Knowing this became the foundation of modern air conditioning design. Air conditioning is increasingly common around the world, and it is typically operated year round in areas that are short of water. Yet this by-product of the air conditioning process, condensate, is commonly wasted. The psychrometric chart is a tool that establishes a useful way of estimating the amount of water that can be reclaimed by the air conditioning process.

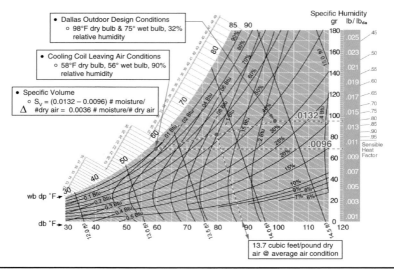

Figure 3.1 Psychrometric chart. (*Courtesy of the Carrier Air Conditioning Corporation*)

Harvesting the Dew

There is always water in the atmosphere. Clouds are, of course, the most visible manifestation of atmospheric water, but even clear air contains water—water in particles that are too small to be seen. One estimate of the volume of water in the atmosphere at any one time is about 3100 mi³ (12,900 km³). This sounds like a lot, but it is only about 0.001 percent of the total Earth's water volume of about 332,500,000 mi³ (1,385,000,000 km³). If all of the water in the atmosphere rained down at once, it would only cover the ground to a depth of 2.5 cm (about 1 in). Yet certain areas of the world have successfully harvested the water in clouds and dew to provide for their water needs.

Harvesting the Fog

Complimentary to dew harvesting is harvesting water from fog (Fig. 5.1). Fog-harvesting installations are operational in the high mountains of South America and in San Francisco Bay, and are viable in coastal and high-altitude sites where clouds consistently form.

Figure 5.1 Morning fog.

Glacier Harvesting

Glaciers historically come and go with the seasons but increasingly are receding a bit each cycle. This causes spring melting to occur later waiting for temperatures to rise at the higher altitudes, which causes the melting process to complete sooner, leaving a shortage of water during the summer growing season. Artificial glaciers take advantage of the seasonal weather cycle to store ice at lower altitudes in winter, which then expands water availability by melting sequentially as temperatures gradually rise at higher altitudes. The illustration in Fig. 6.3 describes how artificial glaciers have been traditionally constructed to provide water to sustain communities in the high desert areas where glaciers occur.

GLACIERS TO ORDER

According to tradition in the Karakoram and Hindu Kush mountains, there are several stages to growing a glacier

Select a site above 4500 m, facing north-west, surrounded by steep cliffs. Ideally it will be rocky, with ice trapped between small boulders of about 25 centimetres diameter

Take 300 kilograms of "female" ice (white, clean and made from snow) and pile it on top of "male" ice (containing stones and soil)

Place gourds of water in the cracks. These will burst when the glacier grows. The contents then freeze and bind the ice together

Cover it with charcoal, sawdust, wheat husk, nutshells or pieces of cloth to insulate the new glacier and leave it for approximately 4 winters

Each season, snowmelt will be trapped by the glacier and refreeze, making it grow. The budding glacier will eventually creep down the hill like a natural glacier

FIGURE 6.3 Schematic of glacier grafting procedure.

Graywater Recycling

Graywater is water recycled from showers and hand sinks that is reclaimed for additional use before discharge as sewage. Common uses are for flushing water closets and urinals, and for landscape irrigation. In this photograph (Fig. 9.4), indoor plants at the Society for the Protection of New Hampshire Forests, in Concord, New Hampshire, are irrigated with graywater.

FIGURE 9.4 Indoor plants at the Society for the Protection of New Hampshire Forests, in Concord, New Hampshire, are irrigated with graywater.

Water Quality

Management of quality, once the water is gathered, is of equal importance to getting water in the first place. Too much treatment is wasteful and too little can present health dangers. Means of sanitizing water include low-tech means such as sand filters, used by the Romans as well as by modern water utilities (Fig. 10.2). Intelligent applications of these technol-

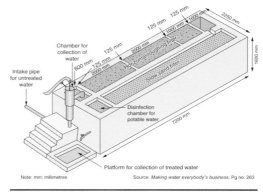

Note: mm: millimetres   Source: *Making water everybody's business*, Pg no. 263

FIGURE 10.2   Slow sand filter.

ogies for treating water can change it to a "fit for use" standard using graduated layers of sand and gravel that tailors the quality of the water according to each specific application.

Aquifer Recharge

Groundwater is normally recharged naturally by rain and snow melt and to a smaller extent by surface water (rivers and lakes) (Fig. 11.6). Recharge may be impeded by human activities including paving that occurs with modern development, or events

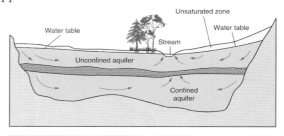

FIGURE 11.6   Groundwater flow. (*Illustration courtesy of United States Geological Survey*)

such as logging that enhance surface water runoff and reduce the soils ability to absorb the water into the aquifer. Pumping of groundwaters, especially for irrigation, also lowers the water tables. Groundwater recharge, which ensures water being retained in the soil is at least equal to what is being extracted, is increasingly being recognized as an important process for sustainable groundwater management.

Aquatic Plants for Waste Processing

The use of aquatic plants to process waste is an attractive alternative to standard waste processing that eventually flushes waste downstream into the watershed (Fig. 12.4). If non-toxic (human) waste is processed, plants like water hyacinths and duckweed generate a nutritious livestock feed. Where heavy metals are present in the effluent, aquatic plants serve to concentrate the metals in the plant, making it possible to reclaim heavy metals for reuse, rather than allowing them to be a detriment to the environment.

FIGURE 12.4   The Lemnaceae family of duckweed occurs worldwide, and can metabolize waste products into nutrients available for livestock feed.

Biological Filters and Constructed Wetlands

Constructed wetlands process building effluent using natural processes to safely treat the waste to a level that is acceptable to the environment. This methodology presents a low-energy consuming option to waste treatment where land is available to infiltrate the waste. This technology also invites the possibility of integrating the wastewater to supplement irrigation of landscaping; making what other-

PLATE 2   Smithsonian Marine Research Station, Bocas Del Toro Panama.

wise would create a concentration of chemicals, such as phosphates and nitrates, into feedstock for plants. In this photo (Plate 2), the attractive landscaping serves as part of an evaporation-transpiration system for processing the graywater for the Smithsonian Marine Research Station in Bocas del Toro, Panama.

Septic Tanks

Widespread in many rural environments, septic tanks are a common means to address the sewage treatment needs of small to medium systems. The septic tank provides settling and preliminary digestion of the solids, allowing the resultant graywater to be infiltrated into the soil. Their effectiveness is largely a function of soil absorptivity and the area available for the graywater to infiltrate into the soil, evaporate into the atmosphere, and transpire through plants (Fig. 15.1).

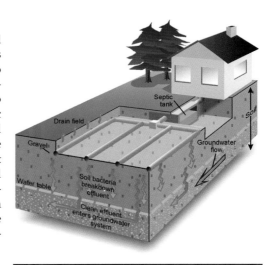

FIGURE 15.1   Septic field schematic.

Composting Toilets

Composting toilets have evolved from a means to bring the privy indoors to a technological means to process waste in a highly sanitary manner. This is done by changing the means of waste digestion from anaerobic (oxygen deprived) digestion common to the privy, to an aerobic (with oxygen) process, with the benefit of better odor and moisture control; and typically using little or no water in the flushing process (Fig. 17.7). Composting toilets may be used as an alternative to flush toilets in situations where there is no suitable water supply or waste treatment facility available or to capture nutrients in human excreta as humanure. The human excrement is normally mixed with sawdust, coconut coir, or peat moss to support aerobic processing, absorb liquids, and to reduce the odor. The decomposition process is generally faster than the anaerobic decomposition used in wet sewage treatment systems such as septic tanks. Composting toilets are in use in many of the roadside facilities in Sweden, and in national parks both in the United States and the United Kingdom.

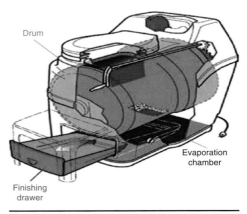

FIGURE 17.7 Sun-Mar composting toilet.

Net Zero Buildings

The waste management paradigm now in existence calls for using water to convey waste away from the point of use, ultimately to rivers and lakes (Fig. 18.2). Whoever would then like to use the water for their use must do so at great care and expense. The goal of the Net Zero building is to tend toward self-sufficiency by incorporating techniques and technologies discussed in this book's chapters, with the end goal being a more long-term sustainable way of managing water and waste which does not deplete the land of its nutrients or contaminate the water we must use to maintain our way of life.

FIGURE 18.2 Closing the sanitation loop to preserve water for the next generation. (*Illustration courtesy of the German Federal Ministry of Economic Cooperation and Development*)

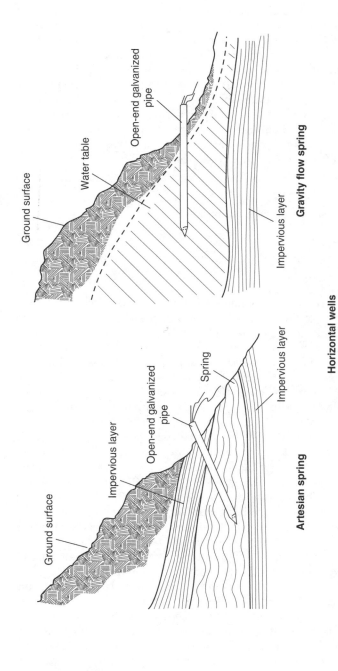

Ground surface

Water table

Open-end galvanized
pipe

Impervious layer

Gravity flow spring

Ground surface

Impervious layer

Open-end galvanized
pipe

Spring

Impervious layer

Artesian spring

Horizontal wells

Figure 2.2 Gravity and artesian flow springs.

17

Benefits of Spring Development Structures

The advantages of spring development structures include increased flow and reduced possibility of spring water contamination. Costs associated with constructing a structure to harvest the spring water are minimal, and maintenance of the system is minimal, requiring only infrequent disinfections and sediment removal. Springs can be developed to augment an existing water supply, should the primary water supply become unreliable. Springs may also be the primary sources themselves, depending upon the water production and the water demand. And contrary to the generally held belief that discharges decline if the springs are touched, the development of natural springs often leads to improved yields.

Basic Design Features

Examples of spring development structures are:

- Spring boxes
- Seepage spring development structures
- Horizontal wells

Of the three examples mentioned above, spring boxes are typically cheapest, require the least skill to construct, and can be made with locally available materials. Spring box technology is simple and can be easily modified to fit just about any situation. If accessibility to the source is an issue, they can be adapted to work in conjunction with other technologies, such as gravity-fed or pumped distribution systems.

Although there are many different designs for spring boxes, they all share common features. A spring box is primarily a watertight collection box constructed of concrete, clay, or brick with one permeable side to allow water to enter the box. The purpose of the spring box is to isolate spring water from surface contaminants such as rainwater or surface runoff. All spring boxes should be designed with a heavy removable cover to provide access for disinfection and maintenance. The weight of the lid should be in the range of 30-40 lb to provide a secure closure and a safety barrier to children.

Spring box design should include an overflow pipe. The overflow should have a screen covering the openings, no greater than a 0.125-inch (3-mm) grid or as otherwise appropriate, for preventing entrance of insects or vermin in the cistern. Provision should also be made for proper handling of overflow water to prevent erosion and possible undermining of the structural support around the tank itself. The grade surrounding the tank should be sloped away from the tank, with the access approximately 12 inches (30 cm) above grade to prevent surface runoff from entering the tank access. Access to the tank by livestock or other animals should be controlled by fencing to prevent contamination and soil compaction that will lead to reduced yield.

There are two thoughts regarding clearing of the vegetation surrounding the tank. Some argue that by removing vegetation from the area surrounding the spring, flow may increase due to reduced water use by vegetation, while others maintain that shallow rooted grasses should be allowed to grow in the area due to their capacity for utilizing surface water before it is able to infiltrate and contaminate spring water. In either case, deep-rooted trees and plants should be avoided, as their root systems in proximity to the spring tank could damage protective structures and reduce spring flow.

Types of Spring Box Construction

There are two basic spring box designs that can be modified to meet local conditions and requirements. The first is a spring box with a single permeable side for hillside collection, and the second design has a pervious bottom for collecting water flowing from a single opening on level ground (Figs 2.3 and 2.4).

The spring box with an open bottom is typically simpler and cheaper to construct because less digging and fewer materials are required. If cement is unavailable, a spring box can be constructed using alternative available materials, such as brick (Fig. 2.5) or large prefabricated concrete drainage pipe.

Commercially manufactured spring boxes are also available (Fig. 2.6).

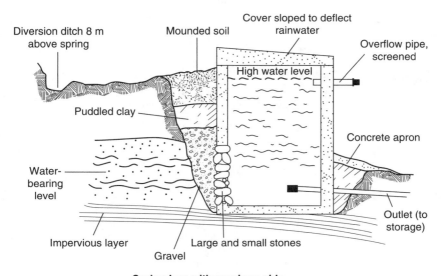

Spring box with pervious side

Figure 2.3   Spring box with single pervious side for hillside collection. (*Courtesy of USAID, 1982, available online at www.lifewater.org*)

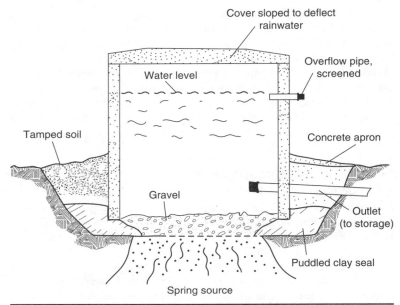

FIGURE 2.4   Spring box with open bottom. (*Courtesy of USAID, 1982, available online at www.lifewater.org*)

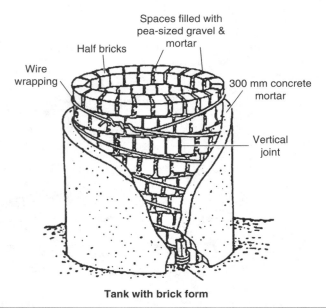

Tank with brick form

FIGURE 2.5   Spring box constructed of brick and concrete. (*Courtesy of USAID, 1982, available online at www.lifewater.org*)

Figure 2.6   Example of commercially available spring box.

Designing the Water Distribution System

Because the site configuration and water needs vary, no one particular design will fit all situations. If, for instance, the spring is located at a higher elevation than the distribution area, and the distance is not too great, it may be preferable to design a spring box that is a storage structure large enough to supply the entire water demand. Such a design could also distribute the water without the need for pumps to pressurize the distribution system. The pressure created by elevating the tank will be 0.43 psi per foot (0.03 kg/cm²) of elevation difference between the water level in the tank and the elevation at the point of end use. Subtracting the friction loss of the volume of water flowing in the pipe will be the delivered pressure at the point of end use.[1]

> **Example:**   If average water level in the storage tank is 10 feet above the point of end use, the static head (pressure with no flow) available will be 4.3 psig (0.30 kg/cm²).
>
> - If there are 10 gallons (38 liters) per minute flowing down 100 feet (30.5 m) of 1-inch (25 mm) polyvinyl chloride (PVC) pipe, the friction loss will be 6.02 feet head loss per 100 feet of pipe (5.9 m/100 m) for 10 gallons per minute. For 100 equivalent feet of pipe,[2] this would result in a pressure loss of 2.6 psig (0.18 kg/cm²).
> - The resultant dynamic pressure at the outlet would be:
>   - 1.7 psig   (4.3 psig − 2.6 psig)
>   - 0.12 kg/cm²   (0.30 kg/cm² − 0.18 kg/cm²)

- If larger pipe was used: if 1.25 inches (32 mm) was used, the friction loss per 100 foot of pipe would be 1.55 feet/100 feet equivalent length of pipe, and the dynamic pressure at point of use would be:
  - 2.7 psig   (4.3 psig – 1.6 psig)
  - 0.19 kg/cm^2 (0.30 kg/cm^2 – 0.11kg/cm^2)

For situations where the groundwater has a high degree of sediment, putting tanks in series will provide some degree of sediment separation in the first tank before overflowing into the second distribution tank. For the settling to occur in the primary tank, the tank must be sized large enough with minimal turbulence so the sediment can settle. As an option to be used in the second tank, run the water inlet to the bottom of the second tank, then have the elbow turned up, and this will result in less agitation of the sediment that has settled to the bottom. Have a surface skimming overflow on the opposite side of where the water inlet will push any floating degree over to the overflow. By so doing, better quality stored water can be maintained in the tank.

Tank Construction

The design chosen for any particular project will depend on local conditions, water yield from the spring versus water requirements, the available materials and the knowledge of the individuals that will be constructing the box. To improve the odds of a successful outcome, it is advisable to generate a dimensioned plan of the spring box (Fig. 2.7) and a map of the area, including the location of the spring. Included with this plan should be the location, distances, and elevation differences of the points of end use (Fig. 2.8). A completely thought-out design is a useful tool for minimizing unanticipated surprises and providing a more accurate list of labor, material, and tools needed. Having such a list at hand will help ensure that all necessary implements will be on hand, delays and setbacks will be avoided, and costs will stay within budget.[3]

Preparing the Site

Once materials and labor are accounted for, the spring needs to be located and the site prepared. The site should be fenced off to protect it from animals, and a diversion ditch needs to be dug approximately 8 meters upslope from the site to divert surface water runoff away from the spring.

Next, dig out the spring until the flow is concentrated from a single source. If the spring is located in a hillside, it may be necessary to dig

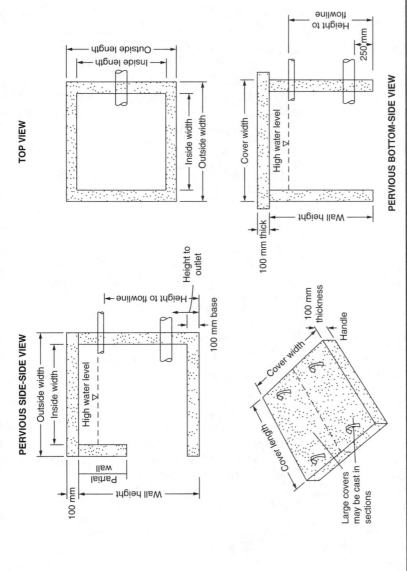

TOP VIEW

Outside length
Inside length

Inside width
Outside width

PERVIOUS SIDE-SIDE VIEW

Outside width
Inside width

High water level

Height to flowline

Height to outlet

100 mm base

Partial wall

100 mm

Wall height

PERVIOUS BOTTOM-SIDE VIEW

Cover width

High water level

Height to flowline

250 mm

100 mm thick

Wall height

Cover width

100 mm thickness

Handle

Cover length

Large covers may be cast in sections

FIGURE 2.7 Dimensional plans of pervious side spring box and pervious bottom spring box. (*Courtesy of USAID, 1982, available online at www.lifewater.org*)

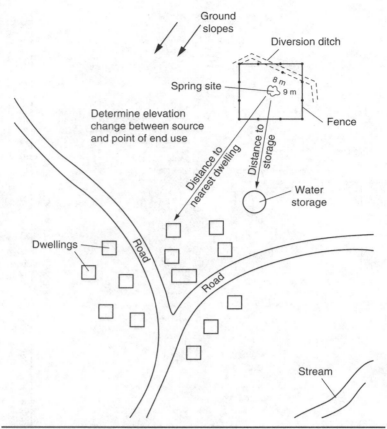

FIGURE 2.8   Location map showing spring location, distances to dwellings, and ground slope. (*Courtesy of USAID, 1982, available online at www.lifewater.org*)

into the hillside far enough to locate the eye of the spring. Look to see if flow from major openings increases or if flow from minor openings decreases or stops. These are signs that the flow is becoming concentrated from a single eye. Remember that the objective is to collect as much water as possible from the spring and that it is generally easier to collect from a single opening than from many. If a single flow source cannot be located because of numerous, separated openings, it will probably be necessary to construct a seep collection system rather than a spring box. Depending on the terrain of the site, it may be necessary to dig a temporary diversion ditch to drain spring water from the excavation site (Fig. 2.9).

Once a single eye is located, dig down until you reach an impervious soil layer. This will make a good waterproof foundation for the spring box. Before installing the spring box, pile stones and gravel

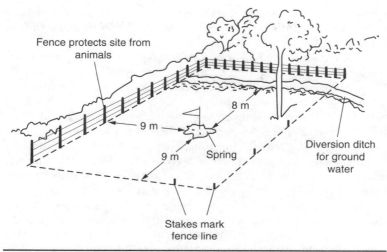

FIGURE 2.9   Excavating the spring site; note temporary diversion ditch. (*Courtesy of USAID, 1982, available online at www.lifewater.org*)

against the spring. This will provide some capacity for sedimentation and will prevent erosion around the spring eye. This will also support the impervious section of the back wall of a pervious-side spring box. If the spring is flowing from a single opening on level ground (pervious-bottom spring box), dig a basin around the spring eye until an impervious layer is reached. Line this basin with rocks and gravel, making sure to cover the spring eye so that water flows through the gravel before entering the spring box.

Constructing the Spring Box

Although concrete construction requires that the concrete remain moist for at least seven days, spring box construction should be done at the peak of the dry season, thereby ensuring that only the most reliable springs are protected. Since the strength of the concrete will increase with curing time, construction of the spring box should start as soon as the proper design for the particular spring is chosen, and it should preferably be the first item to be constructed. The necessity of excavating the spring before construction should not be overlooked. This is so that the type of spring to be protected can be determined and the proper design implemented.

The first step in spring box construction is to ensure that all required materials and tools are available on-site. This will help avoid construction delays. The next step is to build wooden forms. The dimensions of the forms will depend on the dimensions of the spring box, but a good rule of thumb is that the outside dimensions of the forms should be 0.1 meter (4 inches) larger than the dimensions of the spring box.

When constructing a spring box for a spring flowing from a single source on level ground, a form with an open bottom should be built. If a pervious-side spring box is to be constructed, a form needs to be built for a box with an impervious bottom and one permeable side. In order to build a form for a box with a bottom, set the inside form 0.1 meter above the bottom of the floor. This can be done by nailing the inside form to the outside form so that it is left hanging 0.1 meter above the floor. Make holes in the forms to fit the diameter of your overflow and outflow pipes, and place small pieces of pipe in them to ensure that properly sized holes are left in the spring box when the concrete sets. Make sure temporary braces are installed on the outside of the forms and that wire bracing is used between the inside and outside forms. If the forms are not properly braced, the weight of the cement will cause the forms to separate. To brace with wire, drill small holes in the forms and place wire through them to tie them together. Tighten the braces by twisting the wire with a strong stick, thus pulling the forms together[4] (Fig. 2.10).

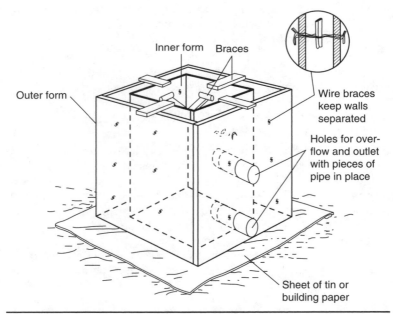

FIGURE 2.10   Form for a poured concrete spring box. (*Courtesy of USAID, 1982, available online at www.lifewater.org*)

Next, set the forms in place, either in the permanent site or nearby, depending on the size of the spring box. If the spring box is so large that installation will be difficult after the concrete has been placed and cured, the forms should be in its permanent site prior to placing the concrete. Make sure that your temporary diversion ditch is functioning properly while the concrete cures.

Prior to placing cement in the forms, the insides of the forms should be coated with old motor oil so that the concrete won't stick to them. It is important to keep in mind that although the spring box walls do not require major enforcement, minor reinforcement around the perimeter of the box is recommended to prevent cracking of the cement. Concrete construction guides should be referred to in order to determine the proper level of reinforcement for your particular spring box.

The tank cover will also require reinforcement, and handles will facilitate cleaning and inspection. The cover should be sloped or concave so that water does not puddle on top of it. For safety reasons, the cover should be heavy enough that a child cannot remove the lid and fall into the tank.

Concrete should be mixed in a proportion of one part cement, two parts sand, and three parts gravel. Only add enough water to make a thick paste, as too much water leads to weak cement. Finally, place the concrete into the forms, tamping as you go to ensure that there are no residual air pockets. This can be done by "vibrating the forms," which can be accomplished by pounding the outside form with a rock or hammer. Smooth out all concrete surfaces to ensure that the cover fits well, and cover with wet canvas, burlap, or straw to prevent the concrete from losing moisture. This protective covering needs to be kept wet throughout the curing process, at least seven days, although longer curing will result in a stronger structure.

Installation

To minimize the possibility of contamination, it is important that the spring box be installed on a solid, impermeable base and that a seal be created between the ground and the spring box so that outside water is unable to infiltrate the box. Place the box so that its permeable section collects the flow of the spring. When installing a hillside collection box, make sure that gravel and stones are piled at the back of it to provide structural support while allowing water to enter the box (Fig. 2.3). Next, create a seal with concrete or puddled clay, a mixture of clay and water, where the spring box comes into contact with the ground. This will ensure that water does not seep in under the box. For hillside spring boxes, backfill the area where the spring enters the box with gravel to the height where the permeable wall ends and the concrete wall begins. Place layers of puddled clay or concrete over the gravel backfill, sloping away from the spring box to divert surface water away from the water source (Fig. 2.4), and then backfill with firmly tamped soil. If puddled clay or concrete is unavailable, soil alone may be used, although it should be at least 2 meters deep to prevent contaminated surface water from reaching the water source. For a level-ground spring box, puddled clay or cement should be placed around the spring box, sloping away from the water source to prevent infiltration.[5]

Remove the pipe pieces used to form the pipe holes in the spring box, and install the outflow and overflow pipes. Seal around the pipes on both sides of the wall to prevent leaks, and secure screening over the pipe openings. Make sure the screen size is small enough to prevent mosquito infestation yet strong enough to deter small animals, and it should be of a durable material so it lasts a long time. Copper or plastic screening works best.

Before completely backfilling the spring box, disinfect the inside and the cover with a chlorine solution and close the box. Remember that all backfill should slope away from the spring box to maximize runoff away from the box.

Seep Collection System

If a single spring eye cannot be located, a seep collection system is a third alternative to a spring box (Fig. 2.11). In a seep collection system, perforated collection pipes are laid in a "Y" shape, in a stone filled trench, perpendicular to the seep flow. This serves to collect and concentrate water, which is then diverted to a spring box or to a storage tank. A covering of puddle clay approximately 4 inches (100mm) thick should cover the stones to prevent surface runoff from infiltrating and contaminating the water source. Designing and constructing a seep collection system is much more difficult and typically more costly than other methods. In addition, collection pipes often clog with soil and rocks, making water collection less efficient and requiring more frequent and intensive maintenance.[6]

Horizontal Well (Horizontal Bore Hole)

Another alternative to a spring box is the development of a horizontal well (Fig. 2.12).

These are particularly useful where the water table is steeply sloped. In a horizontal well, a pipe with a screened or perforated driving point is driven into an aquifer horizontally at a higher elevation than the spring's natural discharge. Often, a headwall will need

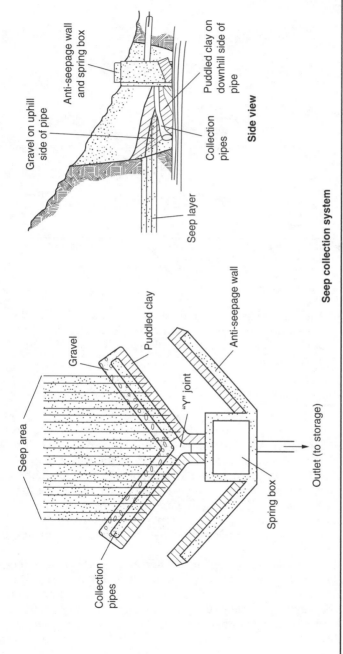

FIGURE 2.11 Seep collection system. *(Courtesy of USAID, 1982, available online at www.lifewater.org)*

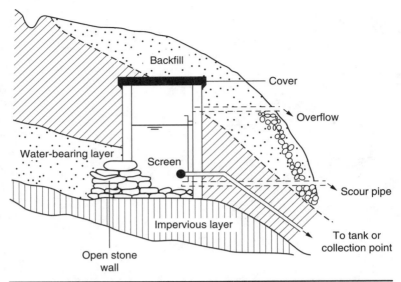

Figure 2.12 Horizontal well placement. (*Courtesy of USAID, 1982, available online at www.lifewater.org*)

to be constructed in order to adequately seal the space outside the pipe. The only requirement of horizontal wells is that the water table be steeply sloped; flat water tables typically won't be under enough pressure to provide adequate flow.

Maintenance

If properly installed, spring boxes require very little maintenance. However, it is recommended that the water quality be checked before they are put into use, and on at least a yearly basis thereafter. It is also a good idea to check that the uphill diversion ditch is adequately diverting surface runoff away from the spring box and that the support fill around the spring box is not eroding. Any necessary animal fencing should be maintained in good repair, to keep livestock from contaminating the water source or compacting the soil around the spring.

For hillside collection boxes, it is important to check that the uphill wall is not eroding and is maintaining structural integrity. The cover should be checked frequently to ensure that it is in place and appears to be watertight. Make sure that water isn't seeping out from the sides or from underneath the spring box, and check that the screening is in place on the overflow pipe.

Once a year, disinfect the system and remove sediment from the spring box. To do so, open the valve on the outlet pipe, allowing the spring box to drain. Remove any accumulated sediment from the box, and wash the interior walls with a chlorine solution. It should be noted that chlorine and chlorine compounds might irritate eyes and skin; proper protective equipment such as gloves, safety glasses, and protective clothing should be worn if available when dealing with chlorine. The solution for washing the spring box should be mixed in a ration of 10 L water with 0.2 L chlorine bleach. After washing the interior of the spring box, chlorine should be added directly to the water in a ratio of 100 parts chlorine per million parts water, and this should be allowed to sit for 24 hours. If it isn't possible to allow the chlorine to sit for 24 hours, two consecutive applications 12 hours apart should provide for adequate disinfection. As mentioned previously, water samples should be analyzed periodically for contamination.

Conclusion

Springs have been developed for over 4000 years as a reliable source of water for domestic, industrial, and agricultural purposes. Spring water is usually remarkably clear; however, it may also be tea-colored if it is tinted after coming in contact with naturally occurring minerals or tannic acids from organic material in subsurface rocks. The rate of flow and the length of the flow path through the aquifer affect the amount of time the water is in contact with the rock, and thus the amount of minerals that the water can dissolve. The quality of water discharged by springs can vary greatly because of factors such as the quality of the water that recharges the aquifer and the type of rocks with which the groundwater is in contact. The quality of the water can also be affected by the mixing of freshwater with pockets of ancient seawater in the aquifer or with modern seawater along an ocean coast.

So should you feel confident about whipping out your canteen and filling it with cool and refreshing spring water? No, you should be cautious. The typical North American spring passes through rock that has a mean annual temperature of approximately 56 degrees Fahrenheit (11 degrees Celsius). The water is crudely filtered in the rock, and the time spent underground allows debris and mud to fall out of suspension. If underground long enough, lack of sunlight causes most algae and water plants to die. However, microbes, viruses, and bacteria do not die just from being underground, nor are any agricultural or industrial pollutants removed.

So although spring water can present a reliable source of water, the quality should be carefully tested before it is assumed to be safe to drink.[7]

Notes and References

1. Friction loss from flow in pipe is traditionally determined using the imperial form of the Hazen-Williams formula:

$$hf = 0.002083 \, L \, (100/C)^{1.85} \times (gpm^{1.85}/d^{4.8655})$$

where: hf = head loss in feet of water
L = length of pipe in feet
C = friction coefficient
gpm = gallons per minute (USA gallons not imperial gallons)
d = inside diameter of the pipe in inches.

2. Equivalent length of pipe is defined as the linear length of straight pipe added to the pressure loss of fittings and valves, expressed in equivalent lengths of straight pipe. For additional information on this topic, consult information on the Hazen-Williams formula, *Crane Catalog No. 60,* or other sources that refer to water pressure loss in piping. Also refer to such tables for determining the friction loss per unit length for materials other than the information on PVC pipe and fittings listed in Appendix I.

3. Hart, W., *Field Engineering in the Developing World,* School of Forest Resources & Environmental Science, Michigan Technological University, April 2003.
4. Hart, W., *Field Engineering in the Developing World.*
5. Hart, W., *Field Engineering in the Developing World.*
6. Hart, W., *Field Engineering in the Developing World.*
7. *How Ground Water Occurs,* Ground Water, U.S. Geological Survey General Interest Publication. Reston, Virginia, 1999

CHAPTER 3

Air Conditioning Condensate Recovery

Introduction

Air conditioning and refrigeration is used around the world for temperature comfort and food preservation. As a by-product of the air being cooled, it is dehumidified, causing water to be removed as condensate. With proper treatment to address biological contaminants, this water can be constructively used.

A typical air conditioning system in a commercial building consists of an air-handling unit that circulates air to the occupied spaces to maintain comfort. As the air returns from the space, it is mixed with outside air, which is necessary to maintain a healthful environment. As the air passes through the air-handling unit, it goes through a cooling coil. The temperature of the air drops, and the humidity, from the added outside air and what is gained in the space, is removed as condensate.

This water is essentially distilled water, low in mineral content, but with an important addition. Air conditioning condensate can amplify Legionella and other airborne bacteria, and it has been shown to be the source of outbreaks in hospitals, motels, and cruise ships. Contamination of air conditioning condensate by Legionella is so common that there are commercially available kits for inhibiting microbial growth in the condensate. Legionella is somewhat unique because low levels are not cause for concern but higher levels are. This is why most natural sources of water are not infective but can become problematic when an amplifying device, such as an air conditioner, is present.

So untreated condensate should be handled in a manner to eliminate any possibilty of creating aerosols that can be inhaled by humans.

Applications for the condensate may include subsurface irrigation or process makeup (such as a cooling tower), where water is treated for biological contamination.

If it is to be used for potable water or for washing, proper disinfection of the water is required. Ultraviolet light, chlorine tablets, ozone injection, and raising water temperature to at least 140 degrees Fahrenheit are examples of methods that can be used to reduce the potential hazard of biological contamination.

Estimation of Condensate Volume

The formula to determine how much cooling coil condensate can be generated at the peak design condition is as follows:

$$\text{Condensate Volume (gpm)} = Q_{air} \times dw_{lb}/8.33 \times V_{da} \qquad \text{Eq. (3.1)}$$

Where:
For specific humidity expressed in $lb_{water}/lb_{dry\ air}$:

$Q_{air} \triangleq$ Air flow (cubic feet per minute)
$dw_{lb} \triangleq$ Difference in specific humidity $(lb_{water}/lb_{dry\ air})$
$S_{Vda} \triangleq$ Specific volume of air (cubic feet/$lb_{dry\ air}$) ≈ 13.8 cubic feet/ $lb_{dry\ air}$
$8.33 = 8.33\ lb_{water}/$gallon

When specific humidity in Eq. (3.1) is expressed in $grains_{water}/lb_{dry\ air}$:

$$\text{Condensate volume (gpm)} = Q_{air} \times dw_{lb}/7000 \times 8.33 \times V_{da} \qquad \text{Eq. (4.2)}$$

$Q_{air} \triangleq$ Air flow (cubic feet per minute)
$dw_{lb} \triangleq$ Difference in specific humidity $(gr_{water}/lb_{dry\ air})$
$V_{da} \triangleq$ Specific volume of air (cubic feet/$lb_{dry\ air}$)
$8.33 = 8.33\ lb_{water}/$gallon
$7000 = 7000\ grains/lb_{dry\ air}$

Calculation Procedure

1. The first step to estimating the volume of condensate that can be reclaimed is determining the difference in the specific humidity $(dS_{V\ lb})$ of the entering and leaving coil condition. Typically, the design outside air condition is known, gathered from historical weather data or obtained from engineering data handbooks such as ASHRAE[1] Fundamentals.

Example:
a. Assuming air flow of 375 cubic feet per minute (cfm) per ton for a 5 ton system would result in 1875 cubic feet per minute.

Assuming that most of the dehumidification results from removing the moisture from the outside air, we are dehumidifying 375 cfm for 20 percent of outside air: (5 tons × 375 cfm/ton × 20%).

b. The design outside air conditions[2] for Dallas, Texas, are 98 degrees dry bulb, 75 degrees wet bulb, 32 percent relative humidity. The Psychrometric Chart (Fig. 3.1) shows that the outside air specific humidity is 0.0132 lb moisture per lb of dry air.

c. The cooling coil discharge condition is presumed to be 58 degrees F dry bulb and 90 percent relative humidity. Also from the Psychrometric Chart, we can see that the coil discharge condition is 0.0096 lb moisture per lb of dry air.

d. The difference in specific volume (d $S_{V\,lb}$) is approximately 0.0036 lb moisture per lb of dry air: (0.0132–0.0096 lb moisture/$lb_{dry\,air}$). (See Fig. 3.1.)

2. The second step is to determine the approximate specific volume (S_{Vda}) of the air. This can also be gathered from the Psychrometric Chart, and for the conditions listed, it is approximately 13.7 cubic feet/$lb_{dry\,air}$.

3. The third step gives us the gallons per minute of generated condensate:

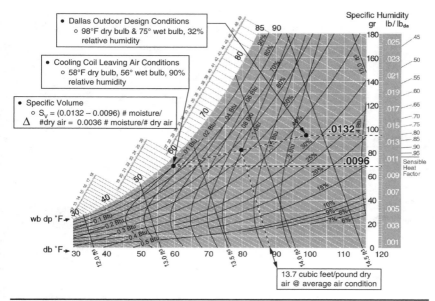

FIGURE 3.1 Psychometric Chart. (See also Color Plates.) (*Courtesy of the Carrier Air Conditioning Corporation*)

Condensate volume (gpm)
$$= Q_{air} \times d \; S_{V\,lb}/8.33 \times V_{da} = (375 \times 0.0036)/8.33 \times 13.7$$

$$= 0.012 \text{ gallons per minute (0.73 gallons per hour)}$$

This is the amount of condensate that can be generated at peak load.

4. To calculate the amount of condensate generated annually, multiply the gpm generated at peak load and the equivalent full load cooling hours.[3] For our Dallas office building example, the equivalent full load cooling hours are between 1350 and 1580 hours.

Assuming this is an average office building, with an interior load of approximately 1.5 watts/square foot, the resultant annual recovered condensate will be about:

Gallons per year = 0.73 gph $\times$ 1465 hr = 1068 gallons per year

By comparison, the approximate amount of condensate that can be recovered from a five-ton air conditioning system, with 20 percent outside air, for an office building with similar interior conditions in other cities would be as follows:

Sample Cities	Equivalent Full Load Cooling Hours (EFLCH) (NOTE 1) @1.5 Watts/sf (NOTE 1)	Dry Bulb/Wet Bulb Design Temperature (NOTE 2)	HR OA @ Design OA Conditions (NOTE 3)	Δ Humidity Ratio with 20% OA (NOTE 3)	Gallons per Hour (NOTE 4)	Gallons per Year (NOTE 5)
Atlanta, Ga	1220	91/74	0.014	0.004800	0.972	1186
Boston, Ma	710	88/72	0.014	0.004000	0.810	575
Dallas, Tx	1465	98/75	0.013	0.003600	0.729	1068
Indianapolis, In	780	89/74	0.015	0.005100	1.033	806
Minneapolis, Mn	465	88/72	0.013	0.003600	0.729	339
Los Angeles, Ca	1475	80/65	0.010	0.000400	0.081	119
Tampa/Miami, Fl	1900	91/77	0.017	0.007500	1.519	2886

TABLE 3.1 Estimated Condensate Production for Various Cities

Table Notes

Note 1: This estimates the annual air conditioning run time if the unit ran at 100 percent of capacity for the equivalent full load cooling hours. Data used is midrange data, as shown in Fig. 3.2, for an office building with an internal load of 1.5 Watts per square foot.

Note 2: The outside air design temperature is a statistically selected value used to properly size air conditioning equipment. The values selected are taken from ASHRAE Fundamentals 2009, Chapter 14, for outside air not exceeding these conditions more than one percent of the time. Units are in degrees Fahrenheit.

Note 3: The humidity ratio (lb moisture/lb$_{dry\ air}$) is the amount of moisture contained in 1 lb of air at the outside air condition stated above. This value is calculated by taking the humidity ratio (HR) of the outside air at the design outdoor air condition minus the HR for air coming off a cooling coil, normally, approximately 58 degree Fahrenheit dry bulb and 90 percent relative humidity. The HR for this condition is 0.0094 lb moisture per 1 lb of dry air.

Figure 3.2 Condensate is centrally gathered at a condensate receiver basin (*Bill Hoffman, P.E., Hoffman & Associates Engineers Photo*)

When bringing in 20 percent outside air, only 20 percent of the difference between these values represents the amount of outside air being dehumidified.

Note 4: This is calculated using the following formula:

$$\text{Gallons/minute} = 4.5/500 \times \text{cfm outside air} \times \Delta \text{ humidity ratio}$$

Note 5: This is calculated using the following formula:

$$\text{Gallons/year} = \text{gallons/minute} \times 60 \text{ minutes/hour} \times \text{EFLCH/year}$$

Calculations assume a room condition of 75 degrees Fahrenheit, 50 percent relative humidity, and 20 percent outside air. For other conditions, the humidity ratios should be changed accordingly.

Application Examples

Because reclaimed air conditioning condensate is high-quality water with low mineral and chemical content, it has many potential uses. But because of the potential bacterial content that commonly exists, care must be taken with its use. Distribution in a fashion that would cause aerosols (e.g., lawn sprinklers) should be avoided due to the possibility of persons in the vicinity being exposed to Legionella bacteria. If the use of air conditioning condensate could expose persons to inhalation of bacteria, then purification of the water should be done prior to use. Examples of common applications are as follows:

- Landscape irrigation (no treatment needed if utilized for subsurface irrigation)
- Swimming pool (with biocide treatment)
- Domestic water (with biocide treatment)
- Cooling tower makeup (can likely run direct to tower without modulating valves) (Fig. 3.3)
- Industrial process makeup

Hybrid systems are being developed that route the condensate into the rainwater catchment cistern. When this is done, the rainwater needs to be considered as part of a gray water system. This means that if used for lawn irrigation, it should be used with subsurface irrigation systems. Because aerosols are a possible result from flushing, if the water is brought into the building for flushing toilets, it needs to be filtered and sanitized before use.

Figure 3.3 Condensate is pumped directly to cooling tower basin (*Bill Hoffman, P.E., Hoffman & Associates Engineers Photo*)

Location	School	Office	Retail	Hospital
Atlanta, GA	690–830	1080–1360	1380–1860	2010–2850
Baltimore, MD	500–610	690–1080	880–1480	1350–2340
Bismarck, ND	150–250	250–540	340–780	540–1290
Boston, MA	300–510	450–970	610–1380	1020–2330
Charleston, WV	430–570	620–1140	820–1600	1260–2560
Charlotte, NC	650–730	1060–1340	1350–1830	1990–2820
Chicago, IL	280–410	420–780	550–1090	870–1780
Dallas, TX	830–890	1350–1580	1660–2090	2320–3100
Detroit, MI	230–360	390–820	530–1170	870–1950
Fairbanks, AK	26–54	64–200	110–320	210–600
Great Falls, MT	130–224	210–490	290–710	500–1210
Hilo, HI	1360–1390	2440–2580	2990–3370	4060–4910
Houston, TX	940–1000	1550–1770	1870–2290	2510–3320
Indianapolis, IN	380–560	560–1000	730–1410	1120–2250
Los Angeles, CA	780–910	1280–1670	1740–2350	2740–3770
Louisville, KY	550–670	770–1250	1000–1720	1480–2690
Madison, WI	210–310	320–640	420–900	680–1490

Table 3.2 Equivalent Full Load Cooling Hours/Year for Various Cities

Location	School	Office	Retail	Hospital
Memphis, TN	700–830	1090–1350	1350–1780	1910–2680
Miami, FL	1260–1300	1980–2150	2350–2740	3110–3890
Minneapolis, MN	200–300	320–610	430–870	680–1420
Montgomery, AL	840–910	1260–1510	1550–1990	2170–2950
Nashville, TN	570–740	830–1280	1030–1710	1490–2620
New Orleans, LA	920–990	1500–1720	182–2240	2500–3280
New York, NY	360–550	540–1040	720–1480	1160–2440
Omaha, NE	310–440	480–820	610–1130	920–1780
Phoenix, AZ	950–1020	1340–1610	1630–2090	2220–3040
Pittsburgh, PA	300–530	440–920	600–1310	960–2160
Portland, ME	190–300	310–630	410–900	700–1520
Richmond, VA	630–730	880–1310	1110–1770	1650–2760
Sacramento, CA	680–850	1080–1430	1460–2020	2250–3180
Salt Lake City, UT	410–710	510–1090	660–1520	1060–2470
Seattle, WA	260–460	440–1200	710–1860	1340–3270
St. Louis, MO	460–550	680–1100	850–1500	1260–2330
Tampa, FL	1050–1110	1800–2000	2170–2580	2910–3710
Tulsa, OK	580–770	830–1300	1030–1730	1470–2630

TABLE **3.2**   Equivalent Full Load Cooling Hours/Year for Various Cities (*Continued*)

Table Notes

Data above was taken from ASHRAE Applications 2007, Chapter 32, Table 8, with the ranges in values determined by internal heat gains ranging between 0.6 to 2.5 Watts/ft^2.

Operating with large temperature setbacks during unoccupied periods (effectively turning the system off) will reduce the cooling EFLCHs by 5 percent. Equations relating to EFLCH for cooling locations other than those listed above can be found in ASHRAE TRP-1120 Final Report (Carlson, 2001).

Conclusion

Condensate from air conditioning units is an often overlooked source of freshwater. The resultant accumulation (which can be just a trickle or sometimes more) can provide a sizable amount of freshwater that can be constructively used to offset the use of potable water. Because of the low mineral content, there is less fouling caused from mineral residue from the evaporation process, making water ideally suited for use in cooling towers and fountains. The single caution for condensate reclaim use would be that aerosols may be created in an occupied space. In this case, treatment for biological elements contained in the water would be appropriate.

As water shortages become more widespread, condensate from air conditioning is gaining increased attention for creative nonpotable and potable applications. If the potential hazard of bacteria from aerosols is addressed, the designer of an air conditioning condensate collection system can utilize this source of water as a viable supplement.

Notes and References

1. American Society of Heating, Air Conditioning, and Refrigeration Engineers (ASHRAE) is a technical research body for engineers and designers active in the design and installation of heating and air conditioning systems.
2. The design air temperature conditions can be obtained from ASHRAE Fundamental 2009, Chapter 14. The design air temperature is the outside air condition that is used for sizing air conditioning systems. This is not the maximum temperature the space may experience, but a statistical high condition such that it will not be exceeded more than a designated percentage of the time (typically between 0.5–5 percent) to avoid over-sizing the equipment. Over-sizing the equipment causes the system to short cycle to cool the space to set point temperature, but does not allow the system to run long enough to properly dehumidify the space. The result of having a cooling coil that is over-sized is a cold, clammy space.
3. Equivalent full load cooling hours are an energy estimating method that approximates the actual operating time, assuming equipment is run continuously at 100 percent capacity. Data shown is from ASHRAE Applications 2007, Chapter 32, Table 8.

CHAPTER 4

Harvesting Water from Dew

D ew harvesting is a viable source of freshwater when other sources of water are not available or as a supplement to a rainwater collection system. Although the water collected from a dew harvesting system is less in volume per event, it can be more reliable than rainwater collection since it is uniformly distributed all year round. As a bonus, the dew harvesting system can be designed to do double duty by also being part of a rainwater collection system.

The key points of the dew harvesting process are as follows:

- Cooling of the condensation surface
- Condensation of the water droplets on the surface
- Gathering the collected water from the condenser surface to the water storage tank

What Is Dew and How Does It Occur?

Dew is water droplets that form on exposed objects in the morning or evening (Fig. 4.1). As the exposed surface cools by radiating its heat to the nighttime sky, atmospheric moisture condenses forming water droplets on the collecting surface. Dew yield increases as the air temperature approaches the dew point temperature, which is the point where the air cannot hold any more moisture (100 percent relative humidity). When temperatures are low enough to allow the moisture to freeze, it is called frost.

Because dew condensation is related to the temperature of surfaces, it is formed most easily on surfaces that are not warmed by the conducted and radiated heat from the ground surface. The preferred weather conditions occur when there are clear night skies, little water vapor in the higher atmosphere, and sufficient humidity in the air where the dew collector is located. Objects where dew formation is commonly found include grass, leaves, railings, car roofs, and bridges.

FIGURE **4.1**   Spider web with dew.

Dew formation is accomplished by two competing effects: heating, by solar radiation, and cooling, mainly by infrared emission. During daylight hours, direct or indirect heating by the sun is greater than radiant cooling, causing evaporation to exceed condensation. During the night, without solar heating, the collecting surface cools allowing condensation to occur. Greenhouse gases present in the atmosphere, like carbon dioxide and especially water vapor, insulate the radiant heat loss and thereby limit the infrared cooling. The radiation exchanges in the temperature of the surfaces where dew forms are all-important in the formation of dew. With a dense cloud cover, radiant cooling can be inhibited to the point that surface temperatures do not drop below dew point and condensation cannot occur. It is for this reason that dew forms best during the clearest nights.

Unlike fog harvesting, where wind assists water harvesting, the turbulence created by wind increases heat transfer from ambient heat sources to the collector surface. The resulting warming of the collector surface can thereby prevent dew from forming.

Historical Applications

Early in the twentieth century, there was interest in high-mass air wells involving stacked beach pebbles and massive masonry construction, but despite much experimentation, this approach proved to be a failure (Fig. 4.2).

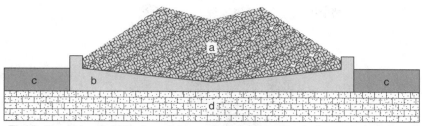

Zibold's dew condenser. (a) is a truncated cone of beach pebbles 20 m in diameter at the base and 8 m in diameter at the top. (b) is a concrete bowl; a pipe (not shown) leads away from the base of the bowl to a collecting point. (c) is ground level and (d) is the natural limestone base

(a)

(b)

FIGURE 4.2   (a) Knapen air well. (b) Zibold air well.[1]

The problem with the high mass collectors was that they could not get rid of sufficient heat during the night to reach dew point temperature at the collector surface.[2] From the late twentieth century onward, there has been much investigation of low-mass radiative collectors, which have proved to be much more successful.

Dr. Girja Sharan, from the Indian Institute of Management, tested a wide range of materials and got good results from galvanized iron and aluminum sheets, but found that sheets of a special 400 micrometers (0.016 inch) plastic foil developed by the OPUR[3] generally worked better and were less expensive. This plastic film, known as OPUR foil, is hydrophilic and is made from polyethylene mixed with titanium oxide and barium sulphate.

Newer materials may make even better collectors. One such material developed at Massachusetts Institute of Technology is inspired by the Namib Desert beetle (*Stenocara*), which survives only on the moisture it extracts from the atmosphere (Fig. 4.3).

The desert beetle has evolved to take perfect advantage of the tiny amount of water available in the desert. The surface of *Stenocara*'s armor-like shell is covered with bumps. The peak of each bump is

FIGURE 4.3   Namib desert beetle.

smooth like glass and attracts water. The slopes of each bump and the troughs in between are covered with wax, which repels water like Teflon. As morning fog sweeps across the desert floor, the water sticks to the peaks of *Stenocara*'s bumps, eventually forming droplets. When the droplets become large and heavy enough, they roll down from the top of the peaks and are channeled to a spot on the beetle's back that leads straight to its mouth. When a sea breeze blows over, the beetle leans into the wind. Tiny droplets of water are attracted to the hydrophilic peaks, where they build up into larger drops, which eventually roll down the beetle's back toward its mouth. Thus, the beetle has a ready source of drinking water even though it seldom rains.[4] Researchers at the Massachusetts Institute of Technology have emulated this capability by creating a textured surface that combines alternating hydrophobic and hydrophilic materials that gathers moisture out of the air from fog, dew and misting rain.[5]

Typical Dew Collection Installation

Unlike fog harvesting systems that require a vertical collection surface, dew collection requires a horizontal surface to take advantage of the nighttime radiant cooling. An efficient dew-condensing surface should be thin and lightweight, made of material with a high emissivity. It should be well insulated underneath and erected against the wind, with sufficient slope for rapid gravity draining of the condensate. Low-mass materials allow heat to be rapidly lost by means of nighttime convective and radiant cooling, thereby allowing the surface temperature of the collector to drop quickly. When the collector surface reaches the dew point temperature of the air, conditions for condensation are created. If the air is also very humid (relative humidity 85 percent or more) and the wind calm, then large amounts of dew condensation will occur.

Dew Collection Research

Dr. Girja Sharan's[6] research on dew collection in his native India showed initial findings that dew water collected from a greenhouse roof could be a significant source of drinkable water. This led to further research with ground and roof mounted systems and the testing of a wide range of materials. Materials studied included commercially available galvanized iron sheeting, aluminum sheeting, fiber-reinforced plastic, polycarbonate sheeting, and the specially made polyethylene film developed by OPUR (Fig. 4.4).[7] This hydrophilic surface, made from polyethylene mixed with titanium oxide and barium sulphate, mimics plant leaves or blades of grass in the manner it permits dew to condense on its surface. Dr. Sharan concluded that collection surfaces using the 400-micrometer (0.016-inch) OPUR foil yielded more harvested water than metals, was more cost-effective, and could easily overlay an existing structure.

FIGURE 4.4   Opur textured surface.

Water gathering performance from specifically designed dew collection systems has been shown to gather between 0.10 –0.50 liter/m² of dew per night. The variance in performance is a function of the quality of the collector surface and local weather conditions. Descriptions of working dew collection systems and their performance results are as follows:

Roof Mounted Systems

Site 1: Some of Dr. Sharan's early data was derived from a test stand built in the town of Kothara, India (Fig. 4.5). located 23° 14' north latitude, 68° 45' east longitude, and 21 meters above sea level.

This initial site consisted of two newly constructed corrugated insulated galvanized roofing panels (3 m by 3 m each), sloping (30 degrees), with one panel oriented east and the other oriented west (Fig. 4.6). Emissivity of the collection surface was measured at 0.23. The results were 1.32 liters/m²/month and 0.10 liter/m² of night dew occurring for the season of October 2005 to May 2006. A key point learned was that the west facing sloped surface collected 35 percent more dew than the east sloped surface because of being in the morning shade longer.

Site 2: Although less than optimal for a dew collection system, galvanized roofing is a popular dew collecting material that is already part of many building structures. Because the incremental investment is usually low, any water collected is seen as beneficial. The second site (Suthari warehouse), located 12 kilometers from the Kothara site, used the uninsulated roof of an existing

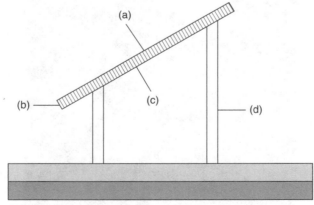

Diagram of a radiative condenser, a device designed to collect dew. (a) radiating/condensing surface, (b) collecting gutter, (c) backing insulation, (d) stand.

Figure 4.5   Kothara test stand schematic.

Figure 4.6   Kothara test stand.

warehouse (Fig. 4.7). Gutters and collection piping were added to the existing 343 m² roof, sloped north and south at a 16 degree slope. The results from this site were 0.5 liters/m²/month and 0.04 liters/m² of dew occurring per night. A key finding of this test site is that the uninsulated roofing, being older with a lower emissivity, produced 40 percent of the water compared to the newer galvanized roofing with an insulated back.

Figure 4.7   Suthari warehouse.

Site 3: OPUR film was used on the third site. The collection surface for Sayara school was a 360 m² existing concrete roof that had specially constructed condensing panels with a selective coating on the foil covering (Fig. 4.8). These panels were 25-mm (1-inch) styrene with 0.2-mm-thick polyethylene film laminated on top, sloped at 15 degrees equally to the north and south. The surface emissivity of the collection surface was measured at 0.83. The results were 1.26 liters/m²/month and 0.10 liter/m² of dew occurring per night.

(a)

(b)

Figure 4.8   (a) Sayara school. (b) Sayara students pumping water from dew collection system.

Site	Kothara (Test Stand)	Suthari (Warehouse)	Sayara (School)
Location (Latitude, Longitude, Meters Above Sea Level)	23° 14' N, 68° 45' E 21 meters a.s.l.	12 km west of Kothara 20 meters a.s.l.	23° 03' N, 69° 03' E 40 meters a.s.l.
Collector Surface Area (m²)	18	343	360
Surface Material	New Corrugated Galvanized Iron (New/Uninsulated)	Existing Corrugated Galvanized Iron (Old/Uninsulated)	Polyethylene w/ Selective Surface (New/Insulated)
Slope (Degrees)	30	16	15
Dates (# Days)	October 2005–May 2006 (243)	October 2005–May 2006 (243)	October 2005–May 2006 (243)
Dew Days/Total (%)	41.8	39.1	42.0
Cumulated Dew (mm)	10.5	4.0	10.1
Cumulated Dew (liters)	189.6	1372	3626
Liters/m²/Dew Night	0.104	0.042	0.100

TABLE 4.1 Roof System Dew Collection Summary

Ground Installed Dew Harvesting

Site 4: The Panandhro site at the Gujarat Mineral Development Corporation (GMDC) is an example of condenser-on-ground installation. This project consists of 10 modules of ridge-and-trough design that have been formed in the earth (Fig. 4.9). The ridges, each 35 m long, are built over gently sloping ground. The ridge is trapezoidal in form (top 50 cm, base 200 cm, with 30 degree sloping sides and a top-to-bottom dimension of 100 cm). The condenser surface is polystyrene insulation on the shaped bare ground covered with a plastic film. This installation measures 850 m² of surface area. This system's performance was comparable to the roof-mounted condensers, with water production of about 1.26 liters/m²/month and 0.10 liters/m² of dew occurring per night.

Although chemical analysis data is not available for water generated from galvanized iron, it is anticipated that this material, already widely used in the world for rain collection, will not impart any chemicals that would be a threat to the health of the user. The cumulated yield on a small two-sided roof of 18 m² area was higher than at Suthari on a per unit area basis—6.24 mm in the season of October 2004 to May 2005 and 10.53 mm from October 2005 to May 2006.

FIGURE **4.9** Panandhro.

Site	Panandhro	Satapar
Location (Latitude, Longitude, Meters Above Sea Level)	23° 14' N, 68° 45' E 21 meters a.s.l.	22° 5' N, 69° 16' E 25 meters a.s.l.
Collector Surface Area (m²)	850	600
Surface Material	Plastic film, with selective surface, on ground-mounted polystyrene insulation	Plastic film, with selective surface, on ground-mounted polystyrene insulation
Slope (Degrees)	30° sloping sides of trench	30° sloping sides of trench
Dates (# Days)	January–May 2007 recorded (68)	Installed April 2007
Dew Days/Total (%)	45	Not available
Cumulated Dew (mm)	16	Not available
Cumulated Dew (liters)	5560	Not available
Liters/m²/Dew Night	0.096	Not available

TABLE **4.2** Ground Surface Installed Collection Summary

Site 5: A similar installation to the Panandhro site was built at a school located in Satapar. This is an example of a condenser-on-ground, plastic film condenser surface with special moisture absorbing selective coatings and an insulation layer between collector film and the bare ground. This installation measures 600 m² of surface area and is owned and operated by a school as the primary source of potable water.

Site	Zadar	Komiza	Bisevo
Location			
(Latitude, Longitude,	43° 08' N,	43° 03' N,	42° 56' N,
	15° 13' E	16° 06' E	16° 47' E
Meters Above Sea Level)	5 meters a.s.l.	20 meters a.s.l.	100 meters a.s.l.
Collector Surface Area (m²)	1	1	15.1
Slope (Degrees)	30	30	30
Dates (# Days)	June 21, 2003– May 31, 2004 (344)	June 24, 2003– July 26, 2004 (399)	April 22, 2005– July 25, 2005 (95)
Dew Days/Total (%)	25	19	59
Cumulated Dew (mm)	13.3	6.1	11.8
Cumulated Dew (liters)	13.3	6.1	180
Liters/m²/Dew Night	0.53	0.32	0.20

TABLE **4.3**   Collection Summary—Croatia

Comparable research was done on 1 m by 1 m test panels in Croatia (Table 4.3). This also incorporated the polyethylene film with the selective surface (OPUR) and has shown improved performance. Further research is needed to determine exactly why the dew harvesting results were better in this case.

The conclusions from the initial roof and surface installed test sites showed that for the east/west oriented roofs, water production was 35 percent more on the west exposure due to longer time in the morning shade after sunrise.

Dew production was 2.5 times better for new roofing surfaces compared to the existing roof surfaces at the Suthari site. This was due to a higher emissivity with the newer roofing material, and in the case of the Sayara school, the application of plastic film with the selective surface coating.

The slope of the roof at 30 degrees seemed to generate slightly more condensate than the 15 degree sloping test site. Research from Kosovo indicated that 30 degrees was an optimum slope that combined the maximum horizontal projection to the night sky with enough slope to drain water from the collector surface.

Selective surface film on the condenser significantly improved dew collection output and was economical. It worked well as a cost-effective retrofit to an existing surface. Production was improved by approximately 40 percent when the collection surface was insulated underneath from heat radiating from the ground surface.[8]

New research is pursuing portable dew collection devices that can be used in mobile applications. The dew collection device, shown in Fig. 4.10, weighs just 400 grams, and under ideal conditions of 50 percent humidity or more, it can develop 1.5 liters of clean water per night.[9]

FIGURE 4.10   Dew collection device.

Other Applications

For agriculture applications, the effects of gravel and sand mulches on dew condensation were studied during the late summer and fall of 1999 in the semiarid loess[10] region of China. The results indicated that there were significant differences in daily dew amount between gravel mulch, sand mulch, and dry loess soil (control). The average dew amount for gravel mulch was 0.071 mm per day with extremes at 0.022 and 0.20 mm per day. The average values for sand mulch and dry loess soil were 0.12 and 0.15 mm per day, respectively. The minimum dew amount was 0.048 mm per day for sand mulch and 0.071 mm/day for dry loess soil, and the maximum dew amount was approximate 0.25 mm per day for both treatments. The results suggest that the lighter mass sand and dry loess soil ground cover are superior for dew collection than the surface stone mulch. Along with gathering moisture, the gravel was also beneficial for inhibiting soil evaporation and erosion.[11]

Conclusions and Recommendations

Although dew harvesting is not seen as a system that will generate large quantities of water, as seen at the Sayara school, it can be a viable option where no other option exists. Because it is available over a longer period of time, it can serve to complement the seasonal rainwater catchment system, while using the same basic structure.

Further research is needed to provide better data on improved collection surfaces and construction techniques, in order to establish effective means and methods to predict how much water can be harvested from a dew collection system. The site-specific variations in collector surface emissivity and water attractiveness, the interaction of weather humidity, temperature, and wind, plus time of exposure to condensing temperatures are all factors that make accurate prediction difficult. Before a significant financial commitment is made to develop a large-scale dew collection system, it is recommended that a test platform (as shown for the Kathara site) be constructed on site first. This will serve to confirm the viability of a dew collection system and to establish water-generating data for scaling up to a larger installation. Based on the research and installations described in this chapter, it can be expected that between 0.10–0.50 liters/m^2/night of clean potable water is an achievable goal.

As dew collection research continues, and with more sophisticated collection surface coatings being developed, it is reasonable to expect dew harvesting system performance and the better predictions about their performance will only improve.

Notes and References

1. Photo: International Organization for Dew Utilization.
2. There is considerable debate as to if these structures are dew collection systems or in fact ancient tombs. (See *The Moisture from the Air as Water Resource in Arid Region: Hopes, Doubts and Facts*, Kogan and Trahtman, versus *Comment on"The moisture from..."* from D. Beysens, et al.) It was left undecided as to why beach stones were specified by Zibold for the construction of this device. Since surfaces covered with seawater have an affinity for absorbing moisture from the air, it causes speculation that perhaps the hydrophilic properties of seawater were the source of producing water, and condensation was never the original intent. More research is needed to confirm or reject the merits of this supposition.
3. International Organization for Dew Utilization (www.opur.com).
4. Harries-Rees, K., "Desert Beetle Provides Model for Fog-Free Nanocoating" *Chemistry World News*, Royal Society of Chemistry, August 31, 2005. (www.rsc. org/chemistryworld/News/2005/August/31080502.asp - observed 2012 June)
5. Trafton, A., *Beetle Spawns New Material*, Professors Michael Rubner and Robert Cohen, MIT Tech Talk, June 14, 2006.
6. Sharan, G., *Harvesting Dew to Supplement Drinking Water Supply in Arid Coastal Villages of Gujarat*, W.P. No. 2007-08-05, Indian Institute of Management, Ahmedabad, India, August 2007.
7. International Organization for Dew Utilization (www.opur.com).
8. Sharan, G., Shaw, R., Milimouk-Melnytchouk, I., Beysens, D., *Roofs as Dew Collectors: I. Corrugated Galvanized Iron Roofs in Kothara and Suthari (NW India)*, September 11, 2006.
9. Alon Alex Gross's dew collector presented at Tuttobene, Milan, as part of his work with the MA Critical Practice program at Goldsmiths College, London, 2008.
10. *Loess* is defined as "a buff to gray windblown deposit of fine-grained, calcareous silt or clay." *American Heritage Dictionary of the English Language, 4th edition*, Houghton Mifflin Company.
11. Cold and Arid Regions Environmental and Engineering Research Institute, Chinese Academy of Sciences, Lanzhou, China.

CHAPTER 5

Water Gathered from Clouds

Harvesting freshwater from clouds is one of the more innovative methods of gathering freshwater, but it hasn't been unproven. Although unconventional, the technology behind cloud water collecting is remarkably simple, and it can be a critical source of water.

When the base of a cloud is in contact with the ground it is called fog. Present research suggests that fog collectors work best in coastal areas where the water can be harvested as the fog moves inland driven by the wind. However, this same technology has the potential to supply water in mountainous areas where stratocumulus clouds are present, at altitudes of approximately 1300 ft (400 m) to 4000 ft (1200 m).

What Is Fog?

Fog forms when the difference between the dry bulb temperature and the condensation temperature (dew point) is generally less than 2.5°C or 4°F. This represents a condition that approaches 100 percent relative humidity. A reading of 100 percent means that the air can hold no additional moisture and is at what is called the saturation, or dew point, condition. Either adding moisture to the air (warm air flowing over a water source) or dropping the ambient air temperature (wind carrying moist air up a mountain slope, dropping temperature with altitude) can increase relative humidity.

Fog can form at lower humidity, and sometimes it does not form with the relative humidity at 100 percent. The air will be supersaturated if moisture is added beyond this point, which presents conditions favorable for condensation in the form of rain or fog.

Fog formation requires all of the elements that normal cloud formation requires, the most important being condensation nuclei, in the form of dust, aerosols, pollutants, etc., for the water to condense upon. When there are exceptional amounts of condensation

FIGURE 5.1   Sea fog in the morning. (See also Color Plates.)

nuclei present, especially hygroscopic (water seeking) particles such as salt, the water vapor may condense below 100 percent humidity.

Sea fog is a fog caused by the peculiar effect of salt. Sea fog is common along coastlines where airborne salt particles are generated from the salt spray produced by breaking waves. These salt particles act as the nuclei for water to condense. Unusual to sea fog is that, due to the hygroscopic nature of salt, condensation can occur with humidity as low as 70 percent. Typically, such lower humidity fog is preceded by a transparent mistiness along the coastline as condensation competes with evaporation, a phenomenon that is typically noticed by beachgoers in the afternoon.

Suitability

Fog typically contains about 0.2 grams of water per cubic meter; the amount of water that can be harvested depends on surface area of the collector, efficiency of collection, and wind speed. Trees serve as natural fog catchers, and an isolated tree typically harvests 10 liters per square meter per day. Forests located in arid areas may drip as much water onto the land as occurs naturally from rainfall.

The keys to the feasibility of artificial fog catchers are meteorology and topography, as well as a community base to maintain the system. First, moisture-laden air must be driven by wind to a geographic barrier, such as a mountain range that intercepts the clouds. These conditions occur most commonly in a place where there are onshore winds and high seaside cliffs so that the air is forced upward into a cooler air mass, which creates appropriate conditions. The same combination can occur with interior mountain chains. Secondly, because the amounts that can be collected are pure in quality but modest in quantity, the more cost-effective fog-collecting site is best located near the point of use.

In order to implement a fog harvesting program, the potential for extracting water from fog first must be investigated. The occurrence of fog can be assessed from reports compiled by government meteorological agencies. To be successful, this technology should be located in regions where favorable climatic conditions exist. Since fog and clouds are carried to the harvesting site by the wind, the interaction of the topography and the wind will be influential in determining the success of the site chosen. The following factors affect the volume of water that can be extracted from fog and the frequency with which the water can be harvested:

- Frequency of fog occurrence is a function of atmospheric pressure and circulation, oceanic water temperature, and the presence of thermal inversions.

- Fog water content is a function of altitude, seasons, and terrain features.

- Design of fog water collection systems takes into account wind velocity and direction, topographic conditions, and the materials used in the construction of the fog collector.

Site Selection

Prior to implementing a fog water harvesting program, a pilot-scale assessment should be undertaken of the proposed collection system and the water content of the fog at the proposed harvesting site. The following factors should be considered in selecting an appropriate site for fog harvesting:

- Global wind patterns: Persistent winds from one direction are ideal for fog collection. The high-pressure area in the eastern part of the South Pacific Ocean produces onshore southwest winds in northern Chile for most of the year and southerly winds along the coast of Peru.

- Topography: It is necessary to have sufficient topographic relief to intercept the fog and clouds; examples, on a continental scale, include the coastal mountains of Chile, Peru, and Ecuador, and on a local scale, isolated hills or coastal dunes.

- Relief in the surrounding areas: It is important that there be no major obstacles to the wind within a few kilometers upwind of the site. In arid coastal regions, the presence of an inland depression or basin that heats up during the day can be advantageous, as the localized low pressure area thus created can enhance the sea breeze and increase the wind speed at which marine cloud decks flow over the collection devices.

- Altitude: The thickness of the stratocumulus clouds and the height of their bases will vary with location. A desirable working altitude is at two-thirds of the cloud thickness above the base. This portion of the cloud will normally have the highest liquid water content. In Chile and Peru, the working altitudes range from 400 m to 1000 m above sea level.

- Orientation of the topographic features: It is important that the longitudinal axis of the mountain range, hills, or dune system be approximately perpendicular to the direction of the wind bringing the clouds from the ocean. The clouds will flow over the ridge lines and through passes, with the fog often dissipating on the downwind side.

- Distance from the coastline: There are many high-elevation continental locations with frequent fog cover resulting from either the transport of upwind clouds or the formation of orographic clouds. In these cases, the distance to the coastline is irrelevant. However, areas of high relief near the coastline are generally preferred sites for fog harvesting.

- Space for collectors: Ridge lines and the upwind edges of flat-topped mountains are good fog harvesting sites. When long fog water collectors are used, they should be placed at intervals of about 4.0 m to allow the wind to blow around the collectors.

- Crestline and upwind locations: Slightly lower-altitude upwind locations are acceptable, as are constant-altitude locations on a flat terrain. But locations behind a ridge or hill, especially where the wind is flowing downslope, should be avoided.[1]

Satellite Imaging

Satellite imaging is a recent scientific development that can be used to pick likely areas for fog harvesting. The spatial coverage provided by satellites can show detailed climatological information on fog and low cloud occurrence. Before the launch of the new Meteosat Second Generation (MSG) weather satellites in early 2004, researchers used satellite images from the polar orbiting NOAA AVHRR satellites that have a pixel resolution of roughly 1 km^2. This allowed for regionally resolved studies of fog distribution. The drawback of these satellites was that they only covered a given geographic area a very few times a day, and the flight times differed from day to day. Moreover, the spectral resolution of the sensors was poor, which meant that fog occurrence could only be studied with sufficient accuracy in night-time images. During the day, the problem of separating images of fog from other clouds and snow has never been truly resolved. The MSG satellites are geostationary and can provide an image every 15 minutes, which means this new technology could potentially overcome the problems associated with the polar orbiting satellites. Although the 1 km^2 pixel resolution of an AVHRR satellite has not been attained yet, the MSG satellites now provide a reasonable spatial resolution that allows researchers to develop a suitable algorithm to retrieve fog and low cloud occurrence data.[2]

Technical Description of a Fog Collector

Fog collectors are a simple concept. The surface of fog collectors is usually made of fine-mesh nylon or polypropylene netting (e.g., shade cloth). It should be made of a relatively fine material that loses heat rapidly, with a black UV-stabilized fabric normally preferred. The fabric should shed its moisture load as rapidly as possible. To more effectively shed the water, vertical fibers are more important than horizontal. The efficiency of various mesh densities was tested. The fog collectors equipped with Raschel netting, with 35–45 percent coverage, mounted in double layer extracts, provided about 30–40 percent of the water from the fog passing through the collection netting (Fig. 5.2).[3] Other materials used with good results are air conditioning filter material and aluminum shade mesh.

The collecting surface is supported vertically by a post at either end and arranged perpendicular to the direction of the prevailing wind (Fig. 5.3). As fog blows through the netting, tiny water droplets are deposited, allowing the water contained in mist or fog to condense on the mesh surface. As the droplets become larger, they run down the net into gutters attached at the bottom. When two layers of collection netting are used, the rubbing of the two layers together in the wind increases water production from the collection systems. From there, water is channeled into cisterns and then, with a hose or pipe, is transported to where it is needed (Fig. 5.4).

Alternatively, the collectors may be more complex structures, made up of a series of collection panels joined together. The number and size of the modules chosen will depend on local topography and the quality of the materials used in the panels. Multiple-unit systems have the advantage of a lower cost per unit of water produced, and the number of panels in use can be changed as climate conditions and demand for water changes.

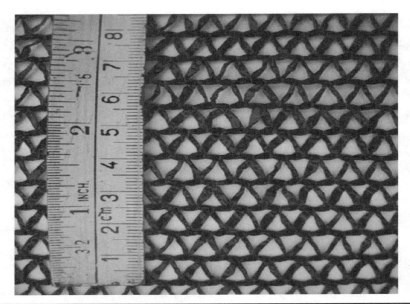

FIGURE 5.2   Raschel-type polypropylene mesh used in the study, with coverage of 45 percent (single layer).

FIGURE 5.3   Fog collection installation.

FIGURE 5.4 Gutter system to transport water from collector.

The collector itself is completely passive, and the water is conveyed to the storage system by gravity. If site topography permits, the stored water can also be conveyed by gravity to the point of use. The storage and distribution system usually consists of a plastic channel or PVC pipe approximately 4 inches (110 mm) in diameter, which can be connected to a ¾-inch (20 mm) to 1-inch (25 mm) diameter water hose for conveyance to the storage site or point of use. Storage can be a closed concrete cistern constructed on-site or an opaque plastic tank. Preventing the stored water from being exposed to sunlight will stop algae from growing on the water surface, which would result in contamination of the water.

For the same collection surface area, performance can be improved with vertical height. As water condenses and flows down the collection material, it gathers more moisture as it falls, thereby improving performance. This improved performance from added height must be traded off against the additional structural support needed to counter the torque from the added wind resistance.

Wind and Site

The role of the wind in fog water collection is important and can be leveraged to obtain optimal water harvesting. Studies by Schemenauer and Cereceda (1994), Schemenauer and Joe (1989), Bridgman et al. (1994), and Marzol (2008, in press) have shown that the optimal wind speed for collecting good water quantities is between 3.5 m/s and 9.0 m/s (8 mph to 20 mph).

Because of the variation in local fog occurrence, a test installation should be provided to confirm the viability of a fog collection system

Figure 5.5   Fog collector test site.

before construction is begun. The length of the test should run at least half a fog season. The results of this test should be compared to the minimum duration fog season that would make this technology cost-effective. In general, fog harvesting has been found to be cost-effective in arid regions when compared to alternative conventional options. An example of a fog collector test site installed in a Moroccan coastal community is shown in Fig. 5.5.

In a mountainous environment, the optimum elevation for the most productive collection of water is obtained from the wetter upper half of a stratoform cloud. For a coastal installation, care must be taken to assure that the water collected will not be contaminated with ocean salt. Collection from the upper elevations of the cloud, no less that 658 ft (200 m) above the ground, will improve the quality of the collected water.[4]

Water Storage

Generally, the storage cistern should be sized to provide at least 50 percent of the expected maximum daily volume of water consumed. But because the occurrence of fog is not perfectly predictable from day to day, it may be necessary to store additional water to meet demands on days when no fog water is collected.

The following is a guideline on how to estimate an appropriate cistern size.

Step 1: Estimating Water Demand

Surveys have shown the average daily water consumption is as follows:

- Per person it's about 4 gal (15 L).
- For goats and sheep it's 0.26 gal (1 L)
- For cows, donkeys, and camels it's about 2.6 gal (10 L).[5]

Step 2: Determining the Maximum Water That Can Be Collected

Harvested water volume (liters/month) = Fog occurrence (days/month) × Average collector efficiency (l/m^2/day) × Collector area (m^2)

Step 3: Calculating the Size of the Tank

The tank needs to be large enough to ensure the following:

- The required volume of water can be collected by the tank.
- The volume of water in the tank will be sufficient to meet demand during the drier months or through periods of low or no fog.
- The simplest way of estimating the cistern size needed to provide water throughout an average year is to use monthly fog data and to assume that at the start of the wetter months the tank is empty. The following formula should then be used for each month:

  $V_t = V_{t-1}$ + Net collected water

  V_t = Theoretical volume of water remaining in the tank at the end of the month

  V_{t-1} = Available volume of water left in the tank from the previous month

  Net collected water = (Water harvested – demand – leakage)

  Demand = (See Step 1)

  Water harvested = (See Step 2)

Leakage (also includes evaporation loss) is estimated between 5–20 percent of volume in the tank (V_t), and it is a function of the time between fog harvesting events and air relative humidity while water is stored.

Starting with the tank empty, $V_{t-1} = 0$. If after any month, the net collected water is positive, then water will be added to V_{t-1} up to the capacity of the tank. If the right side of the equation exceeds the capacity of the tank, then water is lost to overflow, at which time V_{t-1} for the next period equals the volume of the tank (V_t). If after any month, the net collected water is less than the available water (V_t), V_{t-1} for the next period will equal 0.

Tank size is not necessarily based on collecting the entire fog harvest volume per fog occurrence. For example, if the maximum water that can be collected per month is about 40 gal (150 L), and if the water demand is less than this, then some overflow may occur while demand is still met. Optimally, calculations should be repeated using various tank sizes until $V_t \geq 0$ at the end of every month. The greater the values of V_t over the whole year, the greater the security of meeting water demand when fog harvest volume falls below average or when dry periods are longer than normal. However, *the greater the security, the higher the cost of the tank storage.*

Water Purification

Water quality can best be assured by maintaining the cleanliness of the system. Occasional purification of water in the storage tanks may be necessary if the water is used for drinking without first boiling. The normal contamination source could be expected to come from bird or rodent residue. Heavy metal contamination, commonly caused by proximity to industrial or agricultural sites, is best avoided by judicious selection of the fog collecting site. For some recommended solutions to maintain the purity of the stored water, see Chapter 11: Managing Water Quality.

Operation and Maintenance

Operating fog collection technology is very simple after the system and associated facilities are properly installed. Training of personnel to operate the system might not be necessary if the users participate in the development and installation of the required equipment. A very important factor in the successful use of this technology is the establishment of a routine quality control program. This program should address both the fog collection system and the possible contamination of the harvested water, and it should include the following:

- Inspection of support mesh net cable tensions. System misalignment, caused by improper cable tension, can result in water loss from failure to collect the water in the receiving system. It can also cause structural damage to the collector panels.

- Inspection of cable fasteners. Loose fasteners in the collection structure can cause the system to collapse and/or be destroyed.

- Inspection of horizontal mesh net tensions. Loose nets will lead to a loss of harvesting efficiency and can also break easily.

- Maintenance of mesh nets. After prolonged use, the nets may tear. Tears should be repaired immediately to avoid having to replace the entire panel. Algae can also grow on the surface of the mesh net after one or two years of use, accumulating dust, which will cause discoloration in the collected water and offensive taste and odor problems. The mesh net should be cleaned with a soft plastic brush as soon as algal growth is detected.

- Maintenance of collector drains. A screen should be installed at the end of the receiving trough to trap undesirable materials (insects, plants, and other debris) and prevent contamination of water in the storage tank. This screen should be inspected and cleaned periodically.

- Maintenance of pipelines and pressure outlets. Pipelines should be kept as clean as possible to prevent accumulation of sediments and decomposition of organic matter. Openings along the pipe should be built to facilitate flushing or partial cleaning of the system. Likewise, the pressure outlets should be inspected and cleaned frequently to avoid accumulation of sediments. Openings in the system must be protected against possible entry of insects, vermin, and other contaminants.

- Maintenance of cisterns and storage tanks. Tanks must be cleaned periodically with a solution of concentrated calcium chloride to prevent accumulation of fungi and bacteria on the walls. Any holes that admit light into the tank need to be patched to prevent algal growth.

- Monitoring of dissolved chlorine. If bacterial growth is a concern based on observed health issues, periodic doses of chlorine may be advisable. Although chlorine is a universally accepted treatment for bacteria, it has limited effect on cysts and viruses (e.g., *Giardia* and *Cryptosporidium*), which normally come from animal feces. If dosing is necessary, 1 gal (3.8 L) of raw water can be disinfected with 8–16 drops of regular household bleach (visually about ¼ of a teaspoon— double that if the water is cloudy). Agitate and let stand 30 minutes. Long-standing water in tanks will be disinfected with 1 Pt (0.5 L) of household bleach per 1000 gal (3750 L). Immediately after treating, water must initially have a slight smell of chlorine. If it does not, repeat the process. Shock treating with chlorine is preferred to maintaining a constant level to deter the development of chlorine resistant organisms.[6]

Advantages of a Fog Harvesting System

A fog collection system can be easily built or assembled on-site. Installation and connection of the collection panels is quick and simple. Assembly is not labor intensive and requires little skill. Some advantages are as follows:

- No energy is needed to operate the system or transport the water.
- Maintenance and repair requirements are generally minimal.
- Capital investment and other costs are low in comparison with those of conventional sources of potable water used, especially in mountainous regions.
- The technology can provide environmental benefits when used in national parks in mountainous areas or as an inexpensive source of water supply for reforestation projects.
- It has the potential to create viable communities in inhospitable environments and to improve the quality of life for people in mountainous rural communities.
- The water quality is better than from existing water sources used for agriculture and domestic purposes.

Disadvantages of a Fog Harvesting System

- This technology might represent a significant investment risk unless a pilot project is first carried out. The purpose of the pilot project is to quantify the frequency of fog events and to determine the potential water collection yield from the site under consideration.
- Community participation is required in the process of developing and operating the technology in order to reduce installation and operating and maintenance costs.
- If the harvesting area is not close to the point of use, the installation of the pipeline needed to deliver the water can be very costly in areas of high topographic relief.
- The technology is very sensitive to changes in climatic conditions which could affect the water content and frequency of occurrence of fog; a backup water supply to be used during periods of unfavorable climatic conditions is recommended.
- In some coastal regions, fog water has failed to meet drinking water quality standards because of concentrations of chlorine, nitrate, and some minerals from local industry and agriculture. The potential for upwind contamination sources needs to be investigated before construction is initiated.
- Caution is required to minimize impacts on the landscape and the flora and fauna of the region during the construction of the fog harvesting equipment and the storage and distribution facilities.
- Despite technical feasibility, fog catchers have not caught on in many parts of the world, mainly for sociological and

institutional reasons. Although initially fog harvesting systems have been erected and supported enthusiastically, they have been allowed to degrade once funding has run out and/or the initial novelty has worn off.

The Need for Community Support

In applying this technology, it is recommended that the end users fully participate in the construction of the project. Community participation will help to reduce the labor cost of building the fog harvesting system, provide the community with operation and maintenance experience, and most importantly, develop a sense of community ownership and responsibility for the continued success of the project. Government subsidies, particularly in the initial stages, might be necessary to reduce the cost of constructing and installing the facilities. A cost-sharing approach could be adopted so that the end users will pay for the pipeline and operating costs, either in funding or labor to install and maintain. But in the end, the success or failure of the project will rely on the end users maintaining the operational status of the project. As such, the community's involvement with the project is all-important for its success.

Costs

Actual costs of fog harvesting systems vary from location to location. The costs are particularly sensitive to the distance between the ideal location for the fog catcher nets (on top of cliffs) to the village (typically well below the crest or on the coast). In a project in the region of Antofagasta, Chile, the installation cost of a fog collector was estimated to be $8.40 USD/ft^2 ($90 USD/m^2) of mesh. In another project in northern Chile, the cost of a 516 ft^2 (48 m^2) fog collector was approximately $378 USD ($225 in materials, $63 in labor, and $39 in incidentals). This latter system produced a yield of 3.0 l/m^2 of mesh/day. The most expensive item in this system was the pipeline that carried the water from the fog collection panel to the storage tank located in the village.

Maintenance and operating costs are relatively low compared to other technologies. In the project in Antofagasta, the operation and maintenance cost was estimated at $600 USD/year. This cost is significantly less than the more complicated system that involved the pipeline. Operating costs in the latter project were estimated at $4740 USD and maintenance costs at $7590 USD (resulting in a total cost of $12,330 USD/year).

Both the capital costs and the operating and maintenance costs are affected by the efficiency of the collection system, the length of the pipeline that carries the water from the collection panels to the storage areas, and the size of the storage tank. For example, the unit cost for a system with an efficiency of 0.05 gal/ft^2/day (2.0 L/m^2/day) was estimated to be $18.20 USD/1000 gal ($4.80 USD/1000 L/day). If the efficiency was improved to 0.12 gal/ft²/day (5 L/m²/day), then the unit cost would be reduced to $7.20 USD/1000 gal/day ($1.90 USD/1000 L). In the Antofagasta project, the unit cost of production was estimated at $5.34 USD/gal ($1.41/1000 L) with a production of 0.06 gal/ft²/day (2.5 L/m²/day).[7]

Extent of Use

The foggiest place in the world is the Grand Banks off the island of Newfoundland, where the cold Labrador Current from the north meets the much warmer Gulf Stream coming from the south. The foggiest land areas are Menomonie, Wisconsin; Point Reyes, California; Argentina; Newfoundland; and Labrador—all with over 200 foggy days per year. Even in generally warmer southern Europe, thick fog and localized fog is often found in lowlands and valleys. Fog is prevalent in parts of the Po, Arno, and Tiber Valleys in Italy, as well as on the Swiss Plateau that extends beyond the border of Switzerland to its southwestern end in France.

In ancient times, fog water was often collected for domestic and agricultural use. The inhabitants of what is now Israel used to build small, low, circular honeycombed walls around their grapevines so that the mist and dew could precipitate in the immediate vicinity of the plants. In the Atacama desert, which stretches 600 miles (1000 kilometers) from Peru's southern border into northern Chile, both dew and fog were collected by means of a pile of stones arranged so the condensation would drip to the inside of the base of the pile, where it was shielded from the day's sunshine.

The same technique was employed in Egypt, with the collected water stored underground in aqueducts. In Gibraltar, a similar technique was used: a large area on the slope of the rock was covered with cement blocks. As fog and rainwater ran downward, it was collected underground where evaporation was minimized.

On a smaller scale, rain, fog, and dew are collected on enormous granite rocks at Cape Columbine lighthouse at Paternoster on the West coast of South Africa. Low retaining walls have been cemented onto the sloping rock surface to channel the water into a reservoir at the base of the outcrop. The first fog collection installation in South Africa was at Mariepskop in Mpunalanga, in 1969–1970. It was used as an interim measure to supply water to the South African Air Force personnel manning the Mariepskop radar station. Two large fog screens, constructed from plastic mesh and measuring about 92 ft by 11 ft (28m by 3.5m), erected at right angles to each other and to the fog- and cloud-bearing winds, yielded more than 0.27 gal/f². (11 L/m²) per day.

In Chile, the National Forestry Corporation (CONAF), the Catholic University of the North, and the Catholic University of Chile are implementing the technology in several regions, including El Toro, Los Nidos, Cerro Moreno, Travesía, San Jorge, and Pan de Azúcar. The results of the several experiments conducted in the northern coastal mountain region indicate the feasibility and applicability of this technology for a variety of purposes, including potable water and water for commercial, industrial, agricultural, and environmental uses. These experiments were conducted between 1967 and 1988 at altitudes ranging from 1700 ft. to 3100 ft (530 m to 948 m) using different types of fog water collectors.

In Peru, the National Meteorological and Hydrological Service (SENAMHI) has been cooperating with the Estratus Company since the 1960s in implementing the technology in the following areas: Lachay, Pasamayo, Cerro Campana, Atiquipa, Cerro Orara (Ventinilla-Ancón), Cerro Colorado (Villa María de Triunfo), and Cahuide Recreational Park (Ate-Vitarte). In southern Ecuador, the Center for Alternative Social Research (CISA) is beginning to work in the National Park of Machalilla on Cerro La Gotera using the Chilean installations as models.

Despite technical feasibility, fog catchers have not caught on in many parts of the world, mainly for sociological and institutional reasons. The technology was proven in coastal settlements in northern Chile (and later southern Peru and Ecuador), where conditions are ideal for its application: arid conditions on land, local communities very short of water, strong moisture-laden on-shore winds, and seaside cliffs hundreds of meters high. The fog catchers erected and harvested about 0.07 gal/ft²/day (3 L/m²/day). And in the early 1990s, the array produced on average some 270 gal/ft²/day (11,000 L/m²/day), an amount sufficient to provide everyone in the small village nearby with about 8.6 gal (33 L) per day. (Water had previously been trucked 25 mi (40 km) to the village, and given the cost, average use was only 13.7 gal (14 L) per day.)

Fog catcher arrays have also been erected in Mexico, Namibia, Nepal, Oman, and South Africa, with tests sites in several other countries (Fig. 5.6). A review of literature relevant to fog collection, together with field visits, scattered test sites, and a set of key criteria for feasibility, showed that there are more than 20 countries (scattered across all continents) where fog catchers might potentially play a role; many of these countries are chronically or seasonally short of water.

Conclusion

Fog collection is an ancient and proven technology that is a viable source of clean water suitable for agriculture and for domestic use. Problems may occur in the maintenance and ongoing servicing of these systems to keep them operational. Providing sufficient storage volume, to bridge the occurrences when fog is not available, is also an important issue that determines its value as a reliable source of water.

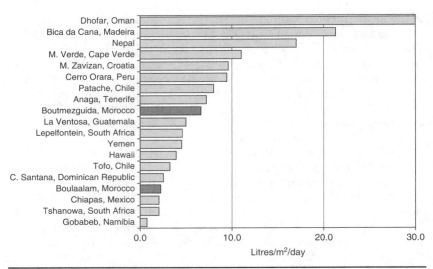

Figure 5.6 Water quantity samples from fog collection.

Figure 5.7 Community spigot for water produced from fog.

"In terms of quality and magnitude of yield, fog harvesting could go a long way to alleviating water shortage problems in the fog-prone mountainous regions of the country. The costs are low. The technology is simple and the source is sustainable for hundreds, even thousands of years."[8]

Notes and References

1. Organization of American States Publication, *Source Book of Alternative Technologies for Freshwater Augmentation in Latin America and the Caribbean,* Chap. 1.3, Fog Harvesting 1999.
2. Cermak, J., *A New Satellite-Based Operational Fog Observation Scheme,* SOFOS, Marburg: Marburger Geographische Gesellscraft, 2007.
3. Schemenauer, R.S., Suit, M., "Alternative Water Supply for Chilean Coastal Desert Villages," *Water Resources Development,* 8(1), 1992.
4. Román, R.L., "Obtención de agua potable por métodos no tradicionales" ("Getting Drinking Water from Nontraditional Methods"), *Ciencia al Día Internacional* (*International Science Day*), Vol 2 (2), May 1999.
5. Marzol, M.V., Megia, J.L.S., *Fog Water Harvesting in Ifni, Morocco, an Assessment of Potential and Demand,* La Laguna and Santa Cruz de Tenerife, 2008.
6. Organization of American States Publication, *Source Book of Alternative Technologies for Freshwater Augmentation in Latin America and the Caribbean,* Chap. 1.3, Fog Harvesting 1999.
7. Organization of American States Publication, *Source Book of Alternative Technologies for Freshwater Augmentation in Latin America and the Caribbean.*
8. Olivier, J., *Fog Harvesting for Water—Clouds on Tap,* University of South Africa's Department of Anthropology, Archaeology, Geography and Environmental Studies.

CHAPTER 6

Glacier Water Reclaim

Introduction

Artificial glaciers are a simple water-harvesting technique suited for high-altitude cold deserts where inhabitants are totally dependent on glaciers as a water source. In this region that seldom sees rain, the glaciers—the fountainheads of water for these regions—are receding at a rapid rate. The snowfall that occurs increasingly melts away quickly and drains into rivers without any beneficial use. The ancient knowledge of artificial glaciers extends the reliability of water supplies to villages in valleys where glacial melt water tends to run out before the end of the growing season.

With winters getting shorter and warmer, natural glaciers are shrinking, and the implementation of artificial glaciers permits the beneficial use of glacial meltwater over a wider period of time than that which occurs naturally now. Natural glaciers at high mountain elevations melt slowly in summer and reach lower elevation villages approximately in midsummer. In contrast, artificial glaciers start melting in spring, conveniently when the first irrigation requirement for crops is needed. As temperatures rise with the approach of spring, the glaciers melt in sequence as the temperature rises at each elevation, allowing water to flow earlier and run later than it occurs without artificial glaciers. With water available over a wider growing season, agriculture productivity increases, which has a positive influence on the local economy.

There are three types of artificial glaciers, which cause similar results, but their construction varies:

- *Aufeis*, or management of mountain stream flow to create multiple layers of ice
- Glacier Grafting, or the enhancement of an existing glacier
- Glacier creation by impoundment of snow and ice

Figure 6.1   Glacier from space.

Artificial Glaciers

Aufeis (German for "ice on top") is a sheetlike mass of layered ice that forms from successive flows of groundwater over an ice surface in freezing temperatures. This form of ice is also called *overflow*, *icings*, or the Russian term, *naled*. According to E. de K. Leffingwell,[1] the term was first used in 1859 by A. T. von Middendorff[2] following his observations of the phenomenon in northern Siberia.

Aufeis accumulates during winter along stream and river valleys in arctic and subarctic environments. It forms by the upwelling of river water behind ice dams or by groundwater discharge. The latter mechanism prevails in high-gradient alpine streams as they freeze solid. Groundwater discharge is blocked by ice, disrupting the steady-state condition and causing an incremental rise in the local water table until discharge occurs along the bank and over the top of the previously formed ice. Successive ice layers lead to *aufeis* accumulations that can be several meters thick (Fig. 6.2). *Aufeis* typically melts out during summer but can form in the same place year after year.[3]

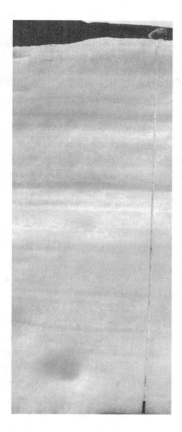

Figure 6.2 An *aufeis*-type artificial glacier is formed as progressive layers of water flowing across an existing ice surface refreeze into multiple layers.

The blockage of the culverts and pipelines can actually help to block flow and lead to the development of more extensive *aufeis*.[4] The research that has been done on *aufeis* has to a large extent been motivated by the variety of engineering problems the ice sheets can cause (e.g., the blocking of drainages and resultant flooding of roads).

Glacier Grafting

Local legend from the 13th century says that when news of Genghis Khan and his marauding Mongol hordes were approaching what is now northern Pakistan, the people there kept the invaders out by blocking the mountain passes by growing glaciers across them. Whether or not these stories are true, the art of glacier growing—also known as glacial grafting—has been practiced for centuries in the mountains of the Hindu Kush and Karakorum ranges.

Legends aside, no one really knows when the first glacier was grown in this region. Inayatullah Faizi, assistant professor in social sciences at the Government Degree College at Chitral in Pakistan's

North-West Frontier Province, cites evidence of a glacier being grown for irrigation purposes as long ago as 1812. However, the first documented reference to the practice does not appear until more than a century later when a British colonial administrator called D. L. R. Lorimer reported it in the 1920s. Though Lorimer described the practice as obsolete, the tradition of glacier growing survived. Today, the skills and know-how needed to grow glaciers are kept alive by a handful of village elders.

In the village of Hanouchal Haramosh in the Karakoram mountains, villagers struggle with their uncertain water supply. Most precipitation falls as snow at altitudes, while the inhabited valleys stay largely dry. To irrigate their fields, villagers rely on snow melting up in the mountains, but by the end of the growing season most of the previous winter's snow has gone. The water dries up, and crop yields suffer.

To successfully graft a glacier, finding the right site is crucial. Usually the glaciers are built in rocky areas prone to ice accumulation at elevations of about 14,800 feet (4500 meters) and where there are small boulders about 10 inches (25 centimeters) across. The rocks serve to protect the ice from sunlight and to trap ice in the gaps between them. Successful glacier grafting commonly occurs on northwest-facing slopes of steep cliffs, atop already advancing glacier slopes. Naturally occurring avalanches and rockfall from the overshadowing cliffs serve to enhance the glacier formation process (Fig. 6.3).

Glacier growers graft glaciers by digging deep into the face of the ice flow to expose "male" ice so that "female" ice can be added, along with large boulders and rocks (Fig.6.4). According to tradition, "male" glaciers are covered in soil or stones and move slowly if at all, while "female" glaciers are whiter, grow faster, and yield more water. In order to grow a glacier, tradition requires equal amounts of both types of glaciers to consummate a successful artificial glacier— just like the birds and the bees, only colder (Fig. 6.5). Just stacking snow or rock alone will not create a sustainable artificial glacier.

The grafted ice is then insulated with charcoal, sawdust, wheat husk, nutshells, or pieces of cloth, and gourds of water are added. Freezing temperature bursts the gourds, spilling the contents, and the process of refreezing bonds the material together. Snowmelt trapped in the young glacier freezes, creating more ice, and cold air pockets moving between the rocks and ice keep the glacier cool. When the newly created frozen mass is heavy enough, it begins to creep downhill, forming a self-sustaining glacier within four years or so. The end result is not quite a true glacier, but growing and flowing areas of ice many tens of meters long, have been reported at the sites of these grafts. When winter arrives, snow bridges the areas between the ice and, over a few years, it forms into a self-sustaining glacier.

GLACIERS TO ORDER

According to tradition in the Karakoram and Hindu Kush mountains, there are several stages to growing a glacier

Select a site above 4500 m, facing north-west, surrounded by steep cliffs. Ideally it will be rocky, with ice trapped between small boulders of about 25 centimetres diameter

Take 300 kilograms of "female" ice (while, clean and made from snow) and pile it on top of "male" ice (containing stones and soil)

Place gourds of water in the cracks. These will burst when the glacier grows. The contents then freeze and bind the ice together

Cover it with charcoal, sawdust, wheat husk, nutshells or pieces of cloth to insulate the new glacier and leave it for approximately 4 winters

Each season, snowmelt will be trapped by the glacier and refreeze, making it grow. The budding glacier will eventually creep down the hill like a natural glacier

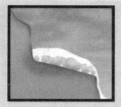

FIGURE 6.3 Schematic of glacier grafting procedure. (See also Color Plates.)

Figure 6.4   Glacier growers will often dig for a meter or more into the face of an existing glacier to find the in-situ "male" ice.

Figure 6.5   A village elder packs the appropriate mix of rock and ice in the shadows of boulders.

Though this practice has its skeptics, there seems to be real merit to it. Researcher Ingvar Tveiten from the Department of International Environment and Development Studies at the Norwegian University of Life Sciences seem to support the locals' methods of glacier farming. Hermann Kreutzmann, a glaciologist at the Free University of Berlin

FIGURE 6.6   Local villagers carrying material to graft a glacier.

in Germany, witnessed a glacier grafting ceremony in Hunza, near Gilgit, in 1985. "It seemed very plausible to me to search for a specific location at the appropriate altitude with a tolerable temperature regime and to place ice there," he says. As ice can absorb and retain water, he reckons that "a substantial amount of ice in a proper location might indeed augment water supplies."

So how well does it work? Villages that have undertaken glacier-growing projects have reported increased water flows. Locals are so convinced of the efficacy of the concept that they continue the heavy lifting required to graft an artificial glacier (Fig. 6.6), and the Pakistani government continues to fund glacier growing efforts as it has for several decades.

Glacier Impoundment[5]

In the cold desert of Ladakh, India,[6] the only source of water is melt from glaciers that naturally occurs in late summer. A great example of how decentralized approaches to water harvesting boost innovation, comes from the pioneering work of Chewang Norphel, a retired engineer of the Indian Department of Rural Development. Norphel developed the concept of impoundment to grow artificial glaciers by combining the benefits of freezing water, as in the *aufeis* process, with impoundment, as in the glacier grafting process. He developed a way to bring glaciers closer to villages by channelizing water from seasonally occurring streams to the shadow area of a mountain, where it can freeze. Thus, the water is stored for future use, rather than being wasted by flowing away to the sea.

At an altitude of more than 14,000 feet, Ladakh has a severe climate and inhospitable terrain, so farmers were previously only able to plant and harvest a single crop each year—wheat, barley, or peas. Since it seldom rains in the area, farmers were heavily dependent on glacier melt to supply water for irrigation. Working against the farmers was the short growing season that began and ended before the bulk of natural glacier meltwater began to flow to the region. Norphel saw firsthand how a more reliable source of irrigation water, if available when critically needed for the growing season, would improve the well-being of the farmers and their community.

What sparked the technique that has now been proven successful was a simple observation. In winter, he noticed water taps were usually kept fully open so the water would run continuously and not freeze. The water that flowed into drains below the tap was essentially wasted, but some of the water surrounding the drain froze. "While there was such a shortage of water at the start of the growing season, I saw a lot of water just running off and getting wasted in winter," said Norphel. "And it was then that it occurred to me, why not try and make artificial glaciers, thereby storing the water for farmers in such a manner that would give them a head start by providing a supply of water at the start of the growing season when most needed." This concept of freezing the water for storage was enhanced by using the simple principle that glaciers melt at different times at different altitudes. *So by providing the water impoundments at different elevations, the water would melt in sequence with approaching warmer weather, thereby extending the time period when water is available.*

Construction

The artificial glacier impoundment is simple and relatively cheap to build. At regular elevation intervals, stone embankments are constructed (Fig. 6.7), which impede the flow of water and make shallow pools (Fig. 6.8). Before the onset of winter, the water flow is distributed by a network of pipes to these small ponds located on the shaded side of a hill, where the winter sun is blocked by a ridge or a mountain slope.

The distributing chamber piping is constructed of a grid of 1.5-inch diameter pipe installed 5 feet on centers for proper distribution of water (Fig. 6.9). Although metal pipes have been used successfully, piping resistant to freezing, such as PEX,[7] is recommended if available. The water flows are made to flow out on a sloping hill face, in small quantities and at low velocity through the distribution pipes, where the water freezes instantly. During the winter, as temperatures fall steadily, the water collected in the small pools continues the

FIGURE 6.7 Retaining walls are constructed to slow water flow down the mountain so it can freeze and create the artificial glacier. (Photos by Nick Pattison, *Scientific American*, May 2010)

FIGURE 6.8 Retaining walls for a storage reservoir, fed by runoff from the glacier higher up the slope. (Photos by Nick Pattison, *Scientific American*, May 2010)

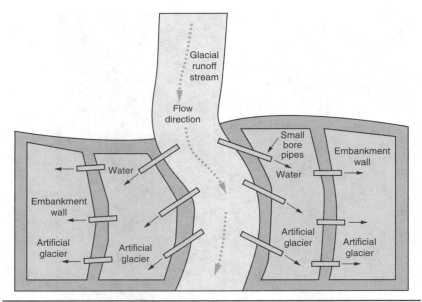

Figure 6.9   Schematic of artificial glacier concept showing diversion piping into ponds that create the artificial glacier.

freezing process. The process of ice formation continues for three to four winter months, and a huge reserve of ice accumulates on the mountain slope, and this becomes the artificial glacier. Once this cycle has been repeated over many weeks, a thick sheet of ice forms that resembles a long, thin glacier (Fig. 6.10).

Figure 6.10   During winter, the glacier formed as planned, storing water that would otherwise have flowed away past the village of Stakmo, India. (Photos by Nick Pattison, *Scientific American*, May 2010)

FIGURE 6.11   Glacial flow during the crucial sowing season—the artificial glacier began melting, releasing much-needed water to the villagers. (Photos by Nick Pattison, *Scientific American*, May 2010)

In spring, the ponds melt sequentially at varying elevations as the warmer weather returns.[8] (Fig. 6.11)

The cropping window, or period in which cultivation can be done, lasts seven months before severe winter arrives in October. Any delay in sowing crops can wipe out a harvest because the crop may not mature in time to beat the cold weather. Today, the residents of Changla can sow their crops on schedule thanks to the artificial glacier that Norphel constructed on the mountain slopes above their village. Phunsok, a farmer from Changla, said Norphel's handiwork has been a big boon for the area. "The true merit of the technology," he said, "lies in the low cost and the minimal maintenance that is required for the upkeep of these artificial glaciers." Sonam Dawa, executive director of the Ladakh Ecological Development Group in Leh, is also enthusiastic about Norphel's achievement. "Artificial glaciers are the cheapest option to the irrigation water needs of this inhospitable cold desert." The novel technique, he added, "offers an elegant solution for the most critical first watering of crops." (Fig. 6.12)

So far Norphel has helped Ladakhi peasants construct five artificial glaciers to increase water supplies in their villages. Although the experiment in Ladakh is still small and site specific, word about its success is spreading and several more are being planned. The largest artificial glacier Norphel has built so far is near the village of Phuktsey. About 1000 feet (300 meters) in length and 150 feet (45 meters) wide, it has an average depth of 4 feet (1 meter) and can supply irrigation

FIGURE 6.12 Greening the Valley. Previously barren fields in Stakmo were turned green by the stored water from the artificial glacier. (Photos by Nick Pattison, *Scientific American*, May 2010)

water to the entire village of about 700 people. Norphel says it was built at a cost of about U.S. $2000.[9]

Some observers think the technique may one day bring relief to many other water-starved villages around the world that face similar conditions. "I have not visited the site, but if it is successful, it is a remarkable achievement," said V. C. Thakur, a geologist and former director of the Wadia Institute of Himalayan Geology in Dehradun, India. "I have never heard of an example like this from anywhere else in the world," he added. In recognition of his achievement, the Far Eastern Economic Review honored Norphel with an Asian Innovation Award in 1999.

Conclusion

The concept of an artificial glacier is simple: During winter, the glacier is formed by the creation of restrictions to seasonal water flow. The restricted water freezes and thereby stores the water that would otherwise have flowed away. When warmer weather comes in the spring, the water melts to become available for use by the local residents.

This technology is applicable in areas that have features like a 15,300 to 17,500 ft (4,666 to 5,333 m) altitude, temperatures as low as −2 to +2 °F (−15 to 120 °C) Celsius during peak winters, and long winter periods of four to five months that assure longer expansion and formation of glaciers.

To maintain such a structure and for the project investment to be sustainable over time, it is crucial that the community being served by this project participate in the associated time, labor, and money. A direct correlation between the individual's labor and the benefits needs to be apparent. In addition, a commitment to the ongoing maintenance of the glaciers needs to be established and a fair method of water distribution arrived at.

Artificial glaciers help in the recharging of groundwater and rejuvenation of springs, the development of pastures for cattle rearing, and the reduction of water sharing disputes. The reliability of the water supply enables farmers to farm with confidence and earn dependable income from their efforts (Fig. 6.13).

Can artificial glaciers help compensate for the disappearance of naturally forming ones? Scientists and aid agencies are studying communities in mountainous regions of India and Pakistan that have a long tradition of using artificial glaciers to enhance their water supplies. As these remote mountain communities come under increasing

FIGURE 6.13 With a more constant water supply, villagers began planting more than one crop per year. (Photos by Nick Pattison, *Scientific American*, May 2010)

pressure from population growth and climate change, researchers and development agencies are starting to take a serious look at how their ancient technologies can better sustain the local water supplies. With observation, and informal interviews with practitioners of glacier growing, there is an increasing acknowledgment among the scientific community as to the merits of this ancient technique. The study of glacier growing, within the context of glaciology, is expanding to determine how natural processes affect the concept of glacier growing and how this concept can find wider application.[10] If their techniques can be popularized, they can bring stability to communities in other areas where climate change has diminished the glaciers and the availability of water.

Notes and References

1. Ernest de Koven Leffingwell (1875–1971) was an arctic explorer, geologist, and Spanish-American War veteran. During the period from 1906 to 1914, Leffingwell spent nine summers and six winters on the Arctic coast of Alaska, where he created the first accurate map of a large part of the Alaskan arctic coastline. He was the first to scientifically describe permafrost and to pose theories about permafrost that have largely proven true. (Wikipedia).
2. Alexander Theodor von Middendorff (18 August 1815-24 January 1894) was a Baltic German zoologist and explorer.
3. Hu, X., and Pollard, W., "The Hydrologic Analysis and Modeling of River Icing Growth, North Fork Pass, Yukon Territory, Canada," *Permafrost and Periglacial Processes*, v. 8: p. 279–294, 1997.
4. Kane, D., 1981, "Physical Mechanics of *Aufeis* Growth," *Canadian Geotechnical Journal*, v. 8: no. 2, p. 186–195, 1981.
5. Based on the pioneering work of Chewang Norphel, Ladakh, India.
6. Ladakh is a region of Jammu and Kashmir, the northernmost state of the Republic of India. It lies between the Kunlun mountain range in the north and the main Great Himalayas to the south and is the most sparsely populated region in Jammu and Kashmir.
7. PEX is an acronym for Cross Linked Polyester Pipe. This pipe is semi-flexible, relatively inexpensive, and has the advantage that water can freeze in the pipes without damaging the pipe integrity.
8. Vince, G., "A Himalayan Village Builds Artificial Glaciers to Survive Global Warming," *Scientific American*, May 24, 2010.
9. Bagla, P., ""Artificial Glaciers" Aid Farmers in Himalayas," September 4, 2001.
10. Tveiten, I., "Glacier Growing—a Local Response to Water Scarcity in Baltistan and Gilgit," Pakistan, 2007.

CHAPTER 7

Harvesting the Rain

Introduction

Rainwater catchment is one of the oldest and simplest ways of expanding the supply of freshwater. All that is needed is a surface to collect the rainwater, a cistern to collect the stored water, and a means to distribute the water. At its simplest, rainwater can be collected in a rain barrel and distributed with a bucket. Larger, more sophisticated systems supply military bases and resort hotels. As the demand for freshwater expands to parts of the world that previously have not experienced water shortages, rainwater collection promises to be increasingly recognized as an approved supplement to conventional water sources.

There are concerns that rain may pick up unhealthy substances as it falls through the atmosphere, runs down a roof, or while being stored. The danger from the first of these, namely atmospheric pollution, seems slight. Measurements of precipitation even in industrialized areas indicate a fairly low absorption of heavy metals from the air and tolerable levels of acidity. Common sense applies here in that it would obviously be unwise to harvest rainwater immediately downwind of, say, a smelter. The probability of finding truly airborne ingestible pathogenic viruses or bacteria is low, and of finding larger airborne pathogens negligible. Interest therefore focuses mainly upon contamination of roofs and the performance of water storage tanks in reducing or increasing pathogens.

Roofs and gutters are made of a variety of materials. For practical purposes, we can exclude discussion of "organic" roofs such as grass, reed, and palm because they yield runoff that is unsuitable for any use other than crop irrigation. The common materials of interest are ceramic, cement, or metal, with plastic roofs being unacceptable since they are neither cheap nor durable. Metal roofs are normally of treated steel or aluminum-coated steel. Corrugated steel roofing employs mild steel protected by hot-dip or electrolytic galvanizing, painting, or "aluminizing." Galvanizing uses zinc compounds that can be leached into the harvested rainwater if the rain is acidic. Although the iron on a rusting roof will enter the runoff, it usually does so in such small quantities that it does not compromise either health or taste. Stainless steel, while an excellent material for rainwater collection, is generally too expensive to use.

Aluminum by itself is subject to corrosion caused by the acidic rainwater.[1] Similarly, use of copper and copper flashing of roofs is not recommended due to copper's reaction to acid, leaching copper into the water and causing a green stain.

If the collection and storage surfaces are to be painted, care should be taken to make sure the paint is approved for potable water use. Even if collected water is not intended to be drunk, watering edible crops with paint not for potable water use is not advised. Bitumen coatings, or shingles made from bitumen, may have some risk to health and/or impart unpleasant taste to the collected water, and it should be avoided. The material that is used to treat shake shingles for moss growth is considered toxic and should not be used even for irrigation. More seriously, water should not be collected where lead, occasionally used to flash around the edges of roofs and roof penetrations, has been used.

One study in Malaysia[2] reported lead levels of up to 3.5 times WHO[3] limits in roof runoff where lead has been used in constructing the roof. Not surprisingly, the safety of water harvested from asbestos roofs (e.g., asbestos-reinforced cement mortar) has been questioned. The consensus is that, although the danger of developing cancer from ingested asbestos is very slight, the danger from inhaled asbestos dust with asbestos sheeting (i.e., from sawing it) is sufficiently high that working without special protection is now generally banned in industrial countries.

Contamination of water might arise from the roofing material itself or from substances that have accumulated on the roof or in the gutter. Metal roofs are comparatively smooth and are therefore less prone to contamination than rougher tile roofs by dust, leaves, bird droppings, and other debris. They may also get hot enough to sterilize themselves. Eliminating overhanging trees, the roosting surfaces for birds and other animals, is recommended. To further improve the quality of the collected water, a roof wash system utilizes the water from the initial part of the rain event to wash away the collected debris before letting water enter the collection tank. Further information on this and other factors in designing a rainwater harvesting system follow.

Estimating Demand

The first step in selection of a rainwater collection system is to determine how much water is needed. The average United States citizen uses approximately 100 gallons (378 liters) per day for drinking, bathing, waste removal (i.e., flushing toilets), and washing clothes. By comparison, United Kingdom citizens use 87 gallons (329 liters) per day, Asians use 22 gallons (83 liters) per day, and Africans use 12 gallons (45 liters) per day.[4] At a minimum, we need about 2/3 gallon (2.4 liters) per day to survive, which represents a vast gulf between water needs and water usage. Of note, for a typical developed world application, only about 12 percent of this total demand actually needs to be potable water.[5] The rest is used for laundry, toilet flushing, and personal hygiene. Supplementing potable water needs with rainwater presents an opportunity to significantly reduce the need for utility-provided water.

Using typical plumbing design data may not provide accurate sizing information for the rainwater harvesting system. Rather than depending on standardized tables that imply an unlimited water supply, the designer should perform his own demand calculation that reflects the use of low-flow fixtures and gives a best estimate of actual fixture usage patterns. This requires that the designer have a detailed understanding of the facility's usage and occupancy patterns.

Table 7.1 shows the format typically used to make a water usage estimate. This format, albeit simpler because of fewer fixture types, can also be used to estimate water demand for commercial facilities. Using a commercial building as an example, the designer can eliminate drinking water from the calculation if it comes from a bottled-water provider. The statistical inclusion of ultra-low flush water closet fixtures, urinals with electronic flush valves that control flush frequency and amount, and separating kitchen water needs (if applicable) creates a more manageable estimated load than provided by traditional calculation techniques.

Special-purpose facilities, such as hotels, add some complexity to the calculations due to the inclusion of showers in the water demand load. When laundry needs, spas, swimming pools, landscape irrigation, and other recreational options are added to the water demand estimate, accurate water demand calculations become more challenging. However, it is important to remember that these latter loads can use lower-quality water and can be separated from the facility's domestic water system.

Water used to irrigate landscaping often equals or exceeds interior water use. Supplemental irrigation water requirements can be greatly reduced by the use of 3 inches or more of top mulch, selecting native plants or plants that thrive in regions with similar climate, and using passive rainwater techniques. Because plant water needs vary greatly depending on soils, climate, plant size, etc., it is recommended that a local plant expert be consulted.

It is not crucial to collect water for 100 percent of demand if replacement water from an alternative source (well, water delivery service, utility) is available. The actual safety factor is a judgment call that depends on replacement water availability and the consequences of running out of freshwater. Due to a number of factors, it may not be feasible to provide for 100 percent of demand without water rationing during periods of low rainfall. It is also important to remember that water demand should best be matched with the corresponding seasonal rain pattern (i.e., the rainy season may not be the tourist season).

Fixture	Flow Rate (per use or min)**	Average # Uses/Day or Min/Day/Person	Daily Demand/Person (gal)	Number of People in Household	Household Total Daily Demand (gal)	Household Total Monthly Demand (gal)	Household Total Yearly Demand (gal)
Toilets	1.6	5.1	8.16	3	24.48	742	8935
Shower (based on 2.5 gal/min)	1.66	5.3	8.80	3	26.39	800	9634
Faucets (based on 2.5 gal/min)	1.66	8.1	13.45	3	40.34	1222	14,723
Dishwasher (1997–2001) (gal/use)	4.5	0.1	0.45	3	1.35	41	493
Clothes washer (1998–2001) (gal/use)	27	0.37	9.99	3	29.97	908	10,939
Total Demand					122.5	3713	44,724

Source: Vickers, A., Handbook of Water Use and Conservation, Waterplow Press, Amherst, MA, 2001.
** Actual Flow (MFR)

TABLE 7.1   Residential Indoor Water Use[1]

Steps to proper selection and sizing of a rainwater harvesting system are as follows:

Step 1: Estimating Rainwater Supply

Once it is known how much water is needed, the next step to sizing a rain harvesting system is to reconcile the roof collection area with the rain occurrence and the cistern storage volume. This is usually an iterative process that balances economics and the geometry of the facility to be served. The starting point to determine the amount of water that can be collected is to assume the collection surface is the horizontal projection of the facility's roof.

To determine how much rain will fall on this collection surface can be found from such sources as the United States National Oceanic and Atmospheric Administration's National Climate Data Center (www.ncdc.noaa.gov/oa/ncdc.html) and the National Weather Service (www.nws.noaa.gov). As a rule of thumb, at least seven years of monthly rainfall records at a location close to the project site should be evaluated. Although average and median rainfall amounts provide some guidance, doing a sensitivity analysis considering the rainwater extremes is also important. After determining the available rainwater density, the water to be collected can be found by multiplying rainfall density (in inches) by the collection surface area (in square feet).

Step 2: Sizing the Collection System

The collection surface is often dictated by architectural constraints, such as roof area, etc. To make an initial approximation of the surface needed to meet the established demand, the following equation can be used:

Surface area (square feet)
= Demand (gallons)/(0.623) × Precipitation density (inches) × (system efficiency)

Note the following:

- 0.623 (gallons/square foot/inch) conversion factor = 7.48 (gallons/cubic foot)/12 (inches per foot). One inch of water covering 1 square foot of surface area = 0.623 gallons.

- Surface area is horizontal projection of roof surface and not actual surface area (measure the area the roof covers, not the actual roof).

- Precipitation density period is consistent with time period being considered (monthly, yearly, etc.).

- This coefficient accounts for collection system loss from leakage, evaporation, roof composition, etc. Roof coefficients are approximately 0.90.

Step 3: Sizing the Storage[6]

Once the area of roof catchment has been determined and the average rainfall has been established, the maximum amount of rain that can be collected can be calculated using the following formula:

$$\text{Runoff (gallons)} = A \times (\text{Rainfall} - B) \times \text{Roof area} \times 0.623$$

- A is the system efficiency. Values of 0.80–0.85 (i.e., 80–85 percent efficiency) have been used as catchall numbers that include system leakage and evaporation.
- B is the loss associated with absorption and wetting of surfaces, and a value of 0.08 inches per month (2.0 inches per year) has been used (e.g., Martin, 1980).
- *Rainfall* should be expressed in inches.
- *Roof area* should be expressed in square feet.
- 0.623 (gallons/square foot/inch) conversion factor = 7.48 (gallons/cubic foot)/12 (inches per foot). One inch of water covering 1 square foot of surface area = 0.623 gallons.

The maximum volumes of rainwater that can be collected from various areas of roof and the range of average annual rainfalls are shown in the appendix at the end of this chapter. This information should only be used as an initial guide. If the maximum volumes are less than the annual water demand, then either the catchment area will need to be increased or water demand will need to be reduced.

Step 4: Sizing the Water Storage Tank (Cistern)

To assure the amount of water stored will meet the demand, the tank needs to be large enough to ensure the following:

- The required volume of water can be collected by the tank.
- The volume of water in the tank will be sufficient to meet demand during the drier months or through periods of low or no rainfall.

The simplest way of estimating the tank size needed to provide water throughout an average year is to use monthly rainfall data and to assume that at the start of the wetter months the tank is empty. The following formula should then be used for each month:

$V_t = V_{t-1} + (\text{Runoff} - \text{Demand})$
V_t = theoretical volume of water remaining in the tank at the end of the month
V_{t-1} = volume of water left in the tank from the previous month

Runoff should be calculated as discussed above ($A = 0.80$, $B = 0.08$ inches). The equation can be explained as follows:

- Starting with the tank empty, $V_{t-1} = 0$.
- If at the end of a period, runoff – demand is a positive number, and V_t exceeds the volume of the tank, then water will be lost to overflow. V_{t-1} for the next period will equal the volume of the tank.
- If runoff – demand is negative, the V_{t-1} is diminished by that amount, indicating that water has been consumed.
- If runoff – demand is equal to or greater than V_{t-1}, then water consumption exceeds the stored value, and $V_t = V_{t-1} = 0$.

Calculations should be repeated using various tank sizes until V_t is greater than 0 at the end of every month. The greater the values of V_t over the whole year, the greater the security of meeting water demand when rainfalls are below average or when dry periods are longer than normal. It should be remembered: *The greater the security, the higher the cost of the tank.*

Step 5: System Adjustment

To optimize performance and cost, going back through the calculation modifying surface area and the cistern storage capacity is recommended.[7]

Acceptable Piping Schematics for Potable and Non-Potable Water

Figure 7.1 shows an above-ground application in a non-freezing environment. In an environment where freezing is possible, the tank should be moved to a heated environment or buried below the frost line, as shown in the following details.

Underground Exterior Cistern for Potable Application

Where carbon filters are used, they may be put downstream of chlorine and ozone disinfection systems, but are recommended to be upstream of ultraviolet disinfection systems. Where soil saturation is a possibility, it is recommended that the combined weight of the tank and ballast must meet or exceed the buoyant upward force of an empty cistern. This buoyant force (in pounds) is equal to the volume of the tank (cubic feet) times 62.4 lb/cubic ft, or tank volume (gallons) times 8.34 lb/gallon of water.

Non-Potable Water

This application is suitable for lawn and plant irrigation or process water makeup. Filters to remove particulate may be added to improve

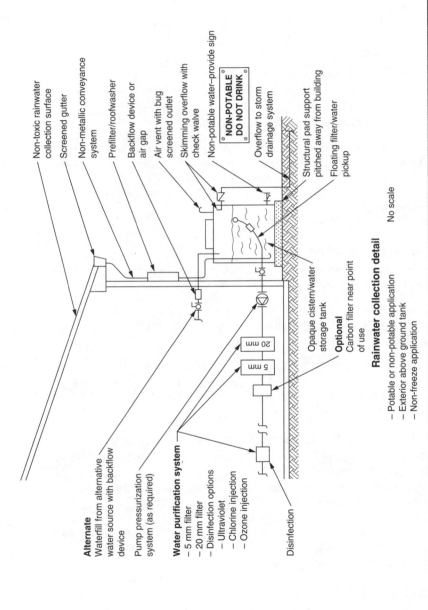

Alternate
Waterfill from alternative water source with backflow device
Pump pressurization system (as required)

Water purification system
– 5 mm filter
– 20 mm filter
– Disinfection options
  – Ultraviolet
  – Chlorine injection
  – Ozone injection

Disinfection

Non-toxic rainwater collection surface
Screened gutter
Non-metallic conveyance system
Prefilter/roofwasher
Backflow device or air gap
Air vent with bug screened outlet
Skimming overflow with check valve
Non-potable water–provide sign

NON-POTABLE
DO NOT DRINK

Overflow to storm drainage system
Structural pad support pitched away from building
Floating filter/water pickup

Opaque cistern/water storage tank

Optional
Carbon filter near point of use

20 mm
5 mm

Rainwater collection detail

No scale

– Potable or non-potable application
– Exterior above ground tank
– Non-freeze application

FIGURE 7.1   Rainwater collection system—above-ground application in a non-freezing environment.

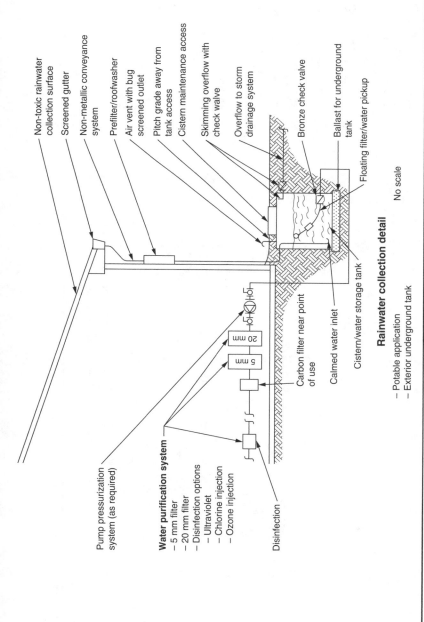

Non-toxic rainwater collection surface

Screened gutter

Non-metallic conveyance system

Prefilter/roofwasher

Air vent with bug screened outlet

Pitch grade away from tank access

Cistern maintenance access

Skimming overflow with check walve

Overflow to storm drainage system

Bronze check valve

Ballast for underground tank

Floating filter/water pickup

Pump pressurization system (as required)

Water purification system
– 5 mm filter
– 20 mm filter
– Disinfection options
  – Ultraviolet
  – Chlorine injection
  – Ozone injection

Disinfection

Carbon filter near point of use

Calmed water inlet

Cistern/water storage tank

20 mm

5 mm

Rainwater collection detail

No scale

– Potable application
– Exterior underground tank

FIGURE 7.2 Rainwater collection system—underground cistern.

97

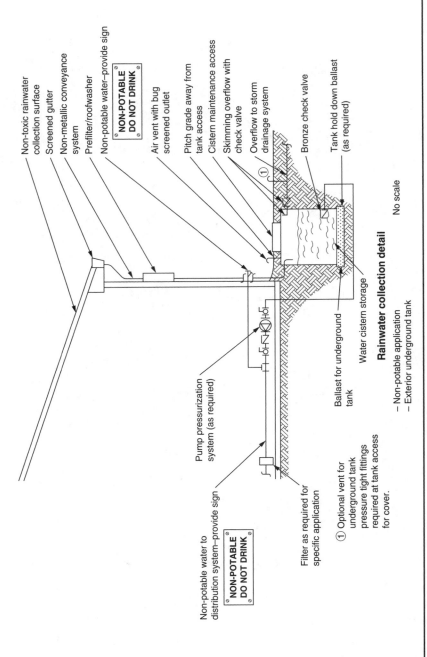

Non-toxic rainwater collection surface

Screened gutter

Non-metallic conveyance system

Prefilter/roofwasher

Non-potable water–provide sign

NON-POTABLE DO NOT DRINK

Air vent with bug screened outlet

Pitch grade away from tank access

Cistern maintenance access

Skimming overflow with check valve

Overflow to storm drainage system

①

Bronze check valve

Tank hold down ballast (as required)

Non-potable water to distribution system–provide sign

NON-POTABLE DO NOT DRINK

Pump pressurization system (as required)

Filter as required for specific application

Ballast for underground tank

① Optional vent for underground tank pressure tight fittings required at tank access for cover.

Water cistern storage

Rainwater collection detail

No scale

– Non-potable application
– Exterior underground tank

FIGURE 7.3 Rainwater collection system—non-potable.

98

water quality in order to avoid problems with sprinkler or process devices. Signage marking water outlets as "Non-Potable, Do Not Drink" are required in a public environment and highly recommended elsewhere.

Freezing Environments

Installing a water storage tank in a heated environment is preferred for an installation subject to freezing. Appropriate signage is necessary to label non-potable water outlets.

Installation Details to Consider

Tank Overflow

There are two processes related to tank overflows: skimming and tank siphoning. The skimming process is quite logical: anything floating on top of the water will be skimmed off when the tank capacity is reached and the tank overflows. What this concept overlooks is that eventually everything that goes into the tank settles to the bottom as sediment, creating what is called the anaerobic zone. In the absence of oxygen, this layer decomposes and creates by-products of methane and SO_2, which add odor and a disagreeable taste to potable water.

A better method is the tank siphon overflow (Fig. 7.5). With this device, when the tank overflows, the overflow water is drawn from the bottom of the tank. This helps to remove what settles to the bottom of the tank, depleting the anaerobic zone that has such a negative effect on the stored water. The key detail in this approach is remembering to drill a hole in the elbow where the siphon leaves the tank to avoid siphoning the entire contents out of the tank.[8]

Water Pickup

For non-potable applications, where the water is used primarily for landscape irrigation, the removal of the water from the cistern can be as close to the bottom as is acceptable—the products of the anaerobic zone provide some benefit to the plants being watered.

For better-quality water, the traditional way of taking water out of the tank is 4 to 6 inches (10.2 to 15.2 cm) above the bottom of the tank. The thinking is that this is as close to the bottom of the tank as is desirable to remove water without getting into the anaerobic zone. For non-potable applications, where the water quality is not critical, or when the bottom of the tank is periodically flushed, this will generally provide acceptable quality water.

Research has shown that the best quality water is between the surface, where the available oxygen promotes aerobic bacterial growth, and the anaerobic zone at the bottom.[9] This is collected using a floating water pickup that uses a float to support the flexible water

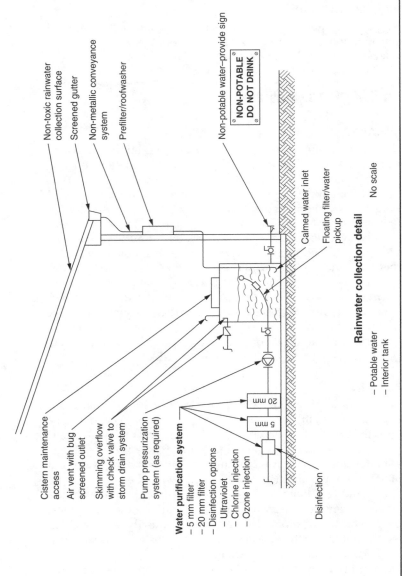

Cistern maintenance access

Air vent with bug screened outlet

Skimming overflow with check valve to storm drain system

Pump pressurization system (as required)

Water purification system
– 5 mm filter
– 20 mm filter
– Disinfection options
  – Ultraviolet
  – Chlorine injection
  – Ozone injection

Disinfection

Non-toxic rainwater collection surface

Screened gutter

Non-metallic conveyance system

Prefilter/roofwasher

Non-potable water–provide sign

NON-POTABLE
DO NOT DRINK

Calmed water inlet

Floating filter/water pickup

20 mm

5 mm

Rainwater collection detail

No scale

– Potable water
– Interior tank

FIGURE 7.4 Rainwater collection system—interior tank.

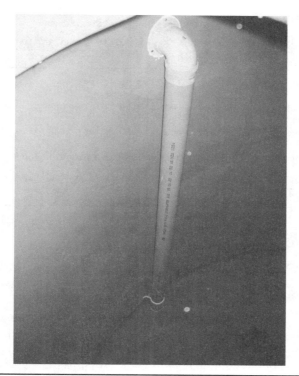

Figure 7.5   Tank siphon overflow.

pickup just below the water surface. This technique is most advisable for potable water applications and for getting the cleanest possible water for use inside a building.

Maintaining Water Quality

Keeping the system clean is the key to maintaining a healthy rainwater collection system, which is the responsibility of the system owner. Rainwater collection and storage facilities require a variety of routine maintenance. Some of the routine maintenance activities that you will need to periodically complete are as follows:

- Removing debris from the roof, leaf guard, gutter, gutter screen, and first flush diverter
- Inspecting and repairing vent screens
- Siphoning sediment from the tank
- Testing the coliform bacteria levels in your untreated-water storage tank

Estimated Roof Contamination Potential		
High Contamination[1]	**Medium Contamination**	**Low Contamination[2]**
0.03"/8 mm	0.01"/2 mm	0.002"/0.5 mm

Notes:
(1) High contamination is considered to have high content of organic debris from animal waste, adjacent trees, and/or airborne contamination.
(2) Low contamination is considered to either have frequent rainfall to keep collection surface clean and/or minimal nontoxic contamination.
(3) Sample calculation: 1000 square foot collection surface, medium contamination:

Gallons = 0.01" rain × 1000 square feet × 0.623 gallons/square foot – inch:
   = 6.23 gallons

TABLE 7.2   Roof Washer First Flush Guidelines

- Disinfecting the untreated-water storage tank if total coliform levels reach 500 colony-forming units per 0.03 gallons (100 mL) (500 CFU/100 mL) or if fecal coliform levels reach 100 CFU/ 100 mL.

Untreated water with total coliform levels above 500 CFU/100 mL or fecal coliform levels rising above 100 CFU/100 mL should not even be used for non-potable purposes until it has been disinfected.

Disinfection can also help prevent heavy biological growth that can foul your treatment system and contaminate your plumbing system. The most convenient way for you to control biological growth in your storage tank is to add a small amount of chlorine bleach periodically. The amount of bleach you will need to add depends on how concentrated the bleach is, how big the storage tank is, and whether you are trying to prevent biological growth or eliminate an existing problem. Particularly for systems that can be publically accessed, periodic testing should be done and a log kept to document that proper rainwater conditions are being maintained suitable for the intended use. Guideline standards for testing rainwater at the point of use are shown in Table 7.4.

Roof Washer

Roof washers are commonly used with the initial water coming off the collection surface before it is allowed to fill the cistern. Commonly used roof wash amounts are indicated below, but they may vary to reflect actual site and seasonal conditions.

There are many different styles of roof wash devices. The simplest versions involve filling a standpipe section of piping that contains adequate volume, which overflows into the cistern once it's full (Fig. 7.6). If sized correctly, research shows this parallel pipe

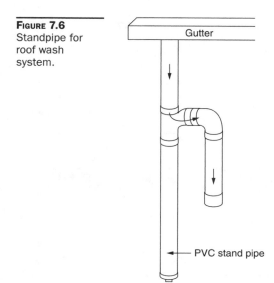

FIGURE 7.6
Standpipe for roof wash system.

style roof washer is very effective, in that there is minimal mixing between the contaminated water and clean water being provided to the cistern.

Another commercially available first flush diverter (Fig. 7.7) attempts to address the mixing issue by using a standpipe and floating ball. Once the standpipe is filled with the pre-wash water, a floating ball seals off the remaining flow, thus preventing the pre-wash water from being mixed with the remaining flow. The remaining rainfall is then diverted to the cistern. This device has a drain at the bottom that allows diverted water to slowly drain after each rainfall event and a clean-out plug to get rid of any accumulated debris.

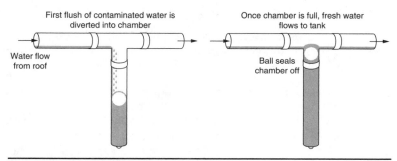

FIGURE 7.7   First flush diverter.

Length: Feet (Meters)	Volume: Gallons (Liters)
11.0 (0.3)	0.7 (2.6)
33.0 (0.9)	2.0 (7.6)
5 15.0 4.6)	3.3 (12.5)
10 (3.0)	6.7 (25.4)
15 (4.6)	15.0 (56.8)

TABLE **7.3a**   4″ PVC Pipe Storage Volume

Length: Feet (Meters)	Volume: Gallons (Liters)
1 1.0 (0.3)	1.5 (5.7)
3 3.0 (0.9)	4.5 (17.0)
5 15.0 (4.6)	7.5 (28.4)
10 10.0 (3.0)	15.0 (56.8)
15 15.0 (4.6)	22.5 (85.2)

TABLE **7.3b**   6″ PVC Pipe Storage Volume

Other commercially available combination pre-filter and roof wash devices are available to help maintain the water quality of the rainwater harvesting system.

The volume of pre-wash for a nominal 4-inch (4.046-inch actual) diameter PVC pipe can be seen in Table 7.3a.

The volume of pre-wash for a nominal 6-inch (15 cm) (6.065-inch actual) diameter PVC pipe can be seen in Table 8.2b.

Conclusion

There are three reasons to consider rainwater catchment as a water source:

1. There is a lack of, or an insufficient quantity, of water available.

2. The water that is available is of poor quality.

3. Collecting rainwater serves to buffer storm runoff and the impact on municipal combination sewers that overflow into the waterways.

Rainwater is nearly universally available as a water source and commonly used as a water source around the world. As it falls from the sky, it is close to pure water. With proper handling, storage, and system maintenance, it can serve as a non-potable substitute for utility-provided water for toilet flushing and other noncritical needs. With adequate treatment, it can meet the needs for both potable and non-

Treatment Level	Non-Potable[1]	Potable
Total Coliform	< 500 CFU/100 mL	0
Fecal Coliform	< 100 CFU/100 mL	0
Turbidity	< 10 NTU	< 0.3NTU
Protozoan Cysts		0
Testing Frequency[2]	Every 12 Months	Every 3 Months

Notes:
1. Suitable for toilet and urinal flushing, washing machine makeup, and other approved applications in occupied space for environments with non-health impaired occupants.
2. Upon failure of total and fecal coliform tests, system shall be recommissioned, which includes cleaning and retesting.

TABLE 7.4 Acceptable Standards of Rainwater Quality[10]

potable requirements. Rainwater catchment is as old as the ages, yet available now to meet modern-day needs.

Notes and References

1. The leaching of aluminum into the water is thought to cause Alzheimer's disease, which causes premature senility.
2. Magyar, M.I., et al., *Lead And Other Heavy Metals: Common Contaminants of Rainwater Tanks in Melbourne*, Monash University, Institute for Sustainable Water Resources, Department of Civil Engineering, Victoria, Australia, 2008. E-mail: mirela.magyar@eng.monash.edu.au.
3. World Health Organization.
4. "The Coming Water Crisis," *Population Reports*, Chapter 1, Volume XXVI: Number 1.
5. American Water Works Association (AWWA) publication.
6. Adapted from Martin, T.J., *Supply Aspects of Domestic Rainwater Tanks*, South Australian Department of Environment, Adelaide, 1980.
7. Boulware, E.W.B., P.E., et al., *Rainwater Catchment Design and Installation Standard*, American Rainwater Catchment Systems Association.
8. This detail developed by Dr. Stanley Abbott, Massey University, Christ Church, New Zealand.
9. Presentation at the American Rainwater Catchment Systems Association (ARCSA) National Convention, October 2010, Austin, Texas by Dr, Peter Coombes, University of New Castle, Newcastle, Australia.
10. Texas Rainwater Standard (as stated in *Harvesting, Storing, and Treating Rainwater for Domestic Indoor Use*, Texas Commission on Environmental Quality, January 2007) and EPA standard for coastal swimming quality water.
11. For metric conversion of the data in the following chapter appendix and any other values, refer to Appendix 2.

Appendix Average Rainfall Data[11]

Location	NORMALS 1971-2000								English Units (inches)					
	YRS	JAN	FEB	MAR	APR	MAY	JUN	JUL	AUG	SEP	OCT	NOV	DEC	ANN
BIRMINGHAM AP, AL	30	5.45	4.21	6.10	4.67	4.83	3.78	5.09	3.48	4.05	3.23	4.63	4.47	53.99
HUNTSVILLE, AL	30	5.52	4.95	6.68	4.54	5.24	4.22	4.40	3.32	4.29	3.54	5.22	5.59	57.51
MOBILE, AL	30	5.75	5.10	7.20	5.06	6.10	5.01	6.54	6.20	6.01	3.25	5.41	4.66	66.29
MONTGOMERY, AL	30	5.04	5.45	6.39	4.38	4.14	4.13	5.31	3.63	4.22	2.58	4.53	4.97	54.77
ANCHORAGE, AK	30	0.68	0.74	0.65	0.52	0.70	1.06	1.70	2.93	2.87	2.09	1.09	1.05	16.08
ANNETTE, AK	30	9.67	8.05	7.96	7.37	5.73	4.72	4.26	6.12	9.49	13.86	12.21	11.39	100.83
BARROW, AK	30	0.12	0.12	0.09	0.12	0.12	0.32	0.87	1.04	0.69	0.39	0.16	0.12	4.16
BETHEL, AK	30	0.62	0.51	0.67	0.65	0.85	1.60	2.03	3.02	2.31	1.43	1.37	1.12	16.18
BETTLES, AK	30	0.84	0.61	0.55	0.38	0.85	1.43	2.10	2.54	1.82	1.08	0.90	0.87	13.97
BIG DELTA, AK	30	0.34	0.41	0.22	0.20	0.77	2.38	2.77	2.11	1.03	0.73	0.59	0.39	11.94
COLD BAY, AK	30	3.08	2.59	2.48	2.30	2.65	2.89	2.53	3.59	4.51	4.54	4.79	4.33	40.28
FAIRBANKS, AK	30	0.56	0.36	0.28	0.21	0.60	1.40	1.73	1.74	1.12	0.92	0.68	0.74	10.34
GULKANA, AK	30	0.45	0.52	0.36	0.22	0.59	1.54	1.82	1.80	1.44	1.02	0.67	0.97	11.40
HOMER, AK	30	2.61	2.04	1.82	1.21	1.07	0.96	1.45	2.28	3.37	2.77	2.87	3.00	25.45
JUNEAU, AK	30	4.81	4.02	3.51	2.96	3.48	3.36	4.14	5.37	7.54	8.30	5.43	5.41	58.33
KING SALMON, AK	30	1.03	0.72	0.79	0.94	1.35	1.70	2.15	2.89	2.81	2.10	1.54	1.39	19.41
KODIAK, AK	30	8.17	5.72	5.22	5.48	6.31	5.38	4.12	4.48	7.84	8.36	6.63	7.64	75.35
KOTZEBUE, AK	30	0.55	0.42	0.38	0.41	0.33	0.57	1.43	2.00	1.70	0.95	0.71	0.60	10.05
MCGRATH, AK	30	1.04	0.74	0.81	0.66	1.02	1.45	2.32	2.75	2.36	1.46	1.46	1.44	17.51
NOME, AK	30	0.92	0.75	0.60	0.65	0.74	1.14	2.15	3.23	2.51	1.58	1.28	1.01	16.56
ST. PAUL ISLAND, AK	30	1.74	1.25	1.12	1.12	1.21	1.41	1.91	2.96	2.79	2.70	2.87	2.13	23.21

TALKEETNA, AK	30	1.45	1.28	1.26	1.22	1.64	2.41	3.24	4.53	4.35	3.06	1.78	1.96	28.18
UNALAKLEET, AK	30	0.40	0.31	0.39	0.35	0.55	1.25	2.15	2.92	2.10	0.89	0.66	0.47	12.44
VALDEZ, AK	30	6.02	5.53	4.49	3.55	3.08	3.01	3.84	6.62	9.59	8.58	5.51	7.59	67.41
YAKUTAT, AK	30	13.18	10.99	11.41	10.80	9.78	7.17	7.88	13.27	20.88	24.00	15.17	15.85	160.38
FLAGSTAFF, AZ	30	2.18	2.56	2.62	1.29	0.80	0.43	2.40	2.89	2.12	1.93	1.86	1.83	22.91
PHOENIX, AZ	30	0.83	0.77	1.07	0.25	0.16	0.09	0.99	0.94	0.75	0.79	0.73	0.92	8.29
TUCSON, AZ	30	0.99	0.88	0.81	0.28	0.24	0.24	2.07	2.30	1.45	1.21	0.67	1.03	12.17
WINSLOW, AZ	30	0.46	0.53	0.61	0.27	0.36	0.30	1.18	1.31	1.02	0.90	0.55	0.54	8.03
YUMA, AZ	30	0.38	0.28	0.27	0.09	0.05	0.02	0.23	0.61	0.26	0.26	0.14	0.42	3.01
FORT SMITH, AR	30	2.37	2.59	3.94	3.91	5.29	4.28	3.19	2.56	3.61	3.94	4.80	3.39	43.87
LITTLE ROCK, AR	30	3.61	3.33	4.88	5.47	5.05	3.95	3.31	2.93	3.71	4.25	5.73	4.71	50.93
NORTH LITTLE ROCK, AR	30	3.37	3.27	4.88	5.03	5.40	3.51	3.15	2.97	3.53	3.81	5.74	4.53	49.19
BAKERSFIELD, CA	30	1.18	1.21	1.41	0.45	0.24	0.12	0.00	0.08	0.15	0.30	0.59	0.76	6.49
BISHOP, CA	30	0.88	0.97	0.62	0.24	0.26	0.21	0.17	0.13	0.28	0.20	0.44	0.62	5.02
EUREKA, CA	30	5.97	5.51	5.55	2.91	1.62	0.65	0.16	0.38	0.86	2.36	5.78	6.35	38.10
FRESNO, CA	30	2.16	2.12	2.20	0.76	0.39	0.23	0.01	0.01	0.26	0.65	1.10	1.34	11.23
LONG BEACH, CA	30	2.95	3.01	2.43	0.60	0.23	0.08	0.02	0.10	0.24	0.40	1.12	1.76	12.94
LOS ANGELES AP, CA	30	2.98	3.11	2.40	0.63	0.24	0.08	0.03	0.14	0.26	0.36	1.13	1.79	13.15
LOS ANGELES C.O., CA	30	3.33	3.68	3.14	0.83	0.31	0.06	0.01	0.13	0.32	0.37	1.05	1.91	15.14
MOUNT SHASTA, CA	30	7.06	6.45	5.81	2.65	1.87	0.99	0.39	0.43	0.87	2.21	5.08	5.35	39.16
REDDING, CA	30	6.50	5.49	5.15	2.40	1.66	0.69	0.05	0.22	0.48	2.18	4.03	4.67	33.52
SACRAMENTO, CA	30	3.84	3.54	2.80	1.02	0.53	0.20	0.05	0.06	0.36	0.89	2.19	2.45	17.93
SAN DIEGO, CA	30	2.28	2.04	2.26	0.75	0.20	0.09	0.03	0.09	0.21	0.44	1.07	1.31	10.77

(Continued)

Location	NORMALS 1971-2000								English Units (Inches)					
	YRS	JAN	FEB	MAR	APR	MAY	JUN	JUL	AUG	SEP	OCT	NOV	DEC	ANN
SAN FRANCISCO AP, CA	30	4.45	4.01	3.26	1.18	0.38	0.11	0.03	0.07	0.20	1.04	2.49	2.89	20.11
SAN FRANCISCO C.O., CA	30	4.72	4.15	3.40	1.25	0.54	0.13	0.04	0.09	0.28	1.19	3.31	3.18	22.28
SANTA BARBARA, CA	30	3.57	4.28	3.51	0.63	0.23	0.05	0.03	0.11	0.42	0.52	1.32	2.26	16.93
SANTA MARIA, CA	30	2.64	3.23	2.94	0.91	0.32	0.05	0.03	0.05	0.31	0.45	1.24	1.84	14.01
STOCKTON, CA	30	2.71	2.46	2.28	0.96	0.50	0.09	0.05	0.05	0.33	0.82	1.77	1.82	13.84
ALAMOSA, CO	30	0.25	0.21	0.46	0.54	0.70	0.59	0.94	1.19	0.89	0.67	0.48	0.33	7.25
COLORADO SPRINGS, CO	30	0.28	0.35	1.06	1.62	2.39	2.34	2.85	3.48	1.23	0.86	0.52	0.42	17.40
DENVER, CO	30	0.51	0.49	1.28	1.93	2.32	1.56	2.16	1.82	1.14	0.99	0.98	0.63	15.81
GRAND JUNCTION, CO	30	0.60	0.50	1.00	0.86	0.98	0.41	0.66	0.84	0.91	1.00	0.71	0.52	8.99
PUEBLO, CO	30	0.33	0.26	0.97	1.25	1.49	1.33	2.04	2.27	0.84	0.64	0.58	0.39	12.39
BRIDGEPORT, CT	30	3.73	2.92	4.15	3.99	4.03	3.57	3.77	3.75	3.58	3.54	3.65	3.47	44.15
HARTFORD, CT	30	3.84	2.96	3.88	3.86	4.39	3.85	3.67	3.98	4.13	3.94	4.06	3.60	46.16
WILMINGTON, DE	30	3.43	2.81	3.97	3.39	4.15	3.59	4.28	3.51	4.01	3.08	3.19	3.40	42.81
WASHINGTON DULLES AP, D.C	30	3.05	2.77	3.55	3.22	4.22	4.07	3.57	3.78	3.82	3.37	3.31	3.07	41.80
WASHINGTON NAT'L AP, D.C.	30	3.21	2.63	3.60	2.77	3.82	3.13	3.66	3.44	3.79	3.22	3.03	3.05	39.35
APALACHICOLA, FL	30	4.87	3.76	4.95	3.00	2.62	4.30	7.31	7.29	7.10	4.18	3.62	3.51	56.51
DAYTONA BEACH, FL	30	3.13	2.74	3.84	2.54	3.26	5.69	5.17	6.09	6.61	4.48	3.03	2.71	49.29
FORT MYERS, FL	30	2.23	2.10	2.74	1.67	3.42	9.77	8.98	9.54	7.86	2.59	1.71	1.58	54.19
GAINESVILLE, FL	30	3.51	3.39	4.26	2.86	3.23	6.78	6.10	6.63	4.37	2.50	2.17	2.56	48.36
JACKSONVILLE, FL	30	3.69	3.15	3.93	3.14	3.48	5.37	5.97	6.87	7.90	3.86	2.34	2.64	52.34

KEY WEST, FL	30	2.22	1.51	1.86	2.06	3.48	4.57	3.27	5.40	5.45	4.34	2.64	2.14	38.94
MIAMI, FL	30	1.88	2.07	2.56	3.36	5.52	8.54	5.79	8.63	8.38	6.19	3.43	2.18	58.53
ORLANDO, FL	30	2.43	2.35	3.54	2.42	3.74	7.35	7.15	6.25	5.76	2.73	2.32	2.31	48.35
PENSACOLA, FL	30	5.34	4.68	6.40	3.89	4.40	6.39	8.02	6.85	5.75	4.13	4.46	3.97	64.28
TALLAHASSEE, FL	30	5.36	4.63	6.47	3.59	4.95	6.92	8.04	7.03	5.01	3.25	3.86	4.10	63.21
TAMPA, FL	30	2.27	2.67	2.84	1.80	2.85	5.50	6.49	7.60	6.54	2.29	1.62	2.30	44.77
VERO BEACH, FL	30	2.89	2.45	4.20	2.88	3.80	6.03	6.53	6.04	6.84	5.04	3.04	2.19	51.93
WEST PALM BEACH, FL	30	3.75	2.55	3.68	3.57	5.39	7.58	5.97	6.65	8.10	5.46	5.55	3.14	61.39
ATHENS, GA	30	4.69	4.39	4.99	3.35	3.86	3.94	4.41	3.78	3.53	3.47	3.71	3.71	47.83
ATLANTA, GA	30	5.03	4.68	5.38	3.62	3.95	3.63	5.12	3.67	4.09	3.11	4.10	3.82	50.20
AUGUSTA, GA	30	4.50	4.11	4.61	2.94	3.07	4.19	4.07	4.48	3.59	3.20	2.68	3.14	44.58
COLUMBUS, GA	30	4.78	4.48	5.75	3.84	3.62	3.51	5.04	3.78	3.07	2.33	3.97	4.40	48.57
MACON, GA	30	5.00	4.55	4.90	3.14	2.98	3.54	4.32	3.79	3.26	2.37	3.22	3.93	45.00
SAVANNAH, GA	30	3.95	2.92	3.64	3.32	3.61	5.49	6.04	7.20	5.08	3.12	2.40	2.81	49.58
HILO, HI	30	9.74	8.86	14.35	12.54	8.07	7.36	10.71	9.78	9.14	9.64	15.58	10.50	126.27
HONOLULU, HI	30	2.73	2.35	1.89	1.11	0.78	0.43	0.50	0.46	0.74	2.18	2.27	2.85	18.29
KAHULUI, HI	30	3.74	2.36	2.35	1.75	0.66	0.23	0.49	0.53	0.39	1.05	2.17	3.08	18.80
LIHUE, HI	30	4.59	3.26	3.58	3.00	2.87	1.82	2.12	1.91	2.69	4.25	4.70	4.78	39.57
BOISE, ID	30	1.39	1.14	1.41	1.27	1.27	0.74	0.39	0.30	0.76	0.76	1.38	1.38	12.19
LEWISTON, ID	30	1.14	0.95	1.12	1.31	1.56	1.16	0.72	0.75	0.81	0.96	1.21	1.05	12.74
POCATELLO, ID	30	1.14	1.01	1.38	1.18	1.51	0.91	0.70	0.66	0.89	0.97	1.13	1.10	12.58
CHICAGO,IL	30	1.75	1.63	2.65	3.68	3.38	3.63	3.51	4.62	3.27	2.71	3.01	2.43	36.27
MOLINE, IL	30	1.58	1.51	2.92	3.82	4.25	4.63	4.03	4.41	3.16	2.80	2.73	2.20	38.04

(Continued)

109

Location	NORMALS 1971-2000							English Units (Inches)					
	YRS	JAN	FEB	MAR	APR	MAY	JUN	JUL	AUG	SEP	OCT	NOV	DEC	ANN
PEORIA, IL	30	1.50	1.67	2.83	3.56	4.17	3.84	4.02	3.16	3.12	2.77	2.99	2.40	36.03
ROCKFORD, IL	30	1.41	1.34	2.39	3.62	4.03	4.80	4.10	4.21	3.47	2.57	2.63	2.06	36.63
SPRINGFIELD, IL	30	1.62	1.80	3.15	3.36	4.06	3.77	3.53	3.41	2.83	2.62	2.87	2.54	35.56
EVANSVILLE, IN	30	2.91	3.10	4.29	4.48	5.01	4.10	3.75	3.14	2.99	2.78	4.18	3.54	44.27
FORT WAYNE, IN	30	2.05	1.94	2.86	3.54	3.75	4.04	3.58	3.60	2.81	2.63	2.98	2.77	36.55
INDIANAPOLIS, IN	30	2.48	2.41	3.44	3.61	4.36	4.13	4.42	3.82	2.88	2.76	3.61	3.03	40.95
SOUTH BEND, IN	30	2.27	1.98	2.89	3.62	3.50	4.19	3.73	3.98	3.79	3.27	3.39	3.09	39.70
DES MOINES, IA	30	1.03	1.19	2.21	3.58	4.25	4.57	4.18	4.51	3.15	2.62	2.10	1.33	34.72
DUBUQUE, IA	30	1.28	1.42	2.57	3.49	4.12	4.08	3.73	4.59	3.56	2.50	2.49	1.69	35.52
SIOUX CITY, IA	30	0.59	0.62	2.00	2.75	3.75	3.61	3.30	2.90	2.42	1.99	1.40	0.66	25.99
WATERLOO, IA	30	0.84	1.05	2.13	3.23	4.15	4.82	4.20	4.08	2.95	2.49	2.10	1.11	33.15
CONCORDIA, KS	30	0.66	0.73	2.35	2.45	4.20	3.95	4.20	3.24	2.50	1.84	1.45	0.86	28.43
DODGE CITY, KS	30	0.62	0.66	1.84	2.25	3.00	3.15	3.17	2.73	1.70	1.45	1.01	0.77	22.35
GOODLAND, KS	30	0.43	0.44	1.20	1.51	3.46	3.30	3.54	2.49	1.12	1.05	0.82	0.40	19.76
TOPEKA, KS	30	0.95	1.18	2.56	3.14	4.86	4.88	3.83	3.81	3.71	2.99	2.31	1.42	35.64
WICHITA, KS	30	0.84	1.02	2.71	2.57	4.16	4.25	3.31	2.94	2.96	2.45	1.82	1.35	30.38
GREATER CINCINNATI AP	30	2.92	2.75	3.90	3.96	4.59	4.42	3.75	3.79	2.82	2.96	3.46	3.28	42.60
JACKSON, KY	30	3.56	3.68	4.38	3.79	5.16	4.67	4.59	4.13	3.77	3.18	4.20	4.27	49.38
LEXINGTON, KY	30	3.34	3.27	4.41	3.67	4.78	4.58	4.81	3.77	3.11	2.70	3.44	4.03	45.91
LOUISVILLE, KY	30	3.28	3.25	4.41	3.91	4.88	3.76	4.30	3.41	3.05	2.79	3.81	3.69	44.54
PADUCAH, KY	30	3.47	3.93	4.27	4.95	4.75	4.51	4.45	2.99	3.56	3.45	4.53	4.38	49.24
BATON ROUGE, LA	30	6.19	5.10	5.07	5.56	5.34	5.33	5.96	5.86	4.84	3.81	4.76	5.26	63.08

LAKE CHARLES, LA	30	5.52	3.28	3.54	3.64	6.06	6.07	5.13	4.85	5.95	3.94	4.61	4.60	57.19
NEW ORLEANS, LA	30	5.87	5.47	5.24	5.02	4.62	6.83	6.20	6.15	5.55	3.05	5.09	5.07	64.16
SHREVEPORT, LA	30	4.60	4.21	4.18	4.42	5.25	5.05	3.99	2.71	3.21	4.45	4.68	4.55	51.30
CARIBOU, ME	30	2.97	2.06	2.57	2.64	3.28	3.31	3.89	4.15	3.27	2.99	3.12	3.19	37.44
PORTLAND, ME	30	4.09	3.14	4.14	4.26	3.82	3.28	3.32	3.05	3.37	4.40	4.72	4.24	45.83
BALTIMORE, MD	30	3.47	3.02	3.93	3.00	3.89	3.43	3.85	3.74	3.98	3.16	3.12	3.35	41.94
BLUE HILL, MA	30	4.78	4.06	4.79	4.32	3.79	3.93	3.74	4.06	4.13	4.42	4.64	4.56	51.22
BOSTON, MA	30	3.92	3.30	3.85	3.60	3.24	3.22	3.06	3.37	3.47	3.79	3.98	3.73	42.53
WORCESTER, MA	30	4.07	3.10	4.23	3.92	4.35	4.02	4.19	4.09	4.27	4.67	4.34	3.80	49.05
ALPENA, MI	30	1.76	1.35	2.13	2.31	2.61	2.53	3.17	3.50	2.80	2.33	2.08	1.83	28.40
DETROIT, MI	30	1.91	1.88	2.52	3.05	3.05	3.55	3.16	3.10	3.27	2.23	2.66	2.51	32.89
FLINT, MI	30	1.57	1.35	2.22	3.13	2.74	3.07	3.17	3.43	3.76	2.34	2.65	2.18	31.61
GRAND RAPIDS, MI	30	2.03	1.54	2.59	3.48	3.35	3.67	3.56	3.78	4.28	2.80	3.35	2.70	37.13
HOUGHTON LAKE, MI	30	1.61	1.25	2.05	2.29	2.57	2.93	2.75	3.72	3.11	2.26	2.14	1.75	28.43
LANSING, MI	30	1.61	1.45	2.33	3.09	2.71	3.60	2.68	3.46	3.48	2.29	2.66	2.17	31.53
MARQUETTE, MI	30	2.60	1.85	3.13	2.79	3.07	3.21	3.01	3.55	3.74	3.66	3.27	2.43	36.31
MUSKEGON, MI	30	2.22	1.58	2.36	2.91	2.95	2.58	2.32	3.77	3.52	2.80	3.23	2.64	32.88
SAULT STE. MARIE, MI	30	2.64	1.60	2.41	2.57	2.50	3.00	3.14	3.47	3.71	3.32	3.40	2.91	34.67
DULUTH, MN	30	1.12	0.83	1.69	2.09	2.95	4.25	4.20	4.22	4.13	2.46	2.12	0.94	31.00
INTERNATIONAL FALLS, MN	30	0.84	0.64	0.96	1.38	2.55	3.98	3.37	3.14	3.03	1.98	1.36	0.70	23.93
MINNEAPOLIS-ST.PAUL, MN	30	1.04	0.79	1.86	2.31	3.24	4.34	4.04	4.05	2.69	2.11	1.94	1.00	29.41
ROCHESTER, MN	30	0.94	0.75	1.88	3.01	3.53	4.00	4.61	4.33	3.12	2.20	2.01	1.02	31.40
SAINT CLOUD, MN	30	0.76	0.59	1.50	2.13	2.97	4.51	3.34	3.93	2.93	2.24	1.54	0.69	27.13

(Continued)

| | NORMALS 1971-2000 | | | | | | | English Units (Inches) | | | | | | |
Location	YRS	JAN	FEB	MAR	APR	MAY	JUN	JUL	AUG	SEP	OCT	NOV	DEC	ANN
JACKSON, MS	30	5.67	4.50	5.74	5.98	4.86	3.82	4.69	3.66	3.23	3.42	5.04	5.34	55.95
MERIDIAN, MS	30	5.92	5.35	6.93	5.62	4.87	3.99	5.45	3.34	3.64	3.28	4.95	5.31	58.65
TUPELO, MS	30	5.14	4.68	6.30	4.94	5.80	4.82	3.65	2.67	3.35	3.38	5.01	6.12	55.86
COLUMBIA, MO	30	1.73	2.20	3.21	4.16	4.87	4.02	3.80	3.75	3.42	3.18	3.47	2.47	40.28
KANSAS CITY, MO	30	1.15	1.31	2.44	3.38	5.39	4.44	4.42	3.54	4.64	3.33	2.30	1.64	37.98
ST. LOUIS, MO	30	2.14	2.28	3.60	3.69	4.11	3.76	3.90	2.98	2.96	2.76	3.71	2.86	38.75
SPRINGFIELD, MO	30	2.11	2.28	3.82	4.31	4.57	5.02	3.56	3.37	4.83	3.47	4.46	3.17	44.97
BILLINGS, MT	30	0.81	0.58	1.12	1.74	2.48	1.89	1.28	0.85	1.34	1.26	0.75	0.67	14.77
GLASGOW, MT	30	0.35	0.26	0.47	0.75	1.72	2.20	1.78	1.25	0.98	0.71	0.39	0.37	11.23
GREAT FALLS, MT	30	0.68	0.51	1.01	1.40	2.53	2.24	1.45	1.65	1.23	0.93	0.59	0.67	14.89
HAVRE, MT	30	0.47	0.36	0.70	0.87	1.84	1.90	1.51	1.20	1.03	0.62	0.45	0.51	11.46
HELENA, MT	30	0.52	0.38	0.63	0.91	1.78	1.82	1.34	1.29	1.05	0.66	0.48	0.46	11.32
KALISPELL, MT	30	1.47	1.15	1.11	1.22	2.04	2.30	1.41	1.25	1.20	0.96	1.45	1.65	17.21
MISSOULA, MT	30	1.06	0.77	0.96	1.09	1.95	1.73	1.09	1.15	1.08	0.83	0.96	1.15	13.82
GRAND ISLAND, NE	30	0.54	0.68	2.04	2.61	4.07	3.72	3.14	3.08	2.43	1.51	1.41	0.66	25.89
LINCOLN, NE	30	0.67	0.66	2.21	2.90	4.23	3.51	3.54	3.35	2.92	1.94	1.58	0.86	28.37
NORFOLK, NE	30	0.57	0.76	1.97	2.59	3.92	4.25	3.74	2.80	2.25	1.72	1.44	0.65	26.66
NORTH PLATTE, NE	30	0.39	0.51	1.24	1.97	3.34	3.17	3.17	2.15	1.32	1.24	0.76	0.40	19.66
OMAHA EPPLEY AP, NE	30	0.77	0.80	2.13	2.94	4.44	3.95	3.86	3.21	3.17	2.21	1.82	0.92	30.22
OMAHA (NORTH), NE	30	0.76	0.77	2.25	3.07	4.57	3.84	3.75	2.93	3.03	2.49	1.67	0.95	30.08
SCOTTSBLUFF, NE	30	0.54	0.58	1.16	1.79	2.70	2.65	2.13	1.19	1.22	1.01	0.80	0.56	16.33
VALENTINE, NE	30	0.30	0.48	1.11	1.97	3.20	3.01	3.37	2.20	1.61	1.22	0.72	0.33	19.52

Location														
ELKO, NV	30	1.14	0.88	0.98	0.81	1.08	0.67	0.30	0.36	0.68	0.71	1.05	0.93	9.59
ELY, NV	30	0.74	0.75	1.05	0.90	1.29	0.66	0.60	0.91	0.94	1.00	0.63	0.50	9.97
LAS VEGAS, NV	30	0.59	0.69	0.59	0.15	0.24	0.08	0.44	0.45	0.31	0.24	0.31	0.40	4.49
RENO, NV	30	1.06	1.06	0.86	0.35	0.62	0.47	0.24	0.27	0.45	0.42	0.80	0.88	7.48
WINNEMUCCA, NV	30	0.83	0.62	0.86	0.85	1.06	0.69	0.27	0.35	0.53	0.66	0.80	0.81	8.33
CONCORD, NH	30	2.97	2.36	3.04	3.07	3.33	3.10	3.37	3.21	3.16	3.46	3.57	2.96	37.60
MT. WASHINGTON, NH	30	8.52	7.33	9.42	8.43	8.21	8.36	8.02	8.08	8.55	7.66	10.49	8.84	101.91
ATLANTIC CITY AP, NJ	30	3.60	2.85	4.06	3.45	3.38	2.66	3.86	4.32	3.14	2.86	3.26	3.15	40.59
ATLANTIC CITY C.O., NJ	30	3.44	2.88	3.79	3.25	3.16	2.46	3.36	4.16	3.02	2.71	2.96	3.18	38.37
NEWARK, NJ	30	3.98	2.96	4.21	3.92	4.46	3.40	4.68	4.02	4.01	3.16	3.88	3.57	46.25
ALBUQUERQUE, NM	30	0.49	0.44	0.61	0.50	0.60	0.65	1.27	1.73	1.07	1.00	0.62	0.49	9.47
CLAYTON, NM	30	0.30	0.27	0.62	0.99	2.08	2.21	2.81	2.69	1.56	0.74	0.54	0.32	15.13
ROSWELL, NM	30	0.39	0.41	0.35	0.58	1.30	1.62	1.99	2.31	1.98	1.29	0.53	0.59	13.34
ALBANY, NY	30	2.71	2.27	3.17	3.25	3.67	3.74	3.50	3.68	3.31	3.23	3.31	2.76	38.60
BINGHAMTON, NY	30	2.58	2.46	2.97	3.49	3.55	3.80	3.49	3.35	3.59	3.02	3.32	3.03	38.65
BUFFALO, NY	30	3.16	2.42	2.99	3.04	3.35	3.82	3.14	3.87	3.84	3.19	3.92	3.80	40.54
ISLIP, NY	30	4.27	3.33	4.76	4.13	3.90	3.71	2.93	4.48	3.39	3.63	3.86	4.13	46.52
NEW YORK C.PARK, NY	30	4.13	3.15	4.37	4.28	4.69	3.84	4.62	4.22	4.23	3.85	4.36	3.95	49.69
NEW YORK (JFK AP), NY	30	3.62	2.70	3.79	3.75	4.13	3.59	3.92	3.64	3.50	3.03	3.48	3.31	42.46
NEW YORK (LAGUARDIA AP), NY	30	3.56	2.75	3.93	3.68	4.16	3.57	4.41	4.09	3.77	3.26	3.67	3.51	44.36
ROCHESTER, NY	30	2.34	2.04	2.58	2.75	2.82	3.36	2.93	3.54	3.45	2.60	2.84	2.73	33.98
SYRACUSE, NY	30	2.60	2.12	3.02	3.39	3.39	3.71	4.02	3.56	4.15	3.20	3.77	3.12	40.05

(Continued)

Location	NORMALS 1971-2000									English Units (Inches)				
	YRS	JAN	FEB	MAR	APR	MAY	JUN	JUL	AUG	SEP	OCT	NOV	DEC	ANN
ASHEVILLE, NC	30	4.06	3.83	4.59	3.50	4.42	4.38	3.87	4.30	3.72	3.18	3.82	3.40	47.07
CAPE HATTERAS, NC	30	5.84	3.94	4.95	3.29	3.92	3.82	4.95	6.56	5.68	5.31	4.93	4.56	57.75
CHARLOTTE, NC	30	4.00	3.55	4.39	2.95	3.66	3.42	3.79	3.72	3.83	3.66	3.36	3.18	43.51
GREENSBORO-WNSTN-SALM-NC	30	3.54	3.10	3.85	3.43	3.95	3.53	4.44	3.71	4.30	3.27	2.96	3.06	43.14
RALEIGH, NC	30	4.02	3.47	4.03	2.80	3.79	3.42	4.29	3.78	4.26	3.18	2.97	3.04	43.05
WILMINGTON, NC	30	4.52	3.66	4.22	2.94	4.40	5.36	7.62	7.31	6.79	3.21	3.26	3.78	57.07
BISMARCK, ND	30	0.45	0.51	0.85	1.46	2.22	2.59	2.58	2.15	1.61	1.28	0.70	0.44	16.84
FARGO, ND	30	0.76	0.59	1.17	1.37	2.61	3.51	2.88	2.52	2.18	1.97	1.06	0.57	21.19
GRAND FORKS, ND	30	0.68	0.58	0.89	1.23	2.21	3.03	3.06	2.72	1.96	1.70	0.99	0.55	19.60
WILLISTON, ND	30	0.54	0.39	0.74	1.05	1.88	2.36	2.28	1.48	1.35	0.87	0.65	0.57	14.16
AKRON, OH	30	2.49	2.28	3.15	3.39	3.96	3.55	4.02	3.65	3.43	2.53	3.04	2.98	38.47
CLEVELAND, OH	30	2.48	2.29	2.94	3.37	3.50	3.89	3.52	3.69	3.77	2.74	3.38	3.14	38.71
COLUMBUS, OH	30	2.53	2.20	2.89	3.25	3.88	4.08	4.62	3.72	2.92	2.31	3.19	2.93	38.52
DAYTON, OH	30	2.60	2.29	3.29	4.03	4.17	4.21	3.75	3.49	2.65	2.72	3.30	3.08	39.58
MANSFIELD, OH	30	2.63	2.17	3.36	4.17	4.42	4.52	4.23	4.60	3.44	2.68	3.76	3.26	43.24
TOLEDO, OH	30	1.93	1.88	2.62	3.24	3.14	3.80	2.80	3.19	2.84	2.35	2.78	2.64	33.21
YOUNGSTOWN, OH	30	2.34	2.03	3.05	3.33	3.45	3.91	4.10	3.43	3.89	2.46	3.07	2.96	38.02
OKLAHOMA CITY, OK	30	1.28	1.56	2.90	3.00	5.44	4.63	2.94	2.48	3.98	3.64	2.11	1.89	35.85
TULSA, OK	30	1.60	1.95	3.57	3.95	6.11	4.72	2.96	2.85	4.76	4.05	3.47	2.43	42.42
ASTORIA, OR	30	9.62	7.87	7.37	4.93	3.28	2.57	1.16	1.21	2.61	5.61	10.50	10.40	67.13
BURNS,OR	30	1.18	1.11	1.24	0.85	1.05	0.66	0.40	0.45	0.50	0.72	1.11	1.30	10.57

EUGENE, OR	30	7.65	6.35	5.80	3.66	2.66	1.53	0.64	0.99	1.54	3.35	8.44	8.29	50.90
MEDFORD, OR	30	2.47	2.10	1.85	1.31	1.21	0.68	0.31	0.52	0.78	1.31	2.93	2.90	18.37
PENDLETON, OR	30	1.45	1.22	1.26	1.13	1.22	0.78	0.41	0.56	0.63	0.99	1.63	1.48	12.76
PORTLAND, OR	30	5.07	4.18	3.71	2.64	2.38	1.59	0.72	0.93	1.65	2.88	5.61	5.71	37.07
SALEM, OR	30	5.84	5.09	4.17	2.76	2.13	1.45	0.57	0.68	1.43	3.03	6.39	6.46	40.00
SEXTON SUMMIT, OR	30	4.71	4.29	3.92	2.38	1.35	0.94	0.35	0.61	1.20	2.93	5.32	5.18	33.18
ALLENTOWN, PA	30	3.50	2.75	3.56	3.49	4.47	3.99	4.27	4.35	4.37	3.33	3.70	3.39	45.17
ERIE, PA	30	2.53	2.28	3.13	3.38	3.34	4.28	3.28	4.21	4.73	3.92	3.96	3.73	42.77
HARRISBURG, PA	30	3.18	2.88	3.58	3.31	4.60	3.99	3.21	3.24	3.65	3.06	3.53	3.22	41.45
MIDDLETOWN/HARRISBURG APT	30	3.18	2.88	3.58	3.31	4.60	3.99	3.21	3.24	3.65	3.06	3.53	3.22	41.45
PHILADELPHIA, PA	30	3.52	2.74	3.81	3.49	3.89	3.29	4.39	3.82	3.88	2.75	3.16	3.31	42.05
PITTSBURGH, PA	30	2.70	2.37	3.17	3.01	3.80	4.12	3.96	3.38	3.21	2.25	3.02	2.86	37.85
AVOCA, PA	30	2.46	2.08	2.69	3.28	3.69	3.97	3.74	3.10	3.86	3.02	3.12	2.55	37.56
WILLIAMSPORT, PA	30	2.85	2.61	3.21	3.49	3.79	4.45	4.08	3.38	3.98	3.19	3.62	2.94	41.59
BLOCK IS., RI	30	3.68	3.04	3.99	3.72	3.40	2.77	2.62	3.00	3.19	3.04	3.77	3.57	39.79
PROVIDENCE, RI	30	4.37	3.45	4.43	4.16	3.66	3.38	3.17	3.90	3.70	3.69	4.40	4.14	46.45
CHARLESTON AP, SC	30	4.08	3.08	4.00	2.77	3.67	5.92	6.13	6.91	5.98	3.09	2.66	3.24	51.53
CHARLESTON C.O., SC	30	3.62	2.62	3.83	2.44	2.77	4.96	5.50	6.54	6.13	3.02	2.18	2.78	46.39
COLUMBIA, SC	30	4.66	3.84	4.59	2.98	3.17	4.99	5.54	5.41	3.94	2.89	2.88	3.38	48.27
GREENV'L-SPARTANB'RG AP, SC	30	4.41	4.24	5.31	3.54	4.59	3.92	4.65	4.08	3.97	3.88	3.79	3.86	50.24
ABERDEEN, SD	30	0.48	0.48	1.34	1.83	2.69	3.49	2.92	2.42	1.81	1.63	0.75	0.38	20.22

(Continued)

115

	NORMALS 1971-2000								English Units (Inches)					
Location	YRS	JAN	FEB	MAR	APR	MAY	JUN	JUL	AUG	SEP	OCT	NOV	DEC	ANN
HURON, SD	30	0.49	0.57	1.67	2.29	3.00	3.28	2.86	2.07	1.80	1.59	0.89	0.39	20.90
RAPID CITY, SD	30	0.37	0.46	1.03	1.86	2.96	2.83	2.03	1.61	1.10	1.37	0.61	0.41	16.64
SIOUX FALLS, SD	30	0.51	0.51	1.81	2.65	3.39	3.49	2.93	3.01	2.58	1.93	1.36	0.52	24.69
BRISTOL-JOHNSON CTY TN	30	3.52	3.40	3.91	3.23	4.32	3.89	4.21	3.00	3.08	2.30	3.08	3.39	41.33
CHATTANOOGA, TN	30	5.40	4.85	6.19	4.23	4.28	3.99	4.73	3.59	4.31	3.26	4.88	4.81	54.52
KNOXVILLE, TN	30	4.57	4.01	5.17	3.99	4.68	4.04	4.71	2.89	3.04	2.65	3.98	4.49	48.22
MEMPHIS, TN	30	4.24	4.31	5.58	5.79	5.15	4.30	4.22	3.00	3.31	3.31	5.76	5.68	54.65
NASHVILLE, TN	30	3.97	3.69	4.87	3.93	5.07	4.08	3.77	3.28	3.59	2.87	4.45	4.54	48.11
OAK RIDGE, TN	30	5.13	4.50	5.72	4.32	5.14	4.64	5.16	3.39	3.75	3.02	4.86	5.42	55.05
ABILENE, TX	30	0.97	1.13	1.41	1.67	2.83	3.06	1.70	2.63	2.91	2.90	1.30	1.27	23.78
AMARILLO, TX	30	0.63	0.55	1.13	1.33	2.50	3.28	2.68	2.94	1.88	1.50	0.68	0.61	19.71
AUSTIN/CITY, TX	30	1.89	1.99	2.14	2.51	5.03	3.81	1.97	2.31	2.91	3.97	2.68	2.44	33.65
AUSTIN/BERGSTROM, TX	30	2.21	2.02	2.36	2.63	5.12	3.42	2.03	2.51	2.88	3.99	3.02	2.53	34.72
BROWNSVILLE, TX	30	1.36	1.18	0.93	1.96	2.48	2.93	1.77	2.99	5.31	3.78	1.75	1.11	27.55
CORPUS CHRISTI, TX	30	1.62	1.84	1.74	2.05	3.48	3.53	2.00	3.54	5.03	3.94	1.74	1.75	32.26
DALLAS-FORT WORTH, TX	30	1.90	2.37	3.06	3.20	5.15	3.23	2.12	2.03	2.42	4.11	2.57	2.57	34.73
DALLAS-LOVE FIELD, TX	30	1.89	2.31	3.13	3.46	5.30	3.92	2.43	2.17	2.65	4.65	2.61	2.53	37.05
DEL RIO, TX	30	0.57	0.96	0.96	1.71	2.31	2.34	2.02	2.16	2.06	2.00	0.96	0.75	18.80
EL PASO, TX	30	0.45	0.39	0.26	0.23	0.38	0.87	1.49	1.75	1.61	0.81	0.42	0.77	9.43
GALVESTON, TX	30	4.08	2.61	2.76	2.56	3.70	4.04	3.45	4.22	5.76	3.49	3.64	3.53	43.84
HOUSTON, TX	30	3.68	2.98	3.36	3.60	5.15	5.35	3.18	3.83	4.33	4.50	4.19	3.69	47.84
LUBBOCK, TX	30	0.50	0.71	0.76	1.29	2.31	2.98	2.13	2.36	2.57	1.70	0.71	0.67	18.69
MIDLAND-ODESSA, TX	30	0.53	0.58	0.42	0.73	1.79	1.71	1.89	1.77	2.31	1.77	0.65	0.65	14.80

PORT ARTHUR, TX	30	5.69	3.35	3.75	3.84	5.83	6.58	5.23	4.85	6.10	4.67	4.75	5.25	59.89
SAN ANGELO, TX	30	0.82	1.18	0.99	1.60	3.09	2.52	1.10	2.05	2.95	2.57	1.10	0.94	20.91
SAN ANTONIO, TX	30	1.66	1.75	1.89	2.60	4.72	4.30	2.03	2.57	3.00	3.86	2.58	1.96	32.92
VICTORIA, TX	30	2.44	2.04	2.25	2.97	5.12	4.96	2.90	3.05	5.00	4.26	2.64	2.47	40.10
WACO, TX	30	1.90	2.43	2.48	2.99	4.46	3.08	2.23	1.85	2.88	3.67	2.61	2.76	33.34
WICHITA FALLS, TX	30	1.12	1.58	2.27	2.62	3.92	3.69	1.58	2.39	3.19	3.11	1.68	1.68	28.83
MILFORD, UT	30	0.73	0.77	1.21	0.99	0.94	0.44	0.76	1.04	0.92	1.12	0.77	0.58	10.27
SALT LAKE CITY, UT	30	1.37	1.33	1.91	2.02	2.09	0.77	0.72	0.76	1.33	1.57	1.40	1.23	16.50
BURLINGTON, VT	30	2.22	1.67	2.32	2.88	3.32	3.43	3.97	4.01	3.83	3.12	3.06	2.22	36.05
LYNCHBURG, VA	30	3.54	3.10	3.83	3.46	4.11	3.79	4.39	3.41	3.88	3.39	3.18	3.23	43.31
NORFOLK, VA	30	3.93	3.34	4.08	3.38	3.74	3.77	5.17	4.79	4.06	3.47	2.98	3.03	45.74
RICHMOND, VA	30	3.55	2.98	4.09	3.18	3.96	3.54	4.67	4.18	3.98	3.60	3.06	3.12	43.91
ROANOKE, VA	30	3.23	3.08	3.84	3.61	4.24	3.68	4.00	3.74	3.85	3.15	3.21	2.86	42.49
OLYMPIA, WA	30	7.54	6.17	5.29	3.58	2.27	1.78	0.82	1.10	2.03	4.19	8.13	7.89	50.79
QUILLAYUTE, WA	30	13.65	12.35	10.98	7.44	5.51	3.50	2.34	2.67	4.15	9.81	14.82	14.50	101.72
SEATTLE C.O., WA	30	5.24	4.09	3.92	2.75	2.03	1.55	0.93	1.16	1.61	3.24	5.67	6.06	38.25
SEATTLE SEA-TAC AP, WA	30	5.13	4.18	3.75	2.59	1.78	1.49	0.79	1.02	1.63	3.19	5.90	5.62	37.07
SPOKANE, WA	30	1.82	1.51	1.53	1.28	1.60	1.18	0.76	0.68	0.76	1.06	2.24	2.25	16.67
WALLA WALLA, WA	30	2.25	1.97	2.20	1.83	1.95	1.15	0.73	0.84	0.83	1.77	2.85	2.51	20.88
YAKIMA, WA	30	1.17	0.80	0.70	0.53	0.51	0.62	0.22	0.36	0.39	0.53	1.05	1.38	8.26
BECKLEY, WV	30	3.23	2.96	3.63	3.43	4.39	3.92	4.78	3.45	3.23	2.64	2.88	3.09	41.63
CHARLESTON, WV	30	3.25	3.19	3.90	3.25	4.30	4.09	4.86	4.11	3.45	2.67	3.66	3.32	44.05
ELKINS, WV	30	3.43	3.20	3.92	3.53	4.77	4.61	4.84	4.26	3.83	2.86	3.42	3.44	46.11

(Continued)

117

Location	YRS	JAN	FEB	MAR	APR	MAY	JUN	JUL	AUG	SEP	OCT	NOV	DEC	ANN
									English Units (Inches)					
NORMALS 1971-2000														
HUNTINGTON, WV	30	3.21	3.09	3.83	3.33	4.41	3.88	4.46	3.88	2.80	2.73	3.32	3.37	42.31
GREEN BAY, WI	30	1.21	1.01	2.06	2.56	2.75	3.43	3.44	3.77	3.11	2.17	2.27	1.41	29.19
LA CROSSE, WI	30	1.19	0.99	2.00	3.38	3.38	4.00	4.25	4.28	3.40	2.16	2.10	1.23	32.36
MADISON, WI	30	1.25	1.28	2.28	3.35	3.25	4.05	3.93	4.33	3.08	2.18	2.31	1.66	32.95
MILWAUKEE, WI	30	1.85	1.65	2.59	3.78	3.06	3.56	3.58	4.03	3.30	2.49	2.70	2.22	34.81
CASPER, WY	30	0.58	0.64	0.90	1.52	2.38	1.43	1.29	0.73	0.98	1.14	0.82	0.62	13.03
CHEYENNE, WY	30	0.45	0.44	1.05	1.55	2.48	2.12	2.26	1.82	1.43	0.75	0.64	0.46	15.45
LANDER, WY	30	0.52	0.54	1.24	2.07	2.38	1.15	0.84	0.57	1.14	1.37	0.99	0.61	13.42
SHERIDAN, WY	30	0.77	0.57	1.00	1.77	2.41	2.02	1.11	0.80	1.38	1.41	0.80	0.68	14.72
GUAM, PC	30	5.58	5.11	4.24	4.16	6.39	6.28	11.66	16.17	13.69	11.88	9.34	6.11	100.61
JOHNSTON ISLAND, PC	30	1.64	1.29	2.01	1.86	1.14	0.87	1.40	2.07	2.46	2.78	4.78	2.70	25.00
KOROR, PC	30	11.20	9.65	8.79	9.45	11.27	17.54	16.99	14.47	11.65	13.41	11.62	12.33	148.37
KWAJALEIN, MARSHALL IS., PC	30	5.12	3.73	3.82	7.63	8.62	8.86	10.24	10.42	11.82	11.46	10.74	7.94	100.40
MAJURO, MARSHALL IS, PC	30	8.09	6.86	8.43	11.30	11.53	11.09	12.41	11.95	11.96	13.73	12.81	11.50	131.66
PAGO PAGO, AMER SAMOA, PC	30	14.02	12.14	11.15	11.16	10.43	5.94	5.76	6.43	7.36	10.03	11.16	13.38	118.96
POHNPEI, CAROLINE IS., PC	30	12.52	9.78	13.96	16.94	19.41	17.06	16.72	16.37	14.94	16.30	14.74	15.87	184.61
CHUUK, E. CAROLINE IS., PC	30	8.58	8.77	8.15	10.94	11.29	12.82	12.45	15.09	13.12	10.69	11.09	10.98	133.97
WAKE ISLAND, PC	30	1.40	1.89	2.38	2.11	1.70	1.95	3.44	5.62	4.82	4.27	2.78	1.87	34.23
YAP, W CAROLINE IS., PC	30	7.24	5.45	6.14	5.58	8.15	13.46	13.25	14.41	13.53	12.25	8.82	9.34	117.62
SAN JUAN, PR	30	3.02	2.30	2.14	3.71	5.29	3.52	4.16	5.22	5.60	5.06	6.17	4.57	50.76

CHAPTER **8**

Solar Water Distillation[1]

Introduction

Ninety-seven percent of the earth's water mass lies in its oceans. Of the remaining 3 percent, five-sixths is brackish, leaving a mere 0.5 percent as freshwater. With pollution, this amount is decreasing. Abundant seawater is a tantalizing resource. Many methods of producing drinking water from the sea have been developed over time; naturally, all of them require energy to separate dissolved salt from the water source. This process is called *desalination*, and its outputs include potable water and usable salts.

Coastal dwellers have a long history of capturing solar energy for the desalination of ocean water. Other communities have used solar distillation to separate freshwater from brackish water, as well. (Brackish water contains large quantities of dissolved solids and is not suitable for human consumption). Solar distillation has also been used to purify bacteria-contaminated water and to generate high-quality water for laboratory and industrial applications.

Documented use of solar stills began in the 16th century; an early large-scale solar still, built in 1872, was used to supply water for an entire mining community in Chile. Mass production of solar distillation devices occurred for the first time during the Second World War, when 200,000 inflatable plastic stills were manufactured for use in life rafts for the U.S. Navy. A simple survival solar still can be constructed with damp ground, hot sun, and a clear plastic sheet (Fig. 8.1).

The purpose of this chapter is to describe the principles behind solar distillation, illustrate how solar radiation can be used to provide potable water, and show examples of current solar distillation installations and applications.

How a Simple Solar Still Operates

Figure 8.2 illustrates a single-basin solar still, a design optimized by solar pioneer Horace McCracken in the 1980s. The principles and

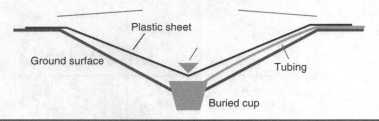

FIGURE 8.1 Field expedient survival still.

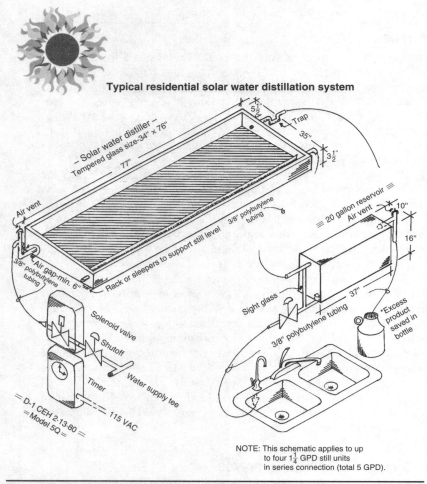

FIGURE 8.2 Single-basin still.

operation of all solar stills are all very similar; the primary design improvements have been to increase distillation efficiency and reduce manufacturing costs. Most of the total cost (life-cycle-cost) for solar distillation is in the initial capital required; operation and maintenance expenditures are relatively low. For large-scale systems, when roof area is unavailable, the cost of land can be an important factor in the capital cost of the system.

In concept, solar distillation is the collection of water evaporated by the heat of the sun. In essence, it is collected *de*w. It has two main steps:

1. The absorption of heat from the sun, which is used to evaporate the feedwater.

2. The condensation of the evaporated water, which can be collected for use.

In practice, the task is to maximize the solar energy absorbed by the feedwater and minimize losses while collecting the condensed water.

A solar distillation panel is designed to transmit solar radiation (called *insolation*) through a transparent cover, usually glass, into a basin containing the feedwater to be distilled. The sunlight is absorbed by a dark surface on the bottom of the basin, a surface which is in contact with the feedwater, and converted to heat. The glass cover traps heat from escaping from the feedwater and prevents it from being reflected back out to space. It utilizes the basic greenhouse effect to produce operating temperatures, which are typically over 160°F, even on cold days (Fig. 8.3).

As the feedwater is heated by the sun, its evaporation rate increases. Water evaporates slowly even at low temperature, but the higher the temperature, the higher the evaporation rate. When the evaporated water (now called *vapor*) comes in contact with a cooler surface, such as the underside of the glass surface, it condenses as water droplets. On clean glass, which can be *wetted* (i.e., it is hydrophilic), drops combine to form a thin film. On tilted glass, this film of condensed vapor, now liquid water, runs down to be collected at the bottom, where it runs into a gutter and is fed into a storage tank. The resulting distilled water is of very high purity.

The composition of seawater and brackish water can vary widely. However, most types of feedwater can be distilled by the sun into potable water that is well suited for human consumption, with only minor differences in production rate. For instance, the production rate from a still fed with seawater is only about 8 percent less than what can be distilled from ordinary tap water, which is already moderately pure. One of the most common residential uses of solar stills is to make drinking water that is higher in quality than most bottled water—using a municipal supply as a feed.

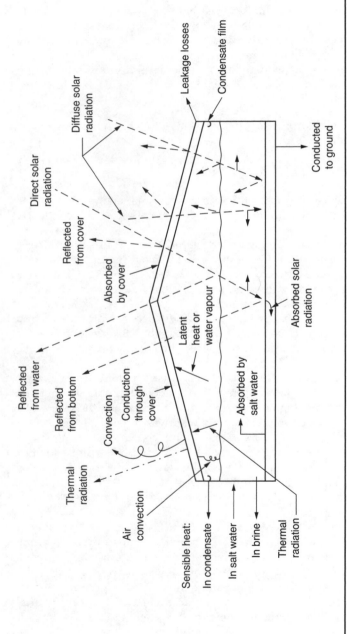

Figure 8.3 How a solar still works.

Salts and almost all dissolved minerals are left behind when evaporated in a solar still. Tap water, brackish water, and seawater feeds all produce good water. Some oils and alcohols will evaporate along with the water, however, and cannot be removed using distillation. This may make the resulting water non-potable.

The basin containing feedwater to be distilled must absorb sunlight efficiently and be well insulated in order to reduce heat loss through the sides and bottom. As feedwater is distilled, it leaves behind dissolved minerals, which create concentrated brine; there must be a convenient method for removing the brine and replacing it with new feed. A considerable amount of solar energy is used up, so to speak, in heating the brine to be removed. So another design consideration is to minimize the amount of leftover brine. An efficient solar still, sized for residential applications, will nominally produce as much as 2.6 gal (10 L)/day during summertime in temperate climates and 0.8 gal (3 L)/day in winter (for Los Angeles). Production can range from a maximum of 1.4 gal/m^2/day to 0.5 gal/m^2/day at the winter solstice. Basic requirements of 1.3 gal (5 L)/person/day can be produced by 17 ft^2 (1.6 m^2) of solar still. A useful design detail is to require 32 ft^2 (3 m^2)/person, which will supply most water needs even in wintertime.

Would a Solar Still Suit Your Needs?

There are a number of approaches to water purification and desalination, such as electrically powered reverse-osmosis or fossil-fuel-derived steam evaporation. Both of these processes use commercially available equipment.

Filtration with a slow sand filter is a simple but effective way to remove commonly found bacteria from contaminated water; however, sand filtration cannot remove dissolved compounds such as salt.

There are several factors to take into account when considering the installation of a solar still. For one thing, human beings need 1 or 2 liters of water a day for survival. The minimum requirement in developing countries is 5 gal (20 L)/day (which includes cooking, cleaning, and washing clothes). For the industrialized world, 50 to 100 gal (200 to 400 L)/day is typical. Some functions can be performed with salty or brackish water, making a typical minimum for basic survival about 5 liters per person per day. Given 5 L/day of required water consumption, an average of 1.1 m^2 of still area should be considered as a baseline for supplying water needs. But it is suggested that 3 m^2 be used to allow for adequate production during wintertime. Therefore, 2 m^2 of solar still is needed for each person served.

Unlike other techniques of desalination, solar stills are more attractive for smaller output. The initial capital cost of a solar still is

roughly proportional to capacity, whereas other methods have significant economies of scale. For the individual household, therefore,
the solar still can still be an economical choice compared with other
desalination methods, when the feedwater requires that minerals
(such as salt) be removed.

Solar stills should normally only be considered for removal of
dissolved salts from water. Although the high temperatures remove
most pathogens and suspended solids, solar stills do not generally
remove all dissolved organics completely. If there is a choice between
brackish groundwater or polluted surface freshwater, it will usually
be cheaper overall to use a slow sand filter or other treatment method
to treat the surface water. If there is no freshwater available, the main
alternatives are desalination, transportation, and rainwater collection.

Energy Requirements for Water Distillation and Performance of Solar Stills

Distillation, as a process for water purification, requires an energy
input. Four components of the required energy can be seen in
Table 8.1.

Reducing energy loss by better insulation of the still and reducing
the rejected brine are the two areas where the best improvements in
efficiency can be achieved. The energy available from incident
sunlight varies from 1 to 8 kWh/m^2/day, based on the latitude and
climate conditions; the amount of this energy that is successfully
collected depends on the design of the still and materials used.

The analysis of still performance and efficiency is very similar to
the standard method for evaluating flat plate thermal solar collectors,
which use sunlight to heat water. This is because the typical still
design (as seen in Fig. 8.1) is very similar to the typical design of a flat
plate thermal collector.

1. Energy to heat feedwater from ambient to operating temperature	280 kJ/kg = 20 BTU/lb
2. Energy for vaporization of water (latent heat)	2260 kJ/kg = 970 BTU/lb
3. Energy lost through panel sides and bottom (estimate)	70 BTU/lb
4. Energy rejected in brine (concentrate after distilled water removal)	120 BTU/lb

TABLE 8.1   Energy Requirements for Distillation

As discussed, the energy input to the system is incident sunlight (the global solar insolation). Much of this energy is lost through the sides and bottom of the basin, and heat can be lost when heating the still equipment. The exact form of the resulting equation is quadratic, and it is based on measurements. The full equation for the output of a conventional solar still is as follows:

$$Q = A \times I \times (0.7300 - 0.6006 \times \Delta T/I - 0.0010 \times \Delta T^2/I)/H$$

Q (gal/day) = estimate of nominal output

I (BTU/ft^2/day) = the global solar insolation for a day

A (ft^2) = the aperture area

H (BTU/gal) = the enthalpy of heating and vaporization, 11,600 BTU/gal at typical conditions

ΔT (°F) = the difference between the still operating temperature and the ambient temperature

For example, consider a standard McCracken design solar still, which uses a common tempered glass size used for sliding doors— 34 by 76 in. Subtracting the edges supporting the glass, the aperture area, which is the area through which sunlight can pass, is 33 by 75 in^2, or 17.1 sq ft.

In April in Los Angeles, the solar insolation is 6.1 kWh/m^2/day = 1945 BTU/sq ft/day, and the average daytime temperature is 62.6°F (typical operating temperature is 180°F). Combining this with the McCracken still aperture area results in a daily output of 2.2 gal/day. Because of hourly variations in effective insolation due to clouds, temperature, angle of the sun's rays, etc., it is more common to estimate production on a daily basis—using average insolation and temperature values.

Daily average insolation and temperature information is available from the National Renewable Energy Laboratory—demonstration values can be seen in Table 8.2. (Hourly average value data for insolation and ambient temperature is also available in some locations. This data can be used to calculate the daily output much more precisely. This kind of accuracy is justified only for larger commercial installations, not for small, residential applications.) Energy to pump and move the water, while relatively minor compared to the energy needed for evaporation, must also be included in the system design. For example, although 2260 kJ/kg is required to evaporate water, to pump 1 kg (2.2 lb) of water through a 20-m (66-ft) head requires an additional 0.2 kJ/kg. This energy must be supplied from an external source, such as an electric pump.

City	Value	Units	Annual	Jan	Feb	Mar	Apr	May	Jun	Jul	Aug	Sep	Oct	Nov	Dec
Charleston, SC	Insolation*	kWh/m²/d	4.6	2.7	3.5	4.7	5.9	6.2	6.2	6.1	5.5	4.7	4.1	3.1	2.5
	Collected** Solar Energy	Btu/day/panel	17,424	9119	12,550	17,741	22,929	24,314	24,395	24,016	21,457	18,000	15,321	10,963	8317
	Day Temp.	F°	69	51.1	54	61.5	68.5	76.2	81.4	84.3	83.4	79.2	70.2	62	54.6
	Output***	Gal/day/panel	1.6	0.8	1.2	1.7	2.2	2.3	2.3	2.3	2.0	1.7	1.4	1.0	0.8
Seattle, WA	Insolation	kWh/m²/d	3.3	1	1.7	2.8	4.1	5.3	5.8	6.1	5.2	3.8	2.2	1.2	0.8
	Collected Solar Energy	Btu/day/panel	11,707	1791	4806	9502	15,066	20,237	22,440	23,778	19,964	13,953	7052	2706	946
	Day Temp.	F°	54.4	41.7	45.4	48	51.9	58	64	68.5	68.7	63.5	55.1	47.1	42
	Output	Gal/day/panel	1.1	0.2	0.4	0.9	1.4	1.9	2.1	2.2	1.9	1.3	0.7	0.3	0.1
Los Angeles, CA	Insolation	kWh/m²/d	4.9	2.8	3.6	4.8	6.1	6.4	6.6	7.1	6.5	5.3	4.2	3.2	2.6
	Collected Solar Energy	Btu/day/panel	18,644	9659	13,060	18,151	23,693	24,997	25,888	28,057	25,532	20,438	15,733	11,423	8813
	Day Temp.	F°	65.4	59.7	60.3	60.5	62.6	64.8	67.8	71.2	72.5	72.1	69.4	64.5	59.8
	Output	Gal/day/panel	1.8	0.9	1.2	1.7	2.2	2.3	2.4	2.6	2.4	1.9	1.5	1.1	0.8

*Source for global insolation data: The Solar Radiation Data Manual for Flat-Plate and Concentrating Collectors. *National Renewable Energy Laboratory.* http://rredc.nrel.gov/solar/old_data/nsrdb/redbook/.

**Estimated collected solar energy calculated by $Q = A \times I \times (0.7300 - 0.6006 \times \Delta T / I - 0.0010 \times \Delta T^2 / I)$ for the solar heat input.

***Water output (gal/day) = Q/H. Where Q is net energy collected and H is the energy required to distill a gallon of water at the daytime temperature listed. (H is the thermodynamic quantity termed enthalpy.)

TABLE 8.2 Average Distilled Water Output for Standard Panel Size (34 by 76 inches), in Three U.S. Cities

Design Objectives for an Efficient Solar Still

Single-basin stills have been much studied and their behavior is well understood. Efficiencies of 35 percent are typical. Daily output as a function of solar irradiation is greatest in the early evening when the feedwater is still hot but when outside temperatures are falling.

For optimal efficiency, there are four key functions that a properly designed solar distillation system should maintain:

1. Absorption of as much of the incoming solar energy as possible
2. A large temperature difference between water vapor and condensing surface
3. A distillation enclosure with low vapor leakage
4. A well-insulated basin to minimize parasitic (unusable) heat loss

A high feedwater temperature can be achieved under the following conditions:

- A high proportion of incoming radiation is absorbed by the feedwater as heat. A low-absorption glazing and a good radiation-absorbing surface on the bottom of the basin are required.

- Heat loss from the enclosure floor and walls is kept low through proper insulation.

- The water level in the evaporator is shallow so there is a minimal water volume to surface area ratio, which will allow the water to be heated more efficiently.

A large temperature difference can be achieved under the following conditions:

- The condensing surface absorbs little or none of the incoming radiation.

- Water condensing on the surface dissipates heat; this water must be removed rapidly by a second flow of water or air, for example, or by condensing at night.

Which Solar Still?

Despite a proliferation of novel still types (see Appendix), the single-basin still is the design most proven in the field. At least 40 single-basin stills with areas greater than 100 m (and up to 9000 m²) were built between 1957 and 1980. Twenty-seven had glass covers and nine had plastic. Twenty-four of the glass-covered stills are still operating in their original form, but only one plastic-covered unit is operational. Hundreds of smaller stills are operating, notably in Africa.

The cost of pure water produced by the distillation process depends on the following:

- The cost of making the still
- The cost of the land
- The life of the still
- Operating costs
- The cost of the feedwater
- The discount rate adopted
- The amount of water produced

The life of a glass still is usually about 20 to 30 years. Operating costs are primarily due to maintenance of the feedwater system. Feedwater is injected each day in a batch manner—not constantly. If a pump is used for the feed, there is an additional electricity cost if the pump is not operated by solar collection. In areas exposed to traffic, breakage and replacement of glazing can be a problem.

While the basic design can take on many variations, the actual shape and concept have not changed substantially from the days of the stills in Las Salinas, Chile, which were built in 1872. The greatest changes have involved the use of new building materials, which have lowered overall costs while providing an acceptably long useful life and better performance.

All basin stills have six major components:

1. A basin
2. A support structure
3. A transparent glazing cover
4. A distillate trough (water channel)
5. A system to regularly flush the brine
6. A system for injecting feedwater

In addition to these, ancillary components may include the following:

- Insulation on all sides of the basin to minimize heat loss. High temperature (300°F) insulation is required. The R-value should be 7 or greater.
- Sealants, to avoid leakage of water.
- Piping and valves, provisions for regular refill.
- Storage tanks.
- An external cover to protect the other components from the weather and to make the still aesthetically pleasing.
- A reflector to concentrate sunlight.

The critical design issues are to minimize side walls and attendant heat loss, to minimize glass support structure, and to maximize the aperture.

The actual dimensions of basin stills vary greatly, depending on the availability of materials, water requirements, ownership patterns, and land location and availability. If the only glazing available is 1 meter at its greatest dimension, the still's maximum inner width will be just less than 1 meter. And the length of the still will be set according to what is needed to provide the amount of square meters to produce the required amount of water. If an entire village were to own and use the still, obviously the total installation would have to be quite large.

Because of the necessity of maintaining as low a water level as possible in the basin, it must be ensured that there is just sufficient fluid left to prevent drying out during operation, which would result in inconvenient precipitation of minerals. This would reduce the effectiveness of the absorption of solar radiation, and therefore decrease the overall efficiency of the still.

Most community size stills are 0.5 to 2.5 meters wide, with lengths ranging up to around 100 meters. Their lengths usually run along an east-west axis to maximize the transmission of sunlight through the equatorial-facing sloped glass.

Residential appliance-type units generally use glass about 0.65 to 0.9 meter wide with lengths ranging from 2 to 2.5 meters. In the United States, two common tempered glass sizes are 34 by 76 inches and 34 by 92 inches. A water depth of 1.5 to 2.5 cm is most common; water depth management is crucial. It is not practical to provide for draining and flushing while maintaining an initial maximum water depth of no more than 2 cm (0.75 inch). This limits the length of a still to about 2.5 meters. Separate panels of these lengths are convenient. However, careful installation of baffles every 2 meters will permit adequate drainage in a lengthy, multi-pane still.

The usual argument for greater depths is that the stored heat can be used at night to enhance production when the air temperatures are lower. Unfortunately, no deep basin has ever attained the 43 percent efficiency typical of a still of minimum water depth. The results to date are clear: the shallower the depth, the better. Of course, if the basin is too shallow, it will dry out and salts will be deposited, which is not good. Note that solar heat can evaporate about 0.5 cm of water on a clear day in summer. By setting the initial charge at about 1.5 cm depth, virtually all of the salts remain in the solution and can be flushed out by the refilling operation.

It is generally best to design an installation with many small modular units to supply the water. This allows the following:

- Units to be added in stages to match increasing demand and available funding

- Smaller and more manageable components to be handled by unskilled persons without expensive mechanical equipment

- Maintenance to be carried out on some units while others continue to operate

Material Requirements of Basin Stills

Although local materials should be used whenever possible to lower initial costs and to facilitate any necessary repairs, keep in mind that solar stills made with cheap, shoddy materials will not last as long as those built with more costly, high-quality material. With this in mind, you must decide whether you want to build an inexpensive and thus short-lived still that needs to be replaced or repaired every few years or build something more durable and lasting in the hope that the distilled water it produces will be cheaper in the long run. Of the low-cost stills that have been built around the world, many have been abandoned. Building a more durable still that will last 20 years or more seems to be worth the additional investment.

The material selection for the components that come in contact with the water represents a key decision in the success of the still. Many plastics will give off a substance that can be tasted or smelled in the product water for periods of anywhere from hours to years. As a general guide, if you are contemplating using any material other than glass or metal in contact with water, you may perform a useful screening test by boiling a sample of the material in a cup of good water for half an hour, then letting the water cool and smelling and tasting it. This is a considerably accelerated test of what happens in the still. If you can't tell any difference between the test water and what you started with, the material is probably safe to use. To get some experience, try this on polyethylene tubing, PVC pipe, and fiberglass resin panel.

Basic Components

This section describes the basic components of the still: basin, support structure, glazing, distillate trough, and insulation. It also discusses the range of materials available for their construction and the advantages and disadvantages of some of those materials.

The Basin

The basin contains the saline (or brackish) water that will undergo distillation. As such, it must be waterproof and dark (preferably black) so that it will better absorb the sunlight and convert it to heat. It should also have a relatively smooth surface so it is easy to clean any sediment from it.

There are two general types of basins. The first is made of a material that maintains its own shape and provides the waterproof containment by itself or with the aid of a surface material applied directly to it. The second type uses one set of materials (such as wood or brick) to define the basin's shape. Into this is placed a second material that easily conforms to the shape of the structural materials and serves as a waterproof liner. No one construction material is appropriate for all circumstances or locations. Table 8.3 lists the various materials and rates them according to properties desirable for this application.

Selecting a suitable material for basin construction is the biggest problem in the solar still industry. The corrosion conditions at the water line can be so severe that basins made of metal, even hose coated with anti-corrosive materials, tend to corrode. Basins made of copper, for example, are likely to be eaten away in a few years. Galvanized steel and anodized uncoated aluminum are likely to corrode in a few months. This is also true of aluminum alloys used to make boats.

Temperature also affects the degree of corrosion that occurs. Many chemical reactions, including those in contact with brackish water, double in rate with each 10°C increase in temperature. Whereas an aluminum boat might last 20 years in seawater at 25°C, if you increase that temperature by 50°, the durability of that aluminum may well be only one or two years.

Porcelain-coated steel lasts only a few years before it is eaten away by corrosion. The special glass used for porcelain is slightly soluble in water, and inside a still it will dissolve away. The typical life of stills equipped with porcelain basins is about five years, although several have been kept operating much longer than that by repairing leaks with silicone rubber.

People have also tried to use concrete because it's inexpensive and simple to work with, but the failure rate has been high because it often develops cracks, if not during the first year, then later on. Concrete and asbestos cement also absorb water. The water may not run right through, but it does soak it up. Everybody knows that satisfactory cisterns and reservoirs are built of concrete, but in a solar still the rules change. Any part of it that is exposed to outside air will permit evaporation. Since it is salt water that is being evaporated, salt crystals will form in the concrete near the surface and break it up, turning it to powder.

What about plastic? Every few years, someone decides that if we could just mold the whole still except for the glass and glass seal out of some plastic such as Styrofoam, it would be so easy and inexpensive. But styrene foam melts at about 70°C. Urethane foam is a little more promising, but it tends to be dimensionally unstable, and if a still is constructed in the inclined tray configuration, the efficiency suffers because the non-wetted portions do not conduct heat to the wetted portions very well.

Material	Durability	Cost	Local Availability	Skills Needed	Cleaning	Portability	Toxicity
Enameled Steel	High	High	Low	Low	High	Medium	Low
EPDM Rubber	High	High	Low	Low	High	High	Low
Butyl Rubber	High	High	Low	Low	High	High	Low
Asphalt Mat	High	Medium	Medium	Medium	Medium	Medium	[a]
Asbestos Cement	High	Medium	Low	Medium	Medium	Medium	High
Black Polyethylene	Medium	Low	Low	Low	Medium	High	Low
Roofing Asphalt on Concrete	Medium	Medium	High	Medium	Medium	Low	[a]
Wood	Low	[a]	[a]	Medium	Medium	Medium	Low
Formed Fiberglass	Medium	Medium	Low	Low	High	Medium	Low

[a] = Unknown or depends upon local conditions.

TABLE 8.3 A Comparison of Various Materials Used in Solar Basin Construction[2]

What about fiberglass? People have spent a lot of time trying to build stills from fiberglass resin formulations. Thus far, they have found the material to be unusable for any part of the still (e.g., the basin or distillate trough) that comes in contact with water, either in liquid or vapor form. Epoxy and polyester resins can impart taste and odor to the distilled water, not just for weeks, but for years. Researchers have found that this problem cannot be eliminated by covering these materials with a coat of acrylic or anything else. The odors migrate right through the coating and make the distilled water unsalable, if not undrinkable. Using fiberglass resin is also not a particularly low-cost approach. Finally, a fiberglass basin or trough that is subjected to hot water for many years develops cracks. Unless researchers find a way to solve these problems, fiberglass remains an unsuitable material.

One alternative is ordinary aluminum coated with silicone rubber. The durability of basins made with this material increased into the 10 to 15 year range. For the hundreds of stills one company sold using this material, the coating was all done by hand. With production roll coating equipment, the basin's durability could probably be increased even more.

Although stainless steel has been used, success has been poor. The degree of "stainless" performance is proportional to the degree of chromium in the alloy. Type 316 stainless steel (which is detectable because it is nonmagnetic) is superior to 304 stainless steel (which is magnetic and more easily corroded). Stainless steel is unusual, in that in order to maintain its "stainless" quality, it needs to maintain an oxide. As a result, stainless steel generally does not perform well under water.

Support Structures

Support structures can form the sides of the still as well as the basin, and they support the glazing cover. As noted earlier, some materials used in forming the basin also form the still support structure, while other still configurations demand separate structures, especially to hold the glazing cover.

It is essential that it be practical to re-level the still (or each section) as the support sags. This is especially important on roof structures. (This is in order to maintain the minimum water depth (1 cm at inlet and 2 cm at outlet) in the basin while providing for a 1 percent slope for drainage.

The primary choices for support structures are wood, metal, concrete, or plastics. In most cases, the choice of material is based upon local availability. Ideally, the frame for the glazing cover should be built of small-sized members so they do not shade the basin excessively.

Wooden support structures are subject to warping, cracking, rot, and termite attack. Choosing a high-quality wood, such as cypress,

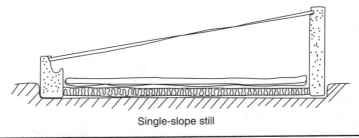

Single-slope still

FIGURE 8.4   Single slope still.

and letting it age may help to alleviate these problems, but if high heat and high humidity prevail inside and outside the still, the still will require frequent repair or replacement. The main advantage of wood is that it can be easily worked with basic hand tools.

Metal may be used for the supports but is subject to corrosion. Since metals are not subject to warping, they can aid in maintaining the integrity of the seals, although the expansion rate of a metal must be taken into account to ensure its compatibility with the glazing material and any sealants used. Use of metal for frame members is practically limited to aluminum and galvanized steel. Both will last almost indefinitely, if protected from exposure. Silicone rubber will not adhere well to galvanized steel, but does so very well to aluminum.

Concrete and other masonry materials may form the sides and glazing support of a still, as well as the membrane. This is more readily possible in a single slope still (Fig. 8.4) than in a double slope still (Fig. 8.5).

Glazing Cover
After the pan, the glazing cover is the most critical component of any solar still. It is mounted above the basin and must be able to transmit a lot of light in the visible spectrum yet keep the heat generated by that light from escaping the basin. Exposure to ultraviolet radiation

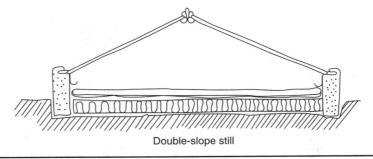

Double-slope still

FIGURE 8.5   Double slope still.

requires a material that can withstand the degradation effects or that is inexpensive enough to be replaced periodically. Since it may encounter temperatures approaching 95°C (200°F), it must also be able to support its weight at those temperatures and not undergo excessive expansion, which could destroy the airtight seals.

Ideally, the glazing material should also be strong enough to resist high winds, rain, hail, and small earth movements, and prevent the intrusion of insects and animals. Moreover, it must be hydrophilic (i.e., wettable).[3] Wetting ability allows the condensing vapor to form as water film on the underside of the glazing cover, rather than as water droplets. Droplets will not run down the inclined glazing to the collection trough, and their formation will reduce the performance of the still for the following reasons:

- Water droplets restrict the amount of light entering the still because they act as small mirrors and reflect it back out.

- A percentage of the distilled water that forms as droplets on the underside will fall back into the basin rather than flow down the glazing cover into the collection trough. Except for temporary conditions at start-up, such a loss of water should not be tolerated. In most cases, if a still operates dry for even a few hours, it will become seriously hydrophobic. That is, the water will condense into droplets on the surface of the glazing.

- When operated dry, the high temperatures achieved vaporize the precipitants in the prior feedwater. A transparent film forms on the glazing, which cannot be removed by ordinary washing; this will seriously impede the effectiveness of the still.

Other factors determining the suitability of glazing material include the cost of the material, its weight, life expectancy, local availability, maximum temperature tolerance, and impact resistance, as well as its ability to transmit solar energy and infrared light. Table 8.4 compares various glazing materials based on these factors.

Of the glazing materials listed in Table 8.4, tempered glass is the best choice in terms of its wettability and its capability to withstand high temperatures. It is also three to five times stronger than ordinary window glass and much safer to work with. One disadvantage of tempered glass is its high cost. While tempered low iron glass, in one series of tests, gave 6 percent additional production, it also added about 15 percent to the cost of the still. Moreover, glass cannot be cut after it has been tempered.

Ordinary window glass is the next best choice, except that it has an oily film when it comes from the factory and must be cleaned carefully with detergent and/or ammonia. If you choose glass as a

Type Glazing Material	Estimated Cost (a) (Dollars/ [ft²])	Weight (lb/[ft²])	Life Expectancy	Maximum Temperature	Solar Transmittance (Percent)	Infrared Light Transmittance (Percent)	Impact Resistance	Wettability	Local Availability
Tempered Low-Iron Glass	3.60	1.6 to 2.5	50+ years	400°–600°F 2040–316°C	91	less than 2	Low	Excellent	No
Ordinary Window Glass	0.95	1.23	50 years	400°F 204°C	86	2	Low	Excellent	Yes
Tedlar	0.60	0.029	5–10 years	225°F 107°C	90	58	Low	Treatable	No
Mylar	?	?	?	?	?	?	Low	Treatable	No
Acrylic	1.50	0.78	25+ years	200°F 93°C	89	6	Medium	Treatable	No
Polycarbonate	2.00	0.78	10–15 years	260°F 127°C	86	6	High	Treatable	No
Cellulose Acetate Butyrate	0.68	0.37	10 years	180°F 82°C	90	?	Medium	?	No
Fiberglass	0.78	0.25	8–12 years	200°F 93°C	72–87	2–12	Medium	Treatable	No
Polyethylene	0.03	0.023	8 months	160°F 71°C	90	80	Low	Treatable	?

(a) Amounts are relative costs, based on U.S. dollars data published between 1980 and 1983.

TABLE 8.4 A Comparison of Various Glazing Materials Used in Building Solar Stills

glazing material, double-strength thickness (i.e., one-eighth of an inch, or 32 millimeters) is satisfactory.

While some plastics are cheaper than either window glass or tempered glass, they deteriorate under high temperatures and have poor wettability. Moreover, under temperature conditions typical of solar stills, the chemicals in plastics are likely to interact with the distilled water, possibly posing a health hazard. Nevertheless, it is a valid choice, certainly for a top-quality appliance-type product.

In a still, the thinner the glass, the better. Thin glass will stay cooler on the inner surface, which helps the water condense faster. In a solar still, the slope of the glass should be between 8° and 12°. Setting the glass at a greater slope increases the volume of air inside the still, reducing the system's efficiency. A lower slope makes it more difficult for the condensed water to run down the glass, and water droplets may just fall back into the basin. Also, the closer the glass is to the water, and the less air space in the still, the more efficient the still will be. According to solar still pioneer Horace McCracken, who designed and built solar stills for more than 30 years, a glass cover that is no more than 2 to 3 inches from the water surface will allow the still to operate efficiently. [4]

A detailed drawing that describes the solar distillation system designed and marketed by McCracken can be seen in Fig. 8.2. This configuration was developed by him after very extensive testing. It minimizes the air space above the water surface in the basin while providing sufficient angle for a water film to drain.

Distillate Trough

The distillate trough is located at the base of the tilted glazing. It serves to collect the condensed water and carry it to storage. It should be as small as possible to avoid shading the basin.

The materials used for the trough must satisfy the general material requirements outlined previously. Those most commonly used include metal, formed materials used in basin construction (with or without plastic liners), or treated materials. Stainless steel is the material of choice, although it is expensive. Common varieties, such as 316, are acceptable. Other metals require protective coatings to prevent corrosion. Aluminum is not supposed to corrode in distilled water, but it seems preferable to rub a coating of silicone rubber over it anyway. Galvanized iron probably will not last more than a few years at most, and copper and brass should not be used because they would create a health hazard. Also, steel coated with porcelain is a poor choice because the glass will dissolve slowly and allow the steel to rust.

Basins lined with butyl rubber or EPDM can have their liners extend beyond the basin to form the trough. This method is inexpensive to implement and provides a corrosion-free channel. No version of polyethylene is acceptable because it breaks up and emits an unpleasant odor and taste. Some people have used polyvinyl chloride (PVC) pipe, slit lengthwise. However, it is subject to significant distortion inside the still, and can give off an undesirable gas, and is subject to becoming brittle when exposed to sunlight and heat. Butyl rubber should be acceptable, but because it is black, the distillate trough becomes an absorber and re-evaporates some of the distilled water (a minor problem).

Ancillary Components

Ancillary components include insulation, sealants, piping, valves, fixtures, pumps, and water storage facilities. In general, it is best to use locally available materials, which are easily replaceable.

Insulation

Insulation, used to retard the flow of heat from a solar still, increases the still's performance. In most cases, insulation is placed under the still basin since this is a large area susceptible to heat loss. The depth of water in the basin is optimized if the water is no greater than ¾ inch at the drainage end and no less than 3/8 inch at the inlet end.

Performance decreases rapidly as the depth of the water in the basin increases. This is due to three causes:

1. Water of greater depth than needed requires more rejected energy as the brine is drained each day.

2. The basin takes longer to heat and may not reach sufficient vaporization temperature until a time when the sun is lower.

3. Increased convection currents in the body of the water result in cooling, which decreases the evaporation rate.

The least expensive insulation option is to build a solar still on land that has dry soil and good drainage. The use of sand helps to minimize solar heat losses and may also serve as a heat sink, which will return heat to the basin after the sun sets and prolong the distillation process.

Insulation, which adds approximately 16 percent to construction costs, may be extruded styrofoam or polyurethane. (Note: Polyurethane in contact with soil will absorb moisture and lose much of its insulation value.)

Sealants

Although the sealant is not a major component of a solar still, it is important for efficient operation. It is used to secure the cover to the frame (support structure), take up any difference in expansion and contraction between dissimilar materials, and keep the whole structure airtight. Ideally, a good sealant will meet all of the general material requirements cited earlier. Realistically, however, it might be necessary to use a sealant that is of lesser quality and has a shorter lifespan but that may be locally available at prices more affordable to people in developing countries. One major drawback of applying low-cost sealants to stills is the frequent labor cost input the stills require to keep them in serviceable condition.

Sealing a solar still is more difficult than sealing a solar water heating panel on two counts: first, an imperfect seal could allow entry to microorganisms, which would contaminate the water; second, applying a sealant that imparts a bad taste or odor to the distilled water will make it unpalatable. Traditional sealants that are locally available include the following:

- Window putty (caulk and linseed oil)
- Asphalt caulking compound
- Tar plastic
- Black putty

A wide variety of other caulk sealants are also available. These include latex, acrylic latex, butyl rubber and synthetic rubbers, polyethylene, polyurethane, silicone, and urethane foam. Most of these will be more costly than traditional varieties, but they may wear longer. Of this group of sealants, molded silicone, or EPDM, clamped in place, seems to be the most promising. Silicone rubber sealant, applied from a tube, is certainly a superior choice, although people have reported a few instances of degradation and seal failure after 5 to 15 years when the seal was exposed to sunlight.

Covering the sealant with a metal strip should extend its life greatly. Researchers are experimenting with an extruded silicone seal, secured by compression.

One final note: Remember that a sealant that works well for windows in a building does not assure that it will work in a solar still, due to higher temperatures, presence of moisture, and the fact that the water must be palatable and unpolluted.

Piping

Piping is required to feed water into the still from the supply source and from the still to the storage reservoir. The general material requirements cited earlier hold true for this component. While stainless steel is preferred, polybutylene is also a satisfactory pipe material. Black polyethylene has held up well for at least 15 years as drain tubing. Nylon tubing breaks up if exposed to sunlight for 5 to 10 years. PVC (polyvinyl chloride) pipe is tolerable, although during the first few weeks of still operation, it usually emits a gas that makes the distilled water taste bad.

Ordinary clear vinyl tubing is unacceptable. While satisfactory for drinking water, this food-grade tubing will degrade from the sun's ultraviolet rays if exposed to sunlight. High-density polyethylene, while used successfully for drinking water and milk bottles, will degrade in the sun, imparting a bad taste to the water. Generally, few plastics can withstand heat and sunlight. Brass, galvanized steel, or copper may be used in the feed system, but not in the product system.

Although a solar still repeatedly subjected to freezing will remain unharmed, drain tubes so exposed may freeze shut unless you make them extra large. Feed tubes can easily be arranged with drain-back provisions to prevent bursting.

Still Operation

A solar still is fed on a batch basis for an hour or two every day. It is necessary to admit some extra water each day to flush out the brine. There is very little pressure available to get the water to drain, so drainage cannot proceed rapidly. To prevent flooding, it's good practice to ensure that the feed rate does not exceed the maximum drainage rate. Though needle valves may be used to restrict the flow, such valves have been found to be unstable over time, generally tending to plug up and stop the flow.

A satisfactory solution that avoids the plugging issue and modulates makeup water flow, when feeding from city water pressure (typically 50 psi+/), is to use a length of small diameter copper tubing of 25 feet or more with a 1/8-inch outside diameter, or 50 feet with a 3/16-inch outside diameter tubing. This serves to restrict inlet water flow so that the feed rate does not exceed this maximum drainage rate. It is good practice to install a filter at the inlet, such as an ordinary hose filter washer (50 mesh or finer stainless steel screen) to prevent the inlet end from plugging.

Storage Reservoir

Storage capacity should be adequate to contain four to five times the average daily output of the still. In selecting materials for the storage reservoir, two precautions should be noted:

1. Distilled water is chemically aggressive, wanting to dissolve a little of practically anything, until it is "satisfied," and then the rate of chemical attack is greatly slowed. What this rate is, in terms of parts per million of different substances, is not well documented. But the practical consequences are that some things, such as steel, galvanized steel, copper, brass, solder, and mortar, are damaged or destroyed in the tank component. This results in the possible contamination of the water.

Stainless steel type 316 is a good choice for use in the storage reservoir. Polypropylene laboratory tanks are satisfactory, but to prevent algal growth, they must either be shielded from sunlight or have an opaque tint as seen in a rainwater catchment cistern. Butyl rubber lining of a structural framework will also work satisfactorily. However, direct contact between distilled water and galvanized steel can last for only a few years, and zinc and iron are added to the water. Concrete can be used, but with the expectation that it will slowly crumble over time. One way to slow the chemical attack on the tank is to introduce limestone or marble chips into the distilled water stream, or into the reservoir itself, to purposely pick up calcium carbonate. Tiny amounts of calcium carbonate are actually beneficial to the human body.

2. Extreme precautions need to be taken to prevent entry of insects and airborne bacteria. Air must leave the reservoir every time water enters it and must re-enter every time water is drawn off. To prevent entry of insects and rainwater, a 50 mesh or finer screen is recommended to cover the vent, and the opening of the vent/screen assembly should be turned downward. If this is ignored and insects enter the basin, then biological growth can occur, thereby compromising the purity of the water until the basin is re-sanitized.

To summarize, materials selected for construction of a solar distillation basin must:

- Contain water without leaking
- Absorb solar energy
- Be structurally supported to hold the water
- Be insulated against heat loss from the bottom and edges

In selecting materials for a solar still, there are almost always tradeoffs. You can save money on materials, but you may lose so much in productivity or durability that the "saving" is not economical.

Conclusion

"The technology for solar water distillation is well developed, precise, and quite mature. Many people have tried repeatedly to make a commercially viable business through sale and installation of solar stills. To my knowledge all have failed. I do not understand why this has been so. From a financial standpoint, compared with purchase of much inferior quality bottled water, solar stills have always been a good investment—if not spectacular, at least solid."[5]

The cost of water transported by vehicles is typically of the same order of magnitude as the cost of water produced by solar stills. A pipeline may be less expensive for very large quantities. Rainwater collection is an even simpler technique than solar distillation in areas where rain is not scarce, but it requires a greater area and usually a larger storage tank. If ready-made collection surfaces exist (such as house roofs), these may provide a less expensive source for obtaining clean water.

For outputs of 1 m³/day (1000 liters) or more, reverse-osmosis or electrodialysis should be considered as an alternative to solar stills. Much will depend on the availability and price of electrical power.

For outputs of 200 m³/day (200,000 liters) or more, vapor compression or flash evaporation will normally be least costly. The efficiency of flash evaporation can be improved by having part of its energy requirement met by solar water heaters. A solar water heater for this application could be described simply as a solar still used in a continuous flow, heating water, but not to the point of evaporation.

The potential market for solar distillation systems exists where (1) the option for solar distillation involves hauling water from a great length, (2) where people have to boil the water that is available to kill germs, and (3) in regions where poor-quality water is available if treated in sufficient quantities to meet the demand. In each of these situations, solar water distillation should be considered as an economical and viable option.[6]

Figure 8.6 Two gallon-per-day still with reflector to improve performance, in operation for about 20 years, showing feedwater makeup—Schultz home, Santa Cruz, California. (Schultz Photo)

Notes and References

1. Credit is due to Mr. Jack Schultz, P.E. who was instrumental in the development of this chapter and a colleague of early solar pioneer Horace McCracken. The information provided in this chapter is largely due to Jack's hands-on experience in designing and installing these systems in the 1980's.
2. McCracken, H., Gordes, J., *Understanding Solar Stills*, VITA, Arlington, Virginia, September 1985. pr-info@vita.org.
3. Also termed "beer-glass clean," the surface needs to be cleaned so there is no soap film to cause the water droplets to bead and drop back into the basin.
4. Ask the Experts: Solar Still By Laurie Stone Apr/May 2009 (#130) pp. 32 Introductory Level
5. Jack Schultz, P.E., Santa Cruz, California. Engineer and former designer and build contractor for Solar Water Distillation System, from 1968 to 1979.
6. *Understanding Solar Stills* by Horace McCracken & Joel Gordes, 9/85, Published by VITA, 1600 Wilson Boulevard, Suite 500, Arlington, Virginia 22209 USA, pr-info@vita.org

Appendix

Multiple-effect basin stills: These have two or more compartments. The condensing surface of the lower compartment is the floor of the upper compartment. The heat given off by the condensing vapor provides energy to vaporize the feed water above (Fig. 8.1A). Efficiency is therefore greater than for a single-basin stills, typically being

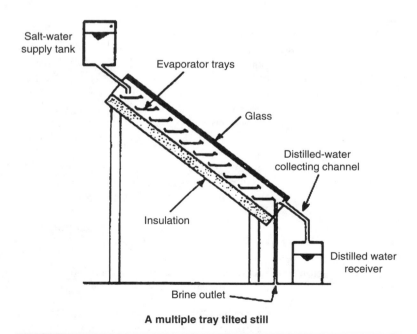

A multiple tray tilted still

Figure 8.1A   Multiple basin solar distillation system.

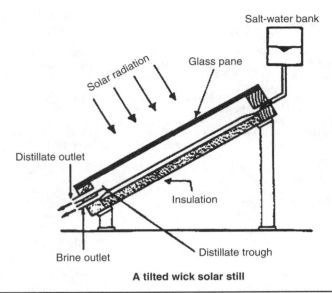

Salt-water bank

Solar radiation

Glass pane

Distillate outlet

Insulation

Brine outlet

Distillate trough

A tilted wick solar still

FIGURE 8.2A   Tilted Wick Solar Still.

35 percent or more, but the cost and complexity are correspondingly higher.

Wick stills: Some designs of wick stills are more efficient than basin stills, and some designs are said to cost less than a basin still of the same output. In a wick still, the feed water flows slowly through a porous radiation-absorbing pad (the wick) (Fig. 8.2A). Two advantages are claimed over basin stills: First, the wick can be tilted so that the feed water presents a better angle to the sun (reducing reflection and presenting a large effective area). Second, less feed water is in the still at any time, so the water is heated more quickly and to a higher temperature.

Emergency still: To provide emergency drinking water on land, a very simple still can be made. It makes use of the moisture in the earth. All that is required is a plastic cover, a bowl or bucket, and a pebble. (Fig. 8.3A & 8.4A)

Hybrid designs: There are a number of ways in which solar stills can usefully be combined with other technologies. (Fig. 8.5A, 8.6A and 8.7A).

Other possible enhancements to a basic solar distillation sytem include

1. Rainwater collection: By adding an external gutter, the still cover can be used for rainwater collection to supplement the solar still output.

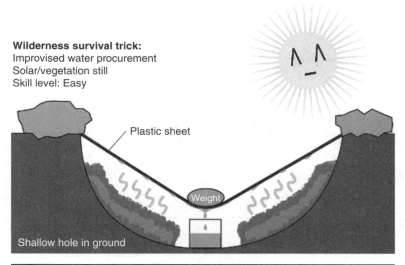

Wilderness survival trick:
Improvised water procurement
Solar/vegetation still
Skill level: Easy

Plastic sheet

Weight

Shallow hole in ground

FIGURE **8.3A** Traditional emergency distillation system.

FIGURE **8.4A** Modern version of a emergency solar distillation system for campers.

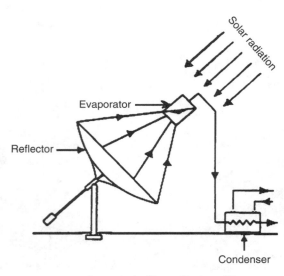

A concentrating collector still

FIGURE 8.5A   Solar focusing solar still.

FIGURE 8.6A   Air supported solar distillation pyramid installed in Gambia Africa.

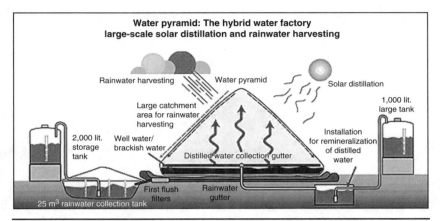

FIGURE 8.7A Schematic of Gambia solar distillation installation.

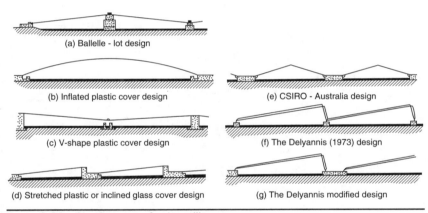

FIGURE 8.8A Simple types of solar stills.

2. Greenhouse-solar still: The roof of a greenhouse can be used as the cover of a still.

3. Supplementary heating: Waste heat from an engine or the condenser of a refrigerator can be used as an additional energy input.

4. Multiple units staged together: see Fig. 8.8A.

CHAPTER 9

Graywater Systems

Introduction

Graywater systems are on-site water treatment systems that gather wastewater, treat it, and return the reclaimed water to the building plumbing system for constructive reuse. Graywater comes from many sources.[1] It is defined as: "Untreated household wastewater that has not come in contact with toilet waste and includes wastewater from bathtubs, showers, washbasins, washing machines and laundry tubs, but does not include wastewater from kitchen sinks, dishwashers or laundry water from the washing of material soiled with human excreta, such as diapers."[2] Although graywater systems can be applied in homes and office buildings, they are also applicable to laundry facilities, car washes, and other such facilities that contain processes that do not require utility-grade potable water. Fig. 9.1 shows a simple graywater system schematic applied to landscape irrigation.

Discussion of Graywater Systems

Only about 20 percent of water used in an average home or commercial office building needs to be potable water, suitable for drinking, food preparation, personal bathing, and washing. The rest can be lower-grade water, for which graywater is an ideal and readily available source.

Benefits of graywater include using less freshwater, sending less wastewater to septic tanks or treatment plants, and fewer chemicals needed to produce the potable water. When graywater is used for landscape irrigation, the phosphorous, nitrogen, and potassium included in soap products, rather than being considered pollutants, can be excellent sources of nutrients when applied to gardens or planter beds.

While shower and lavatory water are perhaps the most popular sources of graywater, other potential sources include condensate gathered from fan-coil units and air-handling units. This condensate is generally free of impurities and can reduce the amount of chemical treatment of makeup water required for the other systems such as

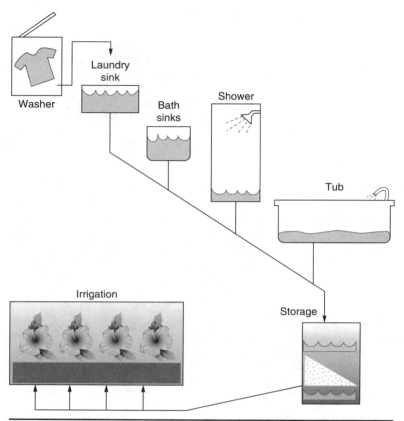

Figure 9.1 Sources of graywater for subsurface irrigation of landscape. (*Source:* Tucson Water, 2009)

cooling towers, boilers, and interior fountains. However, air conditioning condensate, by nature of the air conditioning process, tends to concentrate impurities such as Legionella and bacteria, so it should be used with caution, particularly if the application may create aerosols from spraying. Because this water is distilled water, and therefore low in impurities, use in cooling towers, interior fountains, and boilers, where water treatment is used to control the bacteria, are ideal applications.

Another source of graywater is water gathered from steam boilers, high-efficiency water heaters and boilers, and cooling tower bleed water. Each of these sources has its own chemical content and corrosion properties, which require proper testing and possible treatment before application to their intended reuse. Fig. 9.2 shows a schematic of an industrial application of graywater reuse.

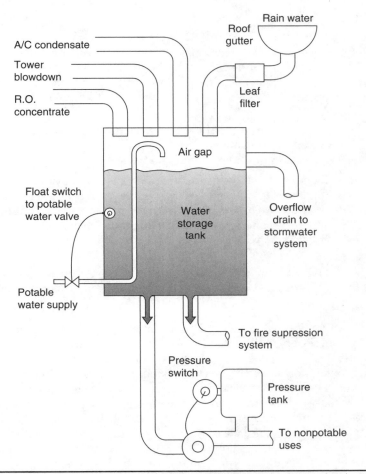

Figure 9.2 Schematic of industrial application of graywater reuse showing multiple sources.

Graywater System Types and Description

Graywater systems vary in their complexity and size and can be grouped according to the type of filtration or treatment they use:

- Direct reuse systems (no treatment): These systems use simple devices to collect graywater from appliances and deliver it directly to the points of use, with no treatment and minimal or no storage.[4]

- Short retention systems: These systems apply a very basic filtration or treatment technique, such as skimming debris off the surface of the collected graywater and allowing particles to settle to the bottom of the tank. They aim to avoid odor and water quality issues by ensuring that the treated graywater is not stored for an extended period.

- Basic physical/chemical systems: These systems use a filter to remove debris from the collected graywater prior to storage while chemical disinfectants (e.g., chlorine or bromine) are generally used to stop bacterial growth during storage.

- Biological systems: These systems introduce an agent, such as oxygen, into the collected graywater to allow bacteria to digest any unwanted organic material. Pumps or plants can be used in this process to aerate the water.

- Biomechanical systems: These systems, the most advanced for domestic graywater reuse, combine biological and physical treatment. These systems remove solid material through settlement and treat organic matter with microbial cultures. Bacterial activity is encouraged by bubbling oxygen through the collected graywater.

- Hybrid systems: These systems either use a mix of the system types detailed above or are combined with rainwater harvesting systems.

Graywater is collected at a central tank and treated to the quality necessary for the intended use. The resultant reclaimed water is distributed using a dedicated piping system. Because graywater generally contains organic matter (such as hair, textile fibers, and skin cells), it cannot be stored for very long. Bacteria in the water quickly break down the organic matter, using up the dissolved oxygen in the process. This measured process is referred to as the biochemical oxygen demand, or BOD. Once the oxygen is depleted, different bacteria take over and the decomposition becomes anaerobic, which can produce smelly methane and hydrogen sulfide gas. For this reason, most graywater systems are designed to minimize retention time by quickly delivering the reclaimed water to its intended use. After treatment, the reclaimed water is distributed back into the plumbing system for reuse, typically for flushing toilets and urinals and for landscape irrigation.

Design Procedure

The type and capacity of the graywater system is calculated using either of the following methods:

- The simplified approach is to use graywater for toilet flushing and/or laundry makeup water in less complex installations, such as single-family residences or for a dedicated process.

- The detailed approach is used where a more sophisticated analysis is necessary for larger applications, such as multi-family, commercial, and industrial facilities.

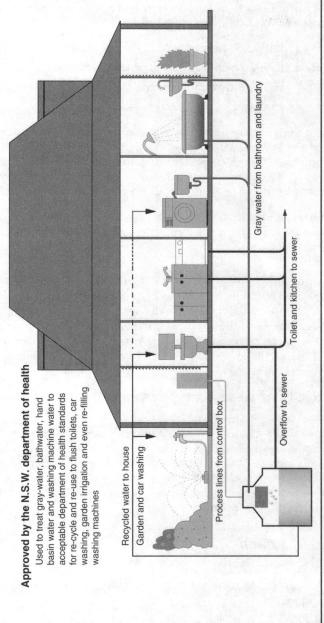

Approved by the N.S.W. department of health

Used to treat gray-water, bathwater, hand basin water and washing machine water to acceptable department of health standards for re-cycle and re-use to flush toilets, car washing, garden irrigation and even re-filling washing machines

Recycled water to house
Garden and car washing

Process lines from control box

Overflow to sewer

Gray water from bathroom and laundry

Toilet and kitchen to sewer

FIGURE 9.3   Residential application of a graywater system.

153

Step 1: Establishing Applications and Sources for Graywater

The first step for a successful graywater system design is to determine an accurate estimate of the demand for the treated graywater. A list of applications of this process, in increasing water quality requirement order, is as follows:

- Toilet and urinal flushing
- External use—non-spray (e.g., subsurface irrigation)
- Laundry use, i.e., washing machine makeup
- External use, i.e., aboveground spray

Knowing the source of the graywater is important to maintaining the safety of the system. Bathroom graywater is preferred, since the level of contamination is consistent and the treatment relatively straightforward. To that end, the following is a list of potential graywater sources prioritized according to level of contamination:

- Showers and baths
- Wash and hand basins
- Washing machines[5]

It is important that the appliances from which graywater is collected are used normally. Where there is a possibility that an appliance is used in an unusual manner, such as frequent use of chemical drain cleaners, the use of the appliance for inclusion in the graywater collection should be reconsidered. Alternatively, the treatment process should be upgraded in the graywater system to accommodate the possible irregularity.

Step 2: Quantifying the Amount of Graywater That Can Be Used

Once the availability and the quality of the graywater are determined, the next step is to quantify the amount of graywater that can be reused. This is important to avoid collecting and treating of graywater that cannot be used. Ideally, the daily amount of reclaimed graywater used will closely match the source amount to minimize storage and to simplify influent treatment.

The Simplified Approach

The simplified approach is based on the following assumptions:

- Relatively constant daily supply of bathroom graywater of 13 gal (50 L)/person
- Relatively constant daily demand of 6.6 gal (25 L)/person for toilet flushing and 4 gal (15 L)/person for laundry

Occupancy	Yield[1]	Average Daily Demand Gallons (Liters)		
# Persons	Gallons (Liters)	Water Closet	Laundry[2]	Other Non-Potable Applications[3]
1	13 (50)	7 (25)	4 (15)	3 (10)
2	26 (100)	13 (50)	8 (30)	5 (20)
3	40 (150)	20 (75)	12 (45)	8 (30)
4	52 (200)	26 (100)	16 (60)	10 (40)
5	63 (250)	33 (125)	25 (75)	13 (50)
6	80 (300)	40 (150)	24 (90)	16 (60)

Notes:
1. Approximate yield from showers, baths, and/or wash and hand basins.
2. The figures are based on average daily demand. Note that, although there are high-efficiency washing machines, an average value for a wash cycle is about 13 gallons (50 liters) per cycle. For best results, use the manufacturer's data for the washing machine being used.
3. Other non-potable uses might include applications such as landscape irrigation and should be quantified accordingly to the specific needs of the plants. It is to be noted that establishing a plant requires approximately twice as much water as ongoing maintenance watering does.

TABLE 9.1 Typical Average Daily Graywater Yield and Demand[6]

When calculating the treatment capacity of the graywater system, the average daily yield and demand per person should be taken into account. See Table 9.1 for approximate graywater yield and demand values that can be used.

The Detailed Approach

The detailed approach assumes that the yield and demand are constant and approximately coincident. The difference in these two events must be accommodated by storage capacity.

Estimating Graywater Yield: Metric Analysis (See Appendix for English Units Conversion)

The following equation should be used to determine the graywater yield:

$$Y_G = n \times [(S \times U_s) + (B \times U_b) + (H_{wb} \times U_{hwb}) + (F_{wb}) + [(W/L) \times U_{wm})]$$

$$(9.1)$$

where: Y_G = the graywater yield in liters (L)
n = the number of persons
S = the average flow rate from the shower in liters per minute/person (L/min)

U_s = the typical usage factor for the shower (minutes)

B = the bath volume to overflow (unoccupied) per person in liters (L)

U_b = the typical usage factor for the bath (% filled × number of fillings/person)

H_{wb} = the mean peak flow rate from taps in liters per minute (L/min)

U_{hwb} = the typical usage factor for the hand wash basins (minutes)

F_{wb} = fixed flow from basin taps used for vessel filling in liters (L)

W_I = the washing machine water consumption per wash cycle in liters (L)

L = the maximum dry wash load recommended by manufacturer in kilograms (kg)

U_{wm} = the typical usage factor for the washing machine (minutes)

Example: The following are some typical values that can be used in Eq. (9.1). Where more precise values are not known, these values may be used for estimation purposes:

n = two people in the master bedroom and one additional person in each of the remaining bedrooms

S = 5 liters/min for low flow shower; 12 liters/min for normal mixer shower; 15 liters/min for high flow showers

U_s = 5.60 for a shower only; 4.37 where there is also a bath (per person)

B = 120 liters to 250 liters (per use)

U_b = 0.5 for a bath only; 0.11 where there is also a shower

H_{wb} = 5 liters/min to 15 liters/min

U_{hwb} = 1.58

F_{wb} = 1.58 liters per person per day

W = 30 liters to 60 liters

L = 4 kg to 10 kg

U_{wm} = 2.1

Note: Washing machine efficiency data is based on a standard 60°C cotton wash. This includes information on water consumption (W) and load capacity (L), which are suitable values for use in Eq. (9.1). Where the performance of the washing machine is unknown, the ratio may be assumed to be 8.17 liters per kilogram.

Estimating Graywater Demand (Interior Building Applications)

The following equation should be used to determine graywater demand where the treated graywater is to be distributed for water closet or urinal flushing and laundry use, as applicable.

$$C = n \times (V_{WC} \times U_{sf})_1 + [(V_{WC} \times U_{sf})_2 + (V_{WC} \times U_{sf})_{n...}] + [(W/L) \times (U_{wm} \times P_{WM})]$$

$$(9.2)$$

where: C = the graywater consumption in liters per time period (L)

n = the number of persons

V_{WC} = the flush volume for a single-flush water closet in liters (L/person/time period)

U_{sf} = the usage factor for a single-flush water closet (# flushes/time period)

W = the washing machine water consumption per wash cycle (see manufacturers data) in liters (L)/time period

L = the maximum dry wash load recommended by manufacturers (see manufacturer's data) in kilograms (kg)

U_{wm} = the usage factor for the washing machine

P_{WM} = the proportion of water consumed by the washing machine to be supplied by non-potable water

Note:

- Where more than one type of water closet is installed, the consumption can be calculated for each water closet$_{(n)}$, or it can be assumed that all such fixtures will be used equally, in which case the standard consumption for each type can be calculated and the results simply averaged.

- Although 24 hours is the common time period for graywater use, the actual time period may vary according to the method for treating the stored graywater.

The following are some typical values that can be used in Eq. (9.2). Where more precise values are not known, these values may be used for estimation purposes:

n = two people in the master bedroom and one additional person in each of the remaining bedrooms

U_{sf} = 4.42

U_{Ff} = 1.46

U_{Pf} = 2.96

W = 30 liters to 60 liters

L = 4 kg to 10 kg

U_{wm} = 2.1

P_{WM} = 1

Graywater Demand (External Application, e.g., Garden Watering)

If the treated graywater is required for external use, demand should be estimated according to the maintenance level of water needed for the applicable landscaping, recognizing that to establish new plants

Figure 9.4   Indoor plants at the Society for the Protection of New Hampshire Forests, in Concord, New Hampshire, are irrigated with graywater. (See also Color Plates.)

will initially take approximately twice the maintenance water volume. This value will depend upon a number of factors, such as soil type, crop type, shading, and topography. Detailed information on specific plant water demand can be obtained from gardening and agricultural sources. Fig. 9.4 shows an indoor application of landscape irrigation for the Society for the Protection of New Hampshire Forests building, Concord, New Hampshire.

Storage Capacity

Depending on the type of graywater system, the optimum storage capacity for treated graywater should be determined by the following factors:

- The peak capacity treatment rate
- The demand, usage, or behavior patterns

It is recommended that storage of treated graywater be minimized to that needed for immediate use. As there is generally a ready supply of untreated graywater, storage equal to a single day's use is normally considered sufficient. Where a combined graywater and rainwater system is used, the storage might need to be increased.

Combined Rainwater/Graywater Systems

Where a graywater reuse or rainwater harvesting system alone cannot provide sufficient water for non-potable use, combining two systems can provide a solution. However, before systems are combined, a thorough assessment should be made of each system individually to determine whether it alone can meet the demand of the intended applications. In a combined graywater system, it is important that all overflows or bypass arrangements discharge into a sanitary sewer and are not allowed to drain to daylight.

Collection

Graywater systems depend significantly on the behavior of the people using the collection appliances, as well as the quality and volume of water collected. Depending on the type of system, graywater can be collected in different ways. It should generally be collected in a separate wastewater drainage system and allowed to flow from collection appliances to the graywater storage vessel by gravity or siphonic action. Where siphonic distribution is not practicable, like in single-story dwellings, pumps need to be considered.

Collection piping should be:

- Dedicated to bathroom graywater.
- Sized and laid out such that the generation of foam is minimized. As air entrainment is a major factor in the generation of foam, turbulence and the use of bends should be minimized.
- Properly identified in terms of components, according to local regulations.
- Free draining to avoid stagnation.

Collection drain piping should be designed and installed to prevent contaminated water from entering the graywater system from other sources. Similarly, it is recommended that hair traps be used to minimize pollutants entering the graywater system, wherever possible. A bypass should be fitted around the graywater system allowing the collected graywater to flow directly to the sewer during periods of maintenance or system isolation. The bypass should not compromise the drainage system.

Storage Tanks and Cisterns

Tanks may be positioned either above- or belowground and should be appropriate to the site. Tanks should be constructed from materials that create watertight structures without encouraging microbial growth. Suitable materials include concrete, glass-reinforced plastic (GRP), polyethylene or polypropylene, and steel coated with non-corrodible materials.

All tanks and cisterns should be installed to allow complete tank drainage to avoid stagnation. Tank inlets should have screened ventilation and fitted lids to prevent the entrance of animals and airborne insects. Tanks and cisterns should be sited so that the stored water does not attain temperatures that could encourage microbial growth, particularly Legionella. Where tanks are positioned near habitable or vulnerable areas, the risk of water leakage should be considered and appropriate protection provided for any potential leaks or flooding.

Aboveground Tanks and Cisterns

Aboveground tanks are particularly cost-effective for retrofit applications. Where they are used, aboveground tanks and cisterns should be installed so as to minimize the potential problems of freezing and warming. Tanks should be opaque to prevent algal bloom. Freezing is an obvious danger. Heating increases biological activity in the stored graywater, and algae contributes to odors resulting from anaerobic digestion in the tank.

Belowground Tanks

Belowground tanks can provide frost protection, prevent undue warming in the summer months, and restrict algal growth owing to the lack of sunlight. Belowground tanks (and their covers) should be sufficiently rigid to resist likely ground and traffic loadings. Where soil saturation is a possibility, it is recommended that the combined weight of the tank and ballast must meet or exceed the buoyancy upward force of an empty storage tank. This buoyant force (lbs) is equal to the volume of the tank (cubic feet) × 62.4 lb/cubic feet, or tank volume (gallons) × 8.34 lb/gallon water.[7]

Treatment

There are many different methods used to filter and treat collected graywater, which range in complexity and level of treatment. The level of treatment should be consistent with the intended use of the collected graywater. This should be considered in order to determine whether filtration or treatment is needed and which method is appropriate (e.g., physical, chemical, or biological).

The collected graywater should only be treated to the extent needed to meet the water quality guidelines (health considerations) of the application being supplied. After choosing the degree of filtration or treatment, the sustainability aspects and the environmental impact should also be considered in order to determine the most appropriate type of filtration or treatment.

Common types of filtration and treatment, either used singly or in combination include the following:

- Sedimentation and flotation, e.g., settlement tanks
- Screening, e.g., large particulate filtration

- Mechanical fine filtration, e.g., membranes;
- Biological treatment, e.g., aeration
- Chemical treatment and disinfection, e.g., chlorine, ozone
- Ultraviolet (UV) disinfection

Different filtration or treatment methods may be used in combination to achieve the necessary results.

Materials and Fittings

The materials selected for the tank and other components should be suitable for the location and temperature ranges anticipated. All components of the graywater system should be capable of withstanding pH levels as low as 5 for the lifetime of the products. Consideration should be given to the environmental impact of materials used.

Backup Water Supply and Backflow Prevention

The backup water supply may be provided from harvested rainwater, where available, or from a potable water supply. Makeup water connections to the graywater storage tank from a potable water supply should be made with either an air gap or other approved backflow device. The backup supply should be fitted with a control mechanism that ensures that the amount of water supplied is adequate for immediate use, but leaves sufficient tank volume for graywater to be added. A warning device should be included in the overflow, set to activate before the water level overflows. The backup supply should be sized to allow it to meet the full demand requirements.

Pumping

If gravity feed of the treated graywater is not possible, selecting a pump to pressurize the distribution system will be necessary. Pumps should be sized so that each pump is capable of overcoming the static lift plus the friction losses of the piping and valves. The pumps should be selected and arranged such that:

- Energy use and noise to the occupied space are minimized.
- Pump cavitation is prevented.
- Air is not introduced into the graywater system.

Pumps should be equipped with dry-run protection, which may be either integral to the pump or provided by an external control device. Expansion tanks and surge-suppressing fittings should be provided to absorb surges created by differential pressure switch control of the pumps and water hammer from quick closing valves.

Pumps for collected graywater should be able to accommodate any solid matter likely to be contained in the graywater. If the pump incorporates any filtering or traps, these should be accessible for maintenance and cleaning. Where graywater drainage cannot be interrupted, duplex pumping devices should be installed, or a means of diverting excess graywater to the sanitary sewer, should be provided.

If installed outside the tank, the pump should have its own self-priming mechanism or a control system which ensures a constant fully primed condition. To eliminate air from the system, the suction line should be laid with a steady gradient upward toward the pump. The pump should be placed in a well-ventilated location and protected from extremes of temperature, with sound-free and vibration-free mountings. A non-return valve (or foot valve) should be provided in the suction line to the pump in order to prevent the water column from draining down. The pressure line of the pump should be supplied with an isolating valve to maintain water in the line should pump maintenance be needed.

If the pump is installed inside the tank, a minimum level of water needs to be maintained above the pump inlet in order to prevent damage by sucking air, sediment, or debris. The immersion depth should be in accordance with the pump manufacturer's instructions. The pump should be removable for maintenance purposes. A non-return (back check) valve should be provided with an isolating valve to enable the non-return valve to be maintained.

Overflow, Bypass, and Drainage

An overflow should be fitted to all tanks and cisterns to allow excess water to be discharged to the sewer system. The overflow should be installed such that backflow from the sewer to the tank is prevented with an appropriate backflow valve. Overflows and tank vents for aboveground tanks and cisterns should be screened to prevent the ingress of insects, vermin, etc.

The capacity of the outlet pipe on the overflow pipe should be capable of draining the maximum inflow without compromising the inlet air gap. Where appropriate, the overflow and bypass should be fitted with a backflow, or anti surcharge valve. The graywater system overflow and any bypass should be connected to the sanitary sewer, and this should be done in a manner to minimize foaming of the effluent as it enters the drainage system.

System Controls

Controls should be incorporated in the graywater system to ensure that users are aware of whether the system is operating effectively.

Optimally, the control unit should do the following to prevent any system failure:

- Make the user aware when any consumable items need replenishment or replacement by a visible or audible warning
- Ensure that the bypass directs untreated water to the sanitary sewer
- Ensure that graywater treatment continues or that graywater is not stored for a period that would allow water quality to deteriorate

In the event of a treatment failure, the control system should ensure that the applications supplied by the graywater system are fed from the backup water supply. Additional monitoring of the overflow, water quality, tank temperature, and other parameters may also be advisable.

Pump controls should do the following:

- Operate the pump(s) to match demand with either variable speed drives or pressure switch control
- Protect the pumps from running dry
- Protect the motor from overheating and electric overload
- Permit manual override

Piping and System Installation

Piping

Piping for graywater and treated graywater should be labeled and/or of contrasting color to differentiate graywater piping from potable water piping. Piping should be sized to provide adequate flow and pressure (e.g., oversized pipes can cause water quality issues from low flows while excessive pressure can cause undue consumption or leakage). Installation should ensure that all components, including tanks, are accessible for future maintenance and/or replacement of consumable parts.

Cistern Installation

Where treated graywater storage cisterns are needed within buildings, they should be installed as for any coldwater cistern with appropriate support, insulation, and means to prevent contamination. The cistern should be supported on a firm, level base capable of withstanding the weight of the cistern when filled with water to the rim. Plastic cisterns should be supported on a flat, rigid platform fully supporting the bottom of the cistern over the whole of its area.

Overflows fitted to storage cisterns should be capable of discharging all inflows into the cistern. In addition, an automatic supply cut-off device activated by an overflow may be installed to minimize damage and the waste of water.

All tanks should be fitted with lids that protect the water from contamination and prevent inadvertent human entry. Where environmental conditions dictate, thermal insulation to the graywater system (i.e., the tank or cistern, treatment, equipment, or piping) should be considered.

Water Quality

It is essential that graywater systems are designed in a way that ensures the water produced is fit for its purpose and presents no undue risk to health. Frequent water sample testing is not necessary; however, observations for water quality should be made to check the performance of the graywater system. Tests should then be undertaken to investigate the cause of any system that is not operating satisfactorily, and any complaints of illness associated with water use from the system should be thoroughly investigated. Testing immediately following the commissioning of systems is not recommended, as systems are generally filled with public mains water in order to facilitate the testing of components, and water quality is therefore not representative of the normal graywater collection.

Water quality should be measured in relation to the guideline values given in Table 9.2 for parameters relating to health risk and Table 9.3 for parameters relating to overall system operation. This provides an indication of the water quality that a well-designed and maintained system is expected to achieve for the majority of operating conditions.

Maintenance

Maintenance of graywater systems is the user's responsibility. Maintenance procedures should be in accordance with the manufacturer's maintenance recommendations. A log should be kept of inspections and maintenance.

Where possible, the graywater system should be drained and flushed with clean water before maintenance to reduce the risk of contamination to maintenance personnel and people in the vicinity. Electricity and all water supplies should be isolated before opening any sealed lids or covers.

Human entry into tanks should be avoided, wherever possible. Where entry is essential, it should only be undertaken by trained personnel with personal protection equipment suitable for confined spaces.

Contaminant	Spray Application	Non-Spray Application		
	Pressure Washing, Aboveground Landscape Irrigation	Urinal and Water Closet Flushing	Below-ground Landscape Irrigation	Washing Machine Use
Escherichia Coli (CFU/100 ml)	Not Detected	250	250	Not Detected
Intestinal (Fecal) Enterococci (CFU/100 ml)	Not Detected	100	100	Not Detected
Legionella pneumophilaa (CFU/100 ml)	10	N/A	N/A	N/A
Total Coliforms (CFU/100 ml)	10	1000	1000	10

Notes:

a. Applicable for high-risk/critical environments (see the following section, Risk Assessment).

b. In the absence of E.coli, intestinal enterococci, and Legionella, there is no need to suspend use of the system if levels of coliforms exceed 10 times the guideline value. However, it might be necessary to include some type of disinfection (e.g., UV or chemical disinfection) to attain the more stringent bacteriological standards suggested, in situations where higher exposure might occur or for high-risk/critical environments.

TABLE 9.2 Graywater Quality Guideline Values for Various Applications,[b] [8]

Contaminant	Spray Application	Non-Spray Application		
	Pressure Washing, Aboveground Landscape Irrigation	Urinal and Water Closet Flushing	Belowground Landscape Irrigation	Washing Machine Use
Turbidity (NTU)	<10	<10	N/A	<10
pH	5–9.5	5–9.5	5–9.5	5–9.5
Residual Chlorine[a]	<2.0	<2.0	<0.5	<2.0
Residual Bromine[a]	0.0	N/A	0.0	N/A

Notes:

a. Where used.

b. In addition to these parameters, all systems should be checked for suspended solids and color. The treated graywater should be visually clear, free from floating debris and not objectionable in color for intended use.

TABLE 9.3 System Monitoring Guidelines,[b] [9]

System Components	Operation	Notes	Frequency
Filters, Membranes, Biological Support Media, and Strainers	Inspection/ Maintenance	Check the condition of the filtration and clean or replace as necessary.	Annually
Biocide, Disinfectant, or Other Consumable Chemical	Inspection/ Maintenance	Check that any dispensing unit is operating appropriately; replenish the chemical supply as needed.	Monthly
UV Lamps (where applicable)	Inspection/ Maintenance	Clean and replace as necessary.	Every 6 Months
Storage Tank/ Cistern	Inspection	Check that there are no leaks, that there has been no buildup of debris, and that all tanks and cisterns are stable and the covers are correctly fitted.	Annually
Storage Tank/ Cistern	Maintenance	Drain down and clean the tanks and cisterns.	Every 5–10 Years or As Necessary
Pumps and Pump Controls	Inspection/ Maintenance	Check for and repair any leaks and check that there has been no corrosion; perform a test to confirm controls and safety limits work appropriately; check gas charge in expansion tank and/or shock arrestors.	Annually
Backup Water Supply	Inspection	Check that the backup supply is functioning correctly and that the air gap/backflow device is functioning correctly.	Annually
Control Unit	Inspection/ Maintenance	Check that the unit is operating appropriately, including the alarm functions where applicable.	Annually
Water Level Gauge (if installed)	Inspection	Check that the gauge indication response is correct for the water level in the supply tank or cistern.	Annually

TABLE 9.4   Maintenance Schedule[10]

System Components	Operation	Notes	Frequency
Piping System	Inspection	Check that there are no leaks, the piping is watertight, and any overflows are clear. This includes the collected and treated graywater supplies, any backwash supply, and the backup water supply.	Annually
System and Piping Supports and Fittings	Inspection/ Maintenance	Adjust and tighten as necessary.	Annually
Markings/ System Identification	Inspection	Check that piping and valve labeling is appropriate and visible.	Annually
Backwash	Inspection/ Maintenance	Check functionality.	Annually

Note:
a. Inspection/maintenance frequencies listed are recommended if there is no other information available from the unit manufacturer.

TABLE 9.4 *(Continued)*

To reduce the buildup of pathogens, systems should be periodically emptied according to manufacturer's suggestions if stored water is unused for an extended period. Small amounts of disinfectants, based on specific system requirements as determined by volume and level of contamination, need to be added to graywater systems to disinfect the collected influent.

Risk Assessment

A risk assessment should be carried out as part of the design process to determine whether the graywater system is appropriate for the intended use. Evaluation of the resultant reclaimed water quality as it applies to the specific environment and the occupants is very important to make sure the system does not present a health or safety hazard. This evaluation includes the following:

- Knowing the health characteristics of the people, including operators, installers, maintainers, and water users, particularly those who might be more affected by poor water quality and/or have compromised immune systems (e.g., children, the ill, or the elderly)

- Knowing the installed environment of the system, including the proximity to domestic and feral animals, birds and fish, plants, as well as water courses and groundwater, tanks, building foundations, drains, paved areas, and gardens.

It is important to note that the likely expectation of the graywater system is that it be entirely self-sustaining; therefore, designing to the capabilities of the maintenance personnel is highly relevant for safe and reliable system operation.

Conclusion

Due to dwindling water supplies and the need to reduce sewer plant overflows into waterways, graywater reuse is gaining popularity. Increasingly, graywater is considered a resource to be separated from the wastewater stream and reused for applications such as toilet flushing, landscape irrigation, air conditioning system makeup, and other applications where utility-provided potable water is not necessary. In this way, potable water is reserved for more vital uses, and the amount of wastewater entering sewers or on-site wastewater treatment systems is reduced; therefore, the day is delayed when municipalities must increase the size of their sewers and sewage treatment plants.

Notes and References

1. *Grey water, gray water, greywater,* and *graywater* all have the same meaning.
2. California Graywater Standards (California Administrative Code).
3. For more in-depth analysis, see Air Conditioning Condensate Recovery, Chapter 4.
4. "It is possible to reuse graywater without any treatment, provided that extended storage is not required. As untreated graywater quality deteriorates rapidly, the collected graywater ideally needs to be reused as soon as it has cooled. Where no treatment is included in the graywater system, applications are generally restricted to subsurface irrigation and non-spray applications and/or use when the general public is not present." British Standard BS 8525-1 Draft, Graywater Systems—Part 1: Code of Practice date of publication.
5. Where washing machines are both taking water from and feeding water into the graywater system, the system needs to incorporate a suitable treatment process.
6. *The Water Efficiency Calculator for New Dwellings,* Communities and Local Government, London: CLG, 2009.
7. Using metric calculations, the buoyancy force (kg) is calculated by multiplying the tank volume in cubic meters (m^3) by 1000 kg/cubic meter.
8. British Standard BS 8525-1 DRAFT, Graywater Systems—Part 1: Code of Practice, p. 28, August 2009.
9. British Standard BS 8525-1 DRAFT, Graywater Systems—Part 1: Code of Practice, p. 29, August 2009.
10. British Standard BS 8525-1 DRAFT, Graywater Systems—Part 1: Code of Practice, p. 38, August 2009.

CHAPTER 10

Managing Water Quality

Maintaining water quality is necessary to maintain the integrity and safety of the water collection system. Should the system be suspected of causing health issues, it likely will be abandoned and fall into disuse. Research has shown that people get accustomed to the normal biological contaminants in their usual water supply.[1] However, maintaining water quality is imperative for the sake of public safety, especially for people not accustomed to local organisms, as well as those with immune system deficiencies due to age or illness.

Bacteria are the second smallest pathogens in water, but they can usually be removed by a 0.2-micron filter. One frequently hears about water being tested for *Escherichia coli*. While strains of this bacterium can be pathogenic, the vast majority is not, and it is, in fact, one of the species required in the intestine for digestion to occur. In fact, without bacteria, we could not live. On the other hand, there are plenty of other bacteria that are happy in the intestine to the detriment of the host (Fig. 10.1).

The smallest parasites are viruses. Pathogenic viruses are seldom found in source water, but the odds increase with the length water is stored, the population density accessing the water, and lack of proper sanitation practices. Ordinary filters *do not* remove viruses—in fact, Louis Pasteur originally defined organisms as "filterable or "non filterable", depending on whether they could pass through a 0.2 micron ceramic filter. This designation resulted from his discovery that organisms existed that were too tiny to be seen by the microscopes of the times.[2]

Filters that have an iodine matrix will kill viruses (or "deactivate" them), and iodine treatment will also work. Boiling is the most reliable way to do away with viruses, and is strongly suggested in third-world countries. The specific viruses you should worry about in water are hepatitis A, rotaviruses, polioviruses, and echoviruses.

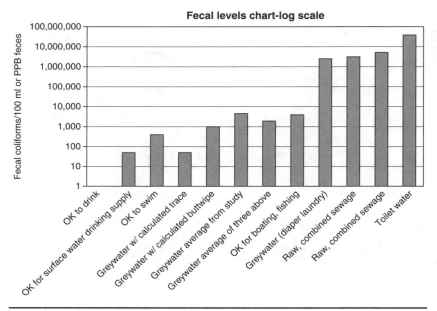

Fecal levels chart-log scale

FIGURE **10.1** Expected water quality from various sources.

All of these will cause diarrhea, intestinal cramps, and discomfort about 48–72 hours after contact. Complications could range from liver damage (from hepatitis), to aseptic meningitis and encephalitis (from echoviruses), and paralysis or death (from polio).

If there are questions about the quality of your source water, the following, although not all-inclusive, are some examples of how water can be sanitized.

Sand Filters[3]

Slow sand filters are a preferred technology worldwide to produce potable water. They are among the oldest water treatment systems and are used in the backyards of remote villages as well as in the most developed cities of the world.

Construction of a sand filter begins with creating a container to hold the filter media. They are typically 1 to 2 meters deep, can be rectangular or cylindrical in cross section, and are used primarily to treat surface water. The length and breadth of the tanks are determined by the flow rate desired by the filters, which typically have a loading rate of 0.1 to 0.2 meters per hour (or cubic meters per square meter per hour).

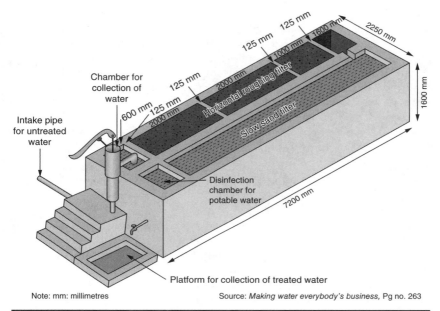

Intake pipe for untreated water

Chamber for collection of water

600 mm
125 mm
2000 mm

125 mm
125 mm
125 mm
125 mm

2000 mm
Horizontal roughing filter
1000 mm
1600 mm

2250 mm

Slow sand filter

1600 mm

Disinfection chamber for potable water

7200 mm

Platform for collection of treated water

Note: mm: millimetres   Source: *Making water everybody's business*, Pg no. 263

Figure 10.2   Slow sand filter. (See also Color Plates.)

Filter Operation

Slow sand filters have a number of unique qualities. Unlike other filtration methods, slow sand filters use biological processes to clean the water, and they are nonpressurized systems. Slow sand filters do not require chemicals or electricity to operate.

Cleaning is traditionally by use of a mechanical scraper, which is usually driven into the filter bed once it has been dried out. However, some slow sand filter operators use a method called "wet harrowing," where the sand is scraped while still underwater, and the water used for cleaning is drained to waste.

Slow sand filters require relatively low turbidity levels to operate efficiently. In summer conditions and in conditions when the raw water is turbid, blinding of the filters occurs more quickly and pre-treatment is recommended.

Unlike other water filtration technologies that produce water on demand, slow sand filters produce water at a slow, constant flow rate and are usually used in conjunction with a storage tank for peak usage. This slow rate is necessary for healthy development of the biological processes in the filter.[4]

While many municipal water treatment works will have 12 or more beds in use at any one time, smaller communities or households may only have one or two filter beds. In the base of each bed is a series of herringbone drains that are covered with a layer of pebbles, which in turn is covered with coarse gravel. Further layers of sand are placed on top followed by a thick layer of fine sand. The whole depth of filter material may be more than 1 meter in depth, the majority of which will be fine sand material. On top of the sand bed sits a supernatant layer of raw, unfiltered water.

How It Works

Slow sand filters work through the formation of a gelatinous layer (or biofilm) called the hypogeal layer or *Schmutzdecke* in the top few millimeters of the fine sand layer. The *Schmutzdecke* is formed in the first 10–20 days of operation[5] and consists of bacteria, fungi, protozoa, rotifera, and a range of aquatic insect larvae. As a *Schmutzdecke* ages, more algae tend to develop and larger aquatic organisms may be present, including some bryozoa, snails, and annelid worms. The *Schmutzdecke* is the layer that provides the effective purification in potable water treatment, while the sand acts as a porous support medium that allows this biological treatment layer to function. As water passes through the *Schmutzdecke*, particles of foreign matter are trapped in the mucilaginous matrix, and dissolved organic material is adsorbed and metabolized by the bacteria, fungi, and protozoa. The water produced from a well-managed slow sand filter can be of exceptionally good quality with 90–99 percent bacterial reduction.[6]

Slow sand filters slowly lose their performance as the *Schmutzdecke* grows and thereby reduces the rate of flow through the filter. Eventually, it is necessary to refurbish the filter. Two methods are commonly used to do this. In the first, the top few millimeters of fine sand is scraped off to expose a new layer of clean sand. Water is then decanted back into the filter and recirculated for a few hours to allow a new *Schmutzdecke* to develop. The filter is then filled to full depth and brought back into service[7] The second method, called "wet harrowing", involves lowering the water level to just above the *Schmutzdecke*, stirring the sand, thereby suspending any solids held in that layer, and then running the water to waste. The filter is then filled to full depth and brought back into service. Wet harrowing can allow the filter to be brought back into service more quickly.[8]

Advantages

As they require little or no mechanical power, chemicals, or replaceable parts, and they require minimal operator training and only periodic maintenance, slow sand filters are often an appropriate technology for poor and isolated areas. Due to their simple design, they may be created as a DIY (do-it-yourself) project (Fig. 10.3), making them an

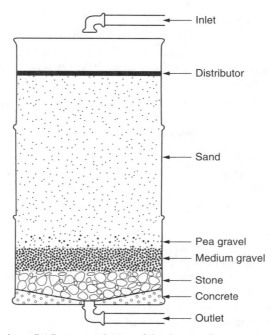

Inlet

Distributor

Sand

Pea gravel
Medium gravel
Stone
Concrete
Outlet

To make a distributor, cut the top of the drum so that
it fits down inside the drum. Drill .5" holes in it spaced 1"
apart. Coat the top with epoxy to protect it from corrosion.

FIGURE 10.3   Home-built slow sand filter.

inexpensive means for water purification in poor areas of the world
when the only option is to make do with the quality of water at hand.

Slow sand filters are recognized by the World Health Organization,[9]
Oxfam, the United Nations[10] and the United States Environmental
Protection Agency[11] as being superior technology for the treatment of
surface water sources. According to the World Health Organization,
"Under suitable circumstances, slow sand filtration may be not only
the cheapest and simplest but also the most efficient method of water
treatment."

Disinfection Methods

Chlorine is one of a group of nonmetallic elements know as halogens.
Chlorine, like ultraviolet (UV) light, is more effective against some
pathogens than it is against others. However, there are several important differences between UV and chlorine, including the following:

- Chlorine is more effective against viruses than it is against
  parasites, while UV is more effective against parasites than
  against viruses.

- UV light can inactivate pathogens in a few tenths of a second, while chlorine requires several minutes to work.

- UV disinfection only works on water that is relatively clear, while chlorine can be used to disinfect water that is relatively cloudy.

- The effectiveness of UV disinfection is unaffected by the pH and temperature of the water, while the effectiveness of chlorine is affected by pH, temperature, and the chlorine concentration in the water.

Experience indicates that you should not rely on chlorine as your primary disinfectant unless you have used a filtration system that has been certified in accordance with ANSI/NSF Standard 53 requirements. But chlorine is an extremely attractive alternative if you have installed such a filtration system because chlorine is extremely effective against viral contaminants.

Chlorine is available in gaseous, solid, and liquid forms. It is recommended that gas chlorine be avoided because it is much more corrosive and hazardous than the solid (calcium hypochlorite) or liquid (sodium hypochlorite bleach) forms.

Many homeowners and public water systems prefer using calcium hypochlorite because, in its solid form, it is more stable than liquid bleach and is available in higher concentrations. Solid calcium hypochlorite is produced in tablet, pellet, and granular forms. Tablets and pellets are frequently used in in-line erosion chlorinators. The calcium hypochlorite tablets or pellets are placed inside the erosion chlorinator; as water flows through the unit, the calcium hypochlorite slowly dissolves and releases chlorine into the water. Granular products are often mixed with water to form a liquid bleach solution that is fed with a metering pump. Due to its high chlorine content, calcium hypochlorite containers should be kept tightly closed and stored away from combustible materials such as oils, fuels, and greases.

You can also use sodium hypochlorite (liquid bleach) as your chlorine source. Liquid bleach is arguably the safest of the chlorine compounds because it is not as concentrated as the other materials and is less likely to release high concentrations of chlorine gas. However, it is much more susceptible to thermal decay, and you should probably not keep more than a 30- to 60-day supply on hand.

Sodium hypochlorite solutions, like solutions of dissolved calcium hypochlorite, are fed with a small metering pump. You should only use chlorine compounds that are certified in accordance with ANSI/NSF Standard 60 requirements. This standard also applies to any other chemical used to treat potable water.

Bleach works well on most harmful viruses and bacteria in water that is *not* heavily polluted, or when *Giardia* or cryptosporidiosis are *not* a concern. The source of these latter concerns is usually animal feces.

For *Giardia* or *Cryptosporidium* parasites, filtration through 0.2 micron is more effective since the amount of bleach needed to kill these parasites makes the water almost impossible to drink. If filtration to 0.2 micron is not practical, boiling the water for 5 minutes, adding 1 minute for every 1000 feet (300 meters) above 10,000 feet (3000 meters), is the best way to ensure safe drinking water.

Disinfection using bleach works best with warm water. Add 1 drop (0.05 mL) of bleach to 1 liter of water, shake, and allow to stand for at least 30 minutes before drinking. For cloudy water, double the amount of bleach or for cooler water. A slight chlorine odor should still be noticeable at the end of the 30-minute waiting period if you have added enough bleach. The longer the water is left to stand after adding bleach, the more effective the disinfection process will be.

Chlorine Tablets

Follow the manufacturer's directions for disinfecting large amounts of water. Always use clean containers designed for storage of food or water. You can use regular household bleach (usually about 5 percent chlorine) or commercial bleach products (usually 10 percent chlorine).

Let the water stand for at least an hour after adding the bleach before you start drinking it. If the water is colder than 10°C or has a pH higher than 8, let the water stand for at least two hours before drinking.

Table 10.1 below shows how much regular household bleach to add to various size water containers *to disinfect relatively clean water.*

Gallons of Water to Disinfect (Metric Equivalent Shown in Brackets)	Amount of Household Bleach (5%) to Add*
1 gal (4.5 liters)	2 drops (0.18 mL)
2.2 gal (10 liters)	5 drops (0.4 mL)
5 gal (23 liters)	11 drops (0.9 mL)
10 gal (45 liters)	22 drops (1.8 mL)
22 gal (100 liters)	0.75 teaspoon (4 mL)
45 gal (205 liters)	1.5 teaspoons (8 mL)
50 gal (230 liters)	1.75 teaspoons (9 mL)
100 gal (450 liters)	3.5 teaspoons (18 mL)
220 gal (1000 liters)	8 teaspoons (40 mL)
500 gal (2200 liters)	6 tablespoons (90 mL)
1000 gal (4550 liters)	6.5 ounces or 12 tablespoons (180 mL)

*Adding household (5%) bleach at these amounts will produce water with about 2 parts per million of chlorine in it (about 0.0002 percent).

TABLE 10.1 Guidelines for Chlorine Water Treatment

If you are treating water from a lake, stream, or shallow well, use twice as much household bleach (5 percent) as indicated in the chart below, and wait twice as long before drinking since it is more likely to contain chlorine-resistant parasites from animal droppings.

While activated carbon filters are reported to remove some chemical contaminants, in general, it is a bad idea to trust any purification system to remove even small quantities of chemicals in water, which could include inorganic contaminants (arsenic and other heavy metals) or organic toxins (such as fertilizers and pesticides).

Iodine

Whenever possible, use warm water (20°C) and let it stand a minimum of 20 minutes after mixing and before drinking. For cold water (5–15°C), increase the waiting time after mixing to 40 minutes. If you are using 2 percent tincture of iodine, use 10 drops (0.5 mL) for every 1 liter of water. With iodine tablets, follow the manufacturer's directions.

Note: Pregnant women should not use iodine drops to purify water, as it may have an effect on the fetus. Iodine should not be used to disinfect water over long periods of time, as prolonged use can cause thyroid problems.

Nano Silver

Silver has had medicinal applications going back centuries. The Phoenicians are said to have stored water, wine, and vinegar in silver bottles to prevent spoiling. In the early 1900s, people would put silver coins in milk bottles to prolong the milk's freshness. Hippocrates, the father of medicine, wrote that silver had beneficial healing and anti-disease properties. In the early 1900s, silver gained regulatory approval as an antimicrobial agent. Prior to the introduction of antibiotics, colloidal silver was used as a germicide and disinfectant. Silver has been used for the treatment of medical ailments for over 100 years due to its natural antibacterial and antifungal properties.

More recently, colloidal silver, impregnated in ceramic water filters, has been integrated into an effective filtering mechanism of contaminated water. A simple, bucket shaped piece of pottery (Fig. 10.4) is constructed of a mix of clay and sawdust. Water is poured through the resultant filter into the accompanying 8 gallon (30-L) plastic or ceramic receptacle, where the water is stored and used to dispense the filtered water.

The process for making the filter is as follows:

1. 60 percent dry pulverized clay (including brick scraps that are not acceptable to bricklayers) and 40 percent screened sawdust are mixed together in a mixer.

2. Water is added to the mix to obtain the correct consistency.

Figure 10.4   Nano Silver Impregnated Clay filter used to filter contaminated water. (*Potters for Peace*)

3. The filters are then formed by hand, turned on a potter's wheel, or press-molded.

4. Filters are fired at 887°C in a brick kiln using wood scraps from industry as the fuel source.

5. Filters are allowed to cool.

6. Filters are soaked for 24 hours to saturate the filter before flow testing.

7. The flow rate of each filter is tested to ensure a rate of between 1 to 2 L/hr–filters outside this range are discarded.

8. Filters are allowed to dry again.

9. 2 mL of 3.2 percent colloidal silver in 250 mL of filtered water is applied with a brush to each filter.

10. Filters are dried and sold.[12]

How it works

Colloidal silver (CS) consists of microscopic particles of ionic silver suspended in water and is applied to the ceramic filters after the firing process. The silver acts as a broad-spectrum germicide, and in some cases, as a bacteriostatic that essentially deactivates bacteria as it passes through the ceramic filter's pores: it thus renders the water potable.

It is believed that CS suffocates bacteria by disabling the particular enzyme that pathogenic bacteria and fungi use for oxygen metabolism. Other pathogens are destroyed by the electric charge of the silver particle, causing their internal protoplast to collapse. Others are rendered unable to reproduce. Parasites are killed while in their egg stage.

The effective filtration rate is between 1.5 and 3.0 L/hr and, if the filter is scrubbed periodically with a soft brush, it will maintain this rate. In one study commissioned by Potters for Peace, the filters removed 100 percent total coliform, fecal coliform and E. Coli, and 94 to 100 percent of fecal Streptococcus. In homes with negative results for hydrogen sulfide producing bacteria and E-coli, no child had had diarrhea in the month prior to the study.[13]

Colloidal Silver is used widely to prevent scumming in water storage tanks, to sterilize and purify water aboard NASA's Space Shuttle and on commercial airline flights.

Developments in nanotechnology have been applied to a water filter that resembles a humble tea bag, have been developed by Professor Cloete of Stellenbosch University. The "tea bag" (Fig. 10.5) is designed to fit in the neck of a standard sized water bottle and be interchangeable. Depending on the quality of water being filtered, it costs between one and five cents to produce a liter of drinkable water.

A German company[14] has developed a product that puts nano-silver particles, impregnated into clay balls, into a water storage tank,

FIGURE 10.5   Nano silver tea bag: South African researchers have created a water-purifying nanotech tea bag that costs half a cent. Portable, instantly effective and with no chance of recontamination.

allowing the silver to leach into the water. When the silver particles come in contact with bacteria and fungus, they adversely affect cellular metabolism and inhibit the growth of bacteria and fungi.

This is an interesting technology that looks to have significant potential. It is proving to be a beneficial application for stored water on boats, recreational vehicles, and rainwater storage in cisterns.

Ultraviolet Light

Since UV is effective only if the microbe is irradiated by the proper wavelength of UV light and for the proper amount of time, filtration is essential for proper sanitation of the water to occur. For typical potable applications of rainwater, the water is first filtered with a 20-micron filter followed by a 5-micron filter. The suspended solids and/or turbidity of the water play a part as does the flow rate through the device. If taste is an issue, then a carbon final filter can be used.

Aside from normal filter maintenance, the UV crystal tube that surrounds the bulb must be kept clean, and the bulb *must* be replaced annually (it is designed to run 24/7 for one year). A choke or regulator must be installed to limit the water flow through the unit, and water quality (transmittance) must be maintained.

There is no test, other than a microscopic culture count, to check the effectiveness. UV does not remove anything, kill anything, or change the chemistry of the water. It simply breaks the DNA of certain types of living cells so they cannot reproduce. In tests, however, some cells have regenerated their ability to reproduce in as little as 45 minutes. UV has no effect on algae, and there is no residual in the treated water. Because there is no residual from a UV system, UV water lamps should be installed downstream of the filters and as close to the end use as possible.

Ozone

Ozone is much different than ultraviolet light in that it is an active oxidizer of microbes, metals, organic, and inorganic compounds. In the process of oxidation, the microbes "burn up" so to speak. Anything that comes in contact with the ozone is oxidized, including algae and phytoplankton, which are not affected by UV. The principal drawback to ozone is that, if bromine is present in the water, bromate is formed and can be toxic. There is a test for presence of ozone, but it is costly, and ozone quickly dissipates so testing is more qualitative rather than quantitative. Residual is also limited unless the temperature of the treated water is lower than 72°F.

Some of the controversy is due to the fact that different types of ozone systems exist, some of which are not intended for the layperson to operate. With older ozone systems, dosing and contact times were tricky to control and excess ozone was released at times.

Newer ozone systems, which are intended for the homeowner, are designed to eliminate these problems by diffusing the ozone thoroughly and producing lower levels of it, so off-gassing doesn't occur. Ozone is reactive on plastic seals and metal pipes, so it should preferably be installed at the point of use, not the point of entry to the house. Due to the limited residual of ozone in water, and therefore its disinfection qualities, there is little protection with extended storage and immediate use is recommended.

Maintenance of ozone units is necessary to assure efficacy. Ozone systems must also be cleaned and flow matched to the rating for the unit. A pressure switch turns the unit on and off with demand and this switch is subject to failure, so that it must be checked and periodically replaced.

Testing for Water Quality

The standards that apply to reclaimed water vary according to the intended purpose of the water. Obviously water for flushing toilets does not need to meet the same standards as for drinking water. However when water is conveyed where splashing is a consequence, the creation of aerosols that can convey airborne bacteria is likely. Therefore in aboveground situations where people are present, even non-potable applications need to have minimum water standards maintained to avoid creating a potential health hazard to passersby. In Table 10.2, guidelines are shown that have been used successfully to maintain safe water quality for both potable and non-potable applications.

Treatment Level	Non-Potable[1]	Potable
Total Coliform	<500 CFU/100 ml	99.9 percent reduction
Fecal Coliform	<100 CFU/100 ml	99.9 percent reduction
Turbidity	<10 NTU	<0.3 NTU
Protozoan Cysts and Viruses		99.99 percent reduction
Testing Frequency[2]	Every 12 Months	Every 3 Months

1. Suitable for toilet and urinal flushing, washing machine makeup, and other approved applications in occupied space for environments with non-health-impaired occupants.
2. Upon failure of total and fecal coliform test, system shall be recommissioned, involving cleaning and retesting for *Giardia* and *Cryptosporidium* before returning system to service.[15]

TABLE 10.2 Water Quality Guidelines

Conclusion

Water disinfection is not a "set it and forget it" situation. No matter what method of disinfection is used, the operator has to be prudent and relentless in the operation and maintenance of the system to ensure proper operation and to avoid problems with water quality.

Notes and References

1. *Rainwater Safe to Drink*, Presentation by Dr. Karin Leder, Monash University, Melborne, Australia, at the American Public Health Association 137th Annual Meeting in Philadelphia on November 2010.
2. Louis Pasteur (1822-1895), The True Master of Microbiology, by King-Thom Chung and Deam Hunter Ferris, http://highered.mcgraw-hill.com/sites/dl/free/0072320419/20534/pasteur.html (Accessed June 3, 2012)
3. United States Environmental Protection Agency (EPA), "Technologies for Upgrading Existing or Designing New Drinking Water Treatment Facilities," Document no. EPA/625/4-89/023, Cincinnati, OH, 1990.
4. United States Environmental Protection Agency (EPA),"Technologies for Upgrading Existing or Designing New Drinking Water Treatment Facilities."
5. National Drinking Water Clearinghouse (U.S.), "Slow Sand Filtration," Tech Brief Fourteen, Morgantown, WV, June 2000.
6. National Drinking Water Clearinghouse (U.S.), "Slow Sand Filtration."
7. Centre for Affordable Water and Sanitation Technology, *Biosand Filter Manual: Design, Construction, & Installation*, July 2007.
8. United States Environmental Protection Agency (EPA), "Technologies for Upgrading Existing or Designing New Drinking Water Treatment Facilities."
9. HDR Engineering, *Handbook of Public Water Systems*, New York: John Wiley and Sons, p. 353, 2001. (Retrieved March 28, 2010)
10. Centre for Affordable Water and Sanitation Technology, *Biosand Filter Manual: Design, Construction, & Installation*.
11. Aquasus, Licher Strasse 19, Reiskirchen, Germany 35447, www.aquasus.de.
12. Investigation of the Potters for Peace, Colloidal Silver Impregnated Ceramic Filter Submitted to Jubilee House Community December 21, 2001 USAID Purchase Order Number: 524-0-00-01-00014-5362
13. Investigation of the Potters for Peace Colloidal Silver Impregnated Ceramic Filter: Intrinsic Effectiveness and Field performance in Rural Nicaragua, December 2001, D.S Lantagne.
14. *Green Plumbing Supplement*, water quality standards as per International Association of Mechanical and Plumbing Officials (IAPMO.org); *Texas Manual on Rainwater Harvesting*, Texas Water Development Board; and the *ARCSA/ASPE Rainwater Design and Installation Standard*, American Rainwater Catchment Systems Association (ARCSA.org).
15. *Rainwater Catchment Design and Installation Standard*, American Rainwater Catchment System Association (ARCSA.org); and *Standard 53*, American Society of Plumbing Engineers (ASPE.org).

CHAPTER 11

Artificial Groundwater Recharge

Introduction

Indirectly related as an Alternative Water Source, is Ground Water Recharge. While it technically does not produce water, artificial groundwater recharge purposefully puts water back into the aquifer for future use, rather than allowing it to flow to the waterways without benefit. Artificial groundwater recharge is designed to increase the natural replenishment or percolation of surface waters into the groundwater aquifers, resulting in a corresponding increase in the amount of groundwater available for extraction.

Although artificial recharging is primarily used to preserve or enhance groundwater resources, it has also been used for many other beneficial purposes, such as conservation of surface runoff and disposal of flood waters, control of salt water intrusion, storage of water to reduce pumping and piping costs, temporary regulation of groundwater abstraction, and water quality improvement through filtration of suspended solids through soils and other materials or via dilution with naturally occurring groundwater.[1]

Other areas in which artificial recharge has been used are in wastewater disposal, waste treatment, secondary oil recovery, prevention of land subsidence, storage of freshwater with saline aquifers, crop development, and stream flow augmentation.[2]

Groundwater is normally recharged by the natural percolation of rain into the soil. With the encroachment of humans on the environment, this process has been impeded with the increased paving of parking and roadways, building construction, and the cutting down of forests. The result is that rainwater, which would normally be added to the aquifer, runs off and flows down drains, creeks, and rivers.

Figure 11.1 1904 Tucson, Arizona, with water table 4 feet below grade. (*Photo courtesy of Arizona Historical Society from the Walter E. Hadsell Collection, Los Angeles, CA*)

The decreased charging of the aquifer is made worse with the increased water demand that accompanies development. Water demand has put more and more emphasis on groundwater extraction. The demand for groundwater to support civilization, especially for farming, is increasing faster than is being replenished; this causes water tables to be driven lower.[3] An example of this occurrence can be seen in the contrast between Figs. 11.1 and 11.2. Figure 11.1 illustrates the Santa Cruz River in Tucson, Arizona, in 1904, with the water table measured in 1946 at 30 feet (9.1 meters) below grade. After the expansion of Tucson, with its expanded agriculture, the 2012 photo shows the Santa Cruz River reduced to a drainage ditch, with a water table now 120 feet (36.5 meters) below grade[4] and dropping.[5]

Groundwater depletion is a problem worldwide, as seen in the graph in Fig. 11.3. It is estimated that about 4500 cubic kilometers of water have been extracted from aquifers between 1900 and 2008. That amounts to 1.26 cm of the overall rise in sea levels of 17 cm in the same period.[6] That 1.26 cm seems insignificant except that groundwater depletion has accelerated massively since 1950, particularly in the past decade. Over 1300 cubic kilometers of the groundwater was extracted between 2000 and 2008, producing 0.36 cm of the total 2.79-cm rise in that time.

Figure 11.2   2007 Tucson, Arizona, with water table 120 feet below grade. (*Photograph courtesy of Bradley Lancaster, www.harvestingrainwater.com*)

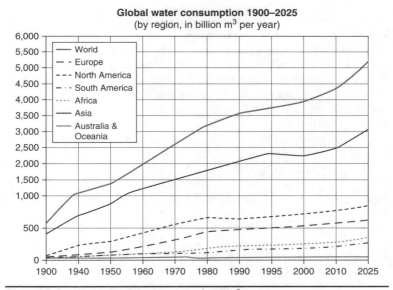

Figure 11.3   Worldwide consumption of water.[7]

In developing countries, the seasonal shortage of drinking water imposes serious health hazards on the rural masses and is responsible for loss of livestock for want of drinking water and fodder. The lowering water tables causes the summer yields of many dug and tube wells to decrease substantially or not produce water at all.[8] In developed countries, seasonal and local water shortages are becoming more common. The overall implications for agriculture and humanity alike are severe.

In general, artificial recharge works because it is effective in minimizing water loss due to evaporation compared with similar surface storage systems. Many environmental problems arising out of surface storage are also avoided using this method. For example, there is generally no loss of agricultural or other lands by inundation as would occur behind a surface storage structure, and the quality of the stored water is maintained and possibly improved since it is less exposed to surface runoff contamination.

As the seriousness of the problem becomes more apparent, people are now beginning to adopt a more enlightened approach to manage the available water in the aquifers. This chapter will discuss successful applications of artificial groundwater recharge and the resulting benefits.

Historical Applications

Groundwater replenishing has been a longtime practice in India, resulting in considerable knowledge about the topic. In India, there are numerous documents, installations and stone inscriptions from as early as A.D. 600 describing ancient kings and other benevolent persons, who considered it their duty to construct retention ponds to collect rainwater and use it to recharge wells. Even today, thousands of such structures exist and are in use for multiple purposes in the southern coastal towns and villages of Tamil Nadu, India, where groundwater is saline.[9]

Similarly, more than 500,000 tanks and ponds, big and small, are dotted all over, particularly in peninsular India. These tanks were constructed thousands of years ago for catering to the multiple uses of irrigated agriculture, livestock, and human uses such as drinking, bathing, and washing. The command area of these tanks has numerous shallow dug wells that are recharged with tank water and accessed to augment surface supplies. Many drinking water wells located within the tank bed and/or on the tank bund (earthen levee) are artificially recharged from the tank to provide clean water supply throughout the year with natural filtering.[10]

The success of these groundwater replenishing programs can be seen in the rivers, which once only flowed during the monsoon months, but now come alive for the full year. Today, with more than 200 johads (Fig. 11.4) in the catchment of the Aravari River, the successful water harvesting and recharging of groundwater upstream has transformed the once seasonal stream into a perennial river. These and other similar programs have been instrumental in achieving productive benefits locally and have led to many such initiatives in other parts of India.

Figure 11.4   A *johad* is a rainwater storage tank, principally used in India, that collects and stores drinking water throughout the year for humans and cattle.

Factors That Affect Infiltration

Factors affecting natural infiltration are precipitation, the absorptive characteristics of the soil, degree of soil saturation, type and quantity of land cover, the slope of the land, and evapo-transpiration.

Precipitation

The greatest factor controlling infiltration is the amount and characteristics (intensity, duration, etc.) of precipitation that falls as rain or snow. Precipitation that infiltrates the ground often seeps into streambeds over an extended period of time, thus a stream will often continue to flow when it hasn't rained for a long time and where there is no direct runoff from recent precipitation.

With changing weather patterns, the trend is that some areas are seeing water shortages, while others are getting too much, resulting in flooding. An additional benefit of artificial groundwater recharge is that, with proper management, rainfall events can serve to moderate these two extremes.

Soil Characteristics

Complementary to the occurrence of rainfall is the ability of the soil to absorb precipitation. Some soils, such as clays, absorb less water at a

Permeability (m/s)											
10^0 10^{-1} 10^{-2} 10^{-3} 10^{-4} 10^{-5} 10^{-6} 10^{-7} 10^{-8} 10^{-9} 10^{-10} 10^{-11}											

Drainage	Good		Poor	Practically impervious
Soil Types	Clean gravel	Clem sands, clean sand & gravel mixtures	Very fine sands, organic & Inorganic silts, mixtures of sand silt & clay, glacial till, stratified clay deposits, etc.	Impervious soils e.g., homogeneous clays below zone of weathering
			"Impervious" soils modified by effects of vegetation & weathering	

It is good to have an idea about the *order of magnitude* for the permeability of a specific soil type.

TABLE 11.1   Permeability and Drainage Characteristics of Soils[11]

slower rate than sandy soils. Soils absorbing less water result in more runoff overland into streams. Soil permeability is a major factor in the design and operation of any groundwater recharge system. The more permeable the soil, the less runoff there will be, and the more easily a given amount of runoff will enter the soil. Table 11.1 shows the interaction between porosity based on soil makeup and permeability. Soil permeability is also closely related to soil health.

Soil Saturation

Like a wet sponge, soil already saturated from previous rainfall can't absorb much more, thus more rainfall will become surface runoff.

Land Cover

Land cover has a great impact on infiltration and rainfall runoff. Vegetation slows the movement of runoff, allowing more time for it to seep into the ground. Agriculture and the tillage of land reduce the ground cover that would otherwise absorb the water and allow the water to drain from the site, often carrying topsoil with it. Impervious surfaces, such as parking lots, roads, and developments, also serve to bypass the absorption process by conveying rainwater directly away from the site before it can be absorbed into the soil. Water, which in natural conditions infiltrated directly into soil, is now lost to streams and rivers before infiltration can occur.

Slope of the Land

To state the obvious, water falling on steeply sloped land runs off more quickly, and infiltrates less, than water falling on flat land.

Evapo-Transpiration

Some infiltration stays near the land surface, which is where plants put down their roots. Plants need this shallow groundwater to grow, and by the process of evapo-transpiration, water is moved back into the atmosphere. Evapo-transpiration is a combination of two events: evaporation and transpiration. Evaporation is what leaves the soil as a function of the level of moisture in the soil and the relative humidity in the atmosphere. Transpiration occurs through the breathing function of plants and varies according to the type of plant and the season in which the plant is growing.[12]

Subsurface Water

To understand the concept of artificial groundwater recharging, it is helpful to understand how the natural process works. As precipitation infiltrates the subsurface soil, it generally forms an unsaturated zone and a saturated zone. In the unsaturated zone, the voids, spaces between grains of gravel, sand, silt, clay, and cracks within rocks contain both air and water. Although a lot of water can be present in the unsaturated zone, this water cannot be pumped by wells because it is held too tightly by capillary forces. The upper part of the unsaturated zone is the soil-water zone. The soil zone is criss-crossed by roots, openings left by decayed roots, and animal and worm burrows. This allows the precipitation to infiltrate into the soil zone. Water in the soil is used by plants in life functions and leaf transpiration, but can also evaporate directly to the atmosphere. Below the unsaturated zone is a saturated zone where water completely fills the voids between rock and soil particles.

How Infiltration Replenishes Aquifers Naturally

Natural refilling of deep aquifers is a slow process because groundwater moves slowly through the unsaturated zone and the aquifer. The rate of recharge is also an important consideration. It has been estimated, for example, that if the Ogallala Aquifer that underlies the High Plains of Texas and New Mexico—an area of slight precipitation—was emptied, it would take centuries to refill it at the present small rate of replenishment.[13] In contrast, shallow aquifers in areas of substantial precipitation, such as those on the coastal plain in southern Georgia (USA) may be replenished almost immediately.

How Groundwater Moves within the Aquifer

Groundwater is water that fills spaces underground, between grains of sand or rock or in subterranean cracks. The area where groundwater occurs is the aquifer. The top level of the aquifer saturated with groundwater is the water table. The most productive aquifers consist of water stored in sand and gravel. These areas are typically ancient floodplains (alluvium) where water, rocks, and sand were deposited at about the same time.

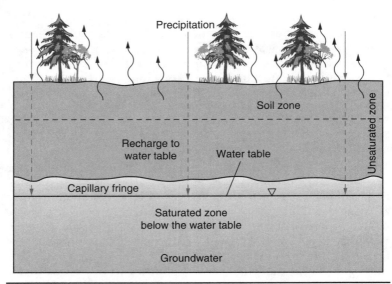

FIGURE 11.5   Schematic of a simple aquifer. (*Illustration courtesy of U.S. Geological Survey*)

Gravity is the force that moves groundwater, generally downward. However, groundwater can also move upward if the pressure in a deeper aquifer is higher than that of the aquifer above it. This often happens where pressurized, confined aquifers occur beneath unconfined aquifers (Fig. 11.5).

As groundwater flows downward in an aquifer, its upper surface slopes in the direction of flow. This slope is known as the hydraulic gradient and is determined by measuring the water elevation in wells tapping the aquifer. For confined aquifers, the hydraulic gradient is the slope of the potentiometric[14] surface. For unconfined aquifers, it is the slope of the water table.

Openings between the subterranean storage areas must be interconnected for water to flow freely. Over the years, water moves slowly downward to the aquifer and also spreads out, moving toward the nearest river basin. At some point, the water will reach solid rock or a layer of clay. When such an impermeable layer is near the surface, water may occur as surface flow or as a spring. If the water table is relatively close to the surface, rivers can flow (Fig. 11.6).

Effects of Water Withdrawal

How aquifers respond when water is withdrawn from a well is an important topic in groundwater hydrology. It explains how a well gets its water, how it can deplete adjacent wells, or how it can induce contamination.

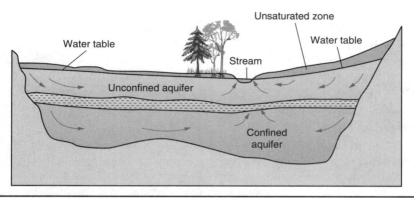

Figure 11.6 Groundwater flow. (See also Color Plates; *illustration courtesy of U.S. Geological Survey*)

When water is withdrawn from a well, the water level in the well drops. When the water level falls below the water level of the surrounding aquifer, groundwater flows into the well. The rate of inflow increases until it equals the rate of withdrawal.

The movement of water from an aquifer into a well alters the surface of the aquifer around the well. It forms what is called a *cone of depression*. A cone of depression is a funnel-shaped drop in the aquifer's surface. The well itself penetrates the bottom of the cone. Within a cone of depression all groundwater flows to the well (Fig. 11.7). The outer limits of the cone define the well's area of influence.

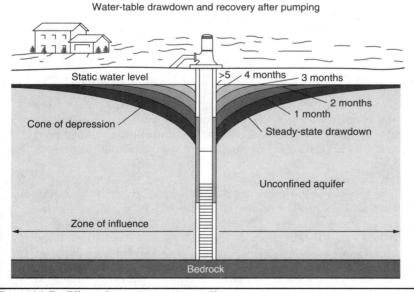

Figure 11.7 Effect of pumping on the aquifer.

Excessive pumping from aquifers causes the water table to recede, and surface flow is no longer possible except during rainy periods. Another impact of pumping is that grains of sand or rock compact as water is withdrawn. With the compacted media, water is much less likely to percolate through the media to refill the spaces, especially at greater depths.[15]

Groundwater Flow

The velocity at which groundwater moves is a function of three main variables: hydraulic conductivity (commonly called permeability), porosity, and the hydraulic gradient. This is important to groundwater replenishing since it determines the rate of water acceptance into the absorption field and the recovery rate of the stored water during drawdown.

The hydraulic conductivity is a measure of the water-transmitting capability of an aquifer. High hydraulic conductivity values indicate an aquifer can readily transmit water; low values indicate poor transmitting ability. In general, course-grained sands and gravels readily transmit water and have high hydraulic conductivities. Fine-grained silts and clays transmit water poorly and have low hydraulic conductivities.

The porosity of an aquifer also has a bearing on its ability to transmit water. Porosity is a measure of the amount of open space in an aquifer. Both clays and gravels typically have high porosities, while silts, sands, and mixtures of different grain sizes tend to have low porosities (Table 11.1).

The velocity at which water travels through an aquifer is proportional to the hydraulic conductivity and hydraulic gradient, and inversely proportional to the porosity. Of these three factors, hydraulic conductivity generally has the most effect on velocity. Thus, aquifers with high hydraulic conductivities, such as sand and gravel deposits, will generally transmit water faster than aquifers with lower hydraulic conductivities, such as silt or clay beds.

Groundwater velocities are typically very slow, ranging from around a centimeter per day to almost a meter per day. However, some very rapid flow can occur in rock with solution cavities or in fractured rock. Very high flow rates (more than 15 m/day) are associated, for example, with some parts of the Columbia River basalt in eastern Washington.[16]

The volume of groundwater flow is controlled by the hydraulic conductivity and gradient and is also controlled by the volume of the aquifer. A large aquifer will have a greater volume of groundwater flow than a smaller aquifer with similar hydraulic properties. But if the cross-sectional area—that is, the height and width—are the same for both aquifers, the aquifer with a greater hydraulic conductivity and hydraulic gradient will produce a greater volume of water.

Artificial Recharge Techniques and Designs

Recharging aquifers has the following objectives:

- To maintain or augment natural groundwater as an economic resource
- To conserve excess surface water underground
- To combat progressive depletion of groundwater levels
- To combat salt and brackish water intrusion into the water table

To achieve these objectives, it is important to plan out the installation process in as scientific a manner as possible. Site investigation involving soil samples should be carried out for selection of the site where the artificial recharge is to occur. The site investigation should do the following:

- Define the sub-surface geology as to soil type, degree of soil saturation, and permeability.
- Determine the presence or absence of impermeable layers or lenses, which can impede percolation
- Define depth of the water table and groundwater flow directions
- Establish the maximum rate of recharge that could be achieved at the site[17]

The type, size, and depth of the structure used will depend on the nature of the rainfall. Tropical regions may have more intense rainfall and need larger capacities to capture the water. In more temperate climates, the rain may come more slowly, giving it more time to be absorbed. These are generalizations, and the requirements of the specific location must be studied when choosing.[18]

Classification of Recharging Methods

Sources and Mechanisms of Recharge

The sources of recharge to a groundwater system include both natural and human-induced phenomena. Natural sources include recharge from precipitation, lakes, ponds, and rivers (including perennial, seasonal, and ephemeral flows), and from other aquifers. Human-induced sources of recharge include irrigation loss from both canals and fields, leaking water mains, sewers, septic tanks, and over-irrigation of parks, gardens, and other public amenities. Recharge from these sources has generally been classified as *direct recharge* from percolation of precipitation and *indirect recharge* from runoff ponding.[19]

Direct Recharge

Direct or diffuse recharge is defined as water added to the groundwater reservoir in excess of soil-moisture deficits and evapotranspiration, by direct vertical percolation of precipitation through the unsaturated zone.

Direct surface techniques are among the simplest and most widely used methods for groundwater recharge. Using these techniques, water is moved from the land surface to the aquifer by means of simple infiltration. The infiltrated water percolates through the unsaturated zone to reach the groundwater table. Through this process, the recharged water is filtrated and oxidized. Direct recharge methods can be grouped into three categories (see Fig. 11.8): (a) when the aquifer is shallow, water may be spread over fields or conveyed to rain gardens, basins, and ditches, where it percolates; (b & c) when an aquifer is situated at greater depths, recharge can be facilitated by flooding pits, trenches, and dug shafts; and (d & e) in cases of high overburden thickness or confining aquifer conditions, recharge can be effected by injecting surface water directly into the aquifer using boreholes or tube wells. Design of pumped wells (Fig. 11.9) is beyond the scope of this book, but additional information can be gathered from the references listed in this chapter's end notes.

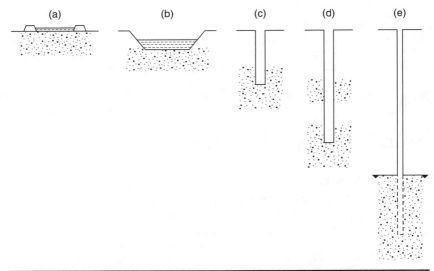

FIGURE 11.8 Recharge systems for increasingly deep permeable materials.[20] Surface basin (rain garden) (*a*); excavated basin (swale) (*b*); trench (*c*); shaft or vadose zone well (*d*); and deep aquifer well (*e*).

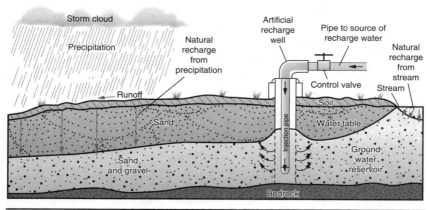

FIGURE 11.9 Illustration of injection well and the influence to the aquifer.

Surface Spreading Techniques

Water spreading is practiced on an increasing scale all over the world and notably in India.[21] In some areas totally reliant upon groundwater, surface storage structures (percolation ponds) are being constructed purely for groundwater recharge.

Percolation tanks (or ponds) are water-harvesting structures constructed across or near streams to impound rainwater and to retain it for a longer time to increase the opportunity time for infiltration. The water storage is expected to induce percolation and replenish the aquifer, which is then exploited through wells located down the gradient.

Check dams, generally constructed for soil conservation, can be considered mini- or micro-percolation tanks from which water is not directly drawn for irrigation but is allowed to percolate into subsurface strata, thus augmenting the groundwater. The aim is to increase the contact area and residence time of surface water over the soil to enhance the infiltration and augment the groundwater storage in phreatic[22] aquifers. The downward movement of water is governed by a host of factors, including vertical permeability of the soil, presence of grass or entrapped air in the soil zone, and the presence or absence of limiting layers of low vertical permeability at depth. Changes brought about by physical, chemical, and bacteriological influences during the process of infiltration are also important in this regard.

Rain gardens, which are shallow retention swales augmented with landscape, combine the benefits of groundwater recharge with aesthetics (Fig. 11.10).

Important considerations in the selection of sites for artificial recharge through surface spreading techniques include:

- The area should have gently sloping land without gullies or ridges.

Figure 11.10 Rain garden integration with Tempe, Arizona, street roundabout manages stormwater runoff while providing the aesthetic benefit of shade and sound abatement. (*Heather Kinkade Photo, www.forgottenrain.com*)

- The aquifer being recharged should be unconfined, permeable, and sufficiently thick to provide storage space.

- The surface soil should be permeable and have a high infiltration rate.

- The unsaturated zone should be permeable and free from clay lenses.

- The water table should be deep enough to accommodate the recharged water so that there is no water logging.

- The aquifer material should have moderate hydraulic conductivity so that the recharged water is retained for sufficiently long periods in the aquifer and can be used when needed.

Examples of surface spreading:

- Recharge pits: 1 to 2 m (3 to 6 ft) wide and up to 3 m (10 ft) deep, filled with rocks or gravel and sand.

- Trenches: Similar in concept to swales, 0.5 to 1 m (2 to 4 ft) wide, 1 to 1.5 m (3 to 6 ft) deep, and 10 to 20 m (38 to 76 ft) long, depending on expected volume of water. These are backfilled with filter materials, commonly with construction debris that includes broken brick, concrete, and sand.

- When the top layer is permeable, surface spreading techniques may be used to spread the water in streams by making check dams, swales, or percolation ponds to retard flow to increase time of percolation into the soil (Fig. 11.11).

FIGURE 11.11   Check dams used in this drainage system to slow storm runoff to increase water absorption into the soil. (*Heather Kinkade Photo, www.forgottenrain.com*)

- Swales: Level shallow trenches that follow the contour of the land and allow water to soak in (Fig. 11.12).
- Rain gardens: Low lying areas, planted with suitable species to provide detention time for the water to be absorbed into the soil, detain rainwater runoff from impervious urban areas

FIGURE 11.12   Containment of parking lot runoff for stormwater management and recharging the aquifer. Note the elevated overflow drains. (*Heather Kinkade Photo, www.forgottenrain.com*)

Figure 11.13   Illustration of an integrated rainwater catchment system with rain garden and overflow into a percolation tank. (*Courtesy the Aquascape Company, www.aquascapeinc.com*)

such as roofs, concrete bitumen, or compacted soil or lawn areas. A full, integrated rainwater retention system can be seen in Fig. 11.13, where roof runoff is first filtered to remove organic debris (1), then conveyed to the water containment cistern (2), where a submersible well pump distributes the water to the garden and water feature (3). When full, this cistern overflows to a similarly constructed cistern (4) that is wrapped with permeable geo-cloth that allows overflow water from the cistern to infiltrate the soil. Only when this cistern is completely filled does water overflow to the storm sewer (5).

Wells

Existing wells may be utilized as a recharge structure provided that a high grade of water is added to the aquifer. The water introduced into the well should pass through filter media to remove any chemical or organic material before being put into the well. The following are types of wells:

- Dug wells
- Recharge wells, specially constructed for recharge—100 to 300 mm (4 to 12 inches) diameter, for recharging the deeper aquifers. Water is passed through filter media to avoid choking of the well.

(a)

(b)

Figure 11.14   A, B, C, D Sequence of installation of a passive rainwater installation utilizing underground vaults to direct stormwater into the aquifer to control soil runoff and support a garden of native plants (*Billy Kniffen Photo, Public Library, Menard, Texas*)

(c)

(d)

FIGURE **11.14** (*Continued*)

- Recharge shafts—for recharging a shallow aquifer located below a clay surface. Shafts of 0.5 to 3 m (10 inches to 10 feet) diameter and 10 to 15 m (33 to 50 feet) deep are made and filled with boulders, gravel, and coarse sand.

- Lateral shafts with bore wells—for recharging shallow and deep aquifers. Shafts of 1.5 to 2 m (5 to 6.6 feet) wide and 10 to 30 m (33 to 98 feet) long are constructed depending upon availability of water with one or two bore wells. Filled with boulders, gravel, and coarse sand.

Recharge through pits and shafts has limited applications, as recharge capacity is low. However, abandoned stone quarries or open (dug) wells located where the water table has dropped below the excavated depths can be used as ready-made recharge pits and shafts.

Indirect Recharge

Indirect recharge results from percolation to the water table following runoff and localization in joints, such as occurs through ponding in low-lying areas and lakes or through the beds of surface water courses. Recharge through such topographic depressions, which are common in the Canadian prairies and Great Plains of the United States, is also known as *depression-focused recharge* and occurs where surface runoff or lateral flow of subsurface moisture accumulates within or beneath such depressions on the landscape. Thus, knowledge of lateral subsurface flow processes becomes important in understanding recharge processes. In arid and semi-arid regions, localized and indirect recharge is often the most important source of natural recharge. Percolation to the water table from streambeds takes two forms, depending on whether there is a saturated connection between the stream and the water table. Where no connection exists (Fig. 11.15a), a situation typical of arid zones where water tables are generally deep, water moves downward from the streambed

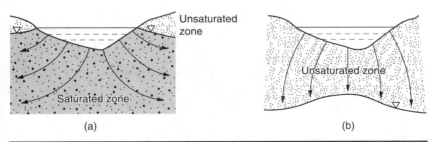

(a)   (b)

FIGURE 11.15 Recharge from streambeds (a) with no hydraulic connection, and (b) with hydraulic connection.

to the water table, forming a groundwater mound which then dissipates laterally away from the stream.

As long as the mound is recharged by unsaturated flow, there is no hydraulic connection between the groundwater and the stream flow, in the sense that the recharge rate is almost unaffected by the groundwater levels. Yet even when the unsaturated condition is present, the stream and aquifer may in fact be hydraulically connected in the sense that further lowering of the regional water table could increase channel loss. At some critical depth to the water table, however, further lowering has no influence on channel loss. At this depth, which depends mostly on soil properties and water stage in the channel, the aquifer becomes hydraulically disconnected from the stream. If the distance from the water table to the stream stage is greater than approximately twice the stream width, the seepage rapidly begins to approach the maximum seepage for an infinitely deep water table. The parameters determining the recharge process are the width, depth, and duration of stream flow, and the hydraulic characteristics of the local material in and below the streambed. The effects of this can be seen in the two photos of the Santa Cruz River in Tucson, Arizona, in Figs 11.1 and 11.2.

Lowering of the water table has a significant impact on vegetation. Beyond certain depths, plants and trees are unable to draw water to survive. Figure 11.1 shows hardwood trees adjacent to the river in Tucson. As the water level dropped to the 2004 level shown in Fig. 11.2, the hardwood trees were been primarily replaced by more tolerant grasses and cacti.

In less arid areas, water table levels tend to rise closer to the streambed. In these situations, a hydraulic connection will usually exist between the stream and the groundwater (Fig. 11.15b), and the recharge rate will decrease as the water table rises. The recharge process will be dominated by horizontal rather than vertical flow and will have a much shorter turnover or transit time than when there is no hydraulic connection. In these less arid environments, there is also likely to be recharge from general catchment percolation, and the mix between the two mechanisms may be hard to predict.

In areas where the potential recharge rate exceeds the rate at which water can flow laterally through the aquifer, the aquifer becomes over-full and available recharge is rejected, a condition known as rejected recharge. In this situation, groundwater pumping in recharge areas can increase the rate of underground flow from the area, and more water can be drawn into the aquifer as induced recharge.[23]

There are a number of other indirect recharge methods. For example, in flood irrigation, excess water percolates to the groundwater table. Similarly, seepage from lake beds, irrigation tanks, streams, and canals recharges groundwater, as does the use of terracing and contour bunds. While recharge from these processes can in some senses be considered accidental, each method's recharge potential can be purposefully enhanced, and a combination of techniques, both direct and indirect, can be used to meet specific terrain and topography conditions and recharge needs.

An experiment by the Uttar Pradesh (India) government to develop a new and practical way to conserve and rejuvenate falling groundwater reserves through the use of floodwater highlights the potential for integrated artificial recharge methods. The Madhya Ganga Canal Project (MGCP), located in the lower Ganga canal, was initiated in 1988. In 2000, the International Water Management Institute (IWMI) carried out a study (Chawla, 2000)[24] on the Lakhaoti branch canal of the MGCP to assess the impact of diversion of surplus Ganga water on groundwater levels and cropping patterns. The Lakhaoti branch is spread over more than 200,000 hectares and covers the districts of Ghaziabad, Bulandsher, and Aligarh in western Uttar Pradesh, and it is bounded by the drainage canals of the Kali and Nim rivers.

According to the study results, the canal project served to raise the groundwater table from 6.6 m to 12.0 m (22 to 39 ft) and brought down the cost of pumping for irrigation by 40 percent. Before the project's successful implantation, monsoon period pumping lowered the groundwater levels, causing severe water shortages during the dry season. Following the guidance of the MGCP who sponsored this project, the seepage from the canals and flooded paddy fields helped recharge the underlying aquifers, which increased water availability, proving the viability of groundwater replenishment.[25]

Recharge for Domestic Supplies

The previous section illustrated how irrigation and irrigation-related storage structures can effectively be used to indirectly recharge groundwater aquifers. This section illustrates the role of groundwater recharging for domestic supplies.

In many parts of India, rural drinking water supply programs often experience a shortage of water from wells because of increased groundwater use for irrigation from boreholes in and around the drinking water bores.[26] The natural recharge studies carried out over different hard-rock terrain indicated that only 5–10 percent of the seasonal rainfall recharges the groundwater (Athavale et al., 1992). The meager annual replenishment of natural recharge to the groundwater (for multiple uses) may not be able to meet the projected demand of 17,000 million cubic meters per year by 2050. Enhancement of recharge to the groundwater has therefore become mandatory in areas where groundwater is the sole source of drinking water supply. The NGRI (National Geophysical Research Institute) recommends that the methodology of artificial recharge and retrieval (ARR) can profitably be used for recharging a well during monsoon season and for drinking water during the dry season.

Cost

The cost of recharge schemes, in general, depends upon the degree of treatment of the source water, the distance over which the source water must be transported, and the stability of recharge structures and resistance to siltation and/or clogging. In general, the costs of construction and operation of the recharge structures (except for the more complex issues associated with injection wells in alluvial areas) are reasonable if the option is transporting the water over great distances to the end user via pipeline or surface transportation.

Maintenance

The effectiveness of artificial groundwater injection is maintained when the source water is adequately cleaned before injection into the aquifer. Accumulated silt and surface debris serves to increase resistance to the water acceptance process. Studies in the semi-arid regions of the Noyyal, Ponani, and Vattamalai (India) river basins showed that percolation rates were as high as 163 mm/day at the beginning of the rainy season, but diminished thereafter mainly due to the accumulation of silt at the bottom of the water percolation tanks.[27] In Punjab, studies of artificial recharge, using injection wells, were carried out in the Ghagger River basin, where using canal water as the primary surface water source showed that the recharge rate from pressure injection was 10 times that of gravity systems and that maintenance was required to preserve efficiency (Muralidharan and Athavale, 1998). In Gujarat, studies of artificial recharge were carried out that showed a recharge rate of 260 m^3/day (68,684 gallons, U.S. liquid) with an infiltration rate of 17 cm/hour (7 in/hour).[28]

Conclusion

Maintaining the aquifer is fundamental to the continuation of human existence. Increasingly, water tables are dropping due to the stress that civilization is imposing on them with domestic and agricultural needs. Groundwater injection serves to complement what occurs naturally by storing water when it is abundant for future use. Recognizing the relationship of water availability and maintaining the water aquifer is an imperative first step.

"Water is our most precious wealth. Left trapped in plastic water bottles that are tossed away, wasted in watering gardens that are meant to grow plants that thrive in dry soil, taken for granted in the home (where) we never imagine life without it, but now, our water needs our consideration more than ever."[29]

Notes and References

1. Asano, T., *Artificial Recharge of Groundwater*, Butterworth Publishers, Boston Massachusetts, 1985.
2. Oaksford, E. T., "Artificial Recharge: Methods, Hydraulics, and Monitoring," in *Artificial Recharge of Groundwater*, T. Asano, pp. 69–127, Butterworth Publishers, Boston Massachusetts, 1985.
3. The Ogallala Aquifer underlies approximately 225,000 square miles in the Great Plains region, particularly in the High Plains of Texas, New Mexico, Oklahoma, Kansas, Colorado, and Nebraska. The depth of the aquifer from the surface of the land varies from region to region. The aquifer has long been a major source of water for agricultural, municipal, and industrial development.
4. Arizona Ground Water Monitoring Site Hydrograph, https://gisweb.azwater. gov/gwsi/Detail.aspx (Accessed February 17, 2012). Well: local ID D-14-13 13CBC, Site ID 321227110574801, Registry ID 619923, Latitude 32° 12' 38.5", Longitude 110° 58' 33.4", Altitude 2368', Water Use—Public Supply, Drill Date March 1, 1946. Depth to groundwater: 1946, 30 ft (9.1 m); 2012, 120 ft (36.5 m).
5. Note: Depths to groundwater vary widely in wells in the Tucson Basin. This Tucson Water well (local ID D-14-13 13CBC) is very close to downtown Tucson and the Santa Cruz River. It is less than 600 yards (547 m) from a hand-dug well on South Main Street (near where the Wishing Shrine is located today), from which in the 1870s Adam Sanders and Joseph Phy obtained water to sell at five cents a bucket. According to "The Lessening Stream: An Environmental History of the Santa Cruz" by Michael F. Logan, "The two entrepreneurs filled an iron tank on a wagon from their well and traveled daily through town selling water. Within twenty-five years municipal water use in Tucson would progress from well water sold by the bucket, to a piped supply tapping the aquifer. When the mains were first opened in September 1882, an almost immediate decline in the water table downstream resulted."
6. Compiled by Leonard Konikov of the U.S. Geological Survey in Reston, Virginia.
7. Church, J. A., et al., "Revisiting the Earth's Sea-Level and Energy Budgets from 1961 to 2008," *Geophysical Research Letters*, Vol. 38: L18601, 8 pp., 2011. doi:10.1029/2011GL048794. Obtained from www.desdemonadespair. net/2011/12/50-doomiest-graphs-of-2011.html.
8. Case studies of locally managed tank systems in Karnataka, Andhra Pradesh, Gujarat, Madhya Pradesh, Gujarat, Orissa, and Maharashtra (India). Report submitted to International Water Management Institute (IWMI), Tata Policy Programme, Anand, India. Shah, T., 1998.
9. *Revisiting Tanks in India*, National Seminar on Conservation and Development of Tanks, DHAN Foundation, New Delhi, India, 2002.
10. *Revisiting Tanks in India*, National Seminar on Conservation and Development of Tanks.
11. Terzaghi, K., Peck, R.B., and Mesri, G., *Soil Mechanics in Engineering Practice, 3rd Edition*, John Wiley & Sons, New York, 1996.
12. For a more extensive discussion of evapo-transpiration, see Hanson, R.L.,"Evapotranspiration and Droughts," U.S. Geological Survey, 1991, in Paulson, R.W., Chase, E.B., Roberts, R.S., and Moody, D.W., Compilers, *National Water Summary 1988–89—Hydrologic Events and Floods and Droughts*: U.S. Geological Survey Water-Supply Paper 2375, pp. 99–104.
13. "The water-level changes from predevelopment (about 1950) to 2005 ranged between a rise of 84 feet and a decline of 277 feet. Area-weighted, average water-level change from predevelopment to 2005 was a decline of 12.8 feet. Approximately 25 percent of the aquifer area had more than 10 feet of water-level decline from predevelopment to 2005; 17 percent had more than 25 feet of water-level decline, and 9 percent had more than 50 feet of water-level decline. Approximately 2 percent of the aquifer area had more than 10 feet of water-level rise from predevelopment to 2005. Water-level changes from predevelopment to 2005 ranged between a rise of 84 feet and a decline of 277 feet.

Area-weighted, average water-level change in the aquifer was a decline of 12.8 feet from predevelopment to 2005, a decline of 0.8 foot from 2003 to 2004, and a decline of 0.2 foot from 2004 to 2005. Total water in storage in the aquifer in 2005 was about 2,925 million acre-feet, which was a decline of about 253 million acre-feet (or 9 percent) since predevelopment (about 1950)." McGuire, V.L., *Changes in Water Level and Storage in the High Plains Aquifer, Predevelopment to 2005*: U.S. Geological Survey Fact Sheet 2007–3029, 2007.

14. A potentiometric surface is based on basic hydraulic principles. For example, if there are two connected storage tanks with one full and one empty, they will gradually fill/drain to the same level. This is because of atmospheric pressure and gravity. A potentiometric surface is the imaginary line where a given reservoir of fluid will "equalize out" if allowed to flow (Younger, P., *Groundwater in the Environment*, Oxford, UK: Blackwell Publishing Ltd., 2007).

15. Perlman, H., *The Water Cycle: Infiltration*, U.S. Department of the Interior, U.S. Geological Survey, March 9, 2012.

16. Chapter 2 of the Washington State Department of Ecology *Ground Water Resource Protection Handbook*, December 1986.

17. For insight in sizing rain gardens and stormwater percolation pits, the reader is invited to use the rainwater quantity estimation techniques described in Chapter 8: Harvesting the Rain. To determine soil percolation rates and the ability for the soil to absorb water, see the methodology used in Chapter 8: Harvesting the Rain and Chapter 16: Septic System Design.

18. The website of the Directorate of Town, Panchayats, India, describes a number of methods of groundwater recharge used to store rainwater and to recharge shallow aquifers: www.tn.gov.in/dtp/rainwater.htm.

19. Other more sophisticated classifications not described herein include localized or focused recharge, preferential recharge, induced recharge, mountain front recharge, and others.

20. Todd, D.K., Mays, L.W., *Ground Water Hydrology*, 3rd Ed., John Wiley & Sons, Inc.

21. Based on paper on artificial recharge in India by Muralidharan and Athavale, National Geophysical Research Institute, CSRI Hyderabad, India, 1998.

22. Defined as soil below water level in which all the pores and inter-granular spaces are full of water.

23. Groundwater Recharge in Sophocleous, M., *Encyclopedia of Life Support Systems* (EOLSS), developed under the auspices of UNESCO, EOLSS Publishers, Oxford, UK, 2004. (www.eolss.net)

24. Shawla, A.S., Ground Water Recharge Studies in Madhya Ganga Canal Project, Consultancy Report to International Water Management Institute (IWMI), Colombo, Sri Lanka, 2000.

25. Sakthivadivel, R., "The Groundwater Recharge Movement in India," in Girodan, M., Villholth, K.G., *The Agricultural Groundwater Revolution: Opportunities and Threats to Development*, 2007.

26. Different villages of Thumbadi watershed in Karnataka state demonstrate how lack of effective zoning and regulation of irrigation wells within a 250 m radius of drinking water wells allows irrigation water supply wells to come into existence, hindering the performance of public water supply wells. Sakthivadivel, R., "The Groundwater Recharge Movement in India."

27. "Importance of Recharging Depleted Aquifers: State of the Art of Artificial Recharge in India," *Journal of Geological Society of India* 51, pp. 429–454.

28. Phadtare, P.N., Tare, S.C., Bagade, S.P., Banerjee, A.K., Srivastava, N.K., Manocha, O.P., Interim Report on Concept, Methodology and Status of Work on Pilot Project for Artificial Recharge, Central Groundwater Board (CGWB), Ahmedabadf, India, 1982.

29. Graef, A., "Fears Grow Over Depletion of the Ogallala Aquifer," Care to Make a Difference, March 31, 2011. (www.care2.com/causes/ogallala-aquifer.html)

CHAPTER 12

Aquatic Plants as a Waste Management System

Introduction

Most people in developed countries have access to relatively pure water, free of pathogens and low in mineral components. However, in developing countries, more than half the population drinks water that has passed through wastewater treatment works. The purification of water for human consumption by recycling wastewater or from treatment of unused surface water, while aesthetically not very acceptable, may be an only recourse in many developing countries. This is particularly true in arid areas, where water is a precious resource. However, care must be taken to effectively treat such water to ensure that health standards are achieved. Aquatic plant–based wastewater treatment is a potential option in such treatment processes.

Conventional versus Aquatic-Plant Treatment

Conventional sewage treatment takes place in two or more stages. A "primary" treatment involves sedimentation, filtration, and screening to remove floating objects, sand, and stones. "Secondary" treatment uses biological methods, including aeration, oxidation, and filtration so that the sludge interacts with air and bacteria to properly decompose. "Tertiary" (or advanced) treatment involves chemical methods such as chlorination, distillation, and depending on what the water is being used for, significant levels of chemical inputs to neutralize pollutants. These high-tech systems generate huge amounts of sludge as a by-product. While some of the treated sludge can be used to fertilize agricultural land, most sludge requires immediate disposal to avoid rapid accumulation. Sludge often needs to be treated as a hazardous waste due to contamination with heavy metals.

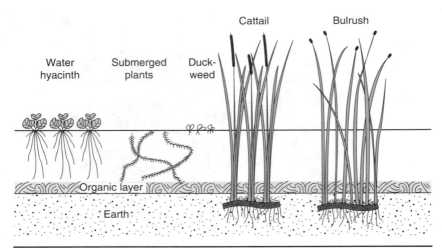

Figure 12.1   Aquatic plants used for wastewater treatment.[2]

In contrast, aquatic plant-based wastewater treatment systems commonly incorporate a stabilization pond, followed by aquatic plant culturing basins. A typical wastewater treatment system using aquatic plants is low tech in comparison to, using minimal chemicals and energy to treat sewage influent. In aquatic plant treatment methodology, the by-product can be used as animal feed, fertilizer, or fiber for paper manufacturing. It can also be composted anaerobically to produce methane gas or used as a feedstock for alcohol production. For industrial wastewater applications, where the objective is to remove or reduce the concentration of organic or inorganic compounds, many of which are toxic, the plants serve to concentrate the heavy metals for more convenient disposal or to recapture the metals possible[1] (Fig. 12.1).

Plants Used for Wastewater Treatment

There are several plants that have the potential for wastewater treatment (Table 12.1). Each has its own benefits and limitations. Plants used to treat raw wastewater and effluents from conventional treatment units need to have the following properties:

1. Rapid vegetative growth rates and productivity, especially when grown in waste enriched waters

2. High nutrient and mineral absorption capabilities

3. To be non-rooted to facilitate harvesting by mechanical means

4. Nutritive value and inorganic content for potential by-products and economic recovery for animal feed, compost, and methane production

Common Name	Scientific Name
Floating Aquatic Plants	
Water Hyacinth	Eichhornia crassipes
Duckweed	Lemna spp.
Lemnaceai Family	Wolffia spp.
Water Primrose	Ludwigia spp.
Water Lettuce	Pistia stratiotes
Water Fern	Salvinia spp.
Water Velvet	Azolla spp.
Alligator Weed	Alternanthera philoxerides
Submerged Aquatic Plants	
Pond Weeds	Potamogeton spp.
Oxygen Weed	Elodea spp.
Coontail	Ceratophyllum spp.
Watermilfoil	Myriophyllum spp.

Note: spp. is the standard abbreviation for "species" (plural). "Species"(singular) is abbreviated "sp."

TABLE 12.1   Potential Aquatic Plants with Wastewater Treatment Potential[3]

5. To be extremely hardy, even in raw sewage

6. To be resistance to disease and insects

The fact that the plants in Table 12.1 have these characteristics, and can be applied to their respective environments, serves to make these plants attractive for wastewater remediation.[4] Because of their performance under differing environments, two plants that have been studied at great length for treating wastewater are water hyacinths and duckweed.[5]

Water Hyacinth Use

The water hyacinth is probably the most intensively studied aquatic plant to date. Successful applications of aquatic plants include use at military facilities (Stennis NASA Space Center, St. Louis, MS) and residential developments (Cedar Lakes Development, North Biloxi, MS). As far back as 1948, research has investigated its biology[6] and use for removal of minerals and pollutants from municipal wastewater.[7]

Several environmental factors affect water hyacinth growth. The most significant is temperature. The optimum water temperature ranges from 68°F (20°C) to 86°F (30°C), with growth ceasing below 40°F (10°C) and above 104°F (40°C). Freezing temperatures will kill the leaves and

FIGURE 12.2   Water hyacinth in bloom.

stems of the plant above the water; however, the plant can regenerate from the rhizomes when weather conditions improve, providing the rhizomes are not frozen. A sustained air temperature lower than 28°F (−2.2°C), however, will usually kill the rhizomes. This generally restricts the area of uniform year-round plant growth to between 32° north and 32° south latitudes.

Water hyacinth use is limited since they are not native plants, having been transplanted from Brazil to North America. As such, they have no natural controls in North America. Consequently, they outcompete other plants and take over aquatic areas quickly. Due to their rapid growth and prolific nature, and tendency to clog slow moving waterways, introduction into the environment can be restricted by local environmental laws (Fig. 12.3).

In a small but growing number of communities (Table 12.2), the water hyacinth is developing an image as a beneficial plant, useful in absorbing and digesting wastewater pollutants and converting sewage effluents into relatively clean water. Thus, the hyacinth shows exciting promise as a natural water purification system, which can be established and maintained at a fraction of the cost of conventional wastewater treatment systems.

Floating aquatic plants provide wastewater treatment, but they change the dynamics of a facultative pond[9] that has algae as the only oxygen producer. In a facultative pond, there is an aerated surface layer that sits on top of an anaerobic layer. The top layer is naturally aerated by the diffusion of oxygen from the atmosphere and from the oxygen release of algae. When floating aquatic plants are placed on

Figure 12.3 Uncontrolled water hyacinth growth can clog waterways if not managed. (Zephyr Photo)

top of this system, they release oxygen above the surface of the water and block the diffusion of oxygen from the atmosphere. As a result, ponds with floating aquatic plants are oxygen deficient and the aerobic zone in the pond is found only near the plant roots (Kadlec and Knight, 1996). Consequently, when a pond is completely covered with water hyacinth, mechanical aeration may be needed to keep the anaerobic bacteria population, and the consequent odors, under control (Wolverton and McDonald, 1978).

To maintain the effectiveness of the wastewater treatment system, the pollution-gorged water hyacinth must be harvested at intervals. But this apparent drawback offers the potential as a money-making asset.

Location	Population	Capacity (Gallons/Day)	Notes
Coral Springs, Florida		150,000	
Lucedale, Mississippi			Methane generation included
Walt Disney World, Lake Buena Vista, Florida		100,000	Methane generation included
Hercules, Georgia		350,000	Has greenhouse cover as freeze protection for winter operation
San Diego, California	2 million	1,000,000	

Table 12.2 Installations of Water Hyacinth Water Sewage Treatment Facilities[8]

The plants can be harvested and dried and used as a component of animal feeds and natural agricultural fertilizers. They have recently been found to absorb a variety of toxins and heavy metals, which in some cases can later be reclaimed. Additionally, the stalks of the water hyacinth have come into use as fiber to make a tear-resistant paper.

When using water hyacinths as a wastewater treatment medium, it should be noted that they can fall victim to a catastrophic event, such as freezing, pests (e.g., coots (*Fulica americana*) and spider mites (*Bryobia praetiosa*)), or a pH imbalance. Thus, the entire population could be wiped out. A reduction of plant cover in a treatment pond can cause a serious decrease in its waste treatment effectiveness for weeks or even months while new plants establish themselves (Kadlec and Knight, 1996). Additionally, water hyacinths do not remove many types of microorganisms common in wastewater, such as fecal coliform bacteria.

Design Parameters for a Water Hyacinth Wastewater Treatment Plant

Wastewater Characteristics

For any aquatic plant–based system, the first requirement is to establish the characteristics of the influent wastewater and the required quality of the effluent. Influent water quality parameters of concern include BOD_5, total suspended solids, nitrogen and phosphorus, temperature, pH, dissolved oxygen, chlorides, and any toxic compounds (particularly important if wastewater is from industrial sources). The values obtained for these parameters should be compared against the range of tolerances for the aquatic plant being considered as the wastewater treatment medium.

Location, Climate, and Temperature

Water hyacinths should not be considered for areas where freezing conditions prevail during the winter, unless provisions are made to protect the plants from freezing or alternative treatment methods are available during the winter. Without such provisions, hyacinths should only be considered for year-round use in southern and coastal areas where temperatures are adequate for growth.

Water hyacinth systems in the United States that are currently treating wastewater are generally located in semitropical or warm temperature climates. The economic benefits of a system in a colder climate need to be investigated, taking into account the need for an enclosure to prevent freezing. The enclosures are typically constructed of rigid plastics or frame-mounted plastic sheets, and they need to provide access for harvesting and cleaning and preferably be removable for warm weather operation.

Site Requirements

The selection of a location for culture basin construction should be based on an evaluation of several site-specific variables. The availability of sufficient land area for the hyacinth basins and any related treatment and disposal systems is the prime consideration. Other factors to be considered are the accessibility of the site, hydrogeological characteristics, and the method and location of final effluent disposal.

Land Requirements

All of the water hyacinth wastewater treatment systems presently operating are less than 4 hectares (10 acres) in surface area, and the majority of the systems are less than 1 hectare (2.5 acres) in surface area.[10]

Many researchers have suggested that ponds ranging from 1 to 2 hectares per 1000 cubic meters/day (10–20 acres per mgd) will provide sufficient area for secondary treatment of primary effluent and advanced treatment of secondary effluents. For treating primary effluent to secondary or better quality, a pond surface area of approximately 1.6 hectares per 1000 cubic meters/day (15 acres per mgd) seems to work reasonably.[11] See a summary of the land requirements in Table 12.3.

Pre-Treatment

Although it is not always necessary, pre-treatment of the influent wastewater prior to application of the aquatic plants may be desirable or even necessary. Pre-treatment may consist of one or more processes, including aeration, screening, grinding, primary settling, and chemical treatment.

Treatment Level	Land Requirements Hectares/1000 m³/day	Land Requirements Acres/mgd
Secondary Treatment and Advanced Treatment of Secondary Effluents	1 to 2	10 to 20
Primary Effluent to Secondary or Better Quality	1.6	15
Polish Secondary Effluent	0.5	5
Nutrient Removal from Secondary Effluent	1.3	12

TABLE 12.3   Land Requirement Summary for Water Hyacinth Wastewater Treatment Systems

Post-Treatment

Post-treatment of the effluent from an aquatic plant culture basin may be necessary to meet the final effluent requirements prior to discharge. Post-treatment may include disinfection, filtration, sedimentation, or other processes. Aeration may be necessary to prevent odor problems resulting from anaerobic conditions in the hyacinth ponds. The addition of chemicals may be required to remove phosphorous remaining in the effluent.

Pond Size, Number, and Configuration

The considerations for pond size, number, and configuration are based on selected harvesting practices and routine maintenance. Optimal design of culture basins is necessary to minimize area requirements, and where potential freezing conditions may exist, the cost of a greenhouse to cover the pond should be considered. To facilitate harvesting, ponds that contain water hyacinths range from 16 to 48 in (0.4 to 1.2 m) deep. By contrast, ponds that harbor duckweed range from 48 to 71 in (1.2 to 1.8 m) in depth.

The physical characteristics of optimal hyacinth culture basins are indicated in Table 12.3.

Organic and Surface Loading Rates

The organic loading rates used for water hyacinth treatment systems are similar to those used for conventional stabilizations ponds. However, the effluent from the plant system may be better in quality than from a stabilization pond. The basic criteria used to design water hyacinth ponds for secondary treatment of primary effluents are the organic uptake rate and the pond surface loading rate. A design value of 5.0×10^{-4} kg BOD_5 per kg wet plant mass per day (5×10^{-4} lb BOD_5 per pound wet plant mass per day) is a recommended value to use.

With water hyacinths, BOD_5 loadings of up to 150 kg per ha per day (13 lb per acre per day) can be used, assuming an average standing crop of 225 metric tons per ha (111 tons per acre). With densely packed hyacinths and 100 percent coverage, surface loading rates of up to 225 kg BOD_5 per ha per day (200 lb BOD_5 per acre per day) may be used.[12] The recommended surface loading rate for ponds were at least 80 kg BOD_5 per ha per day (50 to 200 BOD_5 per acre per day), with a recommended design value of 140 kg BOD_5 per ha per day (125 lb BOD_5 per acre per day).[13]

In reviewing the current full-scale water hyacinth treatment facilities, organic loading rates of less than 30 kg/ha/day would provide satisfactory results when processing raw wastewater. Water hyacinth systems receiving secondary effluents or stabilization pond effluents are more numerous, and a much wider range of organic loading rates have been employed with these systems. Organic loading rates applied

Condition of Hyacinth Field	Loading Rate
Average Standing Crop of 225 metric tons/ha	150 kg BOD_5/ha/day
100% Coverage	225 kg BOD_5/ha/day
80% Coverage	56 to 225 kg BOD_5/ha/day Design Value: 140 kg BOD_5/ha/day
Processing Raw Wastewater	Less than 30 kg BOD_5/ha/day
First Basin in Hyacinth System	31 to 197 kg BOD_5/ha/day

TABLE 12.4   Organic Loading Rates for Water Hyacinth Treatment Systems

to the first basin in hyacinth systems have ranged from 31 kg/ha/day to 197 kg/ha/day. All of these systems produce an effluent which would satisfy the secondary standards of 30 mg/L for BOD_5 and suspended solids. Table 12.4 summarizes the various organic loading rates.[14]

Comparison of several water hyacinth systems indicated that there is a direct correlation between surface loading rates and nutrient removal efficiencies. Effluent requirements for nitrogen and phosphorus concentrations are used to determine the required removal efficiencies for these nutrients and thus the corresponding surface loading rates.

Hydraulic Residence (Detention) Time

A critical factor in aquatic plant pond design, with regard to treatment efficiencies for either secondary or tertiary systems, is the contact time with the root zone of the hyacinth. It is important to allow enough time for adequate sorption, filtration, and nutrient utilizations. Typical detention times for a secondary hyacinth system treating primary effluent range from six to seven days, plus an allowance for sludge storage and peak flows. In a tertiary system, minimum detention times range from one day for a surface loading of 0.2 ha per 1000 m^3/day (2 acres per mgd) to 5.5 days for a surface loading of 1.2 ha per 100 m^3/day (11 acres per mgd).

Hydraulic Loading Rates

Field results say that a hydraulic loading rate of 2000 m^3/ha/day, when treating secondary effluent, will produce an effluent quality that should satisfy advanced secondary standards (BOD_5 less than or equal to 10 mg/L, TSS less than or equal to 10 mg/L, TKN less than or equal to 5 mg/L, and TP less than or equal to 5 mg/L). With nutrient removal as the principal objective, a shallow pond (less than 4 m deep) and a hydraulic loading rate of approximately 500 m^3/ha/day should produce good nitrogen removals (less than or equal to 2 mg/L).

Approximately 50 percent reduction in the total phosphorous could be expected. A reasonable loading rate for hyacinth systems receiving raw wastewater was reported to be approximately 200 m³/ha/day, if nutrient controls was not an objective.

Duckweed Wastewater Treatment

There are probably as many as 100 duckweed sewage treatment plants throughout the world, some of enormous complexity. For effective operation, management has to be well informed about the biochemical and chemical changes owing to such variables as water temperature, light incidence, nutrient levels, and harvesting and replenishment rates, etc. While the necessary knowledge is unique in order to provide an effective water treatment system, it is not over-whelmingly difficult to learn.

The Lemnaceae family of duckweed occurs worldwide, but most diverse species appear in the subtropical or tropical areas. These readily grow in the summer months in temperate and cold regions. Duckweed is a small, fragile, free-floating aquatic plant that occurs in still or slowly moving water, (Fig. 12.4), and it will persist even on mud. Luxurious growth often occurs in sheltered small ponds and ditches or swamps where there are rich sources of nutrients. Duck-weed mats often occur naturally around slow-moving backwaters downstream from sewage works.

Duckweed grows at water temperatures between 6°C (43°F) and 33°C (91°F). However, growth is stressed with temperatures above 30°C (86°F) and is reduced with temperatures below 20°C (68°F). Duckweed survives at pHs between 5 and 9, but grows best over the

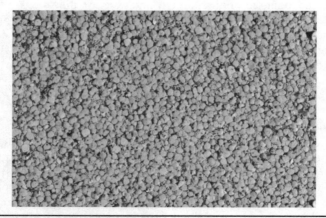

FIGURE 12.4 The Lemnaceae family of duckweed occurs worldwide, and can metabolize waste products into nutrients available for livestock feed. (See also Color Plates.)

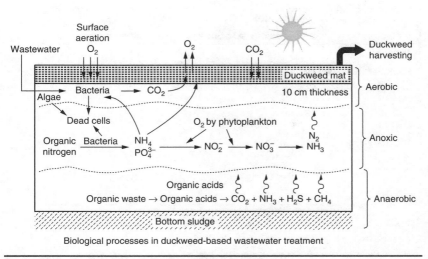

Biological processes in duckweed-based wastewater treatment

FIGURE 12.5 Duckweed biological process schematic.

range of 6.5–7.5.[15] If growing in its optimum range, it can double in volume in one to two days. For a schematic of the duckweed biological process, see Fig. 12.5.

Because duckweed grows so quickly and has a high level of nutrients, being 40 percent protein by weight, in addition to treating wastewater, it can serve as an excellent feed supplement for poultry, livestock, and fish, and can even be served in salads. Like many other plants, the nutritional value reflects the nutrient content of its environment. Duckweed grown in sewage wastewaters that are high in nutrients has been found to have nitrogen and phosphorus concentrations in the plant tissue that range from 5.2–7.2 percent and 1.1–2.8 percent respectively.[16] A study of duckweed grown in sewage, versus pond water without sewage, shows the potential profitable benefits of duckweed as a resource in addition to its water purification benefits (Table 12.5).[17]

Mineral Concentrations

Duckweeds appear to be able to concentrate many macro- and microminerals several hundredfold from water. On the other hand, high mineral levels can depress growth or eliminate duckweed, which grows best on fairly diluted mineral media. There is a mass of data on the uptake by duckweed of micro-elements, which can be accumulated to toxic levels (for animal feed). However, its ability to concentrate trace elements from very diluted media can be a major asset where duckweed is to be used as an animal feed supplement. Trace elements are often deficient in the major feed available to the livestock

(Grams/100 Grams Dry Weight)		
Element	Source #1: Sewage	Source #2: Pond
Nitrogen	5.94	4.43
Phosphorus	1.01	0.36
Potassium	2.13	1.22
Sodium	0.74	—
Calcium	0.88	1.34
Copper	1.41	0.31
Zinc	18.90	2.81
Iron	145.00	14.30
Sulfur	0.85	0.45

TABLE 12.5   Elemental Composition of Duckweed from Two Collection Sites

of small resource poor farmers. For example, in cattle fed mainly straw-based diets, both macro- and micro-mineral deficiencies are present. The ability of duckweed to accumulate minerals from wastewater has made its use as a livestock feed supplement a vital benefit to developing-world farmers that cannot afford the more sophisticated feed supplements available in developed countries.

For an installation to be successful, duckweed needs adequate nutrients and minerals to support growth. Growth is most often affected by the concentrations of ammonia, phosphorous, potassium, and sodium levels. Generally, slowly decaying plant materials release sufficient trace minerals to provide what is needed for growth.

Water Depth

Depth of water required to grow duckweed under warm conditions is minimal, but there is a major problem with shallow ponds in both cold and hot climates where the temperature can quickly move below or above optimum growth needs. However, to obtain a sufficiently high concentration of nutrients and to maintain low temperatures for prolonged optimal growth rate, a balance must be established between volume and surface area. Depth of water is dependent upon the primary purpose for growing duckweed. For agricultural use, anything greater than about 1.6 ft (0.5 m) poses problems for harvesting duckweed. Where water purification is a major objective, it is impractical to construct ponds shallower than about 6.6 ft (2 m) deep.

In practice, depth of water is probably set by the management needs rather than the pool of available nutrients, and harvesting is adjusted according to changes of growth rate, climate changes, and the nutrient flows into the system.

Wastewater Treatment

Urban wastewater treatment systems occur throughout the developing countries but service only a small percentage of the total population. Skillicorn, et al. (1993)[18] argue that duckweed-based wastewater treatment systems provide genuine solutions to the problems of urban and rural human waste management with simple infrastructures and at low cost. For the amount of degradation for phosphorus, sodium, nitrogen, and potassium in the wastewater stream, see Fig. 12.6.

Duckweed wastewater treatment systems are based on stand-alone lagoons. A single or a series of lagoons may be used depending on the size of the treatment plant. The settling tanks need to be dug out once in a while, and two tanks are often needed so that while one is cleaned the other is in use. Following the sedimentation tank is a series of duckweed ponds and, depending on the ultimate water use, possibly a "polishing pond." In the latter pond, sunlight largely removes any pathogens that remain in the water. Generally, these systems require about a 30-day turnover rate of water to be sure of minimum mineral contamination and low bacterial counts in water leaving the works.

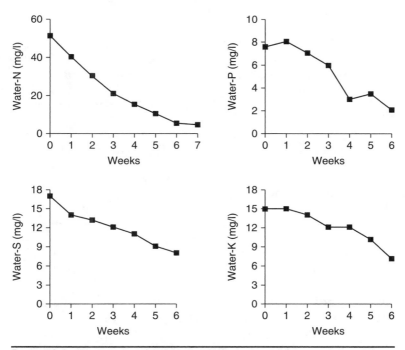

FIGURE 12.6   The uptake of S, P, K, and N by duckweed from sewage water. Duckweed was harvested weekly by placing a piece of wood across the diameter of the tanks and removing half the duckweed.[19]

In most modern sewage works, the ammonia levels have been reduced by a combination of microbial treatment methods. Usually for these systems to effectively grow duckweed, either the de-nitrification step in the treatment works needs to be bypassed, or urea must be applied to provide ammonia so that duckweed growth is vigorous enough to remove the residual phosphorus and other minerals. Even where no de-nitrification is brought about, fertilization with urea in some of the downstream ponds may be necessary to capture as much phosphorous as possible. The potential efficiency of a duckweed treatment plant can be gauged by the fact that effluent water from such an operation can be lower in ammonia, phosphorous, and lower biochemical oxygen demand and turbidity than required by U.S. standards for the Washington D.C. area.

Example

A two-cell lagoon system was installed in the 1970s in Biloxi, Mississippi, using duckweed as the water treatment medium. The mean influent going into the plant is 14,000 gal/day (52.8 m^3/day) from 51 homes.

The two cells operate in series. The first cell has a surface area of 8930 ft^3 (830 m^2) with a depth of 8 ft (2.4 m). The second cell has a surface area of 8070 ft^3 (750 m^3) and a depth of 5 ft (1.5 m). Detention time is 36 days for cell one and 21 days for cell two.

The first cell, principally used for the settlement of solids, utilizes anaerobic digestion to break down the solids. To maintain aerobic conditions at the surface for odor control is a floating 5 hp aerator. Since the natural respiration of the duckweed provides adequate oxygen during daylight, considerable energy savings have been achieved by only needing to run the aerator during night operations.

The second cell, principally used for the removal of suspended solids, contains the duckweed mixture. Due to the thick mat that is created that forms a natural barrier to sunlight, the algae dies off and settles to the bottom of the tank. Water leaving the tank has a yearly mean total suspended solid (TSS) of 18 and a five day biochemical oxygen demand (BOD$_5$) of 15 mg/liter. Periodically, the duckweed produced from this operation is harvested and mixed for use as livestock feed.

At discharge, the effluent flows over a 3 ft (0.9 m) drop to further aerate the water to increase the oxygen level and reduce odor.

Recovery of Biomass

There are thousands of hectares of derelict ponds polluted to eutrophication levels in Bangladesh alone, which could potentially be cleansed of much of their pollutants and resurrected for aquaculture and fish farming at the family farm level. In these systems, the objective would be largely to provide protein of high biological value for the family of small farmers, who often have no animal protein in their diets. To resurrect derelict ponds, the approach might be to first establish duckweed aquaculture as a source of nutrients for terrestrial crop production (e.g., mulches and organic fertilizer). The pond's oxygen levels rise with harvesting of the crop, and fish farming may be introduced either in part of the pond or in adjacent clean water ponds.

A further interesting approach would be to create a market for duckweed locally, as is the case presently in Vietnam, in order to encourage duckweed aquaculture as a cash crop. Undoubtedly, a cash flow from such a market would then stimulate village people to clean up the huge number of polluted ponds. In this case, duckweed collection centers may be established to either sell duckweed directly or after drying for pig, duck, poultry, or even ruminant production through local outlets or to blend duckweed for use in compounded feed (Fig 12.7). The latter is largely imported at great cost in third world countries.[20]

Because proper operation of an aquatic plant–based sewage treatment system requires that there be periodic harvesting of the aquatic plants, there is a possible means of offsetting the cost of wastewater treatment. With intensive harvesting, it is necessary to construct the ponds so that harvesting can be easily accomplished. Resource recovery, however, should be considered carefully, as its inclusion will interfere with the performance or reliability of the aquatic plant systems. Complete removal of the plants will negate the sewage remediation process.

The following is a case history of a sewage treatment facility.

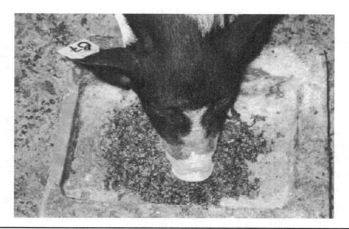

Figure 12.7 Duckweed mixed with feed for animal production.

Natural Treatment for Sewage Treatment Facility

by Dave Maddux

Planting crew in cell 1, planting cattail in mid-June.

Introduction

Talkeetna, Alaska, is a small, unincorporated community of approximately 350 year-round residents and is part of the Matanuska-Susitna Borough located in South Central Alaska. It is about 115 road miles north of Anchorage and about 14 miles off the Parks Highway at the end of the Talkeetna Spur Road. In 1989, a public sewer system and two-stage lagoon facility were constructed to serve the core area of the community. At the time of the proposed improvements described in this report, the system consisted of a force main sewage collection system, which delivered the collected effluent to a lagoon system for treatment and final disposal. The treatment facility consisted of two holding cells and a percolation cell. Approximately 40,000 gal (151 m³) per day of sewage was pumped to the lagoon facility located adjacent to a slough on the Talkeetna River and disposed of through the percolation pit. The stored effluent in the holding cells was typically transferred to the percolation pit twice a year—just after spring breakup and before the onset of winter. Nearly 9 million gal (34,000 m³) gallons of effluent percolated as a slug flow through a thin layer of soil into the shallow groundwater table.

Looking downstream from cell 1, 30 days after planting.

In October 2001, the borough was given a Notice of Violation by the Alaska Department of Environmental Conservation (ADEC) for surpassing water quality discharge standards. Subsequently, the Matanuska-Susitana Borough let an RFP for an upgrade to the system. The upgrade to the town's water and sewer system was funded by the USDA and administered by the Matanuska-Susitna Borough.

The entire project included improving and modifying the existing lagoon flow control system, converting the percolation pit into a third lagoon, installing a new lift station and modifying two existing lift stations, building a constructed wetland treatment system, and camera inspection and cleaning of the existing force main. In addition, a previously nonexistent SCADA monitoring system was installed for remote monitoring and control of all pumps and controls in the lift stations. The prime contractor was Construction Unlimited, and the landscaping contractor was Evergreen Landscaping, both based in Anchorage, Alaska.

After a competitive bid process, a group consisting of Bob Gilfilian, P.E., of Gilfilian Engineering, Dr. Dave Maddux of Applied Wetlands Technology, and Mark Sherman, P.E., of ASCG, Inc., were awarded a contract to design and oversee construction of a surface-flow constructed wetland. This system was a replacement for the percolation pit in the final treatment of effluent before its discharge to the receiving waters of the Talkeetna Slough. The following discussion deals only with the constructed wetland portion of the project.

Benefits of Using a Constructed Wetland

Once the DEC issued a Notice of Violation, it was critical that the problems be remedied as soon as possible. Since Talkeetna is situated in a pristine environment and is considered by many to be the "Gateway to Denali," the town wanted a system that mirrored the natural setting found around the community at the foot of Mt. Denali. Another consideration was the cost of operating and maintaining improvements to the current system.

Selection of a constructed wetland treatment system provided several benefits. When construction was completed, operation and maintenance costs were estimated at $2000 per year, effluent discharge met DEC standards, and the treatment system created an attractive green space that complemented its natural environment.

The cost of the constructed wetland portion was approximately $440,000, which equates to $13 per square foot of constructed wetland surface area. This facility is capable of treating 14.6 million gal (55,300 m^3) of effluent per 145-day treatment season, which easily meets the current needs of the town and allows for future expansion.

Looking from cell 1 at cattails after two months of effluent flow.

Discharge Water Quality Requirements

Prior to construction of the new lagoon and constructed wetland, this facility discharged effluent to a groundwater source and not a surface-water source. As such, water quality criteria was different for percolation pit disposal methods. The Notice of Violation given to the borough prior to new construction related to sludge buildup in the lagoons, percolation cell performance, and groundwater quality issues. Switching to a surface-water discharge was intended to alleviate those issues.

The ADEC sets the discharge permits for treated sewage wastewater discharge to receiving waters in Alaska. There are three main requirements:

1. Reduction of fecal coliforms to a 30-day average of 20 cfu/100 ml
2. Five-day BOD concentration maximum of 65 mg/l
3. A total suspended solids maximum of 70 mg/l (Dissolved oxygen has a minimum of 7 mg/l and a maximum of 17 mg/l.)

Cells 1 and 2 after heavy rain mid-June.

Climatic Conditions Are a Challenge in Alaska

A problem that Alaska faces in using constructed wetlands for wastewater treatment is the shortened treatment season caused by long, cold winters with deep frost penetration. The average winter temperature in Talkeetna is +10.4°F (−12°C), and winter darkness lasts for approximately seven months. This results in a maximum treatment season of 145 days. Ideally, a seasonal storage capacity large enough to hold an entire year's worth of effluent will provide a safety margin to allow this type of system to work in an arctic environment.

Constructed Wetland Design Overview

The existing side-by-side lagoons had surface areas of 2.2 acres each and a working volume of 3.725 million gal (1,400 m³) each. The conversion of the percolation pit to a holding lagoon increased the surface area by 1.1 acres and the holding volume by 1.935 million gal (7,300 m³) for a total holding volume of 9.385 million gal (35,600 m³).

Once the effluent is delivered to the treatment facility, the flow process is a simple gravity-fed discharge from one lagoon to the next. The pumped effluent flows into lagoon #1 and travels the length of the lagoon before being discharged through gravity flow to lagoon #2. The effluent then travels the entire length of lagoon #2 before being discharged through gravity flow into lagoon #3. Finally, the effluent is discharged seasonally to the constructed wetland through a buried 4-in (10.2 cm) HDPE line by opening mechanical distribution header valves.

The constructed wetland is a continuous system comprised of six cells with an operating depth of 12 in (0.3 m). Each cell is separated from the following cell by a deep water trench that is 4 ft (1.2 m) deep and acts as a flow redistribution zone. This zone allows for remixing of the effluent, maximizing an even flow across the entire width of the cell and minimizing the potential for channels (short circuiting) developing through the wetland.

This free-water surface flow system has a surface area of 35,000 ft^2 (3252 m^2) and a volume of 2,618,000 gal (9,849 m^3). Current flow-through is 395 m^3/day (105,000 gpd) with a theoretical hydraulic detention time of 1.86 days. Discharged effluent flows through a V-notched weir to an 8 in (20.4 cm) HDPE buried line that discharges to the Talkeetna River slough, an anadromous fish stream.

Plant Selection

Macrophytes chosen for planting in the constructed wetland were based on four criteria. They had to be indigenous to the area, have an ability to colonize rapidly, be able to withstand high pollutant loads of ammonia, and provide a large surface area for periphyton attachment. Based on these criteria, six species were selected for the project. Cell 1 was planted with *Typha latifolia* (broad-leafed cattail), cell 2 with *Scirpus validus* (soft-stemmed bulrush), cell 3 with *Carex utricularia* (common sedge), cell 4 with *Calla palustris* (calla lily), cell 5 with *Carex aqautilis* (blue-green sedge), and cell 6 with *Carex utricularia* (common sedge).

It was expected that cells 1 and 2 would be subjected to the highest loads of ammonia. Cattails and bulrush were placed in these cells because both species have shown a remarkable ability to thrive while withstanding high pollutant loads that include ammonia. These species also produce a significant volume of biomass, which provides an available carbon source to microbes in the substrate.

Placement of the common sedge in cell 3 was based on the known ability of the sedge to rapidly colonize the open areas and provide a large surface area for periphyton attachment. Both species of sedge used in this constructed wetland do well in polluted water of medium to low concentrations. Cell 3 is far enough along the treatment train that any high concentrations of ammonia should be reduced to levels that will not adversely affect the plants.

Cell 4 was planted with calla lily, which is primarily a floating plant. Calla is slightly rooted at its base in the substrate and sends out thick, floating stems sometimes 10 ft (3.0 m) in length, with large leaves. The remarkable attribute of this plant is the huge amount of subsurface aerial root production along each stem. This species has more biomass in the roots floating in the water column than in the leaves and stems above the waterline. Not only does the plant supply a thick mat of vegetation below the water surface to filter out suspended solids, it also produces a large volume of leaves that decompose rapidly upon senescence and add to the detrital, vegetative mat. Furthermore, since it spreads out and covers the entire water surface, it minimizes the oxygen transfer between the atmosphere and the water surface, providing a near perfect environment for denitrification to occur.

Cells 5 and 6 were planted with blue-green sedge and common sedge respectively, both of which can be found growing along the banks of the receiving water. These two species will grow thick, luxuriant stands of plants, which will effectively filter out seeds and vegetative propagules from the cattail, bulrush, and calla lily. Although these three species are indigenous to the state, they are not indigenous to the local area, and spreading of the species by accident is not desired. These two species of sedge provide final polishing of the effluent before its discharge to the receiving waters.

Site Preparation for the Constructed Wetland

Before construction, the site immediately surrounding the lagoons was a flood plain forest consisting of birch, alder, and spruce trees with a thick tangle of sword ferns, devils club, and a variety of grasses. All vegetation was stripped adjacent to the existing lagoon structures, and a trench was excavated with a bottom width of 32 ft (9.75 m), a length of 1000 ft (305 m), and side slopes of 3:1. After the grade was established and compaction was achieved by roller to 95 percent, a nonwoven, needle-punched polypropylene fabric was placed on top of the compacted soil. A 30-ml polypropylene liner was placed on top of that, then a final layer of geotextile was laid down.

A layer of topsoil 12-in (0.3 m) thick was placed on the liners on the bottom and side slopes of the cells. The soil was manufactured to a minimum specification organic content of 10 percent and a maximum of 20 percent by dry weight of finely chopped, well-mixed organic materials. To keep clumps and rocks out of the mix, the nonorganic portion of the planting soil was required to be less than 0.1 in (2 mm) in size.

Once topsoil was in place, the planting crew began laying out the various planting grids and then using hand-held pluggers to remove soil plugs for planting purposes. Spacing for the various plants was either 12-in, 18-in (0.45 m), or 24-in (0.61 m) centers depending on the species being used.

Plants and the Planting Process

All the plants except calla lily were purchased from a wetland nursery in Montana. The calla was harvested from a lake in Fairbanks and transported by truck to the site the same day they were harvested. The other plants arrived by air and were held until they were needed in a pool designed specifically for wetland plants. Both the cattail and bulrush were planted on 18-in centers requiring approximately 1,760 individual plants of each species. Both species of sedge were planted on 12-in centers, requiring approximately 11,800 plants, and the calla lily was planted on 24-in centers requiring approximately 984 plants.

The planting process was the same for all plants except the calla lily. For the cattail, bulrush, and sedge, a hole was punched in the soil, the soil plug was removed, then approximately two cups of diluted Humi-Zyme®, a liquid fertilizer, was poured into the hole. The plant plug was inserted into the hole, tamped down with some soil, and a final watering of diluted liquid fertilizer was applied. Since the calla lily was harvested by hand from a wild stock, it was not rooted in a tube or pot. Each plant consisted of rhizome-like stalks approximately 18- to 24-in (0.45 to 0.61 m) long with substantial aerial roots along each rhizome. These plants were simply placed on the substrate with a shovel-load of soil placed on the end of the rhizome structure and then tamped down by foot.

After each cell was planted, it was thoroughly watered but not flooded. Once all the plants were in, the cells were flooded to a depth of 1 in (2.5 cm) for 10 days, then the flooding depth was raised to 6 in (0.15 m) for 20 days. The depth was then raised to the operating depth of 12 in (0.30 m) for the following 30 days. No effluent flowed into the constructed wetland during this 60-day acclimation period.

Project Timeline

Construction began on the project in the spring of 2003 with the macrophyte planting taking place in mid-June 2003. Following a 60-day transplant acclimation period, effluent flow through the constructed wetland (cw) began in mid-August 2003. Initial reduction of pollutants was exceptional. Following a six-week discharge period, the flow was shut down for the winter. The following year was the first full treatment season for the constructed wetland and, as can be expected for initial operating seasons, water quality results were mixed.

Discharge structure from lagoon to head of constructed wetland.

Water Quality Results for the First Operating Season

The three main discharge parameters the system is required to meet are total suspended solids (TSS) of 70 mg/L, biological oxygen demand (BOD) of 65 mg/L and fecal coliforms (FC) of 20 cfu/100 ml. The first two parameters were well within the required reduction rates for all sampling events during 2004; however, the FC met discharge standards only once out of four sampling events during 2004.

The effluent exiting the lagoons to the constructed wetland inlet was low in TSS and BOD and the discharge concentrations to the receiving waters averaged well below the ADEC discharge standards. Fecal coliform concentrations exiting the lagoon to the constructed wetland inlet were not as high as was expected except for two sampling events in the fall when concentrations were over 28,000 cfu/100 ml. The FC discharge concentrations to the receiving waters for three sampling events in 2004 were well above the desired ADEC water quality standards.

Outlet of constructed wetland receiving water at full flow.

Problem Areas

Constructed wetland systems usually require a large volume of plants for initial planting purposes. However, this system is small compared to some systems in the lower-48, and only 16,300 plants were required. This still presented a problem because Alaska has no commercial wetland nurseries and although commercial growers in the state were contacted two years prior to the project, none expressed an interest in the wetland plant nursery business. Therefore, most of the plants were purchased from a wetland plant nursery in Montana and air-freighted to Alaska, which added considerable cost to the planting of the constructed wetland.

Another problem that was not anticipated was migration of one of the species due to wind. The calla lily has large leaves which protrude like a flag on the surface of the water. Three days after flooding the cells to 12 in (0.3 m), a stiff wind caught the large leaves like sails and pulled the plants from their mooring, blowing most of them into the upwind end of the cell. This left a large amount of open water available to waterfowl, which can have the effect of increasing the fecal coliform and BOD concentrations in the water.

A drawback from the client's point of view is that constructed wetlands do not show their full potential for pollutant removal until about three years of growth have passed. Unlike engineered systems which are expected to work as soon as they are turned on, biological systems take time to mature. Full plant colonization that allows the various plant and microbe communities to establish themselves and fill all the required niches for efficient pollutant removal takes several seasons.

Lessons Learned

One of the problems this project faced was the very short timeline for construction and implementation of the constructed wetland system. Even a delay of two weeks can spell disaster for a project here in Alaska because the construction season is so short. Allowing a newly established biological system a season of growth before it receives effluent is ideal. However, this project was under pressure to complete planting the constructed wetland in time for it to be functional and ready to receive effluent the same construction season. Therefore, the minimum time allowable for establishment of the plants prior to effluent flow was determined to be 60 days. By mid-August, that establishment period had been reached and effluent flow began.

Part of the performance of the landscaping effort was a guarantee that 90 percent of each species would survive transplant for a period of 60 days. For the most part, this occurred, but two areas of plantings for one species in particular were adversely affected and had to be replanted by the contractor. As of the termination of the 2004 season, those replanted areas still had not reestablished themselves. Either that particular species had a problem with the seed stock used by the greenhouse, or those areas where the planting did not survive had soil that was toxic in some way. Planting procedures followed by the contractor were identical to the successful plantings, so no fault could be found with the contractor.

Acquiring all the necessary permits was somewhat complex because the Talkeetna River Slough is designated an anadromous fish stream. As such, the Alaska Department of Fish and Game became involved in the permitting process, along with the ADEC and the Army Corps of Engineers. Coordination and negotiation with all the agencies involved caused some problems which may have been avoided with more comprehensive permit planning.

Under a competitive bid process, we were unable to select the contractor of our choice, and there was some concern regarding unknown performance. Additionally, this was the first commercial constructed wetland built in Alaska, and therefore no Alaskan contractor had specific experience with these systems. However, we worked with an excellent general contractor, construction crew, and landscape contractor, who all made a potentially difficult project into a straightforward one. The construction schedule for this project worked well, and there were very few changes I would make for a similar project. *For more information, contact Dave Maddux, Ph.D., Wetland Ecologist, Applied Wetlands Technology, P.O. Box 81091, Fairbanks, Alaska 99709, e-mail: davemaddux@wetlandsoptions.com.* ©2004, 1998 Land and Water, Inc.[21]

Conclusion

The use of aquatic plants presents a low-cost way to treat polluted water and serves to preserve the overall quality of the watershed. Several plant species, other than duckweed and water hyacinths, have been used successfully and are worthy of consideration for aquatic plant–based waste remediation systems.

An important benefit to using aquatic plants is that harvested plants, after absorbing the pollutants from the sewage effluent, offer the potential of creating secondary markets for their constructive use. The production of feed and food is a possibility, which would provide for people and their livestock. Processing and product development could evolve from the chemical extraction of protein and perhaps other chemicals including insecticides.

The result of aquatic plant use for sewage water remediation would be overall healthier conditions, particularly for the poorer countries, where energy required, funds needed, and the technical knowledge to maintain more elaborate sewage treatment facilities are not available.[21]

Notes and References

1. Hyde, H., Ross, R., Sturmer, L., *Technology Assessment of Aquaculture Systems for Municipal Wastewater Treatment*, Municipal Environmental Research Laboratory, U.S. Environmental Protection Agency, Cincinnati, Ohio, August 1984.
2. *Constructed Wetlands and Aquatic Plant Systems for Municipal Wastewater Treatment Design Manual*, EPA/625/1-88/022, U.S. Environmental Protection Agency Office of Research and Development, Center for Environmental Research Information, Cincinnati, OH, September 1988.
3. Wolverton, B.C., McDonald, R.C., *Secondary Domestic Wastewater Treatment Using a Combination of Duckweed and Natural Processes*, National Aeronautics and Space Administration Earth Resources Laboratory, NSTL Station, Mississippi, July 1980.
4. Hyde, H., Ross, R., Sturmer, L., *Technology Assessment of Aquaculture Systems for Municipal Wastewater Treatment*.
5. Hyde, H., Ross, R., Sturmer, L., *Technology Assessment of Aquaculture Systems for Municipal Wastewater Treatment*.

6. Penfound, W.T., Earle, T.T., *The Biology of the Water Hyacinth*, Ecological Monographs, 1948.
7. A.C. Robinson et al., *An Analysis of the Market Potential of Water Hyacinth Based Systems for Municipal Wastewater Treatment*, Report No. BCL-OA-TFR-76-5, Battelle Laboratories, Columbus, Ohio, 1976.
8. Stennis Space Center Article, Innovative Partnership Program, "Success at Stennis." (http://technology.ssc.nasa.gov/suc_stennis_water.html accessed 3 June 2012)
9. Facultative pond: The most common type of waste stabilization pond in current use, the facultative pond is designed to remove organic contaminants by natural biodegradation. The upper portion of the pond is aerobic, while the bottom layer is anaerobic, which promotes nitrogen removal. Algae supply most of the oxygen to the upper portion (WHO 2006). Waste stabilization ponds have proved to be a low-cost, sustainable method of wastewater treatment, particularly suited to the socioeconomic and climatic conditions prevailing in many developing countries. No input of external energy or disinfectants is needed (WHO 2006).
10. Middlebrooks, J.E., "Aquatic Plant Processes Assessment," in *Aquaculture Systems for Wastewater Treatment, an Engineering Assessment*, (EPA 430/9-80-007, MCD 68), U.S. Environmental Protection Agency, pp. 43–62, 1980.
11. Reed, S., Bastian, R., Jewell, W., "Engineering Assessment of Aquaculture Systems for Wastewater Treatment: An Overview," *Aquaculture Systems for Wastewater Treatment, an Engineering Assessment* (EPA 430/9-80-007, MCD 68), U.S. Environmental Protection Agency, pp. 1–12, 1980.
12. *Water Hyacinth Wastewater Treatment Design Manual for NASA/National Space Technology Lab*, p. 92, Gee & Johnson Engineers, West Palm Beach Florida, 1980.
13. *Water Hyacinth Wastewater Treatment Design Manual for NASA/National Space Technology Lab*, p. 92, Gee & Johnson Engineers West Palm Beach Florida, 1980.
14. Wolverton, B.C., "Engineering Design Data for Small Vascular Aquatic Plant Wastewater Treatment Systems," Aquaculture Systems for Wastewater Treatment Seminar Proceedings, U.S. Environmental Protection Agency (EPA430-9-80-006), 1979.
15. Leng, R.A., *Duckweed: A Tiny Aquatic Plant with Enormous Potential for Agriculture and Environment*, Animal Production and Health Div.; University of Tropical Agriculture Foundation, Phnom Penh, Cambodia. 1999 *Duckweed*
16. Culley, Jr, D.C., Gholson, J.H., Chisholm, T.W., Stardifer, L.C., Epps, E.A., *Water Quality Renovation of Animal Waste Lagoons Utilizing Aquatic Plants*, Environmental Protection Technology Series, (EPA-600/2-78-153,149), 1978.
17. Wolverton, B.C., McDonald, R.C., *Secondary Domestic Wastewater Treatment Using a Combination of Duckweed and Natural Processes*.
18. Skillicorn, P., Spira, W., Journey, W., *Duckweed Aquaculture—A New Aquatic Farming System for Developing Countries*, p. 76, The World Bank, Washington, D.C., 1993.
19. Stambolie, J.H. & Leng, R.A. 1994. Unpublished observation U.N.E. Armidale NSW, Australia
20. *Duckweed–a potential high-protein feed resource for domestic animals and fish* R A Leng, J H Stambolie and R Bell , Livestock Research and Rural Development, 7(1), October 1995, Centre for Duckweed Research & Development University of New England Armidale, *NSW 2351*
21. Published with permission of *Land and Water Magazine*, 1994. (www.landandwater.com)

CHAPTER 13

Biological Filter and Constructed Wetland Systems

Introduction

Purification of water by biological filters is a process that is as old as the earth, but researchers have been slow to experiment with the treatment of wastewater by biological filters. The design and operation of a biological filter wastewater treatment system is similar to aquatic-based plant based systems in that both use the benefits of plants, but with some subtle differences. Specifically, with a biological filter–based system, there is no surface water exposed to the digestion process. Rather the effluent goes below grade and is biologically digested in much the same manner as a septic field, but with plants incorporated to improve performance.

Discussion

The idea behind all biological methods of wastewater treatment is to introduce contact with bacteria (cells), which feed on the organic materials in the wastewater, thereby reducing its BOD (biochemical oxygen demand) content. In other words, the purpose of biological treatment is BOD reduction. Biological filters are generally applied for secondary treatment applications, which is treatment of the effluent after primary settling and before final discharge.

Typically, wastewater enters the treatment plant with a BOD higher than 200 mg/L, but primary settling has already reduced it to about 150 mg/L by the time it enters the biological component of the system. It needs to exit with a BOD content no higher than about 20–30 mg/L, so that after dilution in the nearby receiving water body (river, lake, etc.), the BOD is less than 2–3 mg/L. Thus, the biological treatment needs to accomplish a sixfold decrease in BOD.

How This Happens

Simple bacteria (cells) eat the organic material present in the waste-water. Through their metabolism, the organic material is transformed into cellular mass, which is no longer in solution but can be precipitated at the bottom of a settling tank or retained as slime on solid surfaces or vegetation in the system. The water exiting the system is then much clearer than when it entered. A key factor in the operation of any biological system is an adequate supply of oxygen. Like humans, cells need not only organic material as food but also oxygen to breathe. Without an adequate supply of oxygen, the biological degradation of the waste is slowed down, thereby requiring a longer residency time of the water in the system. For a given flow rate of water to be treated, this translates into a system with a larger volume that needs more space.

Like all biological systems, operation takes place at ambient temperature. There is no need to heat or cool the water, which saves on energy consumption. Because wastewater treatment operations take much space, they are located outdoors, and this implies that the system must be able to operate at seasonally varying temperatures. Cells come in a mix of many types, and accommodation to a temperature change is simply accomplished by self-adaptation of the cell population.

Filter Types

There are three general types of systems that use the biological filter process to one degree or another as a waste management process: mechanical (commonly a trickling filter), evapo-transpiration (ET), and evapo-transpiration infiltration (ETI). These systems are used individually or in hybrid combination.

Trickling Filter

Trickling filters (TFs) are commonly used to remove organic matter from wastewater for small- to medium-sized communities. The TF is an aerobic treatment system that utilizes microorganisms attached to a medium to remove organic matter from wastewater. This type of mechanical system is common to a number of technologies, such as rotating biological contactors and packed bed reactors (biotowers). These systems are known as attached-growth processes. In contrast, systems in which microorganisms are sustained in a liquid are known as suspended-growth processes.

TFs enable organic material in the wastewater to be adsorbed by a population of microorganisms (aerobic, anaerobic, and facultative bacteria; fungi; algae; and protozoa) that are attached to the medium as a biological film or slime layer (approximately 0.1 to 0.2 mm thick). As the wastewater flows over the medium, microorganisms already in the water gradually attach themselves to the rock, slag, or plastic surface and form a film. The organic material is then degraded by the aerobic microorganisms in the outer part of the slime layer.

As the layer thickens, through microbial growth, oxygen cannot penetrate the medium face, and anaerobic organisms develop. As the biological film continues to grow, the microorganisms near the surface lose their ability to cling to the medium, and a portion of the slime layer falls off the filter. This process is known as *sloughing*. The sloughed solids are picked up by the under-drain system and transported to a clarifier for removal from the wastewater.

Trickling Filter Design

A TF consists of permeable medium made of a bed of fist-sized rocks, slag, or plastic, over which wastewater is distributed to trickle through, as shown in Fig. 13.1. Rock or slag beds can be up to 3 to 8 ft (0.9 to 2.4 m) in diameter and 3 to 8 ft (0.9 to 2.4 m) deep with rock size varying

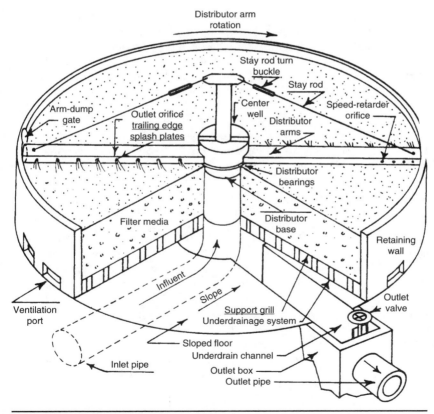

FIGURE 13.1 Typical trickling filter configuration.

from 1 to 4 in (2.5 to 10.2 cm). Most rock media provide approximately 15 ft²/ft³ (149 m²/m³) of surface area with less than 40 percent void space. On the other hand, packed plastic filters (biotowers) are smaller in diameter (20 to 40 ft (4.3 to 12.2 m)) and range in depth from 14 to 40 ft (4.3 to 12.2 m).

These filters look more like towers, with the media in various configurations (e.g., vertical flow, cross flow, or various random packings). Research has shown that cross-flow media may offer better flow distribution than other media, especially at low organic loads. When comparing vertical media with the 60-degree cross-flow media, the former provide a nearly equal distribution of wastewater, minimizing potential plugging at higher organic loads better than cross-flow media. The plastic media also require additional provisions, including ultraviolet protective additives on the top layer of the plastic medium filter and increased wall thickness for the plastic medium packs installed in the lower section of the filter where the loads increase.

After primary settling, the treated wastewater and solids are piped to the trickling filter bed. The effluent is gently sprayed by a rotating arm. Slime (fungi, algae) develops on the rock surface, growing by intercepting organic material from the water as it trickles down. As the water layer passes over the rocks in thin sheets, there is good contact with air, and the cells are effectively oxygenated. Worms and insects living in this "ecosystem" also contribute to removal of organic material from the water. The slime periodically slides off the rocks and is collected at the bottom of the system, where it is removed as sludge.

Water needs to be trickled several times over the rocks before it is sufficiently cleaned. Multiple spraying also provides a way to keep the biological slimes from drying out in hours of low-flow conditions (e.g., at night). Plastic nets are gradually replacing rocks in newer versions of this system, providing more surface area per volume, thereby reducing the size of the equipment.

The design of a TF system for wastewater also includes a distribution system. Rotary hydraulic distribution is usually standard for this process, but fixed nozzle distributors are also being used in square or rectangular reactors. Overall, fixed nozzle distributors are being limited to small facilities and package plants. Recently, some distributors have been equipped with motorized units to control their speed. Distributors can be set up to be mechanically driven at all times or during stalled conditions. In addition, a TF has an underdrain system that collects the filtrate and solids, and it also serves as a source of air for the microorganisms on the filter.

It is essential that sufficient air be available for the successful operation of the TF. It has been found that to supply air to the system, natural draft and wind forces are usually sufficient if large enough ventilation ports are provided at the bottom of the filter and the medium has enough void area.

The following four basic categories of filter design are based on the organic loading of the trickling filter:

1. Low-rate filters: These are commonly used for loadings of less than 40 kilograms of five day biochemical oxygen demand (BOD_5/100 m3/day 25 lb BOD/1000 ft³/day). These systems have fewer problems than other filters with regards to filter flies, odors, and medium plugging because of the lower loading rate. Low-rate filters with a rock medium range in depth from 3 to 8 ft (0.9 to 2.4 m). Most low-rate filters are circular with rotary distributors, but some filters currently in use are rectangular. Both of these configurations are equipped with dosing siphons or periodic pumps to provide a high wetting rate for short intervals between rest periods. A minimum wetting rate of 0.7 gal/ft²/min (0.4 L/m²/sec) is maintained to prevent the high-rate plastic filter medium from drying out. With a rock medium, the filters tend not to be hydraulically limited and have application limits ranging from 0.02 to 0.06 gal/ft²/min (0.01 to 0.04 L/m²/sec).

   The sloughed solids from a low-rate filter are generally well digested, and as a result, these filters yield less solids than higher-rate filters. Secondary quality effluent is readily achievable if the low-rate trickling filter design incorporates filter media with bio-flocculation capabilities or good secondary clarification.

2. Intermediate-rate filters: These filters can be loaded up to 64 kg BOD/100 m³/day (40 lb BOD/1000 ft³/day). In order to ensure good distribution and thorough blending of the filter and secondary effluent, the system should recirculate the trickling filter effluent. The biological solids that slough from an intermediate trickling filter are not as well digested as those using a low-rate filter.

3. High-rate filters: These are generally loaded at the maximum organic loading capabilities of the filter and receive total BOD_5 loading ranging from 64 to 160 kg BOD/100 m³/day (40 to 100 lb BOD/1000 ft³/day). Achieving a secondary quality effluent is less likely for a high-rate filter without a second stage process. As a result, high-rate filters are often used with combined processes.

4. Roughing filters: These are designed to allow a significant amount of soluble BOD_5 to bleed through the trickling filter. Filters of this type generally have a design load ranging from 160–480 kg BOD/100 m³/day (100 to 300 lb BOD/1000 ft³/day).

Filter Type	BOD Removal (%)
Low Rate	80–90
Intermediate Rate	50–70
High Rate	65–85
Roughing Filter	40–65

Source: *Environmental Engineers Handbook*, 1997.

TABLE 13.1 BOD_5 Removal Rates for Various Filter Types

Trickling Filter Performance

Recent efforts have been made to combine fixed film reactors with suspended growth processes to efficiently remove organic materials from wastewater. For example, the combination of a trickling filter with an activated-sludge process has allowed for the elimination of shock loads to the more sensitive activated sludge while providing a highly polished effluent that could not be achieved by a trickling filter alone. Table 13.1 shows the BOD_5 removal rates for the four filter types discussed. Although the TF process is generally reliable, there is still potential for operational problems. Some of the common problems are attributed to increased growth of biofilm, improper design, changing wastewater characteristics, or equipment failure.

Trickling Filter Advantages and Limitations

Advantages:

- Simple, reliable, biological process
- Suitable in areas where large tracts of land are not available for land-intensive treatment systems
- May qualify for equivalent secondary discharge standards
- Effective in treating high concentrations of organics depending on the type of medium used
- Appropriate for small- to medium-sized communities
- Rapidly reduce soluble BOD_5 in applied wastewater
- Efficient nitrification units
- Durable process elements
- Low power requirements
- Moderate level of skill and technical expertise needed to manage and operate the system

Limitations:

- Depending upon the local water quality regulations, additional treatment may be needed to meet more stringent discharge standards.

- Possible accumulation of excess biomass that cannot retain an aerobic condition and can impair TF performance (maximum biomass thickness is controlled by hydraulic dosage rate, type of media, type of organic matter, temperature, and nature of the biological growth).
- Requires regular operator attention.
- Incidence of clogging is relatively high.
- Requires low loadings depending on the medium.
- Flexibility and control are limited in comparison with activated-sludge processes.
- Vector and odor problems.
- Snail problems.

Conclusion

Trickling filters are more conventional systems, commonly used for small- to medium-scale municipal systems. They are effective for first and possibly second stage treatment of effluent, but by themselves may not provide a complete solution to meet the required standards for stream discharge. However, used in combination with other filters, such as constructed wetlands, TF style filters are capable of providing and enhanced sewage treatment performance.

Constructed Wetlands

Constructed wetlands provide simple, effective, and low-cost wastewater treatment when compared with conventional systems.[1] Constructed wetland systems can be used to enhance primary treatment devices such as trickling filters and septic tanks, or in some cases they are used by themselves to adequately treat wastewater. They are low-energy consuming wastewater treatment systems that can also serve as aesthetic enhancements to the community. Types of systems discussed herein are plant rock filters, evapo-transpiration, and recirculating vertical flow systems. A chief advantage of these filtration styles is that, unlike aquatic plant–based systems, the water flow is below grade, and thus does not provide a breeding ground for mosquitoes.

It is important not to use a universal approach to a constructed wetland design. Rather, specific site adjustments need to be made to reflect on-site characteristics and hydraulic and organic loadings. Site factors to consider are soil depth and permeability, seasonal water levels, surface topography, lot size and shape, shading by trees, and owner preferences and attitudes.

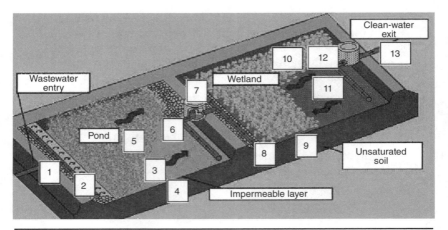

FIGURE 13.2 Constructed wetland schematic.

Although sometimes used to completely treat wastewater, constructed wetland systems are commonly designed so that wastewater enters the constructed wetland from a primary treatment device, such as a septic tank (1). Here, it is distributed evenly across the width of the first cell by a series of plastic valves or PVC tees (2) where microorganisms and chemical reactions break down organic materials and pollutants. The first cell contains gravel (3). A waterproof liner is used on the sides and the bottom of the first cell to conserve water and provide more effective treatment (4). Cattails and bulrushes are usually planted in the first cell (5). The roots of these marsh plants form a dense mat among the gravel. Here, chemical, biological and physical processes take place, which purify the water. Water from the first cell passes into the second cell through a perforated pipe embedded in large stones (6). The water level within each cell is regulated by a swivel standpipe located in concrete tanks at the end of each cell (7). Wastewater in the second cell is distributed evenly across this cell through another perforated pipe (8). Cell 2 has a layer of gravel covered with topsoil and then mulch (9). This cell is planted with a variety of ornamental wetland plants, such as iris, elephant ear, and arrowhead (10). The water in cell 2 eventually seeps into the soil below (11) or passes into another perforated pipe (12) where it is released into a drain field similar to those used with conventional septic tanks (13).

Plant Rock Filters

Plant rock (RF) filters are a type of constructed wetland that uses aquatic plants growing in a rock filled trench, through which

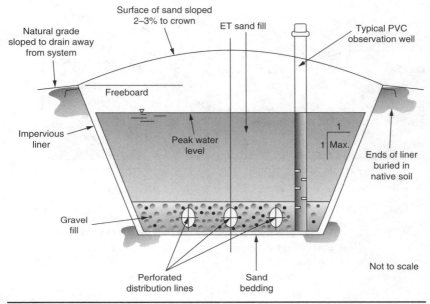

Natural grade
sloped to drain away
from system

Surface of sand sloped
2–3% to crown

ET sand fill

Typical PVC
observation well

Freeboard

Impervious
liner

Peak water
level

1
1 | Max.

Ends of liner
buried in
native soil

Gravel
fill

Not to scale

Perforated
distribution lines

Sand
bedding

FIGURE 13.3   Rock plant filter system schematic.

wastewater flows. Because plant rock filters provide only final treatment, a septic tank or treatment plant must be used before the plant rock filter. This is to provide primary treatment of the wastewater and prevent the plant rock filter from filling with solids. The plant roots and microorganisms living on the plant roots aid in the removal of pathogens and nutrients, making them effective in providing final treatment to wastewater. Additionally, plant rock filters can be useful in the repair of an existing wastewater system that is malfunctioning.

Plant rock filters have an added benefit in that they can be incorporated into the local landscape plan. Properly designed, they can complement a house site with lush green vegetation and colorful flowers while also serving a useful purpose. These systems utilize aquatic plants, which are commonly used in water gardens. A few of the showier plants that can be used include Louisiana Iris, Pickerel Rush, Yellow Canna, Elephant Ears, and Ginger Lilies. With the right combination of plants, they can offer blooms from mid to late spring until frost. Certain plants adapted to this wastewater system will even remain green through the winter months.

A plant rock filter installation can include a single treatment cell or, if soil conditions and space allow, a two-cell with one cell lined and the other washed gravel or stone. The lined cell should be completely lined with a continuous liner of approved material. The liner can be

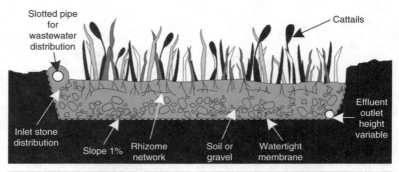

FIGURE 13.4   Rock filter section. (Courtesy of Mississippi State Department of Health Document: Rock Filters, Construction and Maintenance)

either porous or nonporous depending upon soil absorption capability. The lined cell is then filled with gravel. A plant rock filter must have an inlet pipe on one end and an outlet pipe on the opposite end.

The outlet pipe enters a water control structure where the water level in the system can be adjusted so the water is a minimum of one inch below the gravel surface. Water should never be exposed in this system, as this could encourage mosquito breeding; and to maintain an oxygenated environment, no soil material should be placed on top of the gravel.

As commonly done for septic tank–based systems, the tank or treatment plant should be pumped periodically. The pump-out schedule will be determined by how much organic material is loaded into the system. A widely accepted general rule is to have the tank pumped out every three to five years.

Unlike other types of discharge systems, plant rock filters will usually not have a discharge during summer and fall. This is conditional and will differ based on the amount of water used in a household and the number of plants in the system. If for some reason the plant population is not sufficient, there will be a discharge. To minimize any discharge, it's important that plants be placed 12 to 18 in (30.5 to 45.7 cm) apart at planting. Certain species of plants may require an occasional replanting due to insects, disease, or winter freeze to keep discharges minimal. Maintenance of the filter will consist of removing dead foliage after frost. It may also be beneficial to thin plants after several years of use. This will ensure the overall health of the plants and prevent the filter from becoming clogged with excess plant growth.[2]

Rock Filter Design Criteria

An extensive study of rock filters in Illinois[3] found that rock characteristics were very important. Flat rocks, excess fine rocks, soft friable rocks, and rocks of less than 3 in (7.6 cm) in diameter should all be avoided to prevent plugging problems. If stringent ammonia limits prevail, presently designed rock filters may not produce an acceptable effluent.

Compared to mechanical systems, constructed wetlands occupy far more real estate, but if they are integrated into the local landscape design, they can be made aesthetically pleasing. A well-designed system emits no odor, but people should stay away because of the possible danger posed by pathogens. Constructed wetlands have the least energy requirement, since energy is only needed to pump the wastewater to the entrance of the system—from there, gravity and biology do the rest.

The calculations necessary for sizing a rock filter system fall into four stages as follows:

1. *Hydraulic loading calculation:* The hydraulic loading to the RF system, in gallons per day (gpd), is based on the required flow per bedroom for home systems and flow per person or per fixture for other small systems. These rates are often established by each government entity but are generally assumed to be between 120 to 150 gal (454 to 568 L)/day per bedroom for an average home in the United States.[4] Therefore, a three-bedroom house at 120 gpd per bedroom would have a minimum design flow of 360 gpd (1363 L/day).

2. *Organic loading calculation:* The second component to sizing a RF filtration system is to determine the organic loading for the system in pounds BOD per day (lb/d). For a home system, the average daily organic loading per person is approximately 0.17 lb (0.08 kg) BOD per person. It can be assumed that there will be 50 percent BOD removal in the septic tank, and therefore a value of 0.085 lb (0.039 kg) BOD_5 per day per person can be used. Additional organic load reductions can be taken if a septic tank effluent filter or two tanks in the series are used, possibly a total of 70 percent or greater as approved through the appropriate regulatory agency. For other small users, values determined from acceptable engineering measures may be used as approved by the local authorities.

   For a three-bedroom home with a septic tank, with a possible maximum of four occupants, the estimated daily BOD value would be 0.34 lb (0.154 kg)/day.

3. *RF area calculation:* To determine the surface area of the filter bed, multiply surface hydraulic loading criterion in square feet of total surface area per gallon per day (ft²/gpd) by the hydraulic load in gallons per day (gpd).
   - For unrestricted areas—use a surface hydraulic loading criterion of 1.3 ft²/gpd (0.032 m²/L/day).
   - For a restricted small area—use a surface hydraulic loading criterion of 0.87 ft²/gpd (0.022 m²/L/day).

- For cold climates—use a surface hydraulic loading criterion of at least 11.3 ft^2/gpd (0.032 m^2/L/day).

4. Cross-sectional area:
   a. Calculate the cross-section area based on hydraulic loading and organic loading rates. Select the larger value. See Example #1 for details.
   b. Hydraulic loading—calculate area based on the hydraulic loading and Darcy's Law. See Example #1.
      - Use a relatively low hydraulic gradient (assume it is equal to bed slope) up to 1 percent and a conservative long-term hydraulic conductivity (850 feet per day).
      - For a flat bottom (0 percent slope), assume a low hydraulic gradient for the calculation (typically 0.5 percent).
      - For sloping lots, bed slopes of 2 percent or higher can be used to minimize cut and fill.
      - For cells receiving secondary or higher quality wastewater, a higher hydraulic conductivity may be used (up to a tenfold increase, i.e., 8500 ft/day (80 m/day). This will normally be advantageous for larger systems to reduce the total inlet width needed.

It is to be noted that the design values for hydraulic conductivity are still considered conservative. As more experience and data are acquired, the design values may be increased, resulting in improved dimensional flexibility for larger flows. The actual hydraulic gradient is expected to vary with distance down the cell due to the partial, but differential, filling of substrate pore spaces with time. It may be greater than design value in the inlet area and less than design value in the outlet area.

Darcy's Law is used to design the cross-section area to assure that the flow is subsurface, and it is also considered to be a conservative approach. Darcy's Law applies to laminar flow environment. Flow in clean gravel can range from laminar to turbulent, depending on flow rates and gravel sizes. Use of low long-term hydraulic conductivities and small gravel sizes assures a flow environment that is either laminar or in the transitional region, thus assuring practical applicability of Darcy's Law. A continuation of the calculations necessary for sizing the rock filter system, for the home example described above, can be seen in Example #1.

Design Example #1

Objective: To size a residential/small Rock Filter system with septic tank (ST) primary treatment. (English Units)

Assumption:
Assume 3-bedroom house occupied by 4 people.

Note: Calculations are not rounded. Dimensions are rounded (usually to the nearest whole number) after all calculations are completed.

a. Hydraulic Load, Q = 120 gpd per bedroom × 3 rooms = 360 gpd (48.132 ft³/day)

b. Organic Load = 0.17 lb/day/person × 4 people × 50% ST carryover = 0.34 lb/day

c. Hydraulic Loading Criteria = 1.3 ft²/gpd

d. Organic Loading Criteria = 1.0 ft²/0.05 lb BOD/d

Determine RF cell surface area, A_s:

e. A_s = 360 gpd × 1.3 ft²/per gpd = 468 ft²

Site condition: Yard is relatively flat and area is not constrained. The in-situ soil percolation rate is between 90 and 120 min/in.

Therefore, use flat bed data: Assume hydraulic gradient, S = 0.5%, or 0.005, and hydraulic conductivity, K_s = 850 ft/d (conservative rate based on long-term clogging of the gravel).

(1) Determine cell cross-section area, A_x, based on the Hydraulic Load (Q) and Darcy's Law $A_x = Q/(K_s \times S)$, where

K_s = substrate hydraulic conductivity (long term)

S = hydraulic gradient (assume equivalent to bed slope)

A_x = 48.132/(850 × .005)

= 11.325 ft²2

(2) Determine cell cross-section area, A_x, based on Organic Load,

A_x = 1.0 ft²/0.05 lb BOD/day × 0.34 lb BOD/day = 6.8 ft²

(3) Chose the higher of the two values of (1) and (2)

Since A_x is larger by Darcy's Equation, use 11.325 ft².

(4) Using 1.0 ft for cell depth, D (front of cell), determine cell width:

W = A_x/D

= 11.325/ 1.0

= 11.325 ft

(5) Determining the system cell length:

L = A_s/W

= 468/11.325

= 41.325 ft

Rounding to the nearest whole number results in cell dimensions of 11 ft wide and 41 ft long.

Since the soil percolation rate is 120 min/in (47 cm/min) or faster, the wetlands system may be divided into two equally sized cells (each cell 11 ft wide and 20.5 ft long with the second cell unlined) if the local health department does not require a subsequent drain field. If a drain field is required, one lined cell (11 ft wide and 41 ft long) will be more cost-effective.

Substrate fill depth:

- For a one-cell system:

  Depth of the cell inlet: 1.0 ft

  Depth of the cell outlet: 1.0 ft + (0.005 × 41.325) = 1.2066 ft = 14.479 in

- For a two-cell system:

  Depth of the cell inlet: 1.0 ft

  Depth of the cell outlet: 1.0 ft + (0.005 × 68.824/2) = 1.1033 ft = 13.240 in

Use a flat cell bottom (to simplify construction) and a cell depth of 14 inches for a one-cell and 13 inches for a two-cell system. See Fig. 13.5 for a single-cell constructed wetland and Fig. 13.6 for a two-cell constructed wetland.[5]

Rock Filter Conclusion

Overall, rock filters can provide effective BOD and TSS removal most of the time. However, a significant limitation of rock media polishing filters is their inability to meet a consistent 30 mg/L for the BOD and TSS discharge standards. Performance is reduced during winter, especially in regions where the ground freezes during some of the winter months. Therefore, provisions should be made for an alternative solution to anticipate limited performance due to needed maintenance, seasonal freezing, plant disease, or other system calamities.

Evapo-Transpiration Filter

Evapo-transpiration (ET) beds are an alternative to conventional absorption beds and may be appropriate for use where soil and subsurface conditions are not suitable for a conventional soil absorption system. Such unsuitable site conditions for which a lined ET system might be used include very porous soils, fill dirt, karstic limestone, fractured bedrock, insufficient vertical distance to a water table, and other features that would allow unacceptably rapid migration of wastewater effluent to ground or surface water supplies.

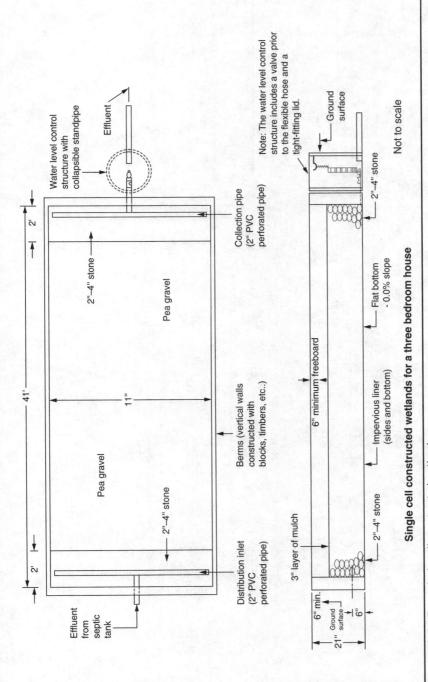

Water level control structure with collapsible standpipe

Effluent

Note: The water level control structure includes a valve prior to the flexible hose and a tight-fitting lid.

Ground surface

2"–4" stone

Not to scale

41'

2'

2'

11"

2"–4" stone

Pea gravel

Pea gravel

2"–4" stone

Collection pipe (2" PVC perforated pipe)

Berms (vertical walls constructed with blocks, timbers, etc..)

Distribution inlet (2" PVC perforated pipe)

Effluent from septic tank

3" layer of mulch

6" minimum freeboard

Impervious liner (sides and bottom)

Flat bottom - 0.0% slope

2"–4" stone

6" min.

Ground surface

6"

21"

Single cell constructed wetlands for a three bedroom house

Figure 13.5 For a single-cell constructed wetland.

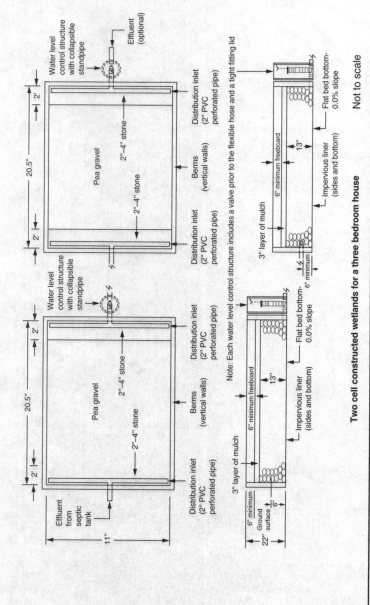

Figure 13.6 For a two-cell constructed wetland.

Because evaporation of moisture from the soil and/or transpiration by plants is the mechanism for effluent disposal in this type of system, they are best suited for arid climates where the net annual evapo-transpiration rate equals or exceeds the annual rainfall and wastewater flow. In these conditions, ET systems can provide a satisfactory means for liquid waste disposal without danger of surface or groundwater contamination.

By contrast, the evapo-transpiration infiltration (ETI) process is a subsurface system designed to dispose of effluent by both evapo-transpiration and infiltration into the soil. The difference is the bed liner: a nonpermeable membrane for the ET system and a permeable membrane for the ETI system. Because ETI beds can take advantage of both plant transpiration as well as infiltration into the soil, they are able to operate in colder seasons when ET systems cannot because of frozen ground and seasonal plant die-off. Year-round ET systems require large surface areas and are most feasible in the areas shown in Fig. 13.7.

Evapo-Transpiration Design Criteria

The inflow rate of domestic wastewater and the evapo-transpiration rate are site-specific parameters. The flow rate can either be estimated based on standard water usage rates[6] or measured with a flow meter. Design guidelines typically incorporate an average wastewater inflow rate for ET bed design. Daily, weekly, or monthly wastewater flow rates can be used for a more site-specific design, particularly at residences where water usage is not consistent. A safety factor to account for peak flows, or future increased site usage, may also be considered in the design flow rate.

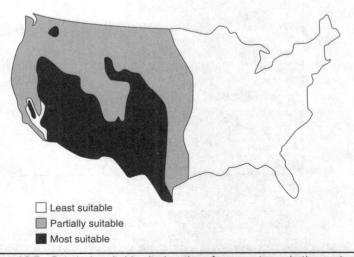

☐ Least suitable
▨ Partially suitable
■ Most suitable

Figure 13.7 Relatively suitable site locations for evapo-transpiration systems.

Evapo-Transpiration Design Parameters

Proper design of an EP or EPI system begins with a water balance calculation. A site-specific water balance is needed to confirm that the evaporation rate exceeds the precipitation occurrence, such that there will be a net transpiration of the moisture entering the bed for the specific location and usage.[7] Periodic exceptions to this principle can be made if there is sufficient storage in the ET/ETI bed to accommodate the disparity between influent coming into the bed and the evapo-transpiration/infiltration going out.

The water balance can be performed for an entire year, on a monthly basis, or a more frequent basis as data is available. Obviously, water balance calculations gain precision as the input variables are better defined and as the time basis is shortened. Often, it is difficult to obtain this data due to either the lack of hard information or constraints on the required research. In this case, a conservative design basis is appropriate. When resources are available, a more detailed evaluation and site-specific design may be possible. An optimal design methodology, where the evapo-transpiration data is superimposed on the precipitation information, can be seen in Fig. 13.8.

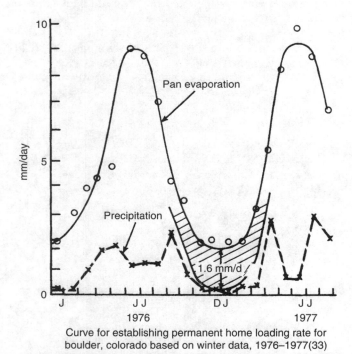

Curve for establishing permanent home loading rate for
boulder, colorado based on winter data, 1976–1977(33)

Figure 13.8 Graph of rainwater occurrence versus evaporation potential for Boulder, CO.

Generally speaking, the water balance can be described as follows:

(Precipitation inflow) – (Wastewater loading)
 – (Evapo-transpiration) = (Accumulation)

- Precipitation inflow (P), wastewater loading (Q), and evapo-transpiration (ET) are site-specific; the development of these parameters is discussed below.

- Accumulation indicates the storage volume required during time periods when ET is less than the total inflow.

The following formula has proven useful in estimating the size of the ET bed:

$$A = n \times Q/(ET - P)$$

where: A = surface area required to evaporate the wastewater
         n = coefficient, which varies from 1 to 1.6
         Q = annual flow volume
        ET = evapo-transpiration rate
         P = precipitation rate

Each of these factors is open to some degree of interpretation. Because these systems are large and expensive, there has been a tendency to minimize their design size and cost, resulting in significant failure rates.

Metric applications of the formula are possible. However, if the formula is based on an estimated daily amount of wastewater, and the net local evapo-transpiration rate is denominated in gallons per day, a conservative modification of the above formula can include a conversion factor of n = 1.6 (12 inches/foot divided by 7.48 gallons/cubic foot), making the formula:

$$A = 1.6 \times Q/(ET - P)$$

Further modification of the constant (P) can be made if an allowance for precipitation runoff is included. But since the data being used is subject to wide variation in actual practice, attempting to achieve too much precision in the calculations for a small system is not advisable.

Precipitation

If more precision is needed in measuring the net precipitation (P_n) that will be included with the wastewater load (Q), the runoff can be estimated using correlations from the Soil Conservation Service, the "Rational Method" formulas, or measurements of a test bed. A conservative ET system design assumes no runoff.

In actuality, there will be some fractional runoff from a sloped surface. For example, with a surface slope of 4 percent, runoff estimates range from a few percent up to 25 percent, depending on land conditions and the precipitation event. Therefore, for a large system, determining the amount of rainwater that is added to the ET load may be appropriate.

When the ground is frozen, there is essentially a 100 percent runoff. However, it has been found that ET beds tend to remain unfrozen when ambient temperature is below 32°F (0°C) due to subsurface microbial activity and the heat content of the wastewater inflow.[8]

Precipitation is primarily measured as rain and snow, with snow being measured as both depth of snow and water content of snow. Precipitation is measured by weather stations operated by various governmental agencies. The amount of precipitation can vary within a small area due to local variations in elevation, terrain, and water bodies. Therefore, precipitation data should be obtained from the nearest weather station that is representative of the site's climatic conditions. If on-site data is collected, a sufficiently long history needs to be analyzed to provide a statistically valid record.

The amount of rain that percolates into an ET bed is estimated from the rainfall event and the runoff estimate, as discussed above. The amount of snow that percolates into the bed depends on weather as well as the run-off estimate. Percolation inflow into an ET bed is often most critical in winter and spring months, when ET rates are at their lowest.

Snow percolation is a function of the water content of the snow (inches of water per inch of snow), sublimation of snow prior to melting, and when the snow melts. Each of these components can be estimated. The water content of snow is variable between location, time of year, and specific snowfall event. During months with only snowfall, the water content per inch of snow can readily be calculated. During months when both rain and snowfall occur, the water-content contribution from snow must be estimated. A more detailed analysis of the impact of snow melt on the evapo-transpiration load of an ET bed can be seen in the work of Garstka, et al.[9] and Croft.[10]

Evapo-Transpiration Rate

The design evapo-transpiration rate (ET) is site-specific. The primary variables that have an impact on the potential ET rate are climate, cover soil, and vegetation. The ET rate is the rate that liquids evapo-transpirate from an ET bed, and it is a function of many variables: air and ET bed temperature, wind and sun exposure, relative humidity, vegetation type and quantity, height of liquid in the bed, and bed fill material. Knowing one or more of these variables, the ET rate can be estimated.[11]

Commonly used parameters to estimate ET rates are pan and lake evaporation. Alternatively, empirical ET rates can be determined from test beds by monitoring total inflow and water levels within the test bed. State or local agencies may provide ET rates for use in the design of bed storage volume and surface area. These rates are typically the yearly average rate and tend to underestimate the ET rate; this provides a safety factor in the design of an ET system. In instances where local guidelines provide ET rates too low to support the use of an ET system, carefully researched ET rates may be high enough to allow the design and use of an ET system.

Historically, ET systems have been used year-round in warm-weather and dry areas or during the summer months in colder and wetter areas. Although biological systems work faster in warm environments, they are successful in colder climates, too. The metabolic rate of the microbes that do the digestion process is directly proportional to the temperature. For every 19°F, their metabolic rate doubles. At 40°F, most microbes are dormant. But wastewater is usually warm, and microbial action generates heat, so treatment may slow during cold seasons, though it rarely, if ever, stops completely.[12] Therefore, expansion of ET systems to year-round use in cold-weather or wet areas may require an increase in ET bed capacity to provide storage during periods when the ET rate is less than the precipitation and wastewater inflow.

In the United States, approximations of the evapo-transpiration data can be gathered from the local Department of Agriculture, the state climatologist, or the regional climate center. Other sources are http://www.stateclimate.org and http://lwf.ncdc.noaa.gov/oa/climate/regionalclimatecenters.html.[13]

Evapo-Transpiration Bed Design

As stated previously, over the course of a year, evapo-transpiration of the bed must exceed the inflow of precipitation and wastewater for the ET system to operate without overflowing. The bed design can incorporate storage for limited periods of time when this is not the case.

In ET systems, an impermeable liner is placed below the bed to prevent infiltration into the water table if the soil conditions do not permit acceptable percolation rates. Impermeable liners are not required in soils with a percolation rate of 4–10 centimeters per second or less. The liner should be 20–40 millimeters in thickness and should be chlorine-free, low-density polyethylene film, as chlorine can leach out from PVC liners, leaving the material brittle and prone to breaking.[14]

The distribution system is placed in 12 inches of gravel (0.75 to 2.5 inches) at the bottom of the bed. Spacing of the distribution pipes is 4 to 12 feet, with lower values preferred for better distribution.

Wicking is accomplished by a 2- to 2.5-foot layer of sand (0.1 millimeter) and a loamy soil-sand mix to raise the water to the surface or a thin layer of soil at the surface.

Both fill material and surface cover affect the efficiency of an ET bed. The fill material should be sand and gravel, selected and placed to facilitate distribution of wastewater within the bed and the wicking of the liquid toward the surface. Surface soil should be good topsoil suitable for supporting the selected vegetation. The surface of the bed is planted with water-tolerant plants suited to the local environment. Surface vegetation should have good water uptake and thrive in the local climate.

When properly operating, the influent to the ET and ETI systems enters through the distribution pipes to the porous bed. As the effluent is drawn up through fine sand media by capillary wicking, it is evaporated or transpired into the atmosphere. In ETI systems, effluent is also allowed to percolate into the underlying soil. Modifications to ET and ETI systems include mechanical evaporating devices and a broad array of different designs and means of distribution, storage of excess influent, wicking, and containment or infiltration prevention.

The most important system variables, which control the movement of wastewater to the surface, are media and the depth to saturated (stored) water. If the water depth exceeds approximately 12 inches, the result can be that the wastewater is stored so deep that the wicking properties of the fill (e.g., the area (voids) through which water must rise to the surface for evaporation) are restricted. This can cause the system to go anaerobic, causing odors. Most published designs are suspect, and therefore conservative assumptions are recommended to allow for seasonal weather variations and variations in wastewater loading.

Experience in ET and ETI system installation has found slight improvements of ET system operation when the use of pure sand for the absorption layer has been exchanged in favor of a loamy soil/sand mix. Professionals say that the absorption rate they get with this mix surpasses that of a straight sand layer. Similarly, the wicks that run from the bottom of the rock media layer up to the soil/sand layer are also made of loamy soil.

Drainage

The area surrounding the ET beds should be graded to prevent precipitation draining into the ET bed. Surface runoff of precipitation depends on several factors: surface soil type and compaction, vegetation cover, surface slope, rate of precipitation, and soil saturation. The bed surface can be graded to enhance precipitation runoff, and swales or culverts can be appropriate drainage modifications to divert any surface water away from the bed.

The contribution of plants, while beneficial, may be seasonal depending upon the growing season, and therefore must not be relied upon if seasonal die-off is a possibility. The use of ET or ETI system will be maximized if the effluent is used for drip landscape irrigation. If this is done, it is recommended that the irrigation system be installed in a shallow trench below the adjacent grade. This is to keep the greywater discharge in the vicinity of the plant, should the wind blow the mulch away.

ET and ETI System Advantages and Limitations

Advantages: ET and ETI systems can overcome site, soil, and geological limitations, or the physical constraints of land that prevents the use of subsurface wastewater disposal systems. One of the most important advantages of constructed wetlands is their potential for providing both a wastewater treatment system and an aesthetically appealing environment if the project is properly landscaped.

Constructed wetlands offer the potential to create habitats for bird and mammal populations, and thus to improve the ecological integrity of a wastewater treatment site. They are also cost-effective. Evidence from a number of experimental sites suggests that the maintenance and labor costs associated with constructed wetlands are competitive, and sometimes significantly lower, than those for conventional treatment facilities dealing with the same volume of wastewater. It should be noted, however, that treatment costs can vary widely depending on the nature of the raw sewage stream (i.e., whether it contains a high proportion of inorganic and trace industrial pollutants from industrial discharges), on climate, and on other local factors.

Limitations

Because ET and ETI are natural processes, their performance is completely controlled by environmental conditions, such as precipitation, wind speed, humidity, solar radiation, and temperature. While a conventional wastewater treatment system is well protected from the elements, disease, and predators, a constructed wetland is vulnerable to all three. In cold climates, climatic variations in particular affect the performance of a natural treatment system. In the spring and during summer thunderstorms, excessive rainfall can flood a wetland, decreasing its treatment effectiveness. Insects and other predators can reduce the populations of key wetland species, thus impairing the treatment function. Transpiration is reduced with winter dieback of vegetation. Simple upsets of temperature, pH, or other factors can affect the health and removal efficiency of wetland plants. These factors create a potential for variability in effluent quality that is unacceptable in a public utility or a private treatment facility subject to legally binding standards for discharge.

ET systems are not suitable where land is limited or where there is irregular terrain. Phosphorus removal is notoriously poor in many constructed wetland systems. The reasons for this are not entirely clear but may be related to reduced oxygenation of the root zone in slow-moving waters. Although constructed wetlands may be able to remove BOD, suspended solids, and nitrogen compounds with reasonable effectiveness, they may not perform well in terms of phosphorus removal. Where phosphorus enrichment is a concern (for instance, because of the potential for eutrophication[15]), constructed wetlands may cause problems rather than solve them. Two typical problems are as follows:

1. *Limited life expectancy:* Although a conventional sewage treatment plant can have a life expectancy of many decades if it is adequately maintained and upgraded, the life expectancy of a constructed wetland system is likely to be much shorter. For one thing, wetlands tend to accumulate suspended solids, so they will fill up with time, gradually reducing their volume and thus their treatment capacity—and their effectiveness. Because of the relative newness of alternative technologies, long-term data for wetland performance is not yet available. Estimates suggest that wetlands may be limited to 15 to 20 years of life, compared to the 25- to 50-year life span possible with conventional facilities.

2. *Possible creation of toxic wetlands:* As wetlands accumulate sediments, they can also accumulate the many pollutants that have a natural affinity for solids. These pollutants include some forms of phosphorus, many heavy metals, and some trace organic pollutants. When the wetland is retired at the end of its useful life, it may not be an inert or innocuous component of the environment but rather a hazardous waste disposal site. Concentrations of hazardous materials in wetland sediments may be too high to permit the use of the site for other purposes, such as recreation. The treatment or removal of those contaminated sediments may therefore add significantly to the costs of building and operating a constructed wetland. But as technology improves, this apparent limitation of pollutant concentration may offer the opportunity to reclaim the heavy metals, rather than being a detriment.[16] To this point, research has shown that using sludge and effluent for crops, such as corn, that rely on phosphorous to grow can be a significant benefit.[17]

Conclusion

ET and ETI systems are potential alternatives for domestic sewage disposal at locations where the annual ET rate exceeds precipitation and wastewater inflows. A site-specific water balance should be used to determine the surface area and storage volume of an ET system if sufficient design and data are available. Careful evaluation of the engineering parameters may extend the applicability of ET systems beyond the conservative guidelines provided by state and local regulatory agencies. In summary, the chief attractions for constructed wetlands as an alternative waste treatment solution are the low cost and ease of operation, which make them very attractive for noncritical applications.

Recirculating Vertical Flow Constructed Wetlands[18]

Recirculating vertical flow (RVF) constructed wetlands are a unique type of constructed wetland that can improve the performance of conventional on-site wastewater treatment. They are placed after the septic tank and before final soil treatment and dispersal. RVF systems are compatible with conventional leach fields, mound systems, drip irrigation, or other approved soil absorption systems, and compared to other systems (e.g., septic, ET/ETI), they require relatively much smaller space.

Who Should Consider RVF Constructed Wetlands?

Building and home owners with space limitations, proximity to environmentally sensitive bodies of water, unavailability of a sanitary sewer, or to improve performance of a failing septic system, can benefit from RVF constructed wetlands. By minimizing the amount of solids and nutrients entering the soil infiltration system, they provide a high level of wastewater treatment. RVFs have been used in the U.S. for many years, but their use as a treatment for residential wastewater is relatively new. The first RVF was installed in Indiana in LaGrange County in 2001. Five such systems are currently in place (as of 2007), and all are performing well.

Design Criteria

The size of the RVF constructed wetland should be based on the expected gallons per day (GPD) of sewage produced. Recommended design parameters for individual residences in the United States are in Table 13.2.

Construction

As a general guideline, the minimum cell size of the RVF constructed wetland is determined on the basis of 0.48 ft² of surface area per gallon of sewage to be daily treated ($0.012 \text{ m}^2/\text{L/day}$). The depth of the constructed wetland cell is from 42 to 48 in (1.1 to 1.2 m), and its design configuration is square.

Like a conventional septic system, wastewater leaving the home should first be collected in a septic tank with a minimum of 48 hours

Residence Bedrooms (#)	Wastewater Daily Flow (Gallons Per Day & Liters Per Day)	Septic Tank Size Volume (Gallons & Liters)	RVF Constructed Wetland Cell Size (Feet & Meters)
1	150 (567.8)	1000 (3785)	8.5 × 8.5 (0.8 × 0.8)
2	300 (1135)	1000 (3785)	12 × 12 (1.1 × 1.1)
3	450 (1703)	1000 (3785)	15 × 15 (1.4 × 1.4)
4	600 (2271)	1250 (4731)	17 × 17 (1.6 × 1.6)
5	750 (2838)	1500 (5678)	19 × 19 (1.8 × 1.8)

TABLE **13.2** Sizing Recommendation for RVF

solids retention time and an effluent filter installed at the tank outlet. The septic tank overflow should be directed to the inlet at the bottom gravel layer of the RVF constructed wetland. A 4-in (10.2 cm) diameter two-row PVC perforated pipe with holes in the 4 and 8 o'clock positions or a three-row pipe with holes at the 4, 8, and 12 o'clock positions is placed across the bottom of the wetland. In early designs, a PVC perforated pipe was also placed at both the inlet and outlet ends of the gravel to distribute and then collect the effluent after it has traveled through the gravel layer at the bottom of the wetland. More recently, plastic soil absorption chambers are often used in place of septic stone in absorption trenches. They have been used as an inlet manifold to distribute effluent across the width of the RVF and have been found to work as well or better than perforated pipe. Chambers have more receiving capacity than traditional 4-inch diameter manifold pipes, and their innovative design facilitates periodic cleanout.

To begin construction, an appropriately sized excavated area is first lined with a 30-mil geo-membrane PVC liner or comparable impermeable material, such as a 45-mil EPDM (ethylene propylene diene monomer) rubber sheet covered with a layer of 13 to 25-mm (½ to 1-inch) diameter gravel. A second layer of impermeable material (PVC or EPDM) is then placed over the most of the top area of the gravel to separate the aerobic from the anaerobic sections of the RVF wetland; this leaves 25 percent of the bottom gravel layer nearest the wetland inlet uncovered. Next, a top layer of 4-mm (¼-inch) diameter gravel (pea gravel) is placed over the membrane and gravel. A set of perforated pressure distribution lines are placed about 6 inches deep in the top of the pea gravel layer in order to uniformly load the wetland with effluent.

Effluent overflowing the septic tank, as well as treated effluent that has passed through the top portion of the wetland, passes through the gravel at the bottom of the wetland and drain to the sump basin. The sump basin consists of a 5-foot long section of 24-inch diameter black corrugated drain tile, installed vertically. Concrete is placed in and around the bottom of the sump basin to seal the tile and prevent the entry of groundwater and the outward seepage of effluent into the surrounding ground. The top of the sump basin should be fitted with a secure, insulated plastic or concrete cover with the bottom of the sump below frost line to prevent freezing. The sump basin holds the recirculation pump, and the sump effluent is distributed over the top of the wetland. The water level in the wetland is maintained with the 4 by 3-inch PVC, and the PVC flexible sewer coupler reducer is set around 20 inches above the wetland bottom.

The effluent pump is controlled by an electronic repeat cycle timer. Each 30-minute period, the timer activates the pump for a 2-minute cycle to pressurize a 1-inch PVC manifold and perforated distribution pipe and distribute effluent uniformly across the top pea gravel layer. The pressure distribution system consists of a closed piping network using 1-inch diameter PVC laterals pipe fed through a manifold by the cycle pump.

The laterals should be placed no more than 2 ft (0.61 m) apart with equally spaced 1/8 in (3.175 mm) holes drilled in the top every two feet and protected with an orifice shield to disperse the effluent. The orifice shields prevent plugging of 1/8-inch openings. The last hole (air relief point) in each lateral should be placed just ahead of the screw-on cap. The manifold and force main pipe must drain back to the sump after each cycle. A ¼-inch pressure relief hole can be drilled in the feed line inside the sump pit to facilitate draining, and a quick disconnect pipe coupling is used to facilitate pump servicing. The manifold and lateral distribution lines are both covered completely by the six inches of pea gravel. The outside edges of the wetland are typically finished with regular leach fieldstone or other locally available material.

The top of the pea gravel layer is planted in rows with plants compatible with the local weather conditions. For the projects constructed in Lagrange, Indiana,[19] river bulrush (*Scirpus fluviatilis*), hard-stemmed bulrush (*Scirpus acutus*), soft-stemmed or great bulrush (*Scirpus validus creber*), prairie cord grass (*Spartina pectinata*), common rush (*Juncus effuses*), dark green rush (*Scirpus atrovirens*), sedges (*Carex spp*), and great spike rush (*Eleocharis palustris*) with a density of one plant per square foot (0.093 m²), and a 1-foot (30-cm) separation between rows was used. These plants have deeper root systems than cattails or bulrushes and function better in constructed wetlands.

Wetland flowering plants, such water iris (*Iris virginica*), swamp milkweed (*Asclepias incarnata*), cardinal flower (*Lobelia cardinalis*), swamp rose mallow (*Hibiscus palustris*), great blue lobelia (*Lobelia siphilitica*), and New England aster (*Aster novae-angliae*) can be planted between the sedges and bulrush. Conventional garden plants have also performed well, such as morning glory vines (*Ipomoea leptophylla*), cheddar bath's pinks (*Dianthus gratianopolitanus*), and ferns.

Landscaping with low flowering plants and a border around the wetland edge of perennial flowers can create the visual effect of a conventional flower garden. When the system is fully operational, it can be walked on since the sewage effluent is well below the surface.

Operation

As sewage effluent leaves the septic tank, it enters the gravel in the bottom portion of the RVF constructed wetland, where it is treated by passing horizontally across the bottom gravel layer. The timer-controlled pump in the sump basin periodically recirculates effluent back to the buried distribution pipe in the top layer of pea gravel. The effluent trickles vertically down through this aerobic upper zone, flows laterally across the impermeable liner separating the two layers of stone, and drops down into the uncovered front portion of the bottom gravel. It also passes horizontally back to the sump basin. As treated effluent builds up in the sump basin, the pump starts another wetland recirculation cycle (timer is in the "on" position), or if the timer is in "off" position (pump is in resting cycle), the overflow is pumped to the soil absorption field.

Maintenance Requirements

Periodic checking of the components is required to assure that the pumps, floats, and controls function as intended. As is customary for septic tanks, they should be pumped (and all solids removed) every three to five years to prevent the overflow of solids. Depending on daily water usage or site-specific circumstances, the tank effluent filter may require more frequent cleaning service. Ideally, the filter should be checked and/or serviced at least annually to maintain peak performance. Cleaning the effluent filter is very simple and usually just involves hosing the solids off the exterior of the filter back into the septic tank with a garden hose. Protective waterproof gloves should be worn when cleaning the filter or performing other maintenance to the on-site system as a safety precaution to ensure there is no direct contact with the wastewater, especially if you have open wounds at the time. As settling of the pea gravel cover occurs, periodic leveling may be necessary.

Green vegetative leaves appear in Indiana by early spring (April–May), grow vigorously throughout the warmer months, and turn brown in late fall or early winter as the plants enter dormancy. This brown vegetative material should remain in the RVF constructed wetland during winter since it provides insulation during the winter months. If removed, it should be done in early spring before new growth is visible. Old growth should not be burned in place since this can damage both growing and dormant plants and possibly even the liner or PVC distribution pipe. Old growth can remain in place for several growing seasons, but it should be removed after three to four seasons by cutting the plants at ground level. Pulling them is not recommended to avoid root damage to other plants. Wetland plants do not require much maintenance, but they should be checked annually as one would any flower bed.

Expected Performance

When compared to a conventional septic tank and soil absorption system, which discharges 100 percent of the septic tank effluent contaminants into the ground, a well-designed, well-constructed, and well-maintained RVF constructed wetland will remove up to 99 percent of the fecal bacteria and E. coli and 80–99 percent of other contaminates—even before the effluent is discharged to the soil absorption field. The physical, chemical, and biological natural filtration treatment processes and the alternating aerobic (oxygen is present) and anaerobic (oxygen is not present) environments present in the constructed wetland layers destroy most pathogens and remove most contaminants. While unusual, the first RVF constructed wetland (15×15 ft (1.4×1.4 m) unit) installed in LaGrange county in 2001 has not discharged effluent to the conventional absorption field during the last three year period (2005–2007) because of water uptake by the plants, the evapo-transpiration process, and the low occupancy of three-bedroom homes (only two people per home).

Summary

Recirculating, vertical flow constructed wetlands are sometimes defined as vegetated re-circulating gravel filters. They treat wastewater by passing sewage through the constructed wetland where it is filtered through the gravel media in the bottom layer, and then recirculate back around the roots and rhizomes several times for more filtration and treatment, before it is finally discharged to the soil absorption area. This simple sewage treatment system is a reasonable economical and effective alternative to conventional wastewater treatments with low maintenance requirements.

Conclusion

Biological filters and constructed wetlands are a viable alternative to conventional sewage treatment systems. With proper integration of traditional engineering wastewater treatment design, with the knowledge of plant and weather interaction, these systems use natural biological processes to provide an efficient alternative to the traditional utility-provided wastewater treatment. These systems are commonly labor and land intensive to install, making them perhaps more applicable to developing countries where the cost of labor is relatively cheap and land, not suitable for agriculture, can be re-purposed.

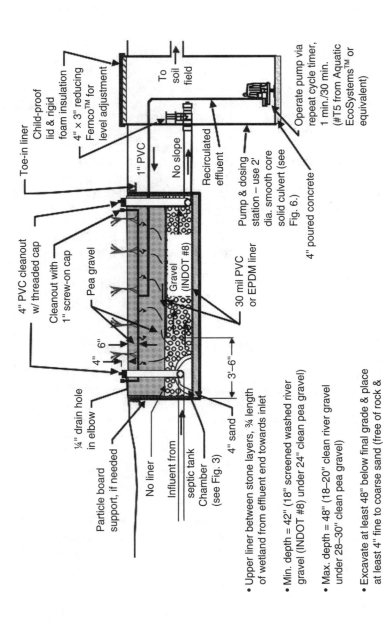

- Upper liner between stone layers, ¾ length of wetland from effluent end towards inlet

- Min. depth = 42" (18" screened washed river gravel (INDOT #8) under 24" clean pea gravel)

- Max. depth = 48" (18–20" clean river gravel under 28–30" clean pea gravel)

- Excavate at least 48" below final grade & place at least 4" fine to coarse sand (free of rock & debris.) Level & compact sand.

Recirculating vertical wetland cross-section

FIGURE 13.9 An elevation view of RVF constructed wetland.

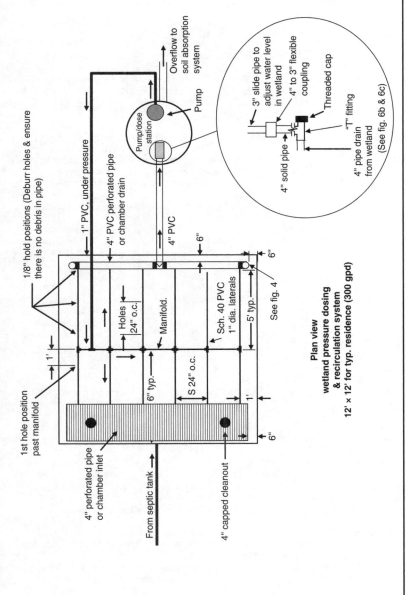

Plan view
wetland pressure dosing
& recirculation system
12' × 12' for typ. residence (300 gpd)

1/8" hold positions (Deburr holes & ensure there is no debris in pipe)

1st hole position past manifold

1" PVC, under pressure

4" PVC perforated pipe or chamber drain

Holes 24" o.c.

Manifold.

Sch. 40 PVC 1" dia. laterals

4" PVC

6"

See fig. 4

6"

6"

1'

6" typ.

S 24" o.c.

5' typ.

1'

6"

4" perforated pipe or chamber inlet

From septic tank

4" capped cleanout

Overflow to soil absorption system

Pump

Pump/dose station

3" slide pipe to adjust water level in wetland

4" to 3" flexible coupling

Threaded cap

"T" fitting

4" solid pipe

4" pipe drain from wetland

(See fig. 6b & 6c)

FIGURE 13.10 Top sectional plan view of a RVF constructed wetland.

263

Notes and References

1. *Wastewater Technology Fact Sheet: Trickling Filters,* EPA 832-F-00-014, September 2000, United States Environmental Protection Agency, Office of Water, Washington, D.C. For additional information on wastewater treatment filters design, see www.epa.gov/owmitnet/mtbfact.htm.

2. Mississippi State Department of Health, Design Standard VII, Plant Rock Filter System excerpt from Steiner, G.R., P.E., Watson, J.T., P.E., *General Design, Construction, and Operation Guidelines, Constructed Wetlands Wastewater Treatment Systems for Small Users Including Individual Residences,* Second Edition, TVA/WM--93/10, Chattanooga, Tennessee, May 1993.

3. Excerpt from *Plant Rock Filters Construction and Maintenance,* Mississippi State Department of Health, General Environmental Services, Jackson, Mississippi.

4. Rock Media Polishing Filter for Lagoons, Environmental Protection Agency, Wastewater Technology Fact Sheet, Office of Water, EPA 832-F-02-023, September 2002.

5. Reduced or increased hydraulic loadings may be approved by the appropriate regulatory agency based on actual usage. A well-engineered system using low-flow plumbing fixtures, as is typically the case in an off-grid application, can be much less than these numbers, commonly about 20–25 gallons per day per person. For a sample water demand estimate, see Appendix B.

6. *Regulation Governing Individual Onsite Wastewater Disposal Design: Plant Rock Filter,* Form No. 309, Standard VII, Mississippi State Department of Health, Jackson, Mississippi, May 1997.

7. Bennett, E.R., Linstedt, K.D., *Sewage Disposal by Evaporation-Transpiration,* EPA-600-2-78-163, U.S. Environmental Protection Agency (EPA) Municipal Environmental Research Laboratory. Office of Research and Development. Cincinnati, Ohio, 1978.

8. Salvato, J.A., "Rational Design of Evapo-Transpiration Bed," *Journal of Environmental Engineering,* Vol 109(3), pp. 646–660, 1983.

9. Bernhart, A.P., *Evapotranspiration—A Viable Method of Reuse (or Disposal) of Wastewater in North America, South of 52nd and 55th Parallel,* pp. 185–195, Individual Onsite Wastewater Systems: Proceeding of the Fifth National Conference, Ann Arbor Science Publishers Inc, 1978.

10. Garstka, W.L., Love, B., Goodell, Bertle, F., "Factors Affecting Snowmelt and Streamflow," U.S. Department of Agriculture Forest Service, March 1958.

11. Croft, A. R., "Evaporation from Snow," *Bulletin American Meteorological Society,* Vol. 25: pp. 334–337, October 1944.

12. Penman, H. L., "Estimating Evaporation," *Transactions American Geophysical Union,* Vol. 27: No. 1, pp. 43–46, February 1956.

13. Steinfeld, C., Del Porto, D., "Growing Away Wastewater," *Landscape Architecture,* pp. 44–53, January 2004.

14. http//:answers@noaa.gov.

15. Steinfeld, C., Del Porto, D., "Growing Away Wastewater."

16. Eutrophication is a condition that generally promotes excessive plant growth and decay, favoring simple algae and plankton over other more complicated plants. It also causes enhanced growth of aquatic vegetation that disrupts normal functioning of the ecosystem, which in turn causes a variety of problems, such as a lack of oxygen and odors from the resultant anaerobic digestion of the effluent.

17. Heathcote, I., "Artificial Wetlands for Wastewater Treatment," excerpt from *Natural Systems for Wastewater Treatment: Manual of Practice FD-16,* Water Pollution Control Federation, Alexandria, VA, 1990.

18. Garcia-Perez, A., Harrison, M., Grant, W., "Recirculating Vertical Flow Constructed Wetland for On-Site Sewage Treatment: An Approach for a Sustainable Ecosystem," *Journal of Water and Environmental Technology,* Vol 9(1), 2011.

19. Information for this section provided from *Recirculating Vertical Flow Constructed Wetlands for Treating Residential Wastewater*, Garcia-Perez, A. (Lagrange County, Indiana, Health Department), Jones, D. (Purdue University), Grant, W. (Lagrange County, Indiana, Health Department), Harrison, M. (Bernardin-Lochmueller), Rural Wastewater RW–4 W, Purdue University Press.
20. Lagrange, Indiana, is located at 41° 38' N, 85° 25' W, with a winter design temperature of –3°F (–20°C).

CHAPTER 14

Blackwater Recycling[1]

Introduction

What is blackwater? Wastewater has many sources. The wastewater that comes from showers, washing machines, and sinks is considered greywater because, while it has particles and contaminants, they are not considered dangerous under normal circumstances. Blackwater recycling is a step beyond greywater recycling, in that everything that goes down the drains, including toilet water and what it carries, is recycled. Because blackwater contains bacteria and pathogens from food particles, feces, and other human body fluids, it is considered hazardous and needs to be treated differently.

Despite many commercial facilities and multi-dwelling apartments having the option to discharge wastewater directly into a local municipal sewer network, on-site blackwater recycling is attractive to many sites for a number of reasons, which include the following:

- Reducing on-site water use by up to 90 percent.

- Proactively demonstrating good environmental management.

- Providing an additional source of water to keep gardens and sports fields green for aesthetic uses in areas that have limited water resources available.

- Reducing the hydraulic impact of a development on a local municipal sewer. This overcomes development restrictions caused when the existing sewer is at its hydraulic capacity and favorably impacts combined sewer overflow[2] common in cities with aging infrastructure.

- Contributing points to building efficiency certification programs such as Green Star (Australia), LEED (USA), BREEAM (UK), and HQE in France.[3]

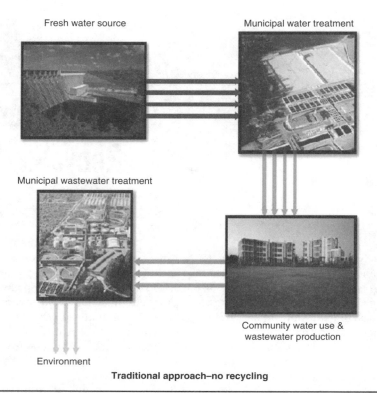

Fresh water source

Municipal water treatment

Municipal wastewater treatment

Community water use &
wastewater production

Environment

Traditional approach–no recycling

FIGURE 14.1 Traditional wastewater treatment process.

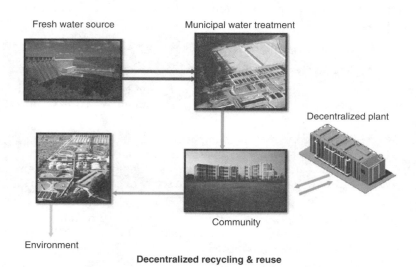

Fresh water source

Municipal water treatment

Decentralized plant

Community

Environment

Decentralized recycling & reuse

FIGURE 14.2 Decentralized systems for recycling and reuse of blackwater.

Blackwater Recycling Systems

The simplest blackwater recycling system is the common septic tank system, where waste is settled and lightly filtered, with the resultant greywater adequate for subsurface irrigation for lawns, plants, and leach fields. More elaborate systems are available to make blackwater clean enough to provide drinkable water. Such systems are expensive, have complex maintenance requirements, and require the people who drink the recycled water to get over the stigma of drinking toilet water.

There are five basic steps to blackwater filtration and treatment: Primary Treatment:

1. Blackwater first flows by gravity to a primary treatment tank. The blackwater sits for 24 hours while settling occurs and an established colony of bacteria works to break down the solids. After the 24-hour period, the settled blackwater is pumped to the secondary treatment tank.

Secondary Treatment:

2. When the settled blackwater is next pumped to the secondary treatment tank, the water is separated into three stages to help with the continuing process.

3. Blackwater aeration stage—the first chamber in the secondary treatment tank begins the aeration stage. This means water and air are injected into the tank at timed intervals so that the tank contents are churned. Bacteria in the tank then settle so they can feed on the sludge in the tank. The water is next moved to the sludge settling chamber.

4. Sludge settling chamber—in the sludge settling chamber, a bacteria biomass mechanism forces sludge downward and the partially treated water upward to be collected and sent next to the irrigation chamber stage.

5. Irrigation chamber—in this last step, the water is treated to tertiary standards by being clarified and chlorinated. At this point in the treatment process, the water looks like normal tap water and is colorless, odorless, and sterile; but it falls short of being suitable for drinking.

After completion of the secondary stage, the recycled water can be piped to applications such as:

- Watering landscaped areas by irrigation
- Flushing toilets
- Water features
- Cooling tower applications
- Car washing

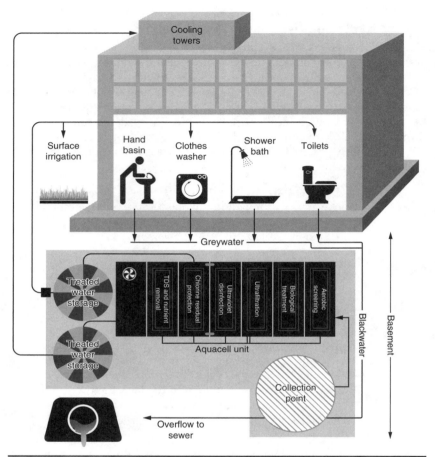

FIGURE 14.3 Blackwater system schematic. (*Graphic courtesy of Aquacell Pty Ltd,* *www.aquacell.com.au*)

A schematic of a blackwater system and its operation can be seen in Fig. 14.3.

Collection point: Water flows from the property to a collection point, where it is pumped into the Aquacell to begin the treatment process.

Screening This first step efficiently reduces insoluble material to a negligible residue. This residue is either discharged to the sewer or it is de-watered and compacted for disposal as solid waste.

Biological treatment: Air is diffused into the water to create optimum conditions for bacteria to consume impurities. A sustainable biomass concentration is maintained, which metabolizes all incoming waste, resulting in negligible sludge.

Ultra-filtration: Ultra-filtration occurs through a special membrane of microscopic pores that prevents particles, bacteria, and viruses from passing through. The membranes are cleaned by air scouring, ensuring no wastewater is produced.

Ultraviolet disinfection: As a precaution, ultraviolet lamps are provided. These act as a barrier, providing additional protection against pathogens.

TDS and nutrient removal: Proprietary technologies are employed for applications such as cooling tower reuse and discharge to sensitive environments.

Chlorination: Finally, a chlorine residual is added to protect the water while in storage and the reticulation system—this is the only time any chemicals are used throughout the treatment process.

Treated water storage: The result is safe water, kept in storage for immediate use in a variety of non-potable applications, including surface irrigation, toilet flushing, and cooling towers.

System Design

The design of a whole house water management system, which includes any water recycling process, must take an integrated water management approach that considers all available sources of water and balances them with available end uses of the reclaimed water. Designers face the challenge of choosing the right water sources and the timing of their occurrence and balancing that with the timing of the end use. All of this is considered while designing the entire system at the right scale for optimum return on investment. A water recycling system has to consider all impacted flows, and must account for environmental, social, and economic risks associated with the system, both on- and off-site. Water source flows include rainwater, storm water, groundwater, drinking water, as well as wastewater. To optimize a whole house water management system, the objective is to make the upstream, on-site, and downstream flows fit together in a balanced and healthy closed loop.

In dealing with complex systems, it is important that an integrated, or "systems thinking," approach be applied. Systems thinking refers to defining a system's boundaries to adequately identify causal relationships and understand the interconnection of these relationships among resources and activities. Advantages of decentralized options are often only apparent when taking a more integrated approach. For example, on-site reuse of blackwater can provide a partially drought-resistant source of landscape irrigation, and can reduce the size of the water main to the building and the waste line from the building. An integrated or systematic approach to water system design links all water-related activities to one another, thereby recognizing the interconnected nature of water and wastewater systems and allowing a concurrent evaluation of a whole system's potential costs and benefits. Any disconnect between supply of the waste water and its use must be made up by storage, supplemental water makeup, or discharge to the public sewers. Minimizing supplemental water makeup and the discharge of wastewater to the sewers is the key to optimizing a successful blackwater reclaim system.[4]

Establishing Water Balance

The first step in evaluating a wastewater reuse system is to establish a numerical account of how much water enters and leaves a site. A water balance sheet should contain detailed information about the amount of water used by each process. Knowing the water balance is crucial for understanding and managing water flows throughout the plant. It also helps to identify equipment with water-saving opportunities and to detect leaks. For a net zero water project, which would be the optimum application of water reuse (in a building with water use that is self–sustaining), the amount of water entering and leaving a site should ideally reflect the natural hydrology of the site.

Bruggen and Braecken[5] offer a "step-by-step method to optimize the water balance" in three steps:

1. Investigate the current water balance in detail.

2. Combine water consuming processes and reuse water where possible for other purposes requiring a lower water quality.

3. Regenerate partial waste streams and reintroduce them into the process cycle.

Developing a water balance includes quantifying all the water sources that can be recycled (e.g., lavatories, showers, toilets and urinals, rainwater catchment) and matching them with applications for the reclaimed water (e.g., landscape irrigation, toilet flushing, cooling tower makeup). Ideally, both water amounts will be approximately equal to avoid storage issues and unnecessary cost associated with system overdesign.

Figure 14.4 provides an overview of possible water source and application opportunities that design teams may select from, in establishing a water balance.

Constructing a spreadsheet, as shown in Fig. 14.5, is a useful way to assist the designer in determining the building water balance by quantifying sources of wastewater to be processed into greywater and opportunities to utilize the reclaimed water.[6]

Figure 14.5 shows how a spreadsheet can be used to assist in achieving the water balance between the source wastewater and what is available for reuse as greywater. Section A of this form estimates the water consumed from a baseline selection of plumbing

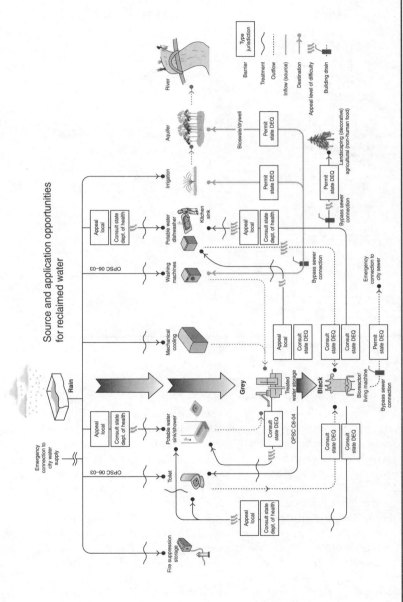

Figure 14.4 Source and application opportunities for wastewater recycling. (*Courtesy of Achieving Water Independence in Buildings, Cascadia Region Green Building Council, www.cascadiagbc.org/lbc/resources/water/oregon*)

Waste Water Reuse Calculator

Non-residential Buildings

Office space	200000	ft2
Occupant ratio	200	ft2 per occupant
Total occupants	1000	
Ratio Male:Female	60%	
Building levels	2	
Roof Area (ft2)	10000	

SECTION A: Base line Fixture Water Consumption

Baseline Case: change occupant values to reflect anticipated occupancy

Fixture Type	Consumption	Daily Uses	Duration	Occupants	Daily Water Use (gal)	
1.6 gpf toilet - male (gallons per flush)	1.6	1	1	600	960	
1.6 gpf toilet - female (gallons per flush)	1.6	5	1	400	1,920	
1.0 gpf urinal - male (gallons per flush)	1	2	1	600	1,200	
Commercial Lavatory Faucet - 0.5 gpm	0.5	3	0.25	1000	375	
Kitchen sink - 2.2 gpm	2.2	1	0.25	1000	550	assume 1 kitchen sink per level
Showerhead - 2.5 gpm	2.5	0.05	8	1000	1,000	
Cooling Tower					0	
Irrigation	Landscaped area, acres	Irrigations per week	Irrigation rate, inches			
	0.25	3	0.5			
			Total Daily Volume		1569	
			x Annual Work Days		7,574	
			Total Annual Usage		1,969,299	Gallons / Year

SECTION B: High Efficiency Fixture Water Consumption   **SECTION C: GREYWATER RECYCLING SAVINGS**

Calculator: To determine estimated savings, insert occupant values (same as Baseline) and
consumption values based on fixtures and fixture fittings installed

Fixture Type	Consumption	Daily Uses	Duration	Occupants	Available Daily Water Source (gal)	Source Water to be Reclaimed	Available for Greywater Recycle Use (gal)
1.28 gpf toilet - male (gallons per flush)	1.28	1	1	600	768		768
1.28 gpf toilet - female (gallons per flush)	1.28	5	1	400	1,536	2,904 blackwater	1,536
0.5 gpf urinal - male (gallons per flush)	0.5	2	1	600	600		600
Commercial Lavatory Faucet - 0.5 gpm	0.5	3	0.25	1000	375		-
Kitchen sink - 2.2 gpm	2.2	1	0.25	1000	550	1,925 greywater	-
Showerhead - 2.5 gpm	2.5	0.05	8	1000	1,000		-
Cooling Tower						from mechanical engineer ==>	-
Irrigation	Landscaped area, acres	Irrigations per week	Irrigation rate, inches				
	0.25	3	0.5				1569
			Total Daily Volume	4,829			4,473
			x Annual Work Days	260			260
			Total Annual Usage	1,255,540	Source Water Gallons / Year		1,163,039 Use Water Gallons / Year
			Annual Savings	713,759	Gallons / Year Saved		1,876,797 Gallons / Year Saved
			% Reduction	-36.2%	from improved fixtures		95% from high efficiency fixtures and greywater reuse

Notes: (1) Values shown in red reflect the maximum consumption values associated with the provisions
called out in the IAPMO Green Plumbing and Mechanical Code Supplement.
(2) If metering faucets are used, insert the flow rate of the faucet in the Consumption column
and insert the cycle time in the Duration column (assume 1 cycle per use).

FIGURE 14.5 Wastewater reuse calculation form.

fixtures. It uses data from the input section at the top of the form to establish the occupancy by gender and count. Then it multiplies the consumption per fixture times daily use times the number of occupants to get daily water use for each fixture. This is done for each fixture to determine the estimated annual consumption for the entire plumbing system. This methodology is consistent with IAPMO Green Plumbing Supplement Appendix A—Method of Calculating Water Savings.

Section B repeats the process above, but for high-efficiency systems to be evaluated as an alternative to the baseline fixtures. The water consumption difference between the two fixture types is used to determine the savings from using high-efficiency plumbing fixtures.

Section C is used to determine the amount of water that can be saved by reprocessing the wastewater into greywater. Net water savings is determined by adding water saved from using high-efficiency fixtures to the wastewater processed into greywater for reuse in toilet and urinal flushing, cooling tower makeup, and landscape irrigation. It is important to compare the total annual source water with the greywater use to determine how close to a balance the building can achieve. If there is an imbalance between the greywater produced and what can be utilized, then a provision for the appropriate makeup water can be recognized and accounted for in the final design.

Fit-for-Purpose Water

The best place to begin any water recycling scheme is with an assessment of the potential sources and reuse applications. Water is often excessively treated for the desired end use, and while safety is always the first priority, it's best to approach each water recycling opportunity with a fit-for-purpose methodology. For example, capturing rainwater off the roof and using it for subsurface irrigation would require no treatment. Rainwater is a low-risk source, and subsurface irrigation is a low-risk application. Greywater contains some pathogens, organics, and contaminants, so it needs some treatment, even if it is being used for subsurface irrigation. More treatment will be required if using greywater for surface irrigation or toilet flushing.

Blackwater is a source containing even higher contaminant levels (see Table 14.1). If it is used for surface irrigation, it is considered a high exposure-risk application due to its potential exposure to humans. This scenario would obviously require a high level of treatment to properly mitigate the associated risks.

All too often, water recycling schemes embark on a treatment agenda that disregards a fit-for-purpose design. Treating the source excessively certainly achieves the water quality objectives adequate for the end use, but overtreatment comes with a cost. For optimal use of resources, both energetic and monetary, it is important that there is a proper evaluation of the level of treatment needed versus the level of risk at the point of use.

Parameter	Influent Water Quality
Biochemical Oxygen Demand (BOD), mg/L	300–600
Suspended Solids, mg/L	300–600
Total Nitrogen, mg/L	70–120
Total Phosphorus, mg/L	20–30
Fecal Coliforms, cfu/100 mL	10^6–10^8
E. Coli, organisms/100 mL	10^6–10^8

TABLE 14.1   Blackwater Contaminant Levels

Parameter	Water Quality—Non-Potable Applications[1]
Total Coliform	< 500 cfu/100 mL
Fecal Coliform	< 100 cfu/100 mL
Turbidity	< 10 NTU
Protozoan Cysts	N/A
Testing Frequency[2, 3]	Every 12 Months

1. As approved by local codes, this quality water is suitable for toilet and urinal flushing, washing machine makeup, and other approved applications in occupied space for environments with non–health impaired occupants.
2. Upon failure of total and fecal coliform test, system shall be recommissioned involving cleaning and retesting.
3. Testing requirements are intended to be a requirement of the system owner to maintain a record of the maintenance and testing history.

TABLE 14.2 Water Quality for Non-Potable Applications[8]

Table 14.2 represents EPA[7] and Board of Health water standards for swimming at public beaches and for non-potable water applications inside buildings, assuming no health-impaired people are present. If the application was to flush toilets in a nursing home, hospital, or clinic, the judgment of the design professional might be for higher-quality water where health-impaired persons are present.

The water quality standard and testing methodology in Table 14.2 is supported by the American Rainwater Catchment Systems Association (ARCSA.org). It exceeds the water quality rainwater use standard in the United Kingdom (BS 8515:2009) as well as Australian Guidelines, and is consistent with Texas Rainwater Standards (as stated in *Harvesting, Storing, and Treating Rainwater for Domestic Indoor Use*, Texas Commission on Environmental Quality, January 2007). It is also consistent with EPA standards for coastal swimming quality water.

Benefits of Blackwater Recycling

The benefits of blackwater recycling are considerable. Aside from constructively utilizing the nutrients found in wastewater, which otherwise would pollute the waterways, the benefits of recycling blackwater include the following:

- Reduction of stress on septic systems—a blackwater recycling system can take some of the stress off older septic systems, which may be close to failing.

- Municipalities can conserve potable water by using recycled blackwater for noncritical applications, such as watering lawns, nonfood crops and gardens, vehicle washing, road cleaning, etc.

- Being a distributed system, backwater recycling systems allow for incremental development, without the up-front cost of a central sewer system.
- When blackwater is treated, it is released back into the environment in a cleaner form and helps preserve natural water habitats.
- Recycled blackwater is rich in nutrients for plants.
- Recycling blackwater can be part of the solution for water shortages experienced in some cities and areas.
- Conserves energy—removing the harmful bacteria and debris from blackwater is costly and requires considerable processing plant energy. By treating the wastewater on-site, treatment and pumping costs are reduced.
- Habitat protection—by recycling blackwater, there is less chance of wastewater seeping into natural habitats.
- Plant growth—the plants that are grown with recycled blackwater rarely need any fertilizer. There are enough nutrients left in the water after treatment that plants can easily feed off of them.

Limitations to Blackwater Recycling

While there are several advantages to working with blackwater recycling, there are also disadvantages to blackwater recycling systems. Some of these drawbacks include the following:

- Cost: Blackwater recycling systems can be expensive—not only to install but to maintain and to fix if something goes wrong. Although higher in cost, because of its near continuous use, blackwater recycling may have a higher return on investment than, say, a rainwater catchment system that can be intermittent in water availability. Although each project needs to be evaluated on its own merits, Fig. 14.6 indicates related payback for various water conservation measures.
- Smell: While many people say there is no discernable smell, there are others who claim that they can smell the system (which is mostly sewage) all the time. If it smells, it usually means that the bacteria are struggling, likely due to something that has been put down the drain that shouldn't have been, such as disinfectant.
- Maintenance Owning a blackwater treatment system normally doesn't pose any serious problems beyond what can be expected

Summary of savings

Conservation approach	% reduction commercial building example	% reduction student housing example
Fixtures	7.9%	3%
Rainwater	20%	?
Greywater	50%	44%
Blackwater	90%	54%

FIGURE 14.6   Relative payback for various water recycling methodologies. (*Courtesy of Aquacell Pty Ltd, www.aquacell.com.au*)

from other commercial building systems. The specialized maintenance, performed every three months or so, is a higher level skill than in-house maintenance staff or local contractors can commonly provide. Generally, it is provided by the system installer. The system owner needs to be careful of what is sent down the drains. Chemicals and anti-bacterial products can destroy the bacteria colony, which has a side benefit of requiring the use of environmentally friendly products.

An order of magnitude of the level of annual maintenance required would be approximately 5 percent of the installed cost. For example, for a $1,000,000 project, annual maintenance cost would need an allocation of around $50,000 per annum. A range of costs would be somewhere around 3 percent for larger projects and 10 percent for smaller projects. The level of maintenance would be such that this would be done best by an individual very familiar with the specific technology.[9]

Other Uses of Blackwater

In situations where there is a water shortage or restricted sewer services available to the site, the concept of sewer mining has been successfully used. This involves the processing of wastewater from the public sewer to meet the greywater needs of a building. Because freshwater needs for drinking and food processing are generally a small portion of a building's water need (approximately 10–20 percent), disinfected rainwater or bottled water can be provided if not available from the public utility. Recycling wastewater by means of sewer mining is a proven methodology that has a future as an answer for combined sewer overflows, undersized water, and sewer infrastructure. It is also a solution in water-deprived environments.

Figure 14.7 shows the various capacities available for commercial microbiological filters, while Fig. 14.8 shows the various components of these devices that are necessary for proper wastewater filtration and treatment.

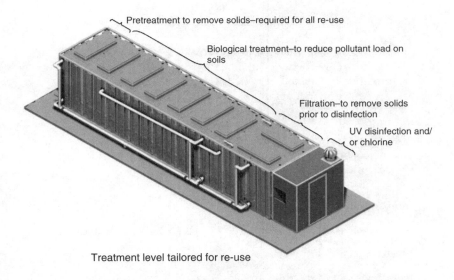

Pretreatment to remove solids–required for all re-use

Biological treatment–to reduce pollutant load on soils

Filtration–to remove solids prior to disinfection

UV disinfection and/ or chlorine

Treatment level tailored for re-use

FIGURE 14.7   Typical microbiological filter and its components. (*Courtesy of Aquacell Pty Ltd, www.aquacell.com.au*)

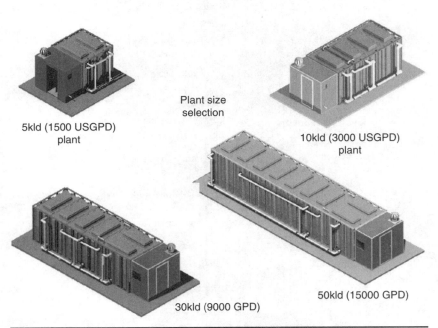

Plant size selection

5kld (1500 USGPD) plant

10kld (3000 USGPD) plant

50kld (15000 GPD)

30kld (9000 GPD)

FIGURE 14.8   Various size of microbiological filter systems. (*Courtesy of Aquacell Pty Ltd, www.aquacell.com.au*)

Conclusion

As technology progresses and improvements to blackwater recycling systems continue, the benefits of on-site waste management will become apparent and system use more widespread. A decentralized approach to waste management, as seen with blackwater systems, offers a solution to overcharged sewer systems, thereby reducing the cost to public utilities. When the ability to recycle a community's waste becomes viable, the wastewater that now is perceived as a liability can be constructively used and treated as an asset, rather than being a contaminant to the water supply.

Notes and References

1. Information provided by Aquacell Pty Ltd, 1/10B Production Place, Penrith NSW 2750 Australia, www.aquacell.com.au.
2. Combined sewer overflow occurs when storm water, commonly piped through the sanitary sewers in older cites, overflows the utility sewage treatment plant, thereby adding diluted waste directly into the stream or river that is used as a water source.
3. Information provided by Aquacell Pty Ltd, www.aquacell.com.au. BREAM (Building Research Establishment Environmental Assessment Method) is a voluntary measurement rating for green buildings that was established in the UK by the Building Research Establishment (BRE). Since its inception, it has grown in scope as well as geographically, being exported in various guises across the globe. Its equivalents in other regions include LEED (Leadership in Energy and Environmental Design) in North America, Green Star in Australia, and HQE in France.
4. Buehrer, M., 2020 Engineering, Bellingham, Washington, 1996.
5. Van der Bruggen, B., Braeken L., "The Challenge of Zero Discharge: From Water Balance to Regeneration," 188.1-3, 2006.
6. The spreadsheet shown is a result of collaboration between the author and Colin Fisher of Aquacell Pty Ltd.
7. United States Environmental Protection Agency (EPA).
8. Water quality and testing procedures for non-potable water applications are as recommended by the International Association of Plumbing and Mechanical Officials (IAPMO.org) *Green Plumbing Supplement*, the ARCSA/ASPE *Rainwater Catchment Design and Installation Standard*, the Texas Water Development Board *Texas Manual on Rainwater Harvesting*, and United States Environmental Protection Agency document # EPA-823-R-03-008 *Recreational Microbial Parameters*, June 2003.
9. Information provided by Colin Fisher, Managing Director for Aquacell Pty Ltd, 1/10B Production Place, Penrith NSW 2750 Australia, www.aquacell.com.au.

CHAPTER 15

Septic System Design

Introduction

Septic systems are on-site household sewage treatment systems that use subsurface drainage fields, in concert with natural processes, to recycle human wastewater. Properly designed systems transmit and store the household effluent below the surface and allow soils and soil microorganisms to clean the wastewater before returning it to the hydrologic cycle.

How Does a Septic System Work?

A septic system is composed of three parts: a septic tank, a drainfield, and soil. The design of the system depends on the number of users, the amount of water used per capita, the average annual temperature, the pump-out frequency of the tank, and the characteristics of the wastewater (Fig. 15.1).

Septic Tank

A septic tank is an underground watertight settling chamber, constructed of concrete, fiberglass, PVC, or plastic, into which raw sewage is delivered through a pipe from plumbing fixtures inside a house or building. Septic tanks can be installed in every type of climate although the efficiency will be negatively affected in colder climates. Even though the septic tank is watertight, it should not be constructed in areas with high groundwater tables or where there is frequent flooding. For an illustration of a typical single compartment septic tank, see Fig. 15.2(a).

The tank should be sized to provide a retention time of 48 hours to achieve moderate treatment of the effluent. The sewage is treated in the tank by separation of solids to form sludge and scum. Generally, the removal of 50 percent of solids, 30 to 40 percent of biochemical oxygen demand (BOD), and a 1-log removal of *E. coli* can be expected

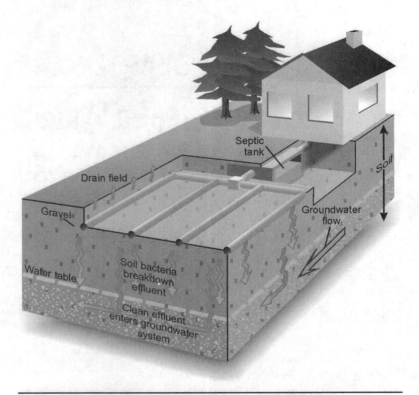

FIGURE 15.1   Septic field schematic. (See also Color Plates.)

in a well-designed septic tank, although efficiencies vary greatly depending on operation, maintenance, and climactic conditions.[1]

Liquid and solid household waste flows into the tank, which is typically belowground. The septic tank primarily functions as a separation chamber and an anaerobic digester. Solid material, more dense than water, sinks to the bottom; and less dense material, such as fats and oils, floats to the top. The liquid in the middle of the tank is termed greywater and exits the tank to flow to the septic drainfield. Settling and anaerobic processes reduce solids and organics, but the treatment is only moderate in the septic tank itself.

A septic tank should preferably have at least two chambers (Fig. 15.2(b)). The first chamber should be at least 50 percent of the total length, and when there are only two chambers, it should be two-thirds of the total length. Most of the solids settle out in the first chamber. The baffle, or the separation between the chambers, is to prevent scum and solids from escaping with the effluent. A T-shaped outlet pipe will further reduce the scum and solids that are discharged.

Liquid flows into the tank and heavy particles sink to the bottom, while scum (oil and fat) floats to the top. With time, the solids that

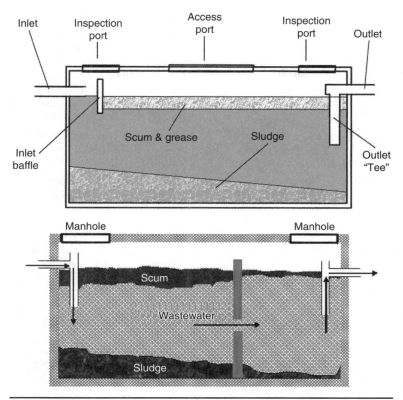

FIGURE 15.2 (a) Typical single chamber septic tank. (b) Typical double chamber septic tank.

settle to the bottom are degraded anaerobically. The system works well where the soil is permeable and not liable to flooding or water logging, provided the sludge is removed at appropriate intervals to ensure that it does not occupy too great a proportion of the tank capacity. Because the rate of accumulation is generally faster than the rate of decomposition, the accumulated sludge must be removed periodically, typically every two to five years, to ensure proper functioning.[2] For a commercial application, more frequent (annual) inspection is recommended.

Drainfield (Also Called a Leachfield)

Drainfields will function properly in a wide range of soils in all climatic regions. In the United States, although septic systems are required in areas where municipal wastewater treatment systems are not available, there are limitations to the capabilities of these systems.

Drainfields will not function hydraulically in soils with high clay or a perched seasonal water table. The high-clay soils will not allow the wastewater to infiltrate fast enough to remain below the soil surface. At the other extreme, soils with extremely high sand content or shallow depth to bedrock do not renovate the wastewater before reaching the aquifer. The soils with a perched seasonal water table near the surface will fill with natural rainwater at some times during the year and will not have the capacity to accept additional wastewater. This will cause the wastewater to surface, where it will run off into streams or may come into direct contact with humans. In some cases, alternative designs that involve pretreatment of wastewater prior to reaching a drainfield may be acceptable if the soil does not provide proper renovation.

Typically, the drainfield consists of three to four trenches that are 18 to 24 in (45.7 to 61.0 cm) deep with a perforated pipe in 12 to 18 in (30.5 to 45.7 cm) or gravel covered by 6 inches of soil. A properly designed drainfield should distribute wastewater over a large enough area to allow it to infiltrate into the soil and be treated.[3] A distribution box is often used to accomplish equal distribution between the septic field trenches.

Soil

The drainfield is sized according to the absorption properties of the soil at each site. Each site should best get a thorough evaluation by a professional prior to installation. Without the assistance of a soil scientist, infiltration can be inferred from the soil texture or crudely determined by a percolation test (see Appendix). The soil must be uniform in color (indicating well-drained conditions) in the area where the drainfield is located. A perched seasonal water table is determined by observing the colors in the soil and knowledge of the depth to the regional aquifer. A grey-colored soil or a soil that has grey mottles (spots) indicated standing water during the rainy season.

Advantages	Disadvantages
Can be built and repaired with locally available materials	Low reduction of pathogens, solids, and organics unless used with leach field
Long service life if properly maintained	Effluent and sludge require secondary treatment and/or appropriate discharge
No problem with flies and odors if used correctly	Requires constant source of water
Moderate operating costs (depending upon water and emptying)	Permeable soil required
No electrical energy required	

TABLE **15.1**   Septic Tank Advantages and Disadvantages

With a properly designed septic system, the wastewater will remain below the soil surface, where it will come in contact with sufficient microorganisms and soil complexing agents to be properly renovated and returned to the hydrologic cycle. The soil acts as a natural filter and contains organisms that help treat the wastewater. The pathogens in the wastewater are typically destroyed by the natural populations of soil microorganisms. Nutrients, such as phosphorus (P), are adsorbed or chemically bonded to the soil particles, which limits mobility.[4]

Operation and Maintenance

Septic tanks should be checked to ensure that they are watertight, and the levels of the scum and sludge should be monitored to ensure that the tank is functioning well. Because of the delicate ecology, care should be taken not to discharge harsh chemicals into the septic tank.

Proper drainfield function requires wastewater with a minimum amount of solid material; therefore, the septic tank should be cleaned out periodically. If the septic tank overfills and solid materials begin exiting the tank with the wastewater, the solids will cause a decrease in the soil permeability, and the drainfield will fail by surfacing. Limiting the flow of water from the household may also be helpful. To lessen water flow, low-flush toilets and other water-conserving measures are commonly used.

Avoid driving or parking over the septic drainfield. Heavy vehicles may crush the pipes within the drainfield or compact the soil, thereby reducing the absorption capability of the soil. Also refrain from flushing nondegradable objects or hazardous chemicals down the toilet. These items may clog the system or disrupt the microbiological treatment of the waste. Building structures over the drainfield or covering the field with hard surfaces such as asphalt will likewise limit the performance of the septic system.

Trees or shrubs over the drainfield may cause problems in some septic systems. These plants have extensive root systems, which seek out the nutrient-rich water, and thus may clog the system.

There is no evidence that septic tank additives improve the performance of septic systems. Additives are costly and may even be harmful to the system.[5]

Anaerobic Baffled Reactor

An anaerobic baffled reactor (ABR) is an improved septic tank because of the series of baffles under which the wastewater is forced to flow (Fig. 15.3). The chief benefit of an ABR is the increased contact time with the active biomass (sludge), which results in improved treatment.

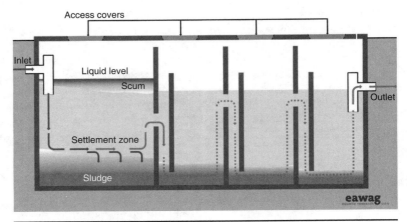

FIGURE 15.3   Anaerobic baffled reactor (ABR).

In operation, the majority of solids are removed in the sedimentation chamber at the beginning of the ABR, which typically represents 50 percent of the total volume. The upflow chambers provide additional removal and digestion of organic matter: BOD may be reduced by up to 90 percent, which is far superior to a conventional septic tank. As sludge is accumulating, de-sludging (tank pump-out) is required every two to three years. Critical design parameters include a hydraulic retention time (HRT) between 48 to 72 hours, up-flow velocity of the wastewater at less than 2 ft (0.6 m)/hr, and an additional number of up-flow chambers (two to three).

Adequacy

This technology is easily adaptable and can be applied at the household level or for a small neighborhood. An ABR can be designed for a single house or a group of houses that are using a considerable amount of water for clothes washing, showering, and toilet flushing. It is mostly appropriate if water use and supply of wastewater are relatively constant. Since the tank is normally installed underground and requires little area, this technology is appropriate for areas where land may be limited. As in the case of a conventional septic system, it should not be installed where there is a high groundwater table, as infiltration will affect the treatment efficiency and contaminate the groundwater.

Typical inflows range from 50 to 500 gpd (2,000 to 200,000 L/day). The ABR will not operate at full capacity for several months after installation because of the long start-up time required for the anaerobic digestion of the sludge. Therefore, the ABR technology should not be used when the need for a treatment system is immediate.

Advantages	Disadvantages
Resistant to organic and hydraulic shock loads	Requires constant source of water
Greywater can be managed concurrently	Effluent and sludge require secondary treatment and/or appropriate discharge
Can be built and repaired with locally available materials	Low pathogen reduction
No problem with flies and odors if used correctly	Requires expert design and construction
Long service life	Permeable soil required
High reduction of organics	Pretreatment is required to prevent clogging
Moderate installation and operating costs (depending upon water and emptying)	
No electrical energy required	

TABLE 15.2   Anaerobic Baffled Reactor Advantages and Disadvantages[8]

To help the ABR to start working more quickly, it can be seeded, i.e., active sludge can be introduced so that active bacteria can begin working and multiplying immediately.[6]

Because the ABR must be emptied regularly, a vacuum truck should be able to access the location. ABRs can be installed in every type of climate although the efficiency will be affected by temperature.[7]

Health Aspects/Acceptance

If septic tanks are used in densely populated areas, on-site infiltration should not be used otherwise the ground will become oversaturated and excreta may rise up to the surface, posing a serious health risk. Instead, the septic tank should be connected to a sewer, and the effluent should be transported to a subsequent treatment or disposal site. Larger multi-chamber septic tanks can be designed for groups of houses and/or public buildings (i.e., schools).

Although the removal of pathogens is not high, the entire tank is below the surface so users do not come in contact with any of the wastewater. Users should be careful when opening the tank because noxious and flammable gases may be released. Septic tanks should have a vent. When pumping the solids from the tank, users generally should not attempt to empty the pit themselves. They should instead use a commercial contractor experienced in this process.

How to Run a Percolation Test[9]

The first step in designing a septic system is to estimate the soil's ability to absorb water. In the absence of a soil scientist to size the drainage field, a percolation test can be used to provide an estimate. The percolation test is designed to determine the suitability of a site for a subsurface private sewage disposal system (i.e., septic system). More specifically, a percolation test measures the ability of the soil to absorb liquid. Septic system designers use the results of percolation tests to properly construct septic systems.

Percolation tests are designed to simulate conditions in a septic system. The percolation test consists of a hole 6 to 12 in (15.2 to 30.3 cm) in diameter dug in the area of the proposed septic system. The depth of this hole varies depending on the soils encountered, but it is generally not greater than 24 in (61.0 cm). The hole is initially filled with water (presoak) in an attempt to saturate the soil, it is allowed to drain away, and then it is refilled with approximately 12 inches of water. The rate at which the water drops in the hole is measured at intervals over a period of time ranging from 30–60 minutes. The uniform slowest rate of drop of the water level over a measured time interval is converted to minutes per inch and used as a basis of design in determining the septic system size. For example, if the water dropped uniformly ¼ inch every five minutes, the rate would be 20 minutes per inch.

Often, the local Authority Having Jurisdiction (AHJ)[10] will provide a simple table that determines the size of the system based on the measured perk rate and the number of bedrooms in the home. The greater the number of bedrooms and the slower the percolation rate, the larger the system required. Commercial systems are sized using the perk rate and projected estimates of water usage in gallons per day.

Part 1: Site Preparation

The following steps outline the procedure for performing a percolation test:

1. Dig at least six test holes. The holes should be:
   - Evenly spaced, approximately 30 to 40 ft (9.2 to 12.2 m) apart, but not less than 30 feet, in the area of the proposed septic field.
   - At least six inches in diameter. Larger holes are acceptable, but will require more water.
   - Dug to a depth of 24 in (0.61 m).
   - No closer than 3 ft (0.91 m) to the 48 in (1.22 m) test hole.
   - No closer than 75 ft (22.9 m) to the nearest water well or proposed water well.

- No closer than 5 ft (1.52 m) to any lot line or easement.
- No closer than 20 ft (6.10 m) to any building.
- Not located in any easement or flood plain area.
- Not located in any area that has previously failed a percolation test.

2. Dig a 48 in (1.22 m) test hole in the lowest part of the test area.

3. The bottom and sides of each 2 ft (0.61 m) test hole may be roughened with a saw blade, knife, or other sharp instrument. It is advisable to roughen those surfaces, which may become smeared with mud during the digging process. Smearing of those surfaces will tend to reduce the seepage rate. Remove all loose soil from the bottom of the holes.

4. Fill each of the 2-foot test holes with water and refill as necessary to maintain a minimum depth of 12 inches for a period of at least four hours. This is the presoak. It must be conducted between 12 noon and 4 o'clock p.m. the day before the test. The local AHJ (Health Department) technicians may spot-check the presoak to make sure it is being done properly. Note: Do NOT put any water in the 48-inch test hole.

Part 2: Testing Procedure

Prepare in advance of the test one wooden stake for each 2-ft (61.0 cm) test hole. Three nails should be pounded into each stake. The first nail should be 3 in (7.62cm) from the bottom of the stake, the second nail must be exactly 6 in (15.24 cm) above the first nail, and the third nail must be exactly 20 in (51 cm) above the second nail.

1. On the morning of the percolation test (the day following the presoak), the test holes should be cleaned out. Any loose soil or silt that accumulated at the bottom of the holes during the presoak should be removed. One of the stakes prepared per the instructions in #4 above should be driven into each hole so that the first nail rests on the bottom and the stake stands by itself.

2. At a time previously arranged between the percolation tester and the AHJ technician, each test hole is filled with water to the level of the second nail on the percolation test stake (which should be exactly six inches). NOTE: Do not wait for the technician to arrive to start the test. After one hour, the percolation tester will measure how far the water has dropped in each test hole. The holes are then refilled with water to the level of the second nail. This process is repeated for at least three more hours. The AHJ technician will take the measurements for the second and all readings after that.

3. In very porous soils, the water in the test holes may seep away in less than an hour. When this happens, the AHJ technician may go to half-hour or even ten-minute readings. A test may also be extended to five or more hours if the last three readings are inconsistent.

Part 3: What Can Go Wrong?

The following is a summary of when and how a percolation test may fail during the presoak:

1. No 48-in (1.22 m) test hole dug
2. Evidence of seasonal high water table within 24 in (0.61 m) of the surface (The test may be continued, but only at the request of the tester.)
3. Improper presoaking—less than 12 in (0.30 m) of water in any test hole at any time
4. Filled lots when the Health Department has not been previously notified
5. Isolation distances or other location problems with the test area
6. Evidence of impropriety

The following is a summary of how and when a percolation test may fail during the test itself:

1. At least half of the test holes fail to drop 1 inch or more in any of the hourly readings.
2. There is evidence of impropriety.

Part 4: Test Results

When the percolation test is completed, most testers are anxious to know the results. The AHJ technician who witnessed the test may be able to say if a test fails, but never if it passes. Percolation test results usually take a few days to process. Testers should advise the AHJ technician during the percolation test about how he or she would like the results of the percolation test reported.

Part 5: Calculations and Sizing the Trench(es)

Percolation test results are calculated using one of the following methods. The method that results in the slowest percolation rate is used.

1. Calculate percolation rate: Divide the time interval by the drop in water level to determine the percolation rate in minutes per inch (MPI).

Example (English Units):
If the drop in water level is 5/8 inch in 30 minutes, the percolation rate is:

$$\frac{30}{5/8} = 30 \times \frac{8}{5} = 48 \text{ MPI} \quad \text{or} \quad \frac{30}{5/8} = \frac{30}{0.625} = 48 \text{ MPI}$$

If the drop is 2½ inches in 10 minutes, then the percolation rate is:

$$\frac{10}{2\,1/2} = \frac{10}{2\,1/2} = 10 \times \frac{2}{5} = 4 \text{ MPI} \quad \text{or} \quad \frac{10}{2\,1/2} = \frac{10}{2\,1/2} = 4 \text{ MPI}$$

Calculate the percolation rate for each reading. When three consecutive percolation rates vary by no more than 10 percent, use the average value of these readings to determine the percolation rate for the test hole. Percolation rates determined for each test hole should be averaged in order to determine the design percolation rate. For reporting the percolation rate, worksheets showing all measurements and calculations should be submitted with the site evaluation report. You can reproduce the blank form on the back page of this folder for use in recording percolation test data.

Note that a percolation test should not be run where frost exists in the soil below the depth of the proposed sewage treatment system.

For our example, let's assume the results of the percolation test indicated the soil would percolation at a rate of 22.3 min/in (8.7 min/cm)

2. Determine the trench bottom area. Table 15.3 shows sewage flows and soil treatment areas. The amount of trench bottom area required depends on the texture of the soil as measured by the percolation rate, the daily sewage flow, and the depth of rock placed below the distribution pipe.

The daily amount of sewage wastes must be estimated in order to size the soil treatment unit. For residences, the daily amount of sewage flow is based on the number of bedrooms and the type of residence. A luxury three-bedroom house likely will generate more sewage than a more modest house. Sewage flows for different types of houses can be estimated from Table 15.3. Using a large sewage flow provides a factor of safety in sizing the soil treatment unit. Also consider future house expansion.

Table 15.3 is used to determine the trench bottom area, for a three-bedroom house, Type I dwelling as follows

- Given that our percolation rate as calculated above is assumed to be 22.3 MPI, from Table 15.3, a three-bedroom Type I dwelling is estimated to generate 450 gal (1.70 m³) of sewage per day.

Estimated Sewage Flows in Gallons (& cubic meters) per Day				
Number of Bedrooms	Type of Residence[a]			
	I	II	III	IV
2	300 (1.14)	225 (0.85)	180 (0.68)	60% of values in Type I,II or III columns
3	450 (1.70)	300 (1.14)	218 (0.82)	
4	600 (2.27)	375 (1.42)	256 (0.97)	
5	750 (2.84)	450 (1.70)	294 (1.11)	
6	900 (3.41)	525 (1.99)	332 (1.26)	
7	1050 (3.97)	600 (2.27)	370 (1.40)	
8	1200 (4.54)	675 (2.55)	408 (1.54)	

[a]**Type I:** The total floor area of the residence divided by the number of bedrooms is more than 800 ft^2 (74.32 m^2) per bedroom, or more than two of the following water-use appliances are installed: automatic washer, dishwasher, water softener, garbage disposal, Jacuzzi, or humidifier in furnace.

Type II: The total floor area of the residence divided by the number of bedrooms is more than 500 ft^2 (46.5 m^2) per bedroom, and no more than two of the water-use appliances are installed.

Type III: The total floor area of the residence divided by the number of bedrooms is less than 500 ft^2 per bedroom, and no more than two of the water-use appliances are installed.

Type IV: No toilet wastes flow into sewage treatment system.

TABLE 15.3   Sewage Flows and Soil Treatment Areas

- The trench bottom area required for a percolation rate in the range of 16 to 30 MPS (6.30 to 11.81 min/cm is 1.67 square feet/gallon (40.98 m^2/m^3) of waste per day.

- Thus, the total required bottom area is 1.67 × 450 = 750 ft^2 (40.98 × 1.7 = 69.7 m^2) square feet for trenches with 6 inches (15.24 cm) of rock below the distribution pipe. If 12 in (30.5 cm) of rock is used as recommended, the trench bottom area can be reduced by 20 percent (see footnote a, Table 15.4). The required trench bottom area is then 0.80 × 750 = 600 ft^2 (55.7 m^2).

The trench bottom area can be reduced by 34 percent for 18 in (.46 m) of rock below the distribution pipe and by 40 percent for the maximum rock depth of 24 in (.61 m). As rock depth increases, the required trench bottom area decreases because more soil is exposed along the trench sidewall, and a greater liquid depth increases the flow through the trench bottom.

The minimum trench width is 18 in; the maximum width is 36 in (.91 m). Using 36-inch wide trenches in the above example, total trench length with 12 inches of rock below the distribution pipe is

Percolation Rate, Minutes per Inch (Minutes/cm)	Soil Treatment Area in Square Feet per Gallon of Waste per Day[a] (m²/m³ per day)
Faster than 0.1[b] (.04)	Soil too coarse for sewage treatment[b]
0.1 to 5 (.04 to 2.0)	0.83 (20.37)
6 to 15 (2.0 to 6.0)	1.27 (31.17)
16-30 (6.0 to 11.8)	1.67 (40.98)
31-45 (11.8 to 17.7)	2.00 (49.08)
46 to 60 (17.7 to 23.6)	2.20 (53.99)
Slower than 60[b]	Refer to information on mounds

[a]For trenches only, the bottom areas may be reduced if more than 6 in (15.24 cm) of rock is placed below the distribution pipe; for 12 in (30.48 cm) of rock below the distribution pipe, the bottom areas can be reduced by 20 percent; a 34 percent reduction for 18 in (45.72 cm); and a 40 percent reduction for 24 in (57.60 cm).
[b]Soil is unsuitable for standard soil treatment units. Refer to information on mounds and alternative systems.

TABLE 15.4   Soil Treatment Areas in Square Feet

200 lineal feet (60.98 m) (600 ÷ 3). It is recommended that the 200 lineal feet should be divided into two or more trenches. The sewage effluent should be distributed between the trenches by means of drop boxes.[11]

Conclusion

Septic tank systems are a proven technology that has provided long-term healthful means of providing waste management. Given proper soil conditions, installation, and maintenance, these systems can be relied upon to provide reliable service for 20 years and more.

Notes and References

1. Mara, D.D., *Low-Cost Urban Sanitation*, Wiley, Chichester, UK, 1996. (Sizing, volume, and emptying calculations and example design solutions, Chapter 6.)
2. Owens, P.R., Rutledge, E.M., *Septic Drain Field Design and Maintenance*, Owens: USDA-ARS, Mississippi State, Mississippi; Rutledge: University of Arkansas, Fayetteville, Arkansas.
3. Design information on sizing drainage fields (also referred to as leach fields) can be found in Chapter 15.
4. The ability of the soil to remove phosphorus is dependent upon the adsorption/precipitation capacity of the soil makeup: *Septic Drain Field Design and Maintenance*.
5. Owens, P.R., Rutledge, E.M., *Septic Drain Field Design and Maintenance*.
6. Owens, P.R., Rutledge, E.M., *Septic Drain Field Design and Maintenance*.

7. Foxon, K.M., Pillay, S., Lalbahadur, T., Rodda, N., Holder, F., Buckley, C.A., *The Anaerobic Baffled Reactor (ABR): An Appropriate Technology for On-Site Sanitation*, 2004.

8. Bachmann, A., Beard, V.L., McCarty, P.L., "Performance Characteristics of the Anaerobic Baffled Reactor," *Water Research*, 19 (1): pp. 99–106, 1985.

9. Reference www.percolationtest.com.

10. The local authority having jurisdiction generally is the local health department, but it may be the regional county engineer. Soil testing procedures to determine the size of a septic field vary for various locals, often requiring a soil analysis by a geotechnical engineer in lieu of a percolation test.

11. Gustafson, D., Machmeier, R.E., "How to Run a Percolation Test," *Biosystems and Agricultural Engineering*, University of Minnesota Extension.

DEPARTMENT OF ENVIRONMENTAL PROTECTION
BUREAU OF WATER STANDARDS AND FACILITY REGULATION

SITE INVESTIGATION AND PERCOLATION
TEST REPORT FOR ONLOT DISPOSAL OF SEWAGE

Application No._____ Municipality _____ County_____

Site Location_____ Subdivision Name _____

☐ SUITABLE   Soil Type _____   Slope _____%   Depth to Limiting Zone _____   Ave. Perc. Rate _____

☐ UNSUITABLE   ☐ Mottling   ☐ Seeps or Ponded Water   ☐ Bedrock   ☐ Fractures   ☐ Coarse Fragments

     ☐ Perc. Rate   ☐ Slope   ☐ Unstabilized Fill   ☐ Floodplain   ☐ Other _____

SOILS DESCRIPTION:

Soils Description Completed by:_____ Date:_____

Inches		Description of Horizon
__0__ TO _____		_____
_____ TO _____		_____
_____ TO _____		_____
_____ TO _____		_____
_____ TO _____		_____
_____ TO _____		_____

PERCOLATION TEST:

Percolation Test Completed by:_____ Date:_____

Weather Conditions: ☐ Below 40°F   ☐ 40°F or above   ☐ Dry   ☐ Rain, Sleet, Snow (last 24 hours)

Soil Conditions: ☐ Wet   ☐ Dry   ☐ Frozen

Hole No.	*** Yes	No	Reading Interval	Reading No. 1: Inches of drop	Reading No. 2: Inches of drop	Reading No. 3: Inches of drop	Reading No. 4: Inches of drop	Reading No. 5: Inches of drop	Reading No. 6: Inches of drop	Reading No. 7: Inches of drop	Reading No. 8: Inches of drop
			10/30								
			10/30								
			10/30								
			10/30								
			10/30								
			10/30								

***Water remaining in the hole at the end of the final 30-minute presoak? Yes, use 30-minute interval; No, use 10-minute interval.

Calculation of Average Percolation Rate:

Hole No.	Drop during final period	Perc. Rate as Minutes/Inch	Depth of Hole
_____	_____ "	_____	_____ "
_____	_____ "	_____	_____ "
_____	_____ "	_____	_____ "
_____	_____ "	_____	_____ "
_____	_____ "	_____	_____ "
_____	_____ "	_____	_____ "

The information provided is the true and correct result of tests conducted by me, performed under my personal supervision, or verified in a manner approved by DEP.

(S_____)
Sewage Enforcement Officer

TOTAL OF MIN/IN →   _____ = _____ **Min**

TOTAL NO. OF HOLES →   _____   **Inch**

☐ White -Local Agency   ☐ Yellow - Applicant   ☐ Pink - Local DEP Office

General Instructions

This form is to be utilized to record the results of site testing for installation of an onlot system. The first section of this form provides general site information, location, and a summary of the observed site conditions. Based on the conditions present, the SEO should check the appropriate suitability block in this section. The type of limiting zone must be noted, such as "mottling," "bedrock," etc.

Soils Description

The name of the individual providing the soils description must be provided, as well as the date of the evaluation. Describe the soil profile by horizons. For each horizon, indicate the depth from the mineral soil surface at which the horizon begins and ends. Indicate the presence and depth of any water seeps or standing water; also describe texture, structure, percentage of coarse fragments, color, indication of mottling, bedrock, masses of loose fragments or gravel or fractures or solution channels, all of which could allow unrestricted downward movement of effluent without treatment; describe any other appropriate information.

Beside the soils description, indicate the depth to limiting zone in inches; if no limiting zone was observed in the excavation, indicate that the limiting zone was greater than the depth of the probe. For example, more than 84".

Percolation Test

The name of the individual conducting the test and the date of the test must be provided. The general conditions under which the test was completed should be indicated by checking the appropriate blocks.

Preparation and initial presoak of the percolation holes must precede the actual test by 8–24 hours. Immediately before conducting the test, two 30-minute presoak periods must be completed. After listing the hole number under the appropriate column, an "X" or checkmark should be placed under the "YES" or "NO" column indicating presence or absence of water in the hole at the end of the final presoak period. Based on that information, the interval between readings should be circled.

The percolation test must be continued in each hole for eight consecutive readings OR until stabilization occurs. This means that the percolation test may continue in some of the holes throughout eight readings, while testing may be discontinued in other holes if stabilization occurs in that particular hole. It is also possible that the interval between readings may differ from one hole to another based on the results of the presoak. Stabilization is defined as "the difference of one-fourth inch or less of drop between the highest and lowest readings of four consecutive readings" in one particular percolation hole.

Upon completion of the percolation test, the final reading of each hole should be recorded in the calculation section and then converted to minutes per inch.

$$\Delta \; LZ = 35'' \qquad \Delta \; LZ = 28'' \qquad \Delta \; LZ = 10''$$
$$::: \; 225 \; min/in \qquad ::: \; 2.1 \; min/in$$

Additional information pertaining to the proper procedures for site investigation and the conduction of a percolation test may be found in Chapter 73, Section 73.12, 73.14, and 73.15.

One copy of this form should be attached to the applicant's copy of the application, one to the sewage enforcement officer's copy, and one to DEP's copy.

CHAPTER 16

Latrines and Privies

Introduction

The proper disposal of human waste (called "night soil" in many parts of the world) is one of the most pressing public health problems in many rural communities. The use of sanitary latrines or privies can be very effective in helping to control disease, which can be spread by water, soil, insects, or dirty hands. While it is necessary to have a sanitary water and food supply, sufficient medical service, and adequate diet to stop disease, the sanitary latrine breaks the disease cycle. Sicknesses that can be controlled by widespread use of sanitary latrines are dysentery, cholera, typhoid, and worms. The human suffering and economic loss caused by these diseases is staggering. In third world countries where proper sanitation is lacking, it has been said that half of the food eaten by a person with intestinal parasites is consumed by the worms.[1]

This chapter discusses the most fundamental of sewage treatment systems: latrines and privies. They are simple to design and construct, yet they provide a significant improvement in a community's level of sanitation.

Latrines and Privies

A pit dug into the ground remains as one of the most simple and used latrine types in the world (Fig. 16.1). There are tens of different types of pit latrine solutions, from a very simple trench, to a rather developed model. Fig. 16.1 shows a model of a basic pit latrine.[2,3]

The pit for the latrine should be dug at least 6ft (2m) deep and approximately 3.3 to 5.0 ft (1 to 1.5 m) in diameter. A lid is constructed from concrete, wood, or other local materials. A hole is made in the lid, where both solid excrement and urine drops to the pit.

Recommendations

The privy shall be constructed to do the following:

- Maintain a minimum of 2 ft (0.61 m) of separation between the bottom of the privy vault and indicators of highest seasonal water level

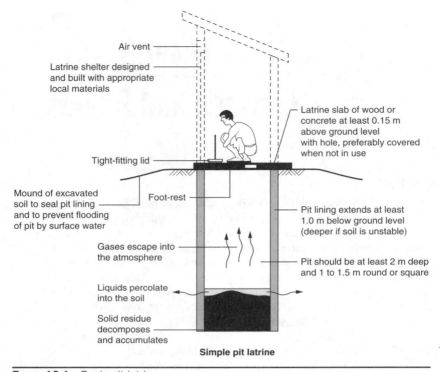

FIGURE 16.1 Basic pit latrine.

- Be located a minimum of 100 ft (30 m) downgrade from any potable water source
- Be located 100 feet from sensitive waters
- Prevent surface water from running into the pit

The pit is used until the pit becomes filled to within 16 in (40.6 cm) of the ground surface, when a new pit should be dug and the old pit filled.[4] It is common not to utilize the nutrient value of the waste from the old pit immediately, but to keep it covered in the old pit from soil removed from the new pit. After a period of time (+/− 1 year), the nutrient value of the excreta can be used as soil enrichment material.

Ventilated Improved Pit Latrine

If the groundwater level is below the bottom of the pit, there are no water sources down grade nearby, and there is no risk of infiltration from rainy season runoffs, ventilated improved pit (VIP) latrines can be used effectively (Fig. 16.2).

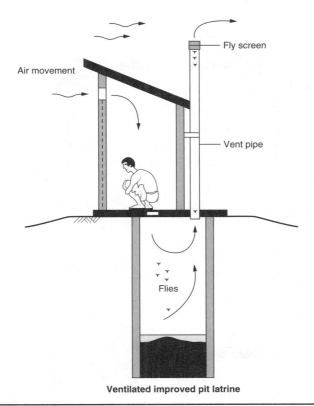

Ventilated improved pit latrine

FIGURE 16.2 Ventilated improved pit latrine (VIP).[5]

To construct the VIP latrine, the pit is dug approximately 6 to 10 ft (2 to 3 m) deep and about 3.3 to 5.0 ft (1.0 to 1.5 m) in diameter and covered with a concrete lid (Fig. 16.3).[6] The hole, approximately 8×12 in (20×30 cm), is made in the lid, so both solid excrement and urine can drop into the pit. To prevent leakage of pathogens to groundwater from excreta, the pit should be constructed at least 6 ft (2 m) above the groundwater surface. If the ground is hard, or groundwater is close to the surface, the pit can be elevated by using the backfill from digging the pit. An additional hole is provided in the concrete lid for a vent pipe, which prevents odor and fly problems. Air circulates through drop hole to the pit and thereafter through the vent pipe to outdoors, which keeps the latrine odorless. Vent pipe diameter should be at least 4 in (100 mm), and the top of the pipe should be 20 in (0.5 m) above the roof to ensure proper air movement.[7]

VIP use can cause diseases spread by insects. Therefore, it is important to pay attention to the coloring of the pit interior and to close the lid when the latrine is not in use. To prevent odor and fly problems, the

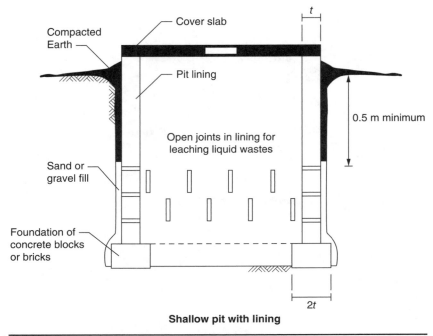

FIGURE 16.3   Shallow pit construction for privy.

condition of the vent pipe and fly screen should be checked on a regular basis. To maintain proper hygiene, it is important to pay attention to cleanliness of the latrine daily.[8]

On the top end of the vent pipe, a fly screen should be installed to prevent insects from entering the latrine. The interior of the latrine pit should be a dark color (e.g., dark blue) to direct the entered flies (from the drop hole) toward the light of the vent pipe. The fly screen will prevent the insects from exiting, and they will eventually die.

Pour-Flush Latrine

The structure of a pour-flush latrine (Fig. 16.4)[9] corresponds in principle to a ventilated improved pit latrine. In pour-flush latrines, there is a U-formed water seal (Fig. 16.5), which prevents flies from entering and acts as an odor seal. The latrine is flushed with a couple of liters of water after every use. Pour-flush latrines can be used when water is needed for cleaning purposes (if enough water is available), if the ground is permeable, and if climate does not present a risk of freezing the water seal. If only one pit is used in the latrine, it can be utilized until the pit is full. Thereafter, a new pit has to be dug or the existing pit emptied before usage can continue.

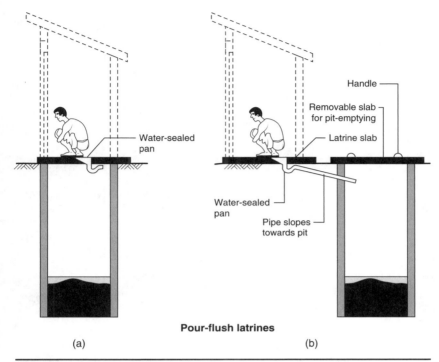

Pour-flush latrines

(a)   (b)

FIGURE 16.4 Pour-flush latrine.

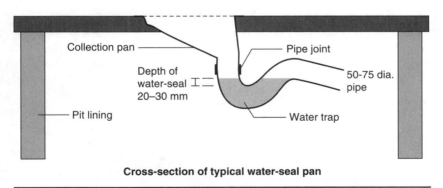

Cross-section of typical water-seal pan

FIGURE 16.4A Cross section of typical water seal pan for pour-flush latrine.

Optionally, the latrine can be constructed initially with two pits (Fig. 16.5).[10] This will allow the excreta to be directed to the second pit, with the help of a valve, after the first pit is full. Then the excreta left in the first pit can be covered and left to decompose while the second pit fills. When the second pit is full, the excreta left in the first

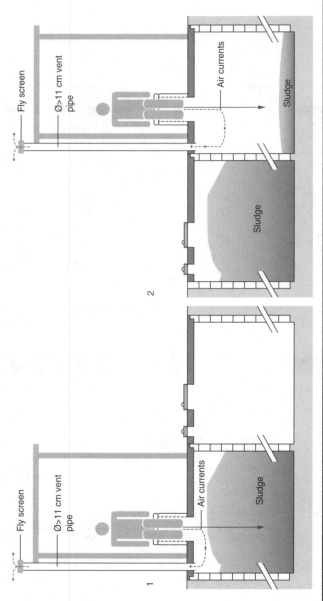

Figure 16.5 Twin privy installation for alternating use when primary pit is filled.

pit will have decomposed and pathogens will have disappeared. At this time, the pit can be emptied safely and the composted material can be used as fertilizer.[11]

Privy with Septic Tank

If water is readily available, be it from rainwater collection or municipal supply, a hybrid system can be constructed that combines multiple waste handling systems. For instance, a flush-style toilet may be incorporated with a septic tank (see Chapter 15), as indicated in Fig. 16.6. This is an improvement over simpler privy versions in that with a septic tank, primary (settling) treatment occurs, and the effluent can then be conveyed to a central sewer, an aquatic plant–based

Pathogen	Survival Time, Days		
	Freshwater and Wastewater	Crops	Soil
Bacteria			
Fecal coliforms[a]	<60 but usually <30	<30 but usually <15	<120 but usually <50
Salmonella (spp.)[a]	<60 but usually <30	<30 but usually <15	<120 but usually <50
Shigella	<30 but usually <10	<10 but usually <5	<120 but usually <50
Vibrio cholerae[b]	<30 but usually <10	<5 but usually <2	<120 but usually <50
Protozoa			
E. histolytica cysts	<30 but usually <15	<10 but usually <2	<20 but usually <10
Helminths			
A. Lumricoides eggs	Many months	<60 but usually <30	Many months
Viruses[a]			
Enteroviruses[c]	<120 but usually <50	<60 but usually <15	<100 but usually <20

[a]In seawater, viral survival is less and bacterial survival is very much less than in freshwater.
[b]*V. cholera* survival in aqueous environments is a subject of current uncertainty.
[c]Includes polio, echo, and coxsackie viruses.
Source: Adapted from Crites and Tchobanoglous, 1998.

TABLE 16.1 Typical Pathogen Survival Times at 20°C to 30°C in Various Environments

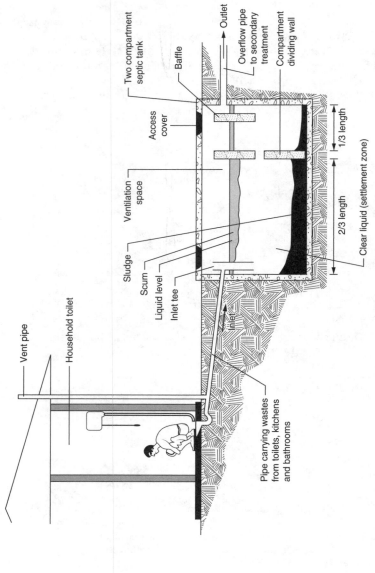

Wastewater treatment using a septic tank

Vent pipe

Household toilet

Pipe carrying wastes from toilets, kitchens and bathrooms

Inlet

Inlet tee

Liquid level

Scum

Sludge

Ventilation space

Access cover

Two compartment septic tank

Baffle

Outlet

Overflow pipe to secondary treatment

Compartment dividing wall

1/3 length

2/3 length

Clear liquid (settlement zone)

Figure 16.6 Wastewater treatment system with septic tank.

system (Chapter 12), or biological filters and constructed wetlands (Chapter 13).

Composting and the Use of the End Product as Fertilizer

Reuse of composted waste can turn what would otherwise be a liability to be disposed of into a beneficial high-end fertilizer. A system that offers that possibility is shown in Fig. 16.7. This illustration shows a urine diverting capability, which serves to lessen the moisture content of the fecal waste, enhancing aerobic digestion and the capability for making compost.

Separately, urine can be collected and also used as a high-end fertilizer. Urine is a well-balanced, nitrogen-rich liquid fertilizer. The average volume ranges from 0.21 to 0.40 gal (0.8 to 1.5 L) per person per day, with the concentration of nutrients in the urine dependent upon food intake. Typically, there are 3–7 grams of nitrogen per liter. Urine phosphorus provides a plant-available form, like mineral fertilizers, and this is particularly important since there are finite limits of mineral phosphates. This places the agriculture industry in a vulnerable "peak phosphate" situation.

If urine is to be separated, it is necessary to provide either a separate toilet bowl or a bench in sitting models (jugs and buckets in squat models can be used). For men-only facilities, the latrine can be equipped with a separate urinal collection system.

To eradicate possible pathogens, urine needs to be stored in closed containers before utilization. If urine is used in household functions (e.g., in the garden or as an additive to the compost), it can be utilized after a couple of days of storage. For urine utilized outside the immediate household, studies in Sweden[12] show that ordinary storage for six months causes E. coli, coliforms, and fecal streptococci to die off, leaving a hygienic quality fertilizer that can be used safely. In storing urine, special attention needs to be paid to the tightness of the containers so that nitrogen and other valuable nutrients are not lost to evaporation.[13,14]

Due to the high ratios of nitrogen to phosphorus and nitrogen to potassium, urine may be added at rates greater than is typical for nitrogen-containing substances as long as salinity is monitored. Urine can be utilized either undiluted or diluted depending on the target. Both forms have their advantages and disadvantages.

Utilization of Undiluted Urine

Urine can be spread undiluted to the soil following immediate watering of a plant. The advantage is that nutrient loss is small since dilution increases evaporation of nitrogen. In garden use,

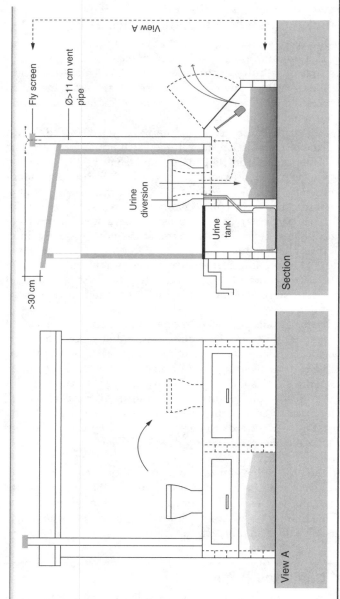

Fly screen

Ø>11 cm vent
pipe

>30 cm

View A

Urine
diversion

Urine
tank

Section

View A

FIGURE 16.7 Alternating use with urine separating system.

Figure 16.8 Using urine as a crop fertilizer.

urine can be poured into the soil with a watering can as fertilizer, and thereafter the soil can be watered with clean water. Watering ensures the absorption of nitrogen to the soil, decreases evaporation, and at the same time various plants have their different water needs met (Fig. 16.8).[15]

A disadvantage of undiluted urine use is that the risk of overfertilization grows, which can lead to the death of the plant. To avoid this risk, urine can be diluted before utilization.[16]

Maintenance

Maintenance of a privy principally involves being aware of the potential health impacts of their use, principally the pollution of the aquifer. Figure 16.9 shows that in dry soil, the pollution can travel 33 ft (10 m) vertically and 10 ft (3 m) horizontally. A privy installation where there is a high water table can amplify the pollution impact on the groundwater and therefore should be installed a safe distance, and preferably down grade, from any source of drinking water.

Conclusion

The essential parts of community sanitation are building and maintenance of sewerage systems, wash-up, and toilet facilities. Where water-based systems are not available, or where water itself is not plentiful, the elemental privy can have a major impact on maintaining the ongoing health and hygiene of the community. Furthermore, effective management of the by-products can provide assistance in the production of food, especially where food production (Figure 16.10) is limited by poor soil conditions.

Outhouses will transmit
pollution 10 feet (3 meters)
vertially and 3 feet (1 meter)
laterlly in dry soil

FIGURE 16.9   Potential impact of privy use with the water table.

FIGURE 16.10   Beneficial life cycle of a well managed privy.

Notes and References

1. Wagner, E.G., Lanoix, J.N., *Excreta Disposal for Rural Areas and Small Communities.* Geneva: World Health Organization, 1958.
2. *Health, Dignity, and Development, What Will It Take?* United Nations Millennium Project, 2005. (Web document, cited June 6, 2005)
3. Harvey, P., Baghri, S., Reed, B., *Emergency Sanitation Assessment and Programme Design.*
4. *Design Standard XIII Non-Waterborne Wastewater Systems, Form No. 309*, February 1997, Mississippi State Department of Health, Jackson, Mississippi 39215–1700.
5. Harvey, P., Baghri, S., Reed, B., *Emergency Sanitation Assessment and Programme Design.*
6. Harvey, P., Baghri, S., Reed, B., *Emergency Sanitation Assessment and Programme Design.*
7. Enwell, Water and Sanitation Project, The Finnish Red Cross.
8. *Sanitation and Cleanliness for a Healthy Environment*, Hesperian Foundation, Berkeley, CA, 2005. (Web document, cited June 17, 2005)
9. Harvey, P., Baghri, S., Reed, B., *Emergency Sanitation Assessment and Programme Design.*
10. Compendium of Sanitation Systems and Technology, Elizabeth Tilley, Christoph Lüthi, Antoine Morel, Chris Zurbrügg and Roland Schertenleib, Publisher: Swiss Federal Institute of Aquatic Science and Technology (EAWAG) (2008)
11. *Sanitation and Cleanliness for a Healthy Environment.*
12. Hoglund C., Stenstrom, T.A., Sundin, A., *Evaluation of Faecal Contamination and Microbial Die-Off in Urine Separating Sewage Systems.* Accepted for Water Quality International 1998, Nineteenth Biennial Conference of the International Association on Water Quality, Vancouver, B.C., Canada, June 21–26, 1998 (Water Science and Technology).
13. Schonning, Stenstrom, *Guidelines for the Safe Use of Urine and Faeces in Ecological Sanitation Systems.* EcoSanRes, Stockholm, 2004.
14. Hoglund, *Evaluation of Microbial Health Risks Associated with the Reuse of Source-Separated Human Urine*, Doctoral thesis, Royal Institute of Technology (KTH), Stockholm, 2001.
15. *Sanitation and Cleanliness for a Healthy Environment*, Hesperian Foundation.
16. (In Finnish) Kuivakäymälän lopputuotteiden käsittely ja käyttö. Käymäläseura Huussi ry, Tampere, 2006.

CHAPTER 17

Composting Toilets

Introduction

Public health professionals are beginning to recognize the need for environmentally sound human waste treatment and recycling methods as an alternative to present day methods that result in increasingly polluted sources of water. In addition to being environmentally beneficial composting toilets provide a solution when public sewers are not available and the soil is not amenable to a septic system installation (Figure 17.1)

Composting is the controlled aerobic (oxygen-using) biological decomposition of moist organic solid matter, in the presence of carbon-based material, to produce a product that can effectively be used as a soil conditioner. In an aerobic system, the microorganisms access free, gaseous oxygen directly from the surrounding atmosphere. The end products of an aerobic process are primarily carbon dioxide and water, which are the stable, oxidized forms of carbon and hydrogen. If the biodegradable starting material contains nitrogen, phosphorus, and sulfur, then the end products may also include their oxidized forms—nitrate, phosphate, and sulfate.[1] In an aerobic system, the majority of the energy in the starting material is released as heat by their oxidization into carbon dioxide and water.[2]

By contrast, an anaerobic system is a digestion process that is conducted in an absence of gaseous oxygen. In an anaerobic digester, gaseous oxygen is prevented from entering the system through physical containment that prevents oxygen from being part of the digestion process. Anaerobes access oxygen from sources other than the surrounding air. The oxygen source for these microorganisms can be the organic material itself or alternatively may be supplied by inorganic oxides from within the input material. When the oxygen source in an anaerobic system is derived from the organic material itself, then the intermediate products (primarily alcohols, aldehydes, and organic acids plus carbon dioxide) are, in the presence of specialized methanogens,[3] converted to the final end products of methane and carbon dioxide with trace levels of hydrogen sulfide. In an anaerobic system, the majority of the chemical energy contained within the starting material is released by methanogenic bacteria as methane.[4]

FIGURE 17.1   Composting toilets provide a solution for off-grid plumbing needs

Anaerobic Digestion	Composting
Digestate	Compost
Carbon Dioxide	Carbon Dioxide
Methane	Heat
Hydrogen Sulfide (trace levels)	

TABLE 17.1   Anaerobic Digestion versus Composting By-Product Summary

Composting systems typically include organisms such as fungi that are able to break down lignin and celluloses to a greater extent than anaerobic bacteria. It is due to this fact that it is possible, following anaerobic digestion, to compost the anaerobic digestate, allowing further volume reduction and stabilization. The emphasis is on the "controlled" system. It is different from the uncontrolled decomposition that occurs in the natural environment: A leaf falls from a tree branch to the forest floor, and microbes transform it into a nutrient form that the tree can consume. The biochemical process is the same. Composting can take place in a matter of hours or years—it depends on the process factors. Essentially, bacteria and other organisms feast on carbon-rich matter and digest it.

In a composting toilet, the objective is to transform potentially harmful residuals—mostly human excrement[5]—into a stable, fully oxidized end product called *humus*. Humus is an odorless brown or black substance resulting from the decay of organic animal and vegetable refuse. Humus builds soil structure and provides a productive environment for plants and essential soil organisms. Compost has also been found to aid in suppression of plant diseases, often reducing or eliminating the need for fungicides.

The humus produced releases plant nutrients gradually, it holds moisture, and it's cheap. The nutrient value of composted yard wastes is modest, amounting to about 10 percent of the nutrients applied as fertilizer. The nutrient content of composted human waste, which is highly nitrogenous thanks to urine, is equal to that of many chemical fertilizers. Although the benefits of composting are extensive, for our purposes, the primary value of composting is its ability to reduce waste.[6]

The Composting Process

Composting typically goes through four stages:

1. Mesophilic phase: Compost bacteria combine carbon with oxygen to produce carbon dioxide and energy. Some of the energy is used by the microorganisms for reproduction and growth; the rest is given off as heat. While a pile of organic refuse begins to undergo the composting process, mesophilic bacteria proliferate, raising the temperature of the composting mass up to 111°F (44°C). This is the first stage of the composting process. These mesophilic bacteria can include E. coli and other bacteria from the human intestinal tract, but they soon become increasingly inhibited by the temperature as the thermophilic bacteria takeover in the transition range of 111°F to 125°F (44°C to 52°C).

2. Thermophilic phase: The second stage of the process begins when the thermophilic microorganisms are very active and produce a lot of heat. This stage can continue to about 158°F (70°C), although such high temperatures are neither common nor desirable in a household plumbing system. This heating stage takes place rather quickly and may last only a few days, weeks, or months. It tends to remain localized in the upper portion of the compost heap where the fresh material is being added. In a batch of compost, the entire composting mass may be thermophilic all at once. This phase serves to dry the mixture, and in so doing, decreases the mass of the compost pile by approximately 90 percent.

3. Cooling phase: After the thermophilic heating period, the compost mix will appear to have been digested, but the coarser organic material will still need further oxidation. This is when the third stage of composting, the cooling phase, takes place.

After the thermophilic stage has been completed, only the readily available nutrients in the organic material will have been digested. The rest takes a longer period (months) to break down.

During this phase, the microorganisms that were chased away by the thermophilic stage of composting migrate back into the compost and continue the digesting process of the more resistant organic materials. Fungi and microorganisms, such as earthworms and sow bugs, also serve to break the coarser elements down into humus.

4. Curing stage: The final stage of the composting process is called the curing, which is the aging and maturing stage of the composting process. A long curing period, optimally a year after the thermophilic stage is completed, adds a safety net for pathogen destruction. Many human pathogens have only a limited period of viability, and the longer they are subjected to the microbiological competition of the composting process, the more likely they are to produce a compost that is not only free of pathogens but is a better soil nutrient for plants. In a composting toilet, time for the curing process is proportional to the hold volume of the toilet.[7]

Factors That Affect the Composting Process

Environmental Factors
How effective the composting operation is depends on the composting environment, as well as pH and food supply. Human excrement and the other organic material that you put in the toilet feed these microbes. The pH is self-regulating if all of the other conditions are satisfactory.

Aeration
To make the composting process work best, the materials being composted should have a loose texture to allow air to circulate freely within the pile. If the material becomes matted down, compacted, or forms too solid a mass, the air will not circulate, and the aerobic organisms will die. Bulking agents may be added, such as wood chips, stale popped popcorn, etc., to increase pore spaces that permit air to reach deep into the biomass and allow heat, water vapor, and carbon dioxide to be exhausted.

Composting is an "unsaturated" aerobic process. This means that the material being composted cannot be immersed in liquid, as that would fill the void spaces (the pores) in the composting mass and prevent oxygen-carrying air from reaching the organisms digesting the food source. If the material becomes saturated, soon the remaining dissolved oxygen in the liquid is consumed. When there is no dissolved or free oxygen, that condition is said to be anoxic. When the anoxic condition persists, anaerobic organisms, which cannot use free oxygen, will take over. The process is called "anaerobiosis" and is typified by offensive odors, such as that of rotten eggs, from sulfides, amines, mercaptans, and flammable methane gas produced by anaerobic bacteria.

To maintain aerobic conditions, the composter must have a ventilation system. If there is an oxygen deficit (a state called hypoxia), the aerobes will die. They will be replaced with anaerobes (organisms that can exist only in the absence of molecular oxygen), which will slow the process and generate odors (thanks to their production of hydrogen sulfide, ammonia, and amines) and potentially flammable methane gas.

The ventilation system in a composter should draw oxygen across and through the compost mix. A key factor for effective ventilation is the surface area-to-volume ratio of the composting substrate (which includes the microbial population). The greater the surface area, the greater the direct contact with oxygen. If the volume of the composting material is greater than the surface area, then oxygen may not reach the microbes, and the process will be limited by this lack of oxygen. Mixing, tumbling, forced aeration, and container design are ways composters provide a good surface-to-volume ratio. However, too much air flow can remove too much heat and moisture, so care must be taken to make sure the composter is not too cool and dry as a result.

Moisture

In order for the microbes in the composter to thrive, the right amount of moisture is needed. Too much water (saturated conditions) will drown them and create conditions for the growth of odor-producing anaerobic bacteria. In optimum conditions, the composting mass has the consistency of a well-wrung sponge—about 45 to 70 percent moisture. When the moisture level drops below 45 percent, it can become too dry for composting.

If the moisture level is higher than 70 percent, leachate will pool at the bottom of the composter, and a drain will be necessary for its removal. Urine and water from micro-flush toilets contribute most of the moisture in a composter and may not be distributed evenly over the mass. (Note that urine-separating toilets are now available to reduce the leachate problem.) Ventilation can be controlled to evaporate the moisture to avoid drowning the microbes. Some modern composting toilets use ventilation and thermostatic control to maintain optimum temperature and moisture for the composting process.

In composting toilets, particularly ones with heaters at the bottom, the upper parts of the biomass may become too dry. If the composting material is too dry, adding water to the compost mass periodically may be necessary. Fresh rainwater, which has little or no dissolved minerals, is best for moisture control, although fresh groundwater will also do as well. In some systems, the drained leachate is re-sprayed over the top to prevent dehydration. This is not the best practice, however, as the concentrated salt and ammonia from urine are toxic to the beneficial bacteria and other organisms that are composting. The use of household exhaust air, particularly the warm, moist vapor exhausted from a clothes dryer, is an innovative way to help maintain the moisture and temperature requirements for the compost mix.

Temperature

The ambient temperature for acceptable biological decomposition is 78° to 113°F (25.5° to 45.0°C). Biological 0 to 41°F (−17.8° to 5°C) (the temperature at which almost no microbial respiration occurs). At this temperature, most microbes cannot metabolize nutrients.

In most composting toilet systems, mesophilic (68° to 112°F (20.0° to 44.4°C)) composting is at work. The heat generated by these microbes is usually lost through the vent stack, so composting toilets rarely reach thermophilic rates (113° to 160°F (45.0° to 71.1°C)), which support thermophilic bacteria. This is the hot composting that takes place at the core of active yard waste composters (that's what generates the steam you see rising from the compost on a cold day). Achieving thermophilic composting would require either heating the composter, which could be expensive, or retaining the heat better by venting it less, which might mean odors and insufficient oxygen. In the highly contained environment of this kind of composter, it's a hard balance to reach.

Moldering toilets support psychrophilic organisms, whose optimum temperature is above 42° to 68°F (5° to 20°C). These are predominately fungi and actinomycete bacteria such as *Streptomyces griseus*, which produces the antibiotic streptomycin. Moldering systems are sized much larger than mesophilic composting systems, which compensates for their reduced processing rate. Moldering is also the last phase after mesophilic and thermophilic processes have completed the early work of degrading sugars, fats, and carbohydrates. As the process cools, fungi and actinomycetous bacteria slowly digest the cellulose and lignin in plant matter, such as wood chips and toilet paper. Performing this balance in modern composting toilets is done automatically with temperature and humidity sensor-controlled fans.

Generally, the rate of processing in a biochemical system is directly proportional to the increase of temperature (within certain limits, the rate doubles with every 18°F (10°C) increase). Conversely, the cooler the process, the slower the composting rate, which may necessitate adjusting the size of the composter to maintain the desired capacity.

A composter in ambient 41°F (5°C) will only accumulate excrement, toilet paper, and additives until the temperature rises. That is why composter manufacturers state their capacities at 65°F (18.3°C) (comfortable room temperature of an average human-occupied space).

Stages of the Composting Process	
Below 42°F (5.6°C)	Little to no active bacterial processing takes place.
Between 42°F and 67°F (5.6°C and 19.4°C)	Psychrophilic (moldering) processing predominantly with actinomycetes and fungi.
Below 68°F, Below 112°F (20°C and 44°C)	Mesophilic bacteria, the typical composting toilet bacteria, are dominant.
Above 112°F (44°C)	Higher temperature increases compost drying and impedes the digestion process.

TABLE 17.2   Stages of the Composting Process

Carbon-to-Nitrogen (C:N) Ratio

Although its significance in composting toilets is often overstated, an important relationship to remember for aerobic bacteria nutrition is the carbon-to-nitrogen ratio—the C:N—of the food source.

Microorganisms require digestible carbon as an energy source for growth, and they need nitrogen and other nutrients, such as phosphorous and potassium, for protein synthesis to build cell walls and other structures (in the same way humans need carbohydrates and proteins). When measured on a dry weight basis, an optimum C:N ratio for aerobic bacteria is 25:1.

Primarily due to the high nitrogen (from urea, creatine, ammonia, etc.) content and low carbon (glucose) content of urine (0:8:1), human urine has a low C:N ratio. Therefore, if the objective is to oxidize all of the nitrogen urinated into the toilet, this would require adding digestible carbonaceous materials on a regular basis. However, the practical fact is that urine, which contains most of the nitrogen, settles by gravity to the bottom of the composter, where it is drained away or evaporated. In either case, the nitrogen passes through the composting mass and is lost to the process. For that reason, adding large amounts of carbon will not help process the nitrogen and will just fill up the composter faster.

The primary reason to add carbon material is to create air pockets in the composting material. Digestible carbonaceous materials include carbohydrates (sugar, starch, toilet paper, popped popcorn), vegetable or fruit scraps, finely shredded black-and-white newsprint, and wood chips. A small handful of dry matter per person per day or a few cups every week is a good rule of thumb to maintain a helpful C:N ratio, absorb excess moisture, and maintain pores in the composting material.

Time

When exposed to an unfavorable environment for an extended period of time, most pathogenic microorganisms will not survive. However, caution is essential when using the compost end product and liquid residual in case some pathogens survive. Table 17.3 gives typical pathogen survival times at 680 to 86°F (20° to 30°C) in various environments.

	Survival Time, Days		
Pathogen	Freshwater and Wastewater	Crops	Soil
Bacteria			
Fecal coliforms[a]	<60 but usually <30	<30 but usually <15	<120 but usually <50
Salmonella (spp.)[a]	<60 but usually <30	<30 but usually <15	<120 but usually <50
Shigella	<30 but usually <10	<10 but usually <5	<120 but usually <50
Vibrio cholerae[b]	<30 but usually <10	<5 but usually <2	<120 but usually <50
Protozoa			
E. histolytica cysts	<30 but usually <15	<10 but usually <2	<20 but usually <10
Helminths			
A. Lumricoides eggs	Many months	<60 but usually <30	Many months
Viruses[a]			
Enteroviruses[c]	<120 but usually <50	<60 but usually <15	<100 but usually <20

[a]In seawater, viral survival is less and bacterial survival is very much less than in freshwater.
[b]V. cholera survival in aqueous environments is a subject of current uncertainty.
[c]Includes polio, echo, and coxsackie viruses.
Source: Adapted from Crites and Tchobanoglous, 1998.

TABLE 17.3 Typical Pathogen Survival Times at 20°C to 30°C in Various Evironments

Process Control

In modern composting toilets, control of the environmental factors (air, heat, and moisture) that affect the composting process is automated to improve operation and make use of these devices as convenient as possible.

Either motorized or manual mixing or turning provides aeration and to move microbial communities into contact with unprocessed composting material.

Blowers and fans are used to remove odors and gases that are the by-products of composting, such as carbon dioxide and water vapor. Fan-speed controllers optimize the efficiency by controlling the speed of the fan according to mix temperature and humidity levels.

Most small manufactured compost toilets have heaters and thermostats to maintain an internal temperature of 90° to 113°F to support the upper mesophilic composting range, while evaporating excess leachate. Evaporation of leachate tends to drop the temperature of the composter through evaporative cooling.

Pumps can be used to move leachate to a management system for harvesting as a high quality fertilizer. Pumps can also be used to spray water over the mass.

Warning indicators and alarms tell the manager when something needs attention.

Composting Toilets

Composting toilets use the composting process to provide a non-water-carriage system that is well suited for (but not limited to) remote areas where water is scarce. They are used in areas with low percolation, high water tables, shallow soil, or rough terrain. A composting toilet is a well-ventilated container that provides the optimum environment for unsaturated but moist human excrement for biological and physical decomposition under sanitary, controlled aerobic conditions. They are used almost anywhere a flush toilet can be used, typically for seasonal homes, homes in remote areas that cannot use flush toilets, or recreation areas. Marine versions have even been adapted for use on boats. Some are large units that require a basement for installation. Others are small self-contained appliances that sit on the floor in the bathroom.

Variations of composting toilets have existed for centuries. A composting (or biological) toilet system serves to contain and process excrement, toilet paper, carbon additive, and sometimes, food waste. This process is similar to a yard waste composter. Unlike a septic system, a composting toilet system relies on unsaturated conditions where aerobic bacteria break down waste. If sized and maintained properly, a composting toilet breaks down waste to 10 to 30 percent of its original volume. Because composting toilets eliminate the need for flush toilets, this significantly reduces water use, while also allowing for the recycling of valuable plant nutrients.

The initial form of toilet was the latrine, which eventually evolved into the classic outdoor privy, common to every rural farmhouse in the mid-1900s. The design concept of the traditional privy is simple: a hole in the ground often enhanced with an enclosure to keep the sun and rain off the occupant, while providing a modicum of privacy. It uses no power or water, is inexpensive to build, and is somewhat easy to relocate if necessary. However, significant downsides include odor in the summer and the necessary trek outside in all weather conditions.

In 1939, Rikard Lindström, a Swedish engineer, developed the original composting toilet by incorporating the major advantages of an outhouse that allowed the concept to be moved indoors. Eventually, in 1962, the system was patented, and in 1964, the first commercial model was constructed out of fiberglass.

The term *clivus multrum* means "sloping chamber" in Swedish and describes the system design (Fig. 17.2). The original system used a single-chamber sloped-bottom design and was constructed

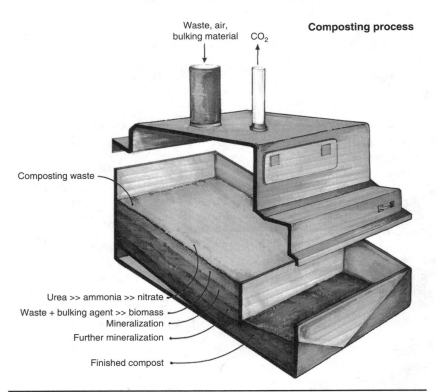

FIGURE 17.2 Schematic of a Clivus multrum composting toilet. (*Courtesy of the Clivus Multrum Company, Lawrence, Massachusetts*)

of concrete. The chief advantage of this style of toilet is the concept of aerobic digestion, in which the primary products of decomposition are relatively odorless carbon dioxide and water vapor as contrasted with the methane, hydrogen sulfide, and ammonia produced under anaerobic digestion, which is common with the standard privy. Designed to accommodate the range of commercial and residential needs, clivus multrum systems are made of polyethylene and can accommodate 18,000 to 65,000 uses per year per composter and a number of space arrangements. Systems come equipped with a fan that eliminates odors in the restroom, a liquid removal pump, an automatic moistening system, and storage for the liquid end product. This style composting system is compatible with both waterless toilets and foam-flush toilets.[8] (Fig. 17.3).

Another "big box" composting toilet that is applicable for a commercial application is the Phoenix manufactured by Advanced Composting Systems (Fig.17.4). These units have horizontal shafts which have tines arrayed radially around the shaft, which both halt the downward progression of the waste and allow for mixing if needed.

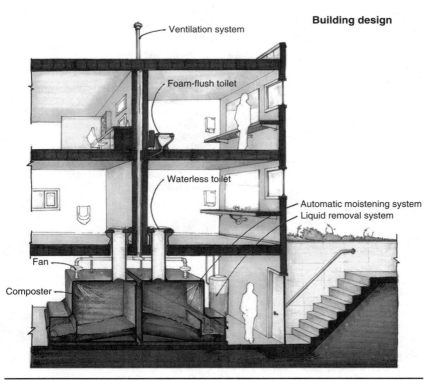

FIGURE 17.3 Typical clivus multrum installation. (*Courtesy of the Clivus Multrum Company, Lawrence, Massachusetts*)

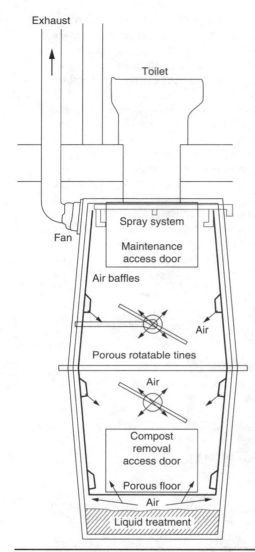

Exhaust

Toilet

Rotatable tines mix waste and control movement of compost to the access area for removal. The Phoenix's vertical design, rotating tines and geometry assure that the entire volume of the tank is used and that only the oldest compost is removed.

Fan

Spray system

Maintenance access door

Air baffles

Air

An energy efficient fan exhausts air from the Phoenix tank to assure odorless operation. Passing behind the side baffles, air makes frequent contact with the compost pile assuring aerobic decomposition. Additional air contact is provided by the porous floor and porous tine shafts.

Porous rotatable tines

Air

Compost removal access door

Porous floor

Air

Liquid treatment

After passing through the compost pile, liquid drains through the porous floor in the bottom of the Phoenix and receives secondary treatment in a stable, well aerated medium. Liquid is periodically sprayed on top of the compost pile to keep the pile moist and to inoculate fresh waste with decomposing organisms.

FIGURE 17.4 Phoenix composting toilet cross section : Illustration courtesy Advanced Composting Systems, 195 Meadows RD, Whitefish, MT 59937, phoenix@compostingtoilet.com

After startup, which involves the addition of wood chips as a carbon source to initiate the composting proeces, additional wood chips are added with each use. The less frequently bulking agent is added, the more mixing becomes important.

Proper composting toilet operation requires that the waste be maintained in an aerobic condition by maintaining a constant oxygen level and temperature within the mix. Common to all composting toilets is the need to maintain a warm environment with the proper level of moisture being critical for proper operation of these units. In a home, any source of air warmer than the basement would help the composting process. To maintain aerobic digestion, the composting toilet uses an exhaust fan and occasional stirring of the waste stack to oxygenate the mix in order to avoid anaerobic digestion. During the digestion process, the waste volume is reduced approximately 70 percent per year, with a humus-like material being removed periodically that can be used as landscape fertilizer. Maintaining aerobic conditions serves to assure that carbon dioxide is the principal by-product and not the noxious hydrogen sulfide and methane odors associated with anaerobic digestion. Properly sized, these units can last several years before pumping out is necessary.

These large box toilets are approved by NSF International for public restrooms and commonly are used for off- grid applications. Figure 17.5 shows a composting toilet installation at a rest station along a trail in a Missouri state park. This installation uses a composting toilet integrated with solar voltaic system that provides power to maintain the mix temperature and to sanitize the collected rainwater used for hand washing.

Klondike County Park, St. Charles, Missouri

FIGURE 17.5 Composting toilet rest facility in Kondike Park, St Charles, Missouri. Photograph courtesy of Advanced Composting Systems, 195 Meadows RD, Whitefish, MT 59937, phoenix@compostingtoilet.com

Residential Composting Toilets

Modern composting toilets are more convenient in their operation, and their appearance is similar to flushing water closets, thus making them an attractive alternative to traditional fixtures. The key to proper operation of composting toilets is accurate sizing for the intended load and the periodic addition of peat moss or wood shavings to the composter to assist the decomposition process.

With modern composting toilets, there are many variations on how the waste pile is mixed. Figures 17.6 and 17.7 show how some manufacturers accomplish this task: Envirolet (Fig. 17.6) uses a raking mechanism that is operated manually. Sun-Mar (Fig. 17.7) uses a barrel to tumble the waste/peat moss/oxygen mix to accelerate decomposition. A third manufacturer, BioLet, uses paddles to keep the mix oxygenated. Whichever system is selected, it should be done after careful investigations to obtain a satisfactory match with the intended application. Reviews of various systems indicate that some systems work better than others in different situations.[9]

The main components of a composting toilet are the following:

- A composting reactor connected to dry or micro-flush toilet(s)

- A screened air inlet and an exhaust system (often fan-forced) to remove odors and heat, carbon dioxide, water vapor, and the by-products of aerobic decomposition

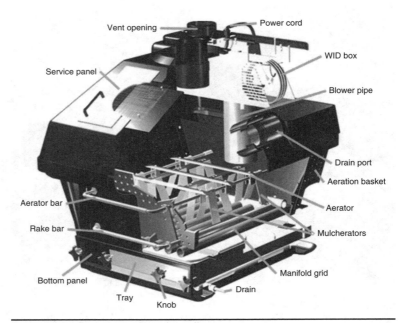

FIGURE 17.6   Envirolet composting toilet.

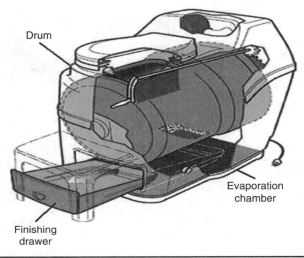

Drum

Evaporation
chamber

Finishing
drawer

FIGURE **17.7**   Sun-Mar composting toilet. (See also Color Plates.)

- A mechanism to provide the necessary ventilation to support the aerobic organisms in the composter
- A means of draining and managing excess liquid and leachate (optional)
- Process controls to optimize and facilitate management of the processes
- An access door for removal of the end product

Two common styles of composting toilets are available: a, self contained, "dry" operating toilet, desirable where freezing is possible (Fig. 17.8) and a dry system with a remote composter (Fig. 17.9). Ultra low water flush versions are available that use a pint of water for the flushing operation (Fig. 17.10). Because the ultra-low-flow toilets use water to assist waste conveyance, the toilet does not need to be directly above the composter, which allows multiple toilets to be installed on a single composter. Where multiple water closets are needed, the water flow in the ultra low flow water closets can convey the waste to a centrally located composter.

The end product from the decomposition process is odorless, has the composition of potting soil, and can be used in gardens (Fig. 17.11). Depending on toilet use, the task of emptying the composter can occur on a quarterly or even annual basis. Available to all are the options of manual operation, 12 volts (battery) for solar power, and 120 volts for utility-based power.

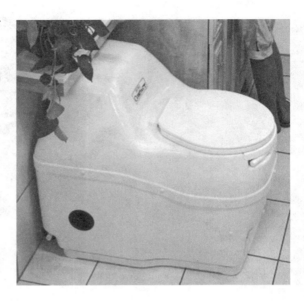

FIGURE 17.8 Dry self-contained composting toilet. (*Courtesy of the Sun-Mar Corporation*)

FIGURE 17.9 Dry remote composting toilet.

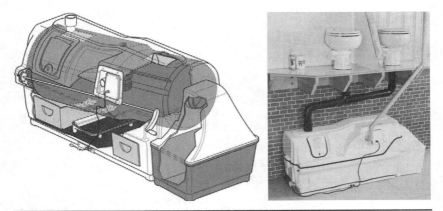

FIGURE 17.10 Remote composter for multiple ultra-low-flush toilets. (*Courtesy of the Sun-Mar Corporation*)

Removal of finished compost from Finishing Drawer

FIGURE 17.11 Finished compost for use in the garden. (*Courtesy of the Sun-Mar Corporation*)

The key limitation of composting toilets appears to be social acceptance of their use as an alternative to the traditional porcelain flush toilet. Because modern composting toilets are comparable to flush toilets in appearance and operation, a little experience usually brings acceptance. Also, in a world concerned about freshwater shortages, an important benefit of composting toilets is reduced water use, with the added benefit of protecting the water that is available by reducing downstream pollution.

Some advantages and disadvantages of composting toilet systems are listed below.

Advantages

- Composting toilet systems do not require water for flushing, and thus, they reduce domestic water consumption.
- These systems reduce the quantity and strength of wastewater to be disposed of on-site.
- They are especially suited for new construction at remote sites where conventional on-site systems are not feasible.
- Composting toilet systems have low-power consumption.
- Self-contained systems eliminate the need for transportation of wastes for treatment/disposal.
- Composting human waste and burying it around tree roots and nonedible plants keeps organic wastes productively cycling in the environment.
- Composting toilet systems can accept kitchen wastes, thus reducing household garbage.
- In many states, installing a composting toilet system allows the property owner to install a reduced-size leachfield, minimizing costs and disruption of landscapes.
- Composting toilet systems divert nutrient- and pathogen-containing effluent from soil, surface water, and groundwater.

Disadvantages

- Maintenance of composting toilet systems requires more responsibility and commitment by users and owners than conventional wastewater systems.
- Removing the finished end product is an unpleasant job if the composting toilet system is not properly installed or maintained.
- Composting toilet systems must be used in conjunction with a greywater system in most circumstances.
- Smaller units may have limited capacity for accepting peak loads.
- Proper operation is vital to satisfactory system operation. Improper maintenance makes cleaning difficult and may lead to health hazards and odor problems.
- Using an inadequately treated end product as a soil amendment may have possible health consequences. Application of compost on edible plants is not recommended.
- There may be aesthetic issues because the excrement in some systems may be in sight.
- Too much liquid residual (leachate) in the composter can disrupt the process if it is not drained and properly managed.

- Most composting toilet systems require a power source.
- Improperly installed or maintained systems can produce odors and unprocessed material.

To be noted, the manufacturers that were listed previously are not the only ones. There are many other manufacturers, each producing composting toilets with different characteristics and performance capabilities.[10]

Operation and Maintenance

Handling raw waste has historically been a problem from a management standpoint. Removing vault or pit type waste from a privy has led to accidental spills and is always a difficult task. This is why managers appreciate the concept of composting human waste.

Management considerations for composting toilets include gathering information on how much maintenance is needed annually, as well as on administration and operation, quality control and assurance, recordkeeping, and training.

In general, operation and maintenance (O&M) for composting toilet systems does not require trained technicians or treatment plant operators. However, regular O&M is of the utmost importance since any system depends on responsible administration. Some required maintenance steps are discussed next.

In cold climates, all composting toilet systems should be heated to levels specified by the manufacturer or designer to maintain performance. Composting toilet systems may require organic bulking agents to be added, such as grass clippings, leaves, sawdust, or finely chopped straw. The agents aid the composting process by providing a source of carbon for the bacteria, as well as keeping the pile porous for proper air distribution. If the facility is used every day, it is recommended to add bulking material at least every other day. Periodic mixing or raking is suggested for single-chamber continuous systems.

The other required maintenance step is occasionally removing the finished end product (anywhere from every three months for a cottage system to every two years for a large central system). If proper composting has taken place, the end product should be inoffensive and safe to handle. Adequate precautions should be taken while handling the humus material.

Design Standards

The standard governing minimum materials, design, construction, and performance of composting toilet systems in the United States is the *NSF International Standard ANSI/NSF 41-1998: Non-Liquid Saturated Treatment Systems*. NSF Standards require the composting unit to be constructed so as to separate the solid fraction from the liquid fraction and produce a stable humus material with less than 200 MPN per gram of fecal coliform. Once the leachate has been drained or evaporated out of the unit, the moist, unsaturated solids must be decomposed by aerobic organisms using molecular oxygen. Bulking agents can be added to provide spaces for aeration and microbial colonization. Key to proper application of a manufactured composting toilet is recognizing at what conditions the toilet was tested to arrive at the cataloged capacity ratings. Optimum composting occurs between 60° and 100°F (15° to 38°C) and between 40 and 70 percent relative humidity. Deviation from these nominal testing conditions requires either adjusting the capacity or the product loading or including a temperature control device within the composting toilet.

Costs

The cost of a composting toilet system depends on the manufacturer and type of design. Although the principle of waste treatment is the same, there are design variations in the containment of the waste, aeration, and other features of the system. The main factors that determine costs are the cost of the equipment, the building foundation, electrical work, and installation labor.

For a year-round home of two adults and two children, the cost for a composting toilet system could range from anywhere between $1200 and $6000 USD, depending on the system. Cottage systems designed for seasonal use range from $700 to $1500 USD. Large-capacity systems for public facility use can cost as much as $20,000 USD or more. However, site-built systems, such as cinder-block double-vault systems are as expensive as their materials and construction labor costs.

In addition to the composting toilet operation is the need to handle the greywater component of the human waste stream. Though much smaller than a conventional system, composting toilet systems will usually require a septic tank and leachfield, a subsurface irrigation system, or a plant-based system to manage greywater.[11] These systems are addressed in other chapters in this book.

Conclusion[12]

Composting toilets use the natural cycle of nature to provide an ecologically beneficial and sanitary means of processing human waste while also producing an end product that serves to fertilize the growth of plants, rather than adding to the pollution of the water supplies. As sources of clean water supplies continue to dwindle, and increased filtration and more chemicals are needed to produce drinking quality water for an expanding population, composting toilets provide a feasible solution to this dilemma by reducing pollution and turning human waste into a constructive asset.

Notes and References

1. *Aerobic and Anaerobic Respiration*, www.sp.uconn.edu, retrieved October 24, 2007.
2. Fergusen, T., Mah, R., *Methanogenic Bacteria in Anaerobic Digestion of Biomass*, p. 49, 2006.
3. meth·an·o·gen (mĕ-thăn′ə-jən, -jĕn′) n., Any of various anaerobic methane-producing bacteria belonging to the family *Methanobacteriaceae*, The American Heritage® Medical Dictionary, Houghton Mifflin Company.
4. *A Comparison of Anaerobic Digestion and Composting*, www.anaerobic-digestion.com, retrieved November 5, 2007.
5. Also referred to as "night soil."
6. Del Proto, D., Steinfeld, C., *Composting Toilet System Book: A Practical Guide to Choosing, Planning and Maintaining Composting Toilet Systems*, Chelsea Green Publishing.
7. Jenkins, J., *The Humanure Handbook*, Third Edition, Chelsea Green Publishing.
8. Foam flush and 1 pint per flush toilets are appropriate where transport of the waste or using the foam or water is necessary to move the waste products to the composting chamber.
9. A good overview of various manufacturers can be seen in Rosenbaum, M., P.E., *Converting "Waste" into Nutrients—Treating Household Organic Waste*, written for the South Mountain Company and Island Cohousing, West Tisbury, MA, June 6, 1998.
10. Additional manufacturers of composting toilets and contact information:
    - Alas Can, Inc., 3400 International Way, Fairbanks, AL, 99701, 907-452-5257. Essentially a house-sized waste treatment system combining greywater and blackwater.
    - Bio-Sun Systems, Inc., RR#2, Box 134A, Millerton, PA, 16936, 717-537-2200, 717-537-6200 fax, contact Allen White. Two models, both suited for high-use areas.
    - Clivus Multrum, Inc., 21 Canal Street, Lawrence, MA 01840-1801, 800-962-8447, 508-794-8289 fax, contact Don Mills.
    - Ecotech, 152 Commonwealth Avenue, Concord, MA 01742-2968, 978-369-9440, 978-369-2484 fax, contact David Del Porto.
    - Carousel: Advanced Composting Systems, 195 Meadows Road, Whitefish, MT 59937, 406-862-3854, 406-862-3855 fax, contact Glenn Nelson. New England Distributor: Tad Montgomery, PO Box C-3, Montague, MA 01351, 413-367-0068.
    - Sun-Mar Corp., 5035 N. Service Road C2, Burlington, Ontario, L7L 5V2, Canada, 800-461-2461, 905-332-1314, 905-332-1315 fax, contact Fraser Sneddon.
    - Phoenix Composting Toilet UK Limited, Regent House, 32 Princes Street, Southport, UK, PR8 1EQ, telephone: + 44 (0)1704 500878.
11. Water Efficiency Technology Fact Sheet: Composting Toilets (EPA832-F-99-066), Environmental Protection Agency, Office of Water, Washington, D.C., September 1999.
12. For an excellent article on homemade bucket-style composting toilets and their operation, see David Omick's articles at http://www.omick.net/composting_toilets/composting_toilets.htm.

CHAPTER 18

Net Zero Water

Introduction

If the principles of the previous chapters are implemented to their optimum benefit, humanity would arrive at a point where water is used efficiently and waste minimized, with no deleterious impact on the environment. The culmination of this effort would be the concept of *net zero water.*

A net zero building incorporates all the ideas presented in this book and integrates them into an interactive system that introduces a responsible means to manage our water sources. Net zero water has the objective of eliminating reliance on publicly provided potable water to meet all the water needs of a building. Similar to net zero energy, which produces energy on-site and does not consume more than it produces, net zero water is a standard that sets out to close the loop of a building's water consumption by producing the water and managing the wastewater on-site. In essence, it applies concepts for working with limited resources to mainstream building systems design.

Discussion

Net zero starts by collecting rainwater that falls on-site and storing it for later potable and non-potable applications. Wastewater is also collected, and it is treated to the level necessary for safe reuse and returned to the building for non-potable applications.

Rainwater, greywater, and blackwater are treated differently to achieve different outcomes. The common methodology today is to use potable water for all water demands—whether it is for toilet flushing, drinking, irrigation, washing, laundry, or other uses. Often overlooked is that the production of potable water is also accompanied by the use of considerable energy and chemicals to achieve potable water standards. In situations where there is no human contact, such as toilet flushing and subsoil irrigation of nonedible crops, fresh potable water is not required to meet the need. Net zero supports treating water only to a fit-for-purpose quality and only to the level needed by the application. The end result is that significant energy and chemicals can be saved along with the water conservation.

Figure 18.1   Nearly half the world will be short of water by 2025.

A Net Zero Example

Many older cities have combined sanitary and storm sewers, whose treatment capacity can be overwhelmed by heavy precipitation. Net zero strives to take what otherwise could be seen as a burden and apply it for constructive use. Since rainwater is cleaner than greywater, less energy is needed to purify the water to drinking quality (Chapter 7). Simple filtration, followed by ultraviolet sterilization, can make rainwater potable. Greywater, which is produced by the bathtub/shower, washing machine, and bathroom sink, can minimally be filtered and disinfected as needed and returned back into the system for toilet flushing, landscape irrigation, and process water makeup (Chapter 9).

Blackwater that comes from toilet flushes, the kitchen sink, and the garbage disposal is the murkiest of wastewater produced with considerably more pathogens and bacteria than greywater. As a result, blackwater is more complicated to treat than greywater, needing extra effort to settle solids, filter the wastewater, and then treat to a biologically safe condition. This can be done using the latest technology of microbiological filters (Chapter 122), which produces greywater for recycling into the building, leaving solids to be addressed by composting or by holding the tank pump out for off-site disposal. Alternately, the blackwater can be handled simply by means of a septic tank (Chapter 15), which can be supplemented with aquatic plants (Chapter12) or biological filters (Chapter 13).

Along with striving to achieve water independence, the net zero water standard also strives to manage storm water runoff to moderate the affects of combined sewer overflows (CSO). Green roofs are one of the most powerful tools to accomplish this, especially when used on a large scale. Studies have shown that a traditional roof's runoff contains high concentrations of pollutants from rainwater, roofing materials, and atmospheric deposition. By capturing and temporarily storing storm water, green roofs can reduce runoff volumes. By reducing runoff, green roofs limit the occurrence of combined sewer overflow events, and thereby diminish the quantity of untreated wastewater entering freshwater bodies.

Green roofs also improve storm water runoff quality. The plants and growing media used in green roofs help decontaminate runoff, loading fewer pollutants into the municipal storm water system. Furthermore, by the reduction of the storm water flow peak, the occurance of flooding or combined sewer overflow is reduced.

A 2005 study by the Casey Trees Endowment Fund, "Re-Greening Washington, DC"[1] showed that a green roof can retain between 65–85 percent of storm water runoff compared to a conventional roof.

Increasingly, jurisdictions have begun to implement storm water utility fee charges, which help raise capital for improved storm water management. The fees assessed to each property owner are typically based on the impervious surface area of the property. For commercial and industrial buildings, the large amount of impervious surfaces often results in larger new storm water fees, which can be a significant cost over a 20 to 40 year period. Storm water fee rebates or reductions can provide a monetary incentive for the use of best management practices.[2]

The importance of water conservation is only going to increase in the near future, and professionals will be needed to respond to this demand. The new zero water program is a unique learning opportunity for architects, engineers, and landscaping and irrigation practitioners to get involved in the sustainable construction industry. This approach promises to significantly reduce the consumption of potable water in buildings, reduce discharge to municipal wastewater systems, and save on municipal energy by lowering the amount of potable water treated at municipal facilities. At the same time, the net zero water concept will develop and efficiently manage limited water resources in order to deliver truly sustainable living systems to the urban environment. Recognition of the connection between water management and energy conservation is emerging as a new opportunity in integrated management systems.

Conclusion

Because of this increasing need for water, net zero water is seen as an up-and-coming concept that is a step forward in relieving the issues many cities face with aging water and wastewater treatment systems. Our present practice of simply flushing our waste downstream to become someone else's problem, once termed linear thinking, is not sustainable. As more and more people compete for the same limited water supply, the increased pollution of our water supply cannot continue into the next generation.

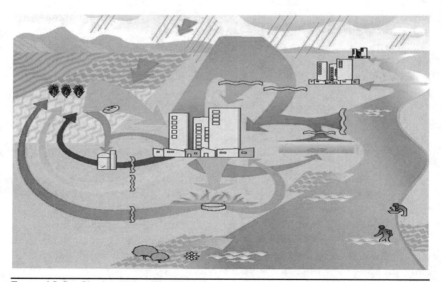

Figure 18.2 Closing the sanitation loop to preserve water for the next generation. (See also Color Plates; *illustration courtesy of the German Federal Ministry of Economic Cooperation and Development*)

Better means of managing our water supply must be found, and it is hoped this book will play a part in that effort. Its purpose is to arouse awareness of viable alternatives to the traditional methods and provide basic knowledge that can be used to make water use more efficient and to reduce contamination now occurring in our drinking water.

Using the design philosophy of net zero water, we will acheive goals necessary for creating a sustainable environment. The end results will be the following:

- The development of a holistic, interdisciplinary approach to the material flow cycle instead of disposal
- Conservation of resources
- Improvement of overall health by minimizing the introduction of pathogens from human excrement into the water cycle
- Promoting safe recovery and use of nutrients, organics, trace elements, water, and energy now seen as a liability
- Preserving soil fertility and improved agricultural productivity

As the above values become commonly supported, a more sustainable society will result, leaving a better world for future generations.

Notes and References

1. *A Green Roof Vision Based on Quantifying Storm Water and Air Quality Benefits.*
2. Excerpt from *Green Building Pro, Net Zero Water—Integrated Water Management for Buildings*, Peck, S.W., GRP, HASLA, founder and president of Green Roofs for Healthy Cities, the North American Industry Association for the green roof and wall industry; and van der Linde, D., Communications and Research Coordinator at Green Roofs for Healthy Cities. www.greenroofs.org.

Pipe Loss Data for Polyvinyl Pipe and Fittings

Feet of Head Pressure Loss per 100' of Polyvinyl Chloride (PVC) Pipe

GPM						Pipe Diameter				
	1/2"	3/4"	1"	1¼"	1½"	2"	2½"	3"	4"	5"
1	2.08	0.51								
2	4.16	1.02	0.55	0.14	0.07					
5	23.44	5.73	1.72	0.44	0.22	0.066	0.038	0.015		
7	43.06	10.52	3.17	0.81	0.38	0.11	0.051	0.021		
10	82.02	20.04	6.02	1.55	0.72	0.21	0.09	0.03		
15		42.46	12.77	3.28	1.53	0.45	0.19	0.07		
20		72.34	21.75	5.59	2.61	0.76	0.32	0.11	0.03	
25			32.88	8.45	3.95	1.15	0.49	0.17	0.04	
30			46.08	11.85	5.53	1.62	0.68	0.23	0.06	0.02
35				15.76	7.36	2.15	0.91	0.31	0.08	0.03
40				20.18	9.43	2.75	1.16	0.40	0.11	0.03
45				25.10	11.73	3.43	1.44	0.50	0.13	0.04
50				30.51	14.25	4.16	1.75	0.60	0.16	0.05
60					19.98	5.84	2.46	0.85	0.22	0.07
70						7.76	3.27	1.13	0.30	0.10
75						8.82	3.71	1.28	0.34	0.11
80						9.94	4.19	1.44	0.38	0.13
90						12.37	5.21	1.80	0.47	0.16

GPM							
100			15.03	6.33	2.18	0.58	0.19
125				9.58	3.31	0.88	0.29
150				13.41	4.63	1.22	0.40
175					6.16	1.63	0.54
200					7.88	2.08	0.69
250					11.93	3.15	1.05
300						4.41	1.46
350						5.87	1.95
400						7.52	2.49
450							3.09
500							3.76

Example Use: 1" pipe, with 5 GPM, would lose 1.72 feet of head pressure per hundred feet.

343

Friction Loss in PVC Fitting = Equivalent Feet of Straight Pipe

PVC Type	Pipe Size								
	½"	¾"	1"	1¼"	1½"	2"	2½"	3"	4"
90° elbow	1.5	2.0	2.25	4.0	4.0	6.0	8.0	8.0	12.0
45° elbow	0.75	1.0	1.4	1.75	2.0	2.5	3.0	4.0	5.0
insert coupling	0.5	0.75	1.0	1.25	1.5	2.0	3.0	3.0	4.0
gate valve	0.3	0.4	0.6	0.8	1.0	1.5	1.6	2.0	3.0
male/female adapter	1.0	1.5	2.0	2.75	3.5	4.5	5.5	6.5	9.5
tee-flow (run)	1.0	1.4	1.7	2.3	2.7	4.3	5.1	6.3	8.3
tee-flow (branch)	4.0	5.0	6.0	7.0	8.0	12.0	15.0	16.0	22.0

Metric Conversion Factors

Multiply	By →	To Get
m³/d	264	gallons per day (gpd)
g/m² – d	8.92	lb/ac – d
kg/m²	0.2	lb/sq ft
kg/ha – d	0.892	lb/ac – d
m³/ha – d	106.9	gpd/ac
m³/m² – d	25	gpd/sq ft
ha (hectare)	2.47	ac (acre)
m (meter)	3.28	ft (feet)
m²	10.76	square feet (sq ft)
m³	264.2	cubic feet
To Get	← by	Divide

Equivalents of Pressure and Head

Pressure or Head	lb/in²	lb/ft²	Atmospheres	kg/cm²	kg/m²	in. water (60°F)	ft water (60°F)	in. mercury (32°F)	mm mercury (32°F)
lb/in²	1	144	0.038046	0.070307	703.067	27.707	2.3039	2.03601	51.7148
lb/ft²	0.0069445	1	0.000473	0.000488	4.88241	0.19241	0.01603	0.014139	0.35913
atmospheres	14.696	2116.22	1	1.0332	10332.27	407.17	33.931	29.921	760.
kg/cm²	14.2234	2048.17	0.96784	1	10000	394.08	32.840	28.959	735.559
kg/m²	0.001422	0.204817	0.0000968	0.0001	1	0.03941	0.003284	0.002896	0.073556
in. water*	0.036092	5.1972	0.002456	0.00253	25.375	1	0.08333	0.073483	1.8665
ft water*	0.433103	62.3668	0.029471	0.03045	304.50	12.	1	0.88180	22.3980
in. mercury†	0.491157	70.7266	0.033421	0.03453	345.316	13.608	1.1340	1	25.40005
mm mercury†	0.0193368	2.78450	0.0013158	0.0013595	13.59509	0.535764	0.044647	0.03937	1

*Water at 60°F
†Mercury at 32°F

Index

Note: Page numbers followed by *f* denote figures; page numbers followed by *t* denote tables; page numbers followed by *n* denote notes and references.

A

A. Lumricoides eggs, 305*t*, 320*t*
ABR. *See* Anaerobic baffled reactor
ADEC. *See* Alaska Department of
    Environmental Conservation
Advanced Composting Systems,
    323–324
Africa, 90
AHJ. *See* Authority Having
    Jurisdiction
Air conditioning condensate, 150
  application examples, 38–40
  applications for, 34
  bacteria and, 33
  commercial kits for inhibiting
    microbial growth in, 33
  cooling tower basin, 39*f*
  equivalent full load cooling
    hours/year for various
    cities, 39–40*t*
  estimate for various cities, 36*t*
  hybrid systems, 38
  introduction to, 33–34
  irrigation, 38
  receiver basin, 37*f*
  swimming pools and, 38
  volume estimate, 34–36
  water distribution, 38
Alaska Department of
    Environmental Conservation
    (ADEC), 221
Alaska Department of Fish and
    Game, 230

Alligator weed, 207*t*
Alluvium, 187
Aluminum, 105*n*1
  sheeting, 47
Alzheimer's disease, 105*n*1
Amebiasis, 9
American Rainwater Catchment
    Systems Association
    (ARCSA), 276
American Society of Heating,
    Air Conditioning, and
    Refrigeration Engineers
    (ASHRAE), 41*n*1
Anaerobic baffled reactor (ABR),
    285–287
  adaptability of, 286
  adequacy of, 286–287
  advantages of, 287*t*
  benefits of, 285
  disadvantages of, 287*t*
  full capacity, 286
  location access, 287
  operation, 286
  schematic, 286*f*
  seeding, 287
  start-up time, 286
Anaerobic system:
  composting toilet compared
    with, 314*t*
  defined, 313
Anaerobiosis, 317
Anaga, Tenerife, 72*f*
Ancient tombs, 55*n*2

Antofagasta, 69–70
Applied Wetlands Technology, 221
Aquatic-plant treatment. *See also*
    specific plants
  conventional compared to,
      205–206
  plants used in, 206–207
  successful, 207
Aquifers. *See also* Ogallala Aquifer
  cone of depression, 189
  confined, 188
  groundwater movement
      through, 187–188
  infiltration replenishing, 187
  pumping, 189*f*, 190
  schematic, 188*f*
  water-transmitting capability
      of, 192
  water withdrawal on, 188–190
Aravari River, 184
ARCSA. *See* American Rainwater
    Catchment Systems
    Association
ASCG, Inc., 221
Aseptic meningitis, 170
ASHRAE. *See* American Society of
    Heating, Air Conditioning,
    and Refrigeration Engineers
Asia, 90
Asphalt caulking compound, 139
Atacama desert, 70
Atlanta, GA, 36*t*, 39*t*
Attached-growth processes, 234
Aufeis, 75, 76–77
  formation of, 76
  research on, 77
Authority Having Jurisdiction
    (AHJ), 288, 289

━━━ **B** ━━━

Bacterial diseases, 9, 169
  air conditioning condensate
      and, 33
  solar water distillation and, 141
Baltimore, MD, 39*t*
Bica da Cana, Madeira, 72*f*
Biochemical oxygen demand
    (BOD), 152
  filter types and, 238*t*

Biological filters, 233
  oxygen supply, 234
  temperature, 234
  types, 234
Bismarck, ND, 39*t*
Black putty, 139
Blackwater, 337
  aeration stage, 269
  applications, 269
  benefits, 276–277
  biological treatment, 270
  chlorination, 271
  collection point, 270
  contamination, 275*t*
  costs, 269, 277
  decentralized systems for, 267*f*,
      271
  defined, 267
  disinfection, 271
  hazards, 267
  incremental development of, 277
  irrigation chamber, 269
  limitations, 277–278
  maintenance, 277–278
  microbiological filter, 279*f*
  nutrient removal, 271
  other uses of, 278
  payback, 278*f*
  plant growth and, 277
  primary treatment, 269
  reasons for, 267
  reuse calculation form, 274*f*
  savings from, 274–275
  scale, 271
  schematic, 270*f*
  screening, 270
  secondary treatment, 269
  sludge settling chamber, 269
  smell, 277
  source and application
      opportunities, 273*f*
  steps, 269
  stigma, 269
  storage, 271
  system design, 271–272
  systems, 269–271
  ultra-filtration, 271
  water balance, 272–275
  water stress and, 276

Bleach, 174–175
  in rainwater catchment, 104
  with warm water, 175
Blue-green sedge, 226, 227
BOD. *See* Biochemical oxygen demand
Boiling, 169
Boston, MA, 36*t*, 39*t*
Boulaalam, Morocco, 72*f*
Boutmezguida, Morocco, 72*f*
Brackish water, 119
  composition of, 121
BREAM. *See* Building Research Establishment Environmental Assessment Method
Broad-leafed cattail, 226
Bromate, 179
Bromine, 179
Building Research Establishment Environmental Assessment Method (BREAM), 280*n*3

C

Calcium hypochlorite, 174
Calla lily, 226
Cape Verde, 72*f*
Carbon-to-nitrogen ratio (C:N), 319
Cardinal flower, 259
Casey Trees Endowment Fund, 337
Center for Alternative Social Research (CISA), 71
Cerro Orara, Peru, 72*f*
Charleston, WV, 39*t*, 126*t*
Charlotte, NC, 39*t*
Cheddar bath's pinks, 259
Chemical fertilizers, 10
Chiapas, Mexico, 72*f*
Chicago, IL, 39*t*
Chile, 72*f*
  fog harvesting in, 69–71
Chlorine, 8. *See also* Bleach
  availability of, 174
  blackwater treated with, 271
  fog harvesting and, 67
  gas, 174
  guidelines, 175*t*
  solid, 174
  tablets, 175–176
  UV differentiated from, 173–174

Chloroform, 8
Cholera, 9, 299
CISA. *See* Center for Alternative Social Research
Cloud water harvesting. *See* Fog harvesting
C:N. *See* Carbon-to-nitrogen ratio
Colloidal silver (CS), 176, 177–178
Commercial buildings, 4
Common rush, 259
Common sedge, 226, 227
Compost, 307
  aeration, 316–317
  defined, 313
  environmental factors, 316
  optimal, 332
  process, 319*t*
  stages, 315–316
Composting toilets:
  advantages, 330
  aeration, 316–317, 321, 325
  anaerobic system compared with, 314*t*
  appearance, 326
  approval of, 325
  barrels, 326
  big box, 323
  BioLet, 326
  blowers, 321
  bulking agents added to, 316, 331
  carbon added to, 319
  carbon-to-nitrogen ration, 319
  Clivus multrum, 322–323, 322*f*, 323*f*
  in cold climates, 331
  cooling phase, 315–316
  costs, 332
  cottage systems, 332
  curing phase, 316
  description, 321–325
  design standards, 331–332
  disadvantages of, 330–331
  dry system with remote, 327, 328*f*
  end product, 327
  Envirolet, 326, 326*f*
  environmental factors, 316
  exhaust fan, 325
  first commercial, 322

Composting toilets (*Cont.*):
  with heaters, 317–318, 321
  initial form of, 322
  installation, 323*f*
  large-capacity, 332
  limitation of, 329
  maintenance, 325, 331
  mesophilic phase, 315, 318
  moisture and, 317–318
  moldering, 318
  objective, 314–315
  off-grid plumbing and, 314*f*
  operation, 331
  oxygen deficit, 317
  paddles, 326
  Phoenix, 324*f*
  process control, 321
  proper, 324–325
  pumps, 321
  raking mechanism, 326
  residential, 326–331
  self-contained dry, 327, 328*f*
  social acceptance of, 329
  in St. Charles, MO, 325*f*
  stages, 315–316
  Sun-Mar, 326, 327*f*
  temperature, 318
  thermophilic phase, 315
  time and, 319
  ultra-low-flush, 329*f*
  ventilation, 317
CONAF. *See* National Forestry
  Corporation
Constructed wetland, 239–257.
  See also Recirculating vertical
  flow constructed wetlands
  advantage of, 239
  benefits of, 222
  design, 225–226
  permits, 228
  plant selection, 226–227
  primary treatment enhanced
    through, 239
  problem areas, 227–228
  schematic, 240*f*
  single-cell, 247*f*
  two-cell, 248*f*
  water quality requirements,
    224–225

Contamination. *See also* Pollution
  blackwater, 275*t*
  fog harvesting, 66
  graywater, 165*t*
  groundwater, 16
  people getting accustomed to,
    169
  rainwater catchment, 90, 102*f*
  Solar water distillation, 139
  sources, 16
  well water, 8
Coontail, 207*t*
Coral Springs, Florida, FL, 209*t*
Croatia, 72*f*
  dew harvesting in, 53
Cryptosporidium, 9, 175
CS. *See* Colloidal silver

D

Dallas, TX, 36*t*, 39*t*
Darcy's Law, 244
Dark green rush, 259
Dawa, Sonam, 85
Desalination, 119
Design air temperature, 41*n*2
Detroit, MI, 39*t*
Dew harvesting, 121
  common objects for, 43
  costs, 55
  in Croatia, 53
  defined, 43–44
  gravel and sand mulches, 54
  ground installed, 51–53
  ground installed collection
    summary, 52
  high-mass air wells, 44
  historical applications of, 44–46
  horizontal surface requirement,
    46
  installation, 46
  key points, 43
  Kothara, India, 48, 49*f*
  low-mass radiative collectors, 45
  materials, 47
  with new roof, 53
  other applications of, 54
  Panandhro site, 51, 52*f*
  performance, 48
  portable, 53, 54*f*

Dew harvesting (*Cont.*):
 recommendations, 54–55
 reliability of, 43
 research, 47–48
 roof mounted systems, 48–51
 roof mounted systems collection
  summary, 51*t*
 roof slope, 53
 Satapar site, 52
 Sayara school, 50, 50*f*
 Suthari warehouse, 48–49, 49*f*
 test platform, 55
 weather conditions for, 43
 yield increase, 43
Dew point condition, 57
Dhofa, Oman, 72*f*
Diarrhea, 170
Disinfection, 8. *See also specific*
  methods of disinfection
 blackwater, 271
 graywater, 161
 methods, 173–180
 rainwater catchment, 104
 spring box, 31
Distilled water, chemical
  aggression of, 140
Dominican Republic, 72*f*
Drainfield, 283–284
 distribution box, 284
 driving on, 285
 proper function of, 285
 soils, 283–285
 trenches, 284
Droughts, 3
Duckweed, 209*t*
 balance between surface and
  volume, 218
 biochemical changes, 216
 biological process schematic, 217*f*
 biomass recovery, 220–221
 as cash crop, 221
 chemical changes, 216
 example, 220
 as feed, 217–218, 221*f*
 harvesting, 219*f*, 221
 lagoons, 219
 Lenacae family, 216
 management, 216
 micro-element uptake, 217

Duckweed (*Cont.*):
 mineral concentrations, 217–218
 nutrients, 221
 pH levels, 216–217
 polishing pond, 219
 pond resurrection with, 220–221
 protein content, 217
 temperature and, 216
 treatment, 219–220
 water depth, 218
Dysentery, 299

━━ **E** ━━

E. histolytica cysts, 305*t*, 320*t*
Echoviruses, 169–170
Ecuador, 71
Egypt, 70
Electric power, 8*f*
Electrodialysis, 142
Elephant Ears, 241
Encephalitis, 170
Energy consumption:
 to generate potable water, 7*f*
 net zero, 335
 in producing clean water, 7
 solar water distillation
  requirements, 124–125, 124*t*
Entamoeba histolytica, 9
Enteroviruses, 320*t*
Escherichia coli, 169
ET. *See* Evapo-transpiration
Eutrophication, 264*n*16
Evapo-transpiration (ET), 246–257
 advantages, 255
 bed liner, 249, 253–254
 climate, 249, 252, 255
 cost, 255
 cover soil, 252
 design criteria, 249
 design parameters, 250–251
 distribution system, 253–254
 drainage, 254–255
 fill material, 254
 flow rate, 249
 groundwater replenishing,
  186–187
 habitats created from, 255
 infiltration, 249
 land limitations, 256

Evapo-transpiration (ET) (*Cont.*):
  life expectancy, 256
  limitations, 255–256
  operation of, 254
  percolation inflow, 252
  precipitation, 251–252
  rate, 252–253
  runoff estimates, 252
  site selection, 249f
  snow, 252
  surface cover, 254
  system variables, 254
  temperature, 253
  toxicity, 256
  in United States, 253
  vegetation and, 252
  water balance, 250
  water depth, 254
Evaporation, 187

━━━ **F** ━━━

Fairbanks, AK, 39t
Faizi, Inayatullah, 77–78
Fecal coliforms, 305t, 320t
Fertilizers:
  chemical, 10
  urine as, 307, 309f
Fiber-reinforced plastic, 47
Fit-for-purpose water, 275–276
Floods, 3, 185
  groundwater replenishing
      through, 200
Fog:
  defined, 57–58
  formation, 57
  frequency, 58
  requirements, 57–58
  sea, 58f
  worldwide, 70
Fog harvesting, 44
  advantages of, 67–68
  altitude and, 60
  in Antofagasta, 69–70
  in Atacama desert, 70
  cable fasteners, 66–67
  in Chile, 69, 70–71
  chlorine and, 67
  cistern size, 64–66
  cisterns, 67
  coastal, 64

Fog harvesting (*Cont.*):
  coastline distance and, 60
  collector drains, 67
  contamination, 66
  cost-sharing approach to, 69
  costs, 69–70
  crestline locations and, 60
  design, 58
  disadvantages of, 68–69
  in Ecuador, 71
  in Egypt, 70
  extent of use, 70–71
  government subsidies, 69
  gutter system, 63f
  installation, 62f
  in Israel, 70
  leakage, 65
  maintenance, 58, 66–67
  mesh densities, 61
  mesh nets, 66–67
  meteorology and, 58
  mountain, 64
  operation of, 66–67
  overflow, 66
  performance improvement, 63
  in Peru, 71
  pilot-scale assessment, 59
  pipelines, 67
  pressure outlets, 67
  problems, 71
  purification, 66
  relief in surrounding areas and,
      60
  satellite imaging and, 60–61
  in series, 61
  site selection, 59–60
  in South Africa, 70
  space for, 60
  storage, 64–66, 67
  suitability of, 58–59
  support, 61, 69
  surface of, 61
  technical description, 61–63
  test site, 64f
  topography and, 58, 59–60
  training, 66
  trees, 58
  upwind locations and, 60
  water quality, 72f
  wind and, 59, 63–64

Food service buildings, 4
Franklin, Benjamin, 5
Freshwater, 3, 119
   on Earth, 5–6
Frost, 43
   percolation test and, 291

━━━ **G** ━━━

Galvanized iron sheeting, 47, 48, 51
Gambia, 146*f*, 147*f*
Giardia lambdia, 9, 175
Giardiasis, 9
Gilfilian, Bob, 223
Gilfilian Engineering, 223
Ginger Lilacs, 241
Glacier grafting, 75, 77–81
   ceremony, 81
   documented, 78
   schematic, 79*f*
   site, 78
   skeptics, 80
Glacier impoundment, 75, 81–86
   applications of, 87
   construction, 82–86
   cost of, 82
   in Ladakh, India, 81–82
   in Phuktsey, India, 85–86
   pipe network, 82
   in Stakmo, India, 84*f*, 86*f*
Glaciers. *See also* Aufeis
   cropping window and, 85
   female, 78
   flow during sowing season, 85*f*
   groundwater recharged by, 87
   maintenance of, 87
   male, 78, 80*f*
   natural, 75
   retaining walls, 83*f*
   schematic, 84*f*
   from space, 76*f*
   stages to growing, 79*f*
   types of, 75
GMDC. *See* Gujarat Mineral Development Corporation
Gobabeb, Namibia, 72*f*
Gravity, 188
Great blue lobelia, 259
Great bulrush, 259
Great Falls, MT, 39*t*

Great spike rush, 259
Green roofs, 337
Graywater, 275, 337
   aboveground tanks, 160
   air elimination, 162
   applications, 149, 154
   backflow prevention, 161
   backup water supply, 161
   basic chemical systems, 152
   basic physical systems, 152
   bathroom, 154
   belowground tanks, 160
   benefits, 149
   biological systems, 152
   biological treatment, 161
   biomechanical systems, 152
   bypass, 162
   chemical treatment, 161
   cistern installation, 163–164
   cisterns, 159–160
   collection, 159
   contamination, 165*t*
   defined, 149
   design procedure, 152–157
   detailed approach, 152, 155–157
   direct reuse systems, 151
   disinfection, 161
   drainage, 159–160, 162
   dry-run protection, 161
   external demand estimations, 157–158
   failures, 163
   fittings, 161
   hair traps, 159
   health characteristics, 167
   hybrid systems, 152
   industrial application of, 151*f*
   installed environment, 168
   internal demand estimations, 156–157
   labeling, 163
   log, 164
   maintenance, 154, 164, 167
   maintenance schedule, 166–167t
   materials, 161
   mechanical fine filtration, 161
   non-return valve, 162
   organic matter in, 152
   overflow, 162
   pathogen reduction, 167

Graywater (*Cont.*):
  payback, 278*f*
  piping, 163
  pollutants, 159
  pumping, 161–162
  quality guideline values, 165*t*
  quantifying use amount, 154
  rainwater catchment combined
      with, 158–159
  residential application of, 153*f*
  retrofit, 160
  reusing, 168*n*4
  risk assessment, 167–168
  screening, 160
  sedimentation and flotation, 160
  short retention systems, 151–152
  simplified approach, 152,
      154–155
  sources, 149–150, 154
  storage capacity, 158
  system controls, 162–163
  system monitoring guidelines,
      165*t*
  system types, 151–152
  tanks, 159–160
  treatment, 160–161
  ultraviolet and, 161
  washing machines and, 168*n*5
  water quality, 154, 164
  yield estimation, 155–156
Groundwater, 7
  contamination, 16
  demand, 182
  depletion, 182
  flow, 189*f*, 190
  glaciers recharging, 87
  movement through aquifer,
      187–188
  sediment, 22
  velocity, 190
Groundwater replenishing.
      See also Infiltration; Water
      spreading; Wells
  Aravari River, 184
  artificial, 191
  benefits of, 181
  classifications, 191–201
  cost, 202
  design, 181, 191

Groundwater replenishing (*Cont.*):
  direct, 192–199
  for domestic supplies, 201–202
  evapo-transpiration, 186–187
  through flooding, 200
  historical applications, 184
  in India, 184, 201, 202
  indirect, 199–201
  land cover and, 186
  maintenance, 202
  mechanisms, 191–192
  natural, 181, 191
  objectives, 191
  plan, 191
  precipitation and, 185
  site investigation, 191
  slope of land and, 186
  soil characteristic and, 185–186
  soil saturation and, 186
  sources, 191–192
  from steambeds, 199*f*
  techniques, 191
  through topographical
      depressions, 199
Gujarat Mineral Development
      Corporation (GMDC), 51

H

Halogens, 173
Hard-stemmed bulrush, 259
Hawaii, 39*t*, 72*f*
Hazen-Williams formula, 32*n*1
Hepatitis, 9, 169
Hercules, GA, 209*t*
Hippocrates, 176
Horizontal well, 28
  placement, 30*f*
  requirements, 30
Houston, TX, 39*t*
HR. *See* Humidity ratio
HRT. *See* Hydraulic retention time
Human waste. *See also* Fecal
      coliforms; Urine
  disposal, 299
  nutrient content, 315
Humidity ratio (HR), 37
Humus, 314–315
Hydraulic conductivity, 190
Hydraulic gradient, 188, 190

Hydraulic retention time (HRT),
    286
Hygroscopic particles, 58
Hypoxia, 317

━━━ **I** ━━━

IAPMO Green Plumbing
    Supplement Appendix A -
    Method of Calculating Water
    Savings, 274
India:
    dew harvesting in, 48, 49f, 50,
        50f, 51, 52, 52f
    glacier impoundment in, 81–82,
        84f, 85–86, 86f
    groundwater replenishing in,
        184, 201, 202
Indianapolis, IN, 36t, 39t
Infiltration, 185–187
    aquifers replenished through,
        187
    evapo-transpiration, 249
Infrared emission, 44
Insects:
    solar water distillation and, 141
    spider mites, 210
    ventilated improved pit latrine
        and, 301–302
Insolation, 121
    daily average, 125
International Water Management
    Institute (IWMI), 201
Intestinal cramps, 170
Iodine, 169, 176
Irrigation:
    air conditioning condensate, 38
    blackwater, 269
    rainwater catchment, 91
Israel, 70
IWMI. *See* International Water
    Management Institute

━━━ **J** ━━━

Johad, 185

━━━ **K** ━━━

Knapen air well, 45f
Kreytzmann, Hermann, 80–81

━━━ **L** ━━━

La Ventosa, Guatemala,
    72f
Ladakh Ecological Development
    Group, 85
Lakes, 6
Latrines. *See also* Pour-flush
    latrine; Privies; Ventilated
    improved pit latrine
    basic, 300f
    new pits, 300
    pit, 299
    sicknesses from, 299
Leachate problem, 317
Leffingwell, E. de K., 76, 88n1
Legionella, 33, 150, 160
Lemnaceai family, 207t
Lindström, Rikard, 322
Liver damage, 170
Loess, 55n10
Lorimer, D. L. R., 78
Los Angeles, CA, 36t, 39t, 126t
Louisiana Iris, 241
Louisville, KY, 39t
Lucedale, MS, 209t

━━━ **M** ━━━

Maddux, Dave, 221
Madhya Ganga Canal Project
    (MGCP), 201
Madison, WI, 39t
Malaysia, 90
McCracken, Horace, 119, 137
Memphis, TN, 40t
Meteosat Second Generation
    weather satellites (MSG),
    60–61
Metric conversion factors, 345
MGCP. *See* Madhya Ganga Canal
    Project
Miami, FL, 36t, 40t
Middendorff, A. T. von, 76, 88n2
Minneapolis, MN, 36t, 40t
Montgomery, AL, 40t
Morning glory vines, 259
Mount Denali, 222
MSG. *See* Meteosat Second
    Generation weather satellites

═══ N ═══

Naled, 76

Namib Desert Beetle, 45–46, 46*f*

Nanotechnology, 178

Nashville, TN, 40*t*

National Forestry Corporation (CONAF), 70–71

National Geophysical Research Institute (NGRI), 201

National Meteorological and Hydrological Service (SENAMHI), 71

National Renewable Energy Laboratory, 125

National Weather Service, 93

Natural Treatment for Sewage Treatment Facility case study, 222–231

Nepal, 72*f*

Net zero energy, 335

Net zero water:
   building towards, 335
   end result of, 339
   example, 336–338
   as learning opportunity, 338
   standards, 337

New England aster, 259

New Orleans, LA, 40*t*

New York, NY, 40*t*

NGRI. *See* National Geophysical Research Institute

Night soil, 299

NOAA AVHRR satellites, 60–61

Norphel, Chewang, 81–82, 85
   honors for, 86

NSF International Standard ANSI/NSF 41-1998: Non-Liquid Saturated Treatment Systems, 331–332

Nutrients:
   blackwater, 271
   duckweed, 219
   in human waste, 315
   sewer loss, 10

═══ O ═══

Oceans, 119

Ogallala Aquifer, 7–8, 187
   depth, 203*n3*
   water removed from, 8

Omaha, NE, 40*t*

OPUR foil, 45, 47, 50

Organic roofs, 89

Oxygen weed, 207*t*

Ozone:
   controversy over, 179
   diffusing, 180
   drawbacks to, 179
   maintenance, 180

═══ P ═══

Parasites, 9

Parking lot runoff, 195*f*

Pasteur, Louis, 169

Patache, Chile, 72*f*

Pathogens, 169. *See also specific pathogens*
   graywater, 167
   survival times, 305*t*, 320*t*
   urine, 307

Percolation test, 295
   calculations, 290–293
   design, 288
   estimated sewage flow, 292*t*
   failure, 290
   frost and, 291
   general instructions, 296
   how to run, 288–293
   procedure, 289–290
   results, 290
   septic simulation, 288
   site preparation, 288–289
   soil description, 296

Permeability, 190

Peru, 71

Phoenix, AZ, 40*t*

Phy, Joseph, 203*n5*

Pickerel Rush, 241

Pipe loss data, 341–344

Pittsburgh, PA, 40*t*

Plants. *See also* Aquatic-plant treatment; *specific plants*
   blackwater, 277
   constructed wetland selection, 226–227
   in recirculating vertical flow constructed wetlands, 259–260
   in rock filters, 241

Polio, 9
Polioviruses, 169
Pollution:
  in graywater, 159
  rainwater catchment and, 89
  sanitation and, 9–10
Polycarbonate sheeting, 47
Polyethylene film, 47
Ponds:
  duckweed resurrecting,
    220–221
  facultative, 232n9
  percolation, 194
  polishing, 219
  water hyacinth, 214
  weeds, 209t
Porosity, 190
Portland, ME, 40t
Potentiometric surface, 188,
  204n14
Potters for Peace, 178
Pour-flush latrine, 302–305, 303f
  structure, 302
  two pit, 303, 305
  water seal pan cross section,
    303f
Prairie cord grass, 259
Pressure and head equivalents,
  348
Privies, 299–300
  design of, 322
  double installation, 304f
  life cycle of, 310f
  maintenance, 309
  with septic tank, 305
  shallow pit construction for,
    302f
  water table and, 310f
Psychometric Chart, 35f

R

Rain gardens, 193, 194f, 195, 196f
  sizing, 204n17
Rainfall data, 105–118
Rainwater catchment, 142, 275,
  336–337
  above-ground application in
    non-freezing environment,
    96f

Rainwater catchment (*Cont.*):
  acceptable piping schematics,
    95–99
  bleach in, 102
  carbon filter, 95
  cistern, underground exterior,
    95, 97f
  cistern sizing, 94–95
  collection sizing, 93
  contamination, 90, 102f
  demand estimates, 90–95
  disinfection, 104
  first flush diverter, 101, 102f
  freezing environments, 99
  graywater combined with,
    158–159
  installation, 99–103
  interior tank, 100f
  introduction to, 89–90
  for irrigation, 91
  maintenance, 103–104
  in Malaysia, 90
  maximum volume, 94
  non-potable water, 95, 98f, 99
  passive, 197–198f
  payback, 278f
  pollution, 89
  pre-wash volume, 103
  roof washer, 101–103
  roofs for, 89–90
  signage, 99
  simplest, 89
  sizing information, 91
  skimming, 99
  solar water distillation and, 144
  special-purpose facilities and,
    91
  standards, 104f
  storage sizing, 94
  supply estimates, 93
  system adjustment, 95
  tank overflow, 99
  tank siphoning, 99, 101f
  water pickup, 99, 101
  water quality, 103–104
Raschel netting, 61, 62f
Rational Method formulas, 251
"Re-Greening Washington, DC,"
  337

Recirculating vertical flow
   constructed wetlands (RVF),
   257–261
   absorption trenches, 258
   construction, 257–260
   design criteria, 257
   elevation view, 261*f*
   expected performance, 261
   first installation, 257
   maintenance, 260–261
   operation, 260
   plants in, 259–260
   pump control, 259
   size of, 257, 258*t*
   top sectional plan view, 262*f*
   vegetation removal, 260–261
Reservoirs, 7
Reverse-osmosis, 142
RF. *See* Rock filter
Richmond, VA, 40*t*
River bulrush, 259
Rivers, 6
Rock filter (RF), 240–246
   area calculation, 243–244
   cross-sectional area, 244
   design criteria, 242–244
   example, 244–246
   fill up of, 241
   hydraulic loading calculation,
      243
   in landscape plan, 241
   limitations, 246
   liner, 241–242
   organic loading calculation, 243
   plants in, 241
   pump-out schedule, 242
   for repair, 241
   rock characteristics, 242–243
   schematic, 241*f*
   section, 242*f*
   size, 243
Roofs:
   dew harvesting mounted on,
      48–51, 51*t*, 53
   green, 337
   new, 53
   organic, 89
   for rainwater catchment, 89–90
   washer, 101–103

Rotaviruses, 169
RVF. *See* Recirculating vertical
   flow constructed wetlands

S

Sacramento, CA, 40*t*
St. Charles, MO, 325*f*
St. Louis, MO, 40*t*
Salmonella, 305*t*, 320*t*
Salt, 58, 123
Salt Lake City, UT, 40*t*
Salt water, 3
San Diego, CA, 209*t*
Sand filters, 170–173
   advantages of, 172–173
   cleaning, 171
   construction of, 170
   do-it-yourself, 172–173, 173*f*
   efficiency of, 171
   herringbone drains, 172
   hypogeal layer, 172
   operation, 171–172
   performance loss, 172
   refurbishing, 172
   simplicity of, 172
   slow, 171*f*
   solar water distillation, 123
   wet harrowing, 171, 172
   working, 172
   WHO on, 173
Sanders, Adam, 203*n*5
Sanitation:
   closed loop, 338*f*
   defined, 12
   future and, 10–11
   pollution and, 9–10
   stress, 2*f*
   UNICEF on, 1
   waste management and, 9
   WHO on, 1, 12
Santa Cruz River, 182, 203*n*5
Satellite imaging:
   fog harvesting and, 60–61
   spatial resolution, 61
Saturated zone, 187
Saturation condition, 57
Schmutzdecke, 172
Schultz, Jack, 143*n*1
Seattle, WA, 40*t*, 126*t*

Seawater, 119
  composition of, 121
Sediment:
  groundwater, 22
  spring box removal, 31
Seep collection system, 24, 28, 29*f*
SENAMHI. *See* National
  Meteorological and
  Hydrological Service
Septic system:
  components, 281–285
  design, 281–285
  field schematic, 282*f*
  percolation tests simulating, 288
Septic tank, 281–283
  acceptance of, 287
  advantages of, 284*t*
  chambers, 282, 283*f*
  cleaning, 285
  disadvantages of, 284*t*
  health aspects, 287
  maintenance, 285
  materials, 281
  operation, 285
  primary functions, 282
  privies with, 305
  size, 281
  sludge removal, 283
Sewers:
  development of conventional,
    10
  improvement to, 9
  network, 13
  nutrients lost in, 10
  overflow, 280*n*2, 337
  population migration and, 13
Sharan, Girja, 45
Sherman, Mark, 221
Shigella, 305*t*, 320*t*
Shigellosis, 9
Silver:
  colloidal, 176, 177–178
  filter, 176–177
  medicinal applications of,
    176–179
  nano tea bag, 178–179*f*
Snow, 252
Society for the Protection of New
  Hampshire Forests, 158*f*

Sodium hypochlorite, 174
Soft-stemmed bulrush, 226, 259
Soil:
  characteristics, 185–186
  drainage characteristics, 186*t*
  drainfield, 283–285
  evapo-transpiration cover, 252
  groundwater replenishing and,
    185–186
  night, 299
  percolation test, 296
  permeability of, 186*t*
  saturation, 186
Soil Conservation Service, 251
Solar radiation, 44, 121
Solar water distillation:
  ancillary components, 128,
    138–141
  bacteria and, 141
  brine rejection, 124
  Charleston average, 126*t*
  components, 128, 130–138
  contamination, 139
  cost, 121, 123–124
  depth, 129, 138
  design objectives, 127
  diagram, 122*f*
  distillation trough, 137–138
  documentation of, 119
  double slope, 134*f*
  drainage, 138
  durability, 130
  efficiency, 127
  emergency still, 144, 145*f*
  energy requirements for,
    124–125, 124*t*
  enhancements, 144, 147
  equation for output, 125
  feedwater, 123
  feedwater temperature, 127
  field survival, 120*f*, 145*f*
  in Gambia, 146*f*, 147f
  glass, 128
  greenhouse-solar still, 147
  hybrid designs, 144
  insects and, 141
  insulation, 124, 138
  Los Angeles average, 126*t*
  market potential, 142

Solar water distillation (*Cont.*):
  material requirements, 130
  modular units, 129–130
  multiple-effect basin stills, 143–144, 143*f*
  operation of simple, 119, 121, 123
  panel design, 121
  piping, 139–140
  pure water cost, 128
  rainwater catchment and, 144
  re-leveling, 133
  residential, 120*f*, 129
  sand filtration, 123
  sealants, 138–139
  Seattle average, 126*t*
  single slope, 134*f*
  size, 129
  solar focusing still, 146*f*
  steps, 121
  still operation, 140
  storage reservoir, 140–141
  supplementary heating, 147
  support structure, 133–134
  types, 123–124, 127–130, 147*f*
  wick stills, 144, 144*f*
Solar water distillation basin, 130–133
  color, 130
  corrosion, 131
  materials, 131, 132*t*, 133
  multiple, 143–144, 143*f*
  shape, 131
  single, 120*f*
  surface, 130
  types, 131
Solar water distillation glazing, 134–137
  material composition, 136*t*
  materials, 135, 137
  strength, 135
  wetting ability, 135
South Africa, 70, 72*f*
Spider mites, 210
Spring box:
  backfilling, 28
  brick and concrete, 20*f*
  commercially manufactured, 19, 21*f*

Spring box (*Cont.*):
  common features of, 18
  concrete mixture for, 27
  construction, 25–27
  delays, 25
  dimensioned plan of, 22, 23*f*
  disinfection, 31
  diversion ditch, 22, 24, 30
  excavation of, 25*f*
  fencing off, 22
  foundation, 24
  hillside, 27–28
  installation, 27–28
  location map, 24*f*
  location of, 21
  maintenance, 30–31
  materials, 19
  newly completed, 16*f*
  open bottom, 20*f*, 26
  overflow pipe, 18
  with pervious side, 19*f*, 26
  purpose of, 18
  sediment removal from, 31
  single flow source, 24
  site preparation, 22, 24–25
  tank construction, 22
  tank cover, 27
  types of, 19–21
  vegetation surrounding, 19
  washing solution, 31
  water distribution system, 21–22
  water quality check, 30
Spring box forms, 25–26, 26*f*
  motor oil coating on, 27
  setting, 27
  vibrating, 27
Spring development:
  benefits of, 18
  costs, 18
  design features, 18–19
  examples, 18–19
Springs:
  artesian, 15, 17*f*
  artesian fissure, 15
  artesian flow, 16
  categories, 15
  color of, 31
  defined, 15
  depression, 15

Springs (*Cont.*):
  fracture and tubular, 15, 16
  gravity, 15, 17*f*
  protection, 16
  water quality, 31
Stabilization, 297
Storm water:
  fee rebates, 337
  runoff management, 337
  utility fee charges, 337
  waste management and, 12–13
Stratocumulus clouds, 57
Streptomycin, 318
Subsurface water, 187
Swamp milkweed, 259
Swamp rose mallow, 259
Swimming pools, 38

━━━ **T** ━━━

Tampa, FL, 36*t*, 40*t*
Tar plastic, 139
Texas Rainwater Standards, 276
TF. *See* Trickling filter
Thakur, V. C., 86
THMs. *See* Trihalomethanes
Transpiration, 187
Transport costs, 142
Trickling filter (TF), 234–235
  activated-sludge process
    combined with, 238
  advantages of, 238
  air availability, 236
  common uses of, 234
  configuration, 235*f*
  cross-flow media, 236
  design, 235–237
  effectiveness of, 239
  fixed nozzle distributors, 236
  high-rate, 237
  intermediate-rate, 237
  limitations of, 238–239
  low-rate, 237
  multiple spraying, 236
  organic loading of, 237
  performance, 238
  plastic, 236
  rock media, 236
  rotary hydraulic distribution, 236
  roughing, 237

Trihalomethanes (THMs), 8
Tucson Water well, 203*n*5
Tulsa, OK, 40*t*
Tveiten, Ingvar, 80
Typhoid, 9, 299

━━━ **U** ━━━

Ultraviolet (UV):
  chlorine differentiated from,
    173–174
  effectiveness of, 179
  graywater and, 161
  maintenance of, 179
United Kingdom, 90
United Nations Children's Fund
    (UNICEF), 1
United States, 14*n*5, 90, 253. *See also*
    specific places in United States
United States National Oceanic
    and Atmospheric
    Administration's National
    Climate Data Center, 93
Unsaturated zone, 187
Urine:
  average person's, 307
  diluting, 309
  as fertilizer, 307, 309*f*
  pathogens, 307
  separating system, 308*f*
  storage, 307
  utilization of undiluted, 307–309
Use profile:
  for residential facilities, 5*f*
  for various buildings, 4*f*
UV. *See* Ultraviolet

━━━ **V** ━━━

Vapor, 121
Vegetation. *See also* Plants
  evapo-transpiration and, 252
  recirculating vertical flow
    constructed wetlands and
    removal of, 260–261
  spring box and, 19
  water table lowered by, 200
Ventilated improved pit latrine
    (VIP), 300–302, 301*f*
  construction, 301
  insects and, 301–302
  leak prevention, 301

Vibrio cholerae, 305*t*, 320*t*
VIP. *See* Ventilated improved pit latrine
Viral agents, 9, 169

W

Walt Disney World, 211*t*
Washing machines:
  efficiency data, 156
  graywater and, 168*n*5
  water consumption, 92*t*
Waste management:
  alternative, 12–14
  sanitation and, 9
  storm water and, 12–13
Water availability, 3, 205
Water balance:
  blackwater, 272–275
  evapo-transpiration filter, 250
  spreadsheet, 272
  step-by-step method for, 272
Water consumption, 90–91
  in Africa, 90
  in Asia, 90
  developing countries minimum, 123
  dishwasher, 92*t*
  faucets, 92*t*
  to generate electric power, 8*f*
  industrialized world, 123
  per animal, 65
  per person, 65
  residential indoor, 92*t*
  shower, 92*t*
  toilet, 92*t*
  in United Kingdom, 90
  United States, 14*n*5, 90
  washing machine, 92*t*
  worldwide, 183*f*
Water cycle, 3–5, 3*f*
Water distribution, 4*f*
  air conditioning condensate, 38
  spring box, 21–22
Water fern, 207*t*
Water hyacinth, 209*t*
  aeration, 211
  basic criteria for, 214
  in bloom, 210*f*
  catastrophic events and, 212

Water hyacinth (*Cont.*):
  climate, 212
  configuration, 214
  in Coral Springs, Florida, 211*t*
  depth, 214
  design parameters, 212–216
  environmental factors, 209–210
  environmental laws restricting, 210
  harvesting, 211–212
  in Hercules, GA, 211*t*
  hydraulic loading rates, 215–216
  hydraulic residence time, 214
  installations, 211*t*
  land requirements, 213, 213*t*
  location, 212
  in Lucedale, MS, 211*t*
  natural controls, 210
  organic loading rates, 214–215, 215*t*
  paper from, 212
  pond size and, 214
  post-treatment, 214
  pre-treatment, 213
  in San Diego, CA, 211*t*
  site requirements, 213
  surface loading rates, 214–215
  systems comparison, 215
  temperature and, 209–210, 212
  uncontrolled, 211*f*
  in Walt Disney World, 211*t*
  water quality and, 212
Water iris, 259
Water lettuce, 209*t*
Water primrose, 209*t*
Water quality:
  constructed wetland, 224–225
  expected, 170*f*
  fog harvesting, 72*f*
  graywater, 154, 164
  guidelines, 180
  for non-potable applications, 276*t*
  rainwater catchment, 103–104
  spring box, 30
  springs, 31
  testing, 180
  water hyacinth and, 210

Water sources, 7–9
  blackwater, 273*f*
  graywater, 149–150, 154
  groundwater replenishing,
    191–192
Water spreading:
  check dams, 193, 194, 195*f*
  examples, 194–196
  parking lot runoff, 195*f*
  percolation ponds, 194
  percolation tanks, 193
  rain gardens, 193, 194*f*, 195, 196*f*
  recharge pits, 194
  site selection, 193–194
  swales, 194, 195
  trenches, 194
Water stress, 2*f*, 202
  blackwater and, 276
  health hazards, 183
  livestock loss, 183
Water table:
  privies, 310*f*
  vegetation and, 202
Water velvet, 209*t*
Watermilfoil, 209*t*
Wells, 7. *See also* Horizontal well;
    specific types of wells
  contamination of, 8

Wells (*Cont.*):
  dug, 196
  Knapen air, 45*f*
  lateral shafts, 201
  recharge, 198
  recharge shafts, 201
  Tucson Water, 205*n*5
  types, 198
  Zibold air, 45*f*
WHO. *See* World Health
    Organization
Wind:
  fog harvesting and, 59, 60, 63–64
  global patterns, 59
Window putty, 139
Wishing Shrine, 203*n*5
World Health Organization
    (WHO):
  on sand filters, 173
  on sanitation, 1, 12
Worms, 299

Yellow Canna, 241
Yemen, 72*f*

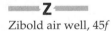

Zibold air well, 45*f*

76676

Y0-BZB-302

Seeing

A

W

T

Publication of this volume was made possible in part by a grant
from the National Endowment for the Humanities.

The University Press of Kentucky
Scholarly publisher for the Commonwealth,
serving Bellarmine University, Berea College, Centre College of Kentucky,
Eastern Kentucky University, The Filson Historical Society, Georgetown
College, Kentucky Historical Society, Kentucky State University,
Morehead State University, Murray State University, Northern Kentucky
University, Transylvania University, University of Kentucky, University of
Louisville, and Western Kentucky University.
All rights reserved.

Editorial and Sales Offices: The University Press of Kentucky
663 South Limestone Street, Lexington, Kentucky 40508-4008
www.kentuckypress.com

08 07 06 05 04   5 4 3 2 1

The Library of Congress has cataloged the hardcover edition as follows:

McEuen, Melissa A., 1961–
    Seeing America : women photographers between the wars / Melissa A.
McEuen.
        p.   cm.
    Includes bibliographical references and index.
    ISBN 0-8131-2132-9 (cloth : alk. paper)
    1. Women photographers—United States—Biography. 2. Documentary
photography—United States—History—20th century. 3. Photography—
United States—History—20th century. I. Title.
TR139.M395   1999
770'.92'273—dc21
[B]   99-17219

Paper ISBN 0-8131-9094-0

This book is printed on acid-free recycled paper meeting
the requirements of the American National Standard
for Permanence in Paper for Printed Library Materials.

Manufactured in the United States of America.

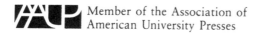 Member of the Association of
American University Presses

For Family,

the McEuens
and
the Stantons

Contents

Acknowledgments   ix

Introduction   1

1   Documentarian with Props
    Doris Ulmann's Vision of an Ideal America   9

2   Portraitist as Documentarian
    Dorothea Lange's Depiction of
    American Individualism   75

3   A Radical Vision on Film
    Marion Post's Portrayal of Collective Strength   125

4   Of Machines and People
    Margaret Bourke-White's Isolation
    of Primary Components   197

5   Modernism Ascendant
    Berenice Abbott's Perception of
    the Evolving Cityscape   251

Conclusion   291

Notes   299

Bibliography   331

Index   349

Acknowledgments

THE PLASTIC CAMERA MY PARENTS GAVE ME on my eighth birthday had only two features, a shutter button and a neck strap. That such a simple box could create a world of images amazed me. For several years, I proudly displayed my black and white pictures as squares of captured time, shots of ordinary people going about their daily lives in my small western Kentucky hometown. Many years later I began probing the meanings of photographs, inspired by my graduate school mentor, Burl Noggle. He directed an early paper I wrote on "realities" in pictures taken in the 1920s. From there I expanded my arguments about visual images, first in a master's thesis and then in a doctoral dissertation, all the while sustained by Burl's patience and encouragement.

My dissertation lies at the core of this book, and I want to acknowledge the financial support I received to complete both projects. The T. Harry Williams Fellowship in History at Louisiana State University allowed me a year's leave, the valuable time necessary for sustained research. A generous faculty research grant from Georgetown College made it possible for me to spend a summer in Washington, D.C., and a Jones Faculty Development Grant from Transylvania University funded another research trip in 1997.

I received kind assistance from many cooperative and patient archivists, curators, and staff members at the Archives of American Art; the Art Department of Berea College; Special Collections at Berea College; the Bancroft Library of the University of California, Berkeley; the J. Paul Getty Museum; the University of Kentucky Art Museum; the University of Louisville Photographic Archives; the National Archives; the New York Historical Society; the Southern Historical Collection at the University of North Carolina, Chapel Hill; Special Collections at the University of Oregon; and the South Carolina Historical Society. Especially helpful were Lisa Carter, at the University of Kentucky Special Collections; Therese Thau Heyman, at the Oakland Museum; and Amy Doherty, at George Arents Research Library, Syracuse University. My unending gratitude goes to Beverly Brannan, Curator in the Prints and Photographs Division at the Library of Congress. Over the years we have discussed ideas, read each other's manuscripts, and shared numerous stories about women who chose photography as a profession. Beverly's enthusiasm and wide-ranging knowledge in the field are blessings.

The comments offered by colleagues and members of the audience at several meetings, including those of the American Studies Association; the American Historical Association, Pacific Coast Branch; the Louisiana Historical Association; and the 1997 Doris Ulmann Symposium at the Gibbes Museum of Art, helped me to shape my positions on several issues. Fellow participants at the 1995 National Endowment for the Humanities Institute, "The Thirties in Interdisciplinary Perspective," directed by John and Joy Kasson at the University of North Carolina, prodded me with thought-provoking questions that strengthened the book. I extend special thanks to Robert Snyder for his insights on my work and for taking a chance on a young scholar by inviting me to contribute to a special issue of *History of Photography* that he edited. Those who have read all or parts of the manuscript and whose invaluable comments have enhanced it beyond measure include Jessica Andrews, Peter Barr, Robert Becker, Beverly Brannan, James Curtis, Gaines Foster, the late Sally Hunter Graham, Philip W. Jacobs, Wendy Kozol, Heather Lyons, Richard Megraw, Mary Murphy, Daniel Pope, Janice Rutherford, Charles Thompson, and Alan Trachtenberg. An anonymous reader for the University Press of Kentucky offered excellent suggestions. In the Social Science Division Office at Transylvania University, Linda Denniston helped me tremendously and usually on short notice.

Those familiar with liberal arts colleges devoted to undergraduate education know that time for research and writing is precious; there are no teaching assistants or graders, and teaching loads are heavy. So I remain awed by the example of my former Georgetown College colleague, Steven May, who has gracefully balanced his roles as an award-winning teacher, a faculty leader, and a prolific scholar for thirty years. His sound advice and hearty encouragement were extremely important when I was starting out in the academic world.

Loving friends have been with me at the times I needed them most—I am lucky to be able to share secrets and an occasional breakfast, lunch, or afternoon tea with Sharon Brown, Barbara Burch, Regina Francies, and Mary Jane Smith. As always, the warm embrace of my parents, Bruce and Peggy McEuen, and my brothers, Kevin McEuen and Kelly Brown McEuen, has been constant and life-sustaining. The family that I married into about the time I began writing this book also have given me their unconditional support. No one has lived with this project more than my husband, Ed Stanton. As a scholar, he knows the rigors of academia and so has carefully protected my solitude. As a companion and lover, he keenly understands what the most essential things in life are and has passionately safeguarded our time to enjoy them together. He not only made this book possible but allowed its creator to thrive in the sweetest Eden imaginable.

Introduction

WHEN *LIFE* PHOTOGRAPHER MARGARET BOURKE-WHITE drafted an essay for *Popular Photography* magazine in the fall of 1939, she reminded readers and fellow photographers, "It is the thoughts that live in your head that count even more than the subjects in front of your lens."[1] Her judgment alerted every creator of visual images and every subsequent observer of those pictures to the vital reality that understanding the substance of a photograph requires understanding the person behind the camera. The whole range of ideas, prejudices, and desires that a photographer harbors is as significant as what he or she chooses to frame.

This book examines the lives and work of five American women who distinguished themselves as professional photographers in the years between the world wars. They are tied together by their passion for viewing people and places in the United States and, more importantly, by a common desire for their visual images to make a difference, serve a purpose, or influence what Americans thought about themselves or other people or distant locales or new ideas. As a result of these motivations, all five photographers ultimately embraced the most popular vehicle for socially conscious expression in the 1930s—documentary. Each woman then molded the genre to advance her own agenda, at the same time reshaping the visual form itself, even creating ameliorative possibilities for it. By freely allowing personal prejudices to permeate their gazes on the world, Doris Ulmann, Dorothea Lange, Marion Post, Margaret Bourke-White, and Berenice Abbott revealed the malleable nature of documentary photography. This examination of their lives and their pictures attempts to illuminate the primary impulses that drove photographers to use their cameras to send highly charged political and social messages in an age when most people believed that pictures did not lie but rather substantiated what was questionable or clarified what was imperceptible.[2]

The following study focuses on women photographers. Why women exclusively? After having set out to delve into New Deal politics and the photography it inspired in the thirties, I soon reached the same conclusions as Chicago gallery owner Edwynn Houk, who in 1988 planned a photography exhibit that would display a solid cross-section of twenties and thirties pictures. Nearly all of the final selections for the show, he realized, were photographs taken by women. Houk found the results intriguing and concluded, "Without attempting to focus on women artists, the Gallery nevertheless came to represent the works of many women by offering the best and most significant images produced in photography during the twenties and thirties." Similarly, a substantial number of the most penetrating visual studies I viewed in the early stages of my research were created by women. Their photographs seemed endless. Yet the scholarly literature on them was scant compared to that based on their male contemporaries. Perhaps worse, women were poorly represented or omitted completely from the best-known photography anthologies. Given their marginalization in the scholarship on photography, I grew even more curious about the photographers themselves. Why did they take up camera work initially? What led them to become professional photographers? What obstacles did they face or what freedoms did they enjoy because they were women in the profession? What were they trying to accomplish? How did they feel about the use of their photographs by employers or gallery owners or others? What political or cultural connections did they make with their visual imagery, or did they care at all about these matters? The most important questions, in my opinion, probed the inextricable relationship between the photographers' lives, the conceptual frameworks they built around their subjects, and the final images they produced. After pursuing the answers to these questions, I saw that they revealed the rich texture of American culture and a web of ideologies that circulated in the first half of the twentieth century. So what had begun as a project narrowly defined as political history became women's history and then developed into a larger examination of American history and culture. What kept appearing in my imagination was the superb title that editors Linda Kerber, Alice Kessler-Harris, and Kathryn Kish Sklar gave to a 1995 essay collection dedicated to Gerda Lerner—*U.S.*

History as Women's History. It seemed an appropriate description for my own discoveries regarding the development of documentary photography through the lens of its female practitioners. The composite analysis finally showed, as their essay collection did, "a vision of U.S. history as women's history quite as much as it is men's history."[3]

Analyzing photographs and evaluating aesthetic philosophies proved to be complementary to the demands of a feminist theoretical framework, which encouraged deep probing into the photographers' backgrounds, including what they thought their work did for them on a personal level. Historically, the photography profession provided an attractive alternative to the constrictive boundaries of nineteenth-century domestic existence, which still affected many women in the first decades of the twentieth century. In 1902 Myra Albert Wiggins stated, "Nothing has revealed human nature, given me a chance to travel, [and] given me valued acquaintances and friends as much as photography." And for women who desired a sense of independence, the vocation allowed "an individual working alone . . . [to] achieve something." As a low-ranking profession in the nineteenth century, photography was considered an acceptable pursuit for members of politically marginalized groups, particularly women. Those who engaged in taking pictures did not threaten powerful elements in the hegemonic structure, because photography was not steeped in tradition, as were the fields of law, medicine, and academia. Successful careers in photography did not depend upon attendance at august institutions, where women were rarely if ever admitted. But as early as 1872, the Cooper Union offered photography courses to women in New York City, hoping to prepare them for employment as assistants in the rapidly developing field. Pictures taken by American women were exhibited at the 1876 U.S. Centennial Exhibition in Philadelphia and accounted for a significant part of the Paris Exhibition in 1900. As camera equipment became less bulky and more inexpensive in the early twentieth century, an individual wanting to experiment with photography needed little capital. Many women were able to borrow cameras from friends or relatives or use the equipment owned by the studios where they retouched negatives, made prints, or posed models for well-established photographers.[4]

Photography opened doors for women, perhaps at no time more widely than in the years between the world wars. Exhibit curator Paul Katz noted that this generation of female photographers "wanted careers—public lives that would be more like a marriage with the world. Photography offered that possibility. In their quest they were aided by the vast increase in photographically illustrated publications and the creation, as a result, of new fields such as photojournalism and advertising photography. The needs of editors tended to override sexual prejudices, and the relatively low status of the profession as an art form made it easier for women to enter." Katz contends that "the sheer number of women who found a vocation in photography proclaims a social revolution . . . as emblematic of the age as the feats of Amelia Earhart and Gertrude Ederle."[5] Finding a vocation in photography did not necessarily guarantee a comfortable life, though. In the present study, only one of the five women, Doris Ulmann, never had to worry about money. Personal wealth sustained her career and her expensive habits. In contrast, Dorothea Lange saw her immediate family members, including her young children, scatter in different directions when the Depression began; Berenice Abbott took on a variety of odd jobs to support her career; Margaret Bourke-White had outstanding accounts at nearly every major department store in New York City during the 1930s; and Marion Post once admitted having said "yes" to any man who asked her out so that she could have at least one good free meal that day.[6] Despite their sporadic economic hardships, female photographers in the twenties and thirties received recognition as equals of, even superiors to, their male colleagues and competitors.

If there were so many women working in photography during this period, then why single out these five—Berenice Abbott, Margaret Bourke-White, Dorothea Lange, Marion Post, and Doris Ulmann? What makes them so compelling? Chiefly, all were prolific photographers who turned to documentary expression in the interwar years. Here the term *documentary* is defined broadly, not as a distinctive and recognizable style that focuses on specific subjects (especially since 1930s documentary was expressed in various styles using all kinds of subjects), but instead as a touchstone measuring two elements: first, the photographer's role as both recorder and participant in the cultural dramas in which she engaged, and second, the extent of her desire to have her images used for larger social or political purposes. For this reason photographers such as Laura Gilpin and Imogen Cunningham, whose reputations were made primarily as art photographers during the 1920s and 1930s, are not included here. And although Tina Modotti has been labeled a documentarian, her oeuvre is largely Mexican, which puts her photography outside the geographical parameters of this study, namely the United States. Beyond my desire to focus on photographers who considered themselves documentarians of some sort and who completed all or most of their work in the United States, I wanted to show the tremendous range of documentary styles exhibited by women photographers, which in turn would foster a discussion about their contributions in shaping the genre and its role in public life. To accomplish this, I chose five individuals who carried out extensive "fieldwork" in the discipline by traveling to unfamiliar surroundings or uncharted territory in order to survey American life. Each produced perceptive views on the astounding variety of occupations, values, and leisure activities in the nation between the world wars, and in the process they made considerable contributions to historical photography. Finally, each cultivated a distinctive style woven from the skeins of her aesthetic sensibilities, her personal politics, and the pressing social and cultural forces of her time.

Together, the five women produced a corps of visual images that covers an impressively broad spectrum in tastes, methods, and perspectives, all of which fit comfortably under the large umbrella of documentary photography. That these women worked during such a critical time in the nation's history simply augments their professional achievements. When their pictures are viewed collectively and examined across time, patterns emerge that show the development of documentary as a medium of expression. The life of socially conscious visual expression in the 1920s and 1930s may be plotted along the paths taken by Ulmann, Lange, Post, Bourke-White, and Abbott. Beginning with Ulmann's studio-in-the-field approach in the mid-1920s, documentary then experienced modifications by Lange, who fashioned slightly more informal portraits than Ulmann did while on the road. Post turned the medium into a forum for radical political views, exposing racism and class stratification in the United States through her angles on social situations and her telling backdrops. Bourke-White attempted to infuse documentary with the high-style modernism of innovative advertising photography. But not until Abbott systematically utilized a different kind of modernist aesthetic in her large-scale project "Changing New York" did documentary and modernism coexist harmoniously on photographic paper. Despite the apparent incongruity of a marriage between documentary and modernism, Abbott managed to combine the two forces almost seamlessly.[7] Her calculated juxtapositions of old monuments with new architectural creations showed layers of the past stacked up next to the present and the foreseeable future, an array of generations realized in two-dimensional form.

Beyond their diverse stylistic preferences, these five photographers posited certain nationalist ideals by pursuing subjects that they believed would highlight American cultural strength and in turn promote greater social awareness or change. In each woman's prescriptive works, themes emerge that connect present circumstances with eventual consequences. Ulmann perceived American ingenuity and continuity overwhelmingly in rural Appalachian craftspeople, whereas Abbott found characteristic "Americanness" in urban growth and renewal. Bourke-White pictured sophisticated machine technology as the nation's greatest hope for a promising future, while Lange illuminated the steadfastness and survivalist spirit of its ordinary people as the country's most reliable resources. Post idealized the notion of collective cooperation as a means of alleviating the most deeply rooted social problems in the United States. Over a twenty-year span, the five women analyzed here articulated in pictures the principal cultural forces that manipulated American thought and action in the critical years between the world wars.

More than anything else, this is a study of visual images as the tangible results of personal motivations and historical forces. I began my research on this project by following James Borchert's prescription for evaluating visual evidence, a charge to "cast as wide a net as possible." He maintains that scholars may more easily determine "bias" if they look at a substantial number of pictures. The virtue of quantity also provides clues as to what surrounding evidence a photographer may have purposefully left out. To that end, photographic series of subjects, including whole jobs and complete assignments rather than isolated images, form the visual evidence base of this study. Consequently, the historian's task involves interpreting the visual thinking of the photographer. Thomas Schlereth suggests that historians of visual imagery attempt to "get inside the mind of the photographer." To accomplish this rather difficult task, I examine the ways each photographer prepared for fieldwork, dealt with local officials, approached her subjects, described her perceptions of various jobs, and handled her superiors, such as supervisors and editors. In the process of contextualizing each woman's life, I attempt to show that a photographer born in the 1880s was more greatly swayed by her training in the 1910s than by the stock market crash, and that another, who was a teenager in the 1920s, viewed Americans differently than her institutional colleague who had been an established portraitist in that same decade. The more familiar historical markers, such as the 1929 stock market crash and presidential election years, appeared to me to be artificial guidelines, since social and cultural changes in the United States did not necessarily parallel economic and political shifts. It took time for some photographers to recognize the enormity of the Great Depression and its effects on the nation; only after witnessing hunger and despair firsthand did they seek out "the people" as their principal subject. And although picturing the "common" man and woman is often interpreted as a requisite function of documentary expression in the 1930s, there were American photographers like Doris Ulmann experimenting with these subjects in the 1920s and even earlier. Historian David Peeler has written that "one of the more enduring American myths is that social art of the thirties, with all its intensity and commentary, was completely divorced from a frivolous and self-indulgent twenties culture."[8] The fluidity of artistic, ideological, and cultural trends in the interwar years led me to construct a narrative organized to enhance the historical contexts in which these photographers worked. For that reason, I have chosen a biographical approach for ordering my analyses. Although such a schema does present the possibility of thematic discontinuity, its advantages outweigh the conceivable impediments.

In his provocative study of American modernism and its purveyors in the South, Daniel Singal defended his use of a biographical framework by noting that "[a] sociologist may be trained in the most advanced social science theory, or a novelist may be steeped in the literature of his times, but in each case the beliefs and perspectives actually absorbed and utilized will depend on the constellation of formative experiences the person has undergone." Likewise, a photographer's vision emerges from the mélange of past experiences, present emotions, careful calculations, and technological processes that come together at the moment in time and space when a scene is framed through the lens and recorded on a glass plate or a strip of film. Since images cannot be separated from their creators' intentions, they are treated as such in this text. Photo scholar Allan Sekula has pointed out that "every photographic image is a sign, above all, of someone's investment in the sending of a message." Such messages cannot be fully understood if the photographer is cast on the periphery by researchers. The most significant recent scholarship on American photography has shown the primacy of examining the sources of images, their creators, in order to understand more clearly the messages being sent. I have built upon the exceptional work of Alan Trachtenberg, whose book *Reading American Photographs: Images as History, Mathew Brady to Walker Evans* displays a method for examining photographs as cultural texts while keeping the photographer's responses and motivations near the center of the analysis. James Curtis provides yet another revisionist model in *Mind's Eye, Mind's Truth: FSA Photography Reconsidered*, a work based on primary meanings of photographs, with the creator's intentions and biases always at the forefront. Halla Beloff wrote in *Camera Culture* that "the camera and the film link a photograph concretely with a machine, and yet we understand that a human intelligence, and sensitivity, and a human need have made us that picture."[9] In the 1920s and 1930s, those human intelligences, sensitivities, and needs manifested themselves powerfully through the vehicle of documentary photography. The fruits of the documentary visions cultivated by Doris Ulmann, Dorothea Lange, Marion Post, Margaret Bourke-White, and Berenice Abbott are rich representations of the intricate workings of American culture in the years between the wars.

Documentarian with Props

DORIS ULMANN'S VISION OF AN IDEAL AMERICA

One picture . . . cannot express an individual.
—Doris Ulmann

A FEW WEEKS BEFORE HER DEATH at age fifty-two, Doris Ulmann wrote, "Personally, I think there is always more value in doing one thing thoroughly and as well as possible than in spreading over a large area and getting just a little of many things."[1] The specific reference was to her current photography project, but the statement also clearly defined the approach she had taken in her twenty years behind the camera. Spending hours with each subject, posing and reposing, Ulmann ultimately created a composite image of the person or object on which she focused. Her method of painstakingly observing each portrait sitter remained the hallmark of her in-depth studies. Beginning as a photographer who posed wealthy, educated, and privileged individuals in New York City, she later broadened her focus to create images of rural Americans. She chose ethnically distinctive enclaves that interested her and carefully studied individuals within those groups. Combining an interest in human psychology, a nostalgia for an idealized American past, and the finest available training in photography,

Ulmann produced some of the most penetrating character studies of Americans in the 1920s and 1930s.

That she realized her portraits could serve a social purpose places her squarely within the documentary tradition of American photography. She reached this conclusion well past the midpoint of her career, embarking upon new and extensive projects despite debilitating physical frailties. Although her style and her equipment remained virtually unchanged for twenty years, Ulmann's camera eye shifted significantly three times: in 1919, when the hint of publishing success ensured her status as a professional photographer; in the mid-1920s, after her marriage legally ended, her mentor died, and she suffered a crippling fall; and in 1933, when she began a comprehensive survey of southern Appalachian handicrafts to illustrate a colleague's written text on the subject. At each juncture Ulmann embraced subjects that she felt deserved the attention of the public and, most of all, required a photographer's interpretative eye (her own) to grasp and hold that attention. The faces and scenes she rendered reflect her desire to create photographic records that not only would illuminate personalities and lifestyles but also would expose ideal worlds—worlds created by the good intentions and active imaginations of Ulmann and her upper-middle-class counterparts. Their interests led them to grapple with the myriad changes wrought by a modern, industrialized, and increasingly urbanized nation.

Ulmann's family background and educational pursuits set the stage for the work she found most satisfying as a professional photographer. She was born in 1882 into a wealthy Jewish family, her father having immigrated to the United States from Bavaria in the 1860s. Supported by a successful textile manufacturing business, the Ulmann family lived in New York City's heart, Manhattan. The urban environment provided the cosmopolitan influences that shaped Ulmann's initial aesthetic tastes and values. She cultivated many interests that she would continue to enjoy for the rest of her life, from literature to theater to modern dance. Her New York public school education was supplemented by excursions abroad with her father, Bernhard Ulmann. In 1900 she enrolled in teacher training at the Ethical Culture School, an institution founded by Felix Adler, who was an optimistic reformer driven by humanistic impulses and a great need to sponsor and help the burgeoning working classes. His progressive institution functioned according to the Ethical Culture Society's motto, "Deed not creed," thus setting it apart from other contemporary reform efforts that were heavily infused with religious messages and influences.[2] The Ethical Culture School appealed to several constituencies, including successful immigrants seeking to Americanize their children and provide them with a living conscience sufficient to embrace problems posed by the new industrial order in the United States.

With hopes of becoming an educator, Ulmann spent four years at Ethical Culture, during the same period that a young teacher named Lewis Hine went there to teach biology. At the insistence of the school's superintendent, Hine ended up taking students on several field trips to Ellis Island to photograph newly arrived immigrants. He also began offering lessons in photography, where Ulmann probably had her first contact with him. Their mutual devotion to Ethical Culture's philosophies gave them common ground on which to build their respective photographic achievements. For Hine, the task began almost immediately, as he published both words and pictures addressing society's problems.[3] For Ulmann, the reform impulse lay dormant for nearly twenty years, awakening when she realized that her camera work could transcend its status as a hobby and could make a difference in distinctive communities in the United States. To accomplish her goals, she embraced an element of Hine's approach that had become one of the hallmarks of his socially charged visual images: a focus upon individual faces, not the masses. Hine portrayed dignity in his subjects, despite their horrid living and working conditions in mills and mines and sweatshops. Connecting people intimately with their work, especially that accomplished by their hands, Hine created portraits that bespoke his appreciation for individual laborers. Ulmann's photography in Appalachia and the Deep South in the 1920s and 1930s mirrored Hine's imagery in its emphasis on the individual life, the character of manual labor, and the maintenance of human dignity.

But long before she created the photographs that made her famous, Ulmann spent several years studying. Columbia University proved to be a significant influence in Ulmann's young adulthood. Here she pursued the two subjects that would direct her life's work, psychology and photography; here also she met Dr.

Charles Jaeger, the man she eventually married. Ulmann's attraction to psychology, a relatively new social science, led her to pursue a teaching career at Teachers College, Columbia University. She joined hundreds of single young women who filled the social science departments at major universities in the early twentieth century. Their interests in philosophy and pedagogy, particularly educational psychology, caused them to seek vocational avenues where their scholarship could be directly applied. Many of these women became teachers or ran urban settlement houses or rural settlement schools, carving out socially acceptable careers for themselves as independent women working alone or in single-sex groups.[4] Although Ulmann never pursued those vocations, she later became closely acquainted with a number of women who did.

While a student at Columbia, Ulmann also took courses in law, but she developed such a distaste for the field that she abandoned it after one term. She felt that "a welter of legal technicalities" smothered the human element. In 1914 Ulmann began serious study of photography at Teachers College with the acclaimed instructor Clarence H. White. She had already taken a few classes with White soon after he arrived in New York City, but her true dedication to the art form began in 1914. She joined a legion of students under White's mentorship, many of them women who later enjoyed high-profile careers as professional photographers, including Margaret Bourke-White, Dorothea Lange, and Laura Gilpin.[5] Ulmann, known as one of White's "most devoted pupils," later taught at the master's photography school. Given the time and energy she put into developing her art, it seems unusual that Ulmann claimed to have taken up photography as "an excuse for doing something with her hands when her mind was tired." But she was known to suffer from any number of simultaneous physical ailments, including stomach ulcers (which she had developed as a child), arthritic pain, and a general nervousness that led her to seek solace in activities that would calm her. Her physical weaknesses combined with society's expectations of a woman reared in the nineteenth-century bourgeois tradition kept Ulmann from venturing out too far away from her Manhattan home with her camera. But these limitations would soon be eased by the companions she cultivated. In 1915 Ulmann's professional interests and personal interests intersected. She married orthopedic surgeon Charles

Jaeger, who was a friend and physician of Clarence White, an instructor of orthopedic surgery at Columbia University, and himself a photography buff.[6]

Because of their shared interest in photography, husband and wife often traveled to picturesque settings—coastal villages in Maine, Massachusetts, and the Carolinas—with hopes of finding appropriate subject matter for their respective visual studies. The two soon became active leaders in the Pictorial Photographers of America, a group that continued a forty-year-old tradition of creating naturalist-inspired scenes altered by manipulations in the darkroom. At the turn of the century, pictorialism had been supported by gallery owner and photographer Alfred Stieglitz, who served as the inspiration for a number of artists and artistic movements. One of those movements was a branch of pictorialism called Photo-Secession, whose practitioners, such as Clarence H. White, sought to create symbolic art. Stieglitz believed photography should be considered an art and nothing more, an end in itself, certainly not an extension of the muckraking journalists' stories designed to arouse social change. So at the same time Lewis Hine shaped his style employing the camera for reform purposes, Stieglitz had initiated a movement in New York that sought to keep the camera from becoming such an instrument.[7] These were the preeminent standards and approaches in American photography at the time Ulmann was developing her own camera eye.

These two powerful forces in photography—the reform impulse of Lewis Hine and the artistic-pictorialist focus of Alfred Stieglitz and Clarence H. White—are clearly traceable in Ulmann's aesthetic sense. She did not claim to have copied any particular photographic style, but the dominant philosophies of the era are revealed in the thousands of images that make up the Ulmann oeuvre. Reflecting the standards set by Hine's work, Ulmann focused on the unknown individual whom society judged more often by ethnic, religious, or economic affiliations than by personal merits. In a 1917 study she initially titled *The Back Stairs* but later recast as *The Orphan* (fig. 1), Ulmann captured a small, darkhaired child amid the symbols of urban poverty. The child plays barefooted among broken stones, discarded wood pieces, and other debris. Additional messages about the child's existence may be detected in the rickety rail accompanying the stairs to her home

Fig. 1. Doris Ulmann. *"The Back Stairs." 1917.* Audio-Visual Archives, Special Collections and Archives, University of Kentucky Libraries.

and the empty barrel she leans on. Despite the instability and emptiness characteristic of the child's environment, Ulmann portrays her as an angelic figure—a tender face in profile, her tiny body robed in white play clothes made brighter by the natural light. A certain universality in the child's forlorn look, much like a paper-cut silhouette, ensures her status as an innocent in the midst of social disarray.

Like Hine, Ulmann eschewed evaluation according to the group standard, although she did find individuals in certain groups more fascinating to photograph than those in others. Among the groups she studied early in her career were physicians at Columbia University and writers in New York City; she later expanded her examination of groups to include fishermen in Massachusetts, Dunkards in Pennsylvania, Shakers in New York, mountaineers in Appalachia, Gullah African Americans in South Carolina, and

Creoles in New Orleans. Seeking out particular "types" that could be categorized, Ulmann proceeded to isolate particular individuals within a community who possessed intriguing physical characteristics or who worked at unusual occupations. That was a practice not uncommon in the early twentieth century; Lewis Hine had selected extraordinary persons from bands of workers and ethnic groups to propel his arguments about the need for labor reform.[8]

From the Stieglitz association's artistic philosophy, a viewpoint seemingly adversarial to that of Hine, Ulmann co-opted ideas she could assimilate into her own aesthetic. The sensorial appeals achieved by the early pictorialists' romantic imagery also pervaded Ulmann's photographs. The relationship between soft backgrounds and sharply defined foreground foci allowed for a play of textures that remained the single most continuous thread in Ulmann's images throughout her years as a professional photographer. She began her photographic studies with the requisite nature scenes that pictorialists often sought. In one 1917 experiment with light and shadows, Ulmann focused on a barren tree without leaves, its branches and its spindly, dark shadow set against a white building. Other similar treatments of trees, architectural structures, clouded skies, and snowy landscapes are representative of Ulmann's early pictorialist-inspired vision.[9] In one composition (fig. 2), gradations of light combined with myriad textures to form the sensory depths of the photograph, from the softly focused leaves in the lower left section of the frame to the harder lines of the main vine. Geometric patterns are emphasized as the vertical plane is determined by the strongest vine trunk, which divides the frame. The smaller arm cuts across the horizontal plane of the photograph, and the planks on the wooden structure provide subtle reminders that balance has been achieved in these perpendicular relations. Ulmann continued throughout her career to seek the play of light and dark and shades of gray in similar natural settings and in her portraiture. A photograph taken in the late teens or early twenties and entered in a local exhibition was a picturesque landscape she entitled "III Clouds over the Mountain." Her choice of subject matter and textual characteristics reflect the influence of Stieglitz's approach. Clouds were a subject that Stieglitz had obsessively embraced during World War I.[10] To further emphasize

Fig. 2. Doris Ulmann. *Scene at Georgetown Island, Maine, ca. 1918-19.* Photography Collections, J. Paul Getty Museum, Los Angeles, California.

her attachment to this particular artistic philosophy and style, Ulmann joined no other photography collectives but the local pictorialist group and the Pittsburgh Salon of Photography. She generally avoided camera clubs, unions, and similar groups, even though she lived in a city that boasted the most exciting photography community in the nation and, perhaps, the world.

As a student and friend of Clarence White's, Ulmann drew ideas from the Photo-Secession movement, which suggested that one had to assume the mantle of "artist" in order to create art with a camera. Ulmann inculcated this idea, believing herself an artist and thus carrying out many of the same processes a person working with canvas and brush had to master. Nowhere was this more evident than in her oil pigment printing, a method requiring careful brush strokes of lithographic ink on water-soaked prints. Since no two oil pigment prints were exactly alike, they were more like paintings than most photographic prints produced in the 1920s. Although David Featherstone has suggested that Ulmann accumulated "raw sociological data," her finished prints reveal an artistry that transcends mere collection of evidence. She employed light to its fullest effect, sought figured shadows, and focused on patterns, objects, hands, and faces. But these constituted only a portion of her work. The painstaking printing processes Ulmann performed required as much time and manipulation as her choice and recording of subjects. It is this conscious creation and re-creation of her subject matter in the darkroom that keeps her work from constituting simply a mass of empirical data on which historians or other observers can hang hypotheses about particular subcultures in the United States. Those seeking to use her photographs as clear windows through which to view American culture must consider Ulmann's self-professed biases and her conscious deliberation over positioning subjects and using props. From her earliest work in the portrait studio, she sought complete control over her attempts to "express an individual."[11]

In 1919 Ulmann published her initial work as a professional photographer, a handsome portfolio entitled *Twenty-Four Portraits of the Faculty of Physicians and Surgeons of Columbia University*. As Mrs. Charles Jaeger, she had gained entrance into the prestigious circle of physicians to which her husband belonged. In the years immediately following, she developed a reputation as an outstanding portraitist. In 1922 her second major collection, *A Book of Portraits of the Medical Faculty of Johns Hopkins University,* was published in Baltimore. A marked difference in the publication information recorded in these two collections suggests Ulmann's direction on a path independent of her husband. The author listing she chose for the 1919 Columbia University collection was "Mrs. Doris U. Jaeger," the signature she had most commonly used on her early prints. By 1922 she had begun to use her family name, Ulmann, on her published work and on her exhibition prints. The exhibition entry entitled "III Clouds over the Mountain" reveals her new professional name, along with her impressive studio address—Doris Ulmann, 1000 Park Avenue, New York City. To sever her past ties with Jaeger, Ulmann returned to some of her early prints and erased the original signature that bore his name, replacing it with the name she had reclaimed.[12] Even though they had studied photography together and had been prime motivators behind the local Pictorial Photographers of America chapter, Ulmann sought to strike out on her own. Her act of wiping out her husband's name implies her dedication to a life and profession not only separate from his but also not tainted by her previous relationship to him. However, her title remained vague. She rarely used one on letters, notes, or prints, but her principal traveling companion in later life, John Jacob Niles, referred to her as "Miss Doris Ulmann." Ulmann's brother-in-law, Henry Necarsulmer, insisted after her death that although she had divorced her husband and resumed her family name, she had been married and thus "was known as *Mrs.* Doris Ulmann." Even in the early 1930s, a divorced woman traveling hundreds of miles in rural America with a male companion, especially one of whom her family did not approve, was inconsistent with the demands of upper-class New York social circles.[13] In Ulmann we see a complex woman whose quiet demeanor and upper-crust sophistication were matched by a need to exert personal control over her present, her past, and even her legacy.

The two published books of physicians' portraits provide the necessary clues to understanding Ulmann's use of her skills as a photographer and her need to control her work. With these collections Ulmann exhibited a strong desire to assume and complete whole projects, comprehensive surveys focused on particular ends. These projects set the stage for future work by giving her a taste of the kind of material she would find most satisfying throughout her career as a professional photographer—theme-centered studies, built on extensive series of images rather than on a single mesmerizing frame. Outside the portrait studio, Ulmann's meticulous examinations of groups and communities helped her shape her niche in photography. One of her more thorough early examinations of a particular place and people—the fishing village of Gloucester, Massachusetts—included scenes of boats, pier buildings, equipment for the trade, and the characters who made their living on the sea.[14]

As a result of Ulmann's drive to embrace complete projects, she carried out each step of the photographic process herself. She handled the glass plates, mixed the chemicals, developed the negatives, made the prints, and mounted the finished photographs. And she preferred to keep her creative secrets to herself, allowing no one to assist or interrupt the magical process that unfolded in her darkroom, a converted bathroom. Only years later, after her health seriously deteriorated, did she allow anyone to help her in the darkroom. She even refused to allow other eyes to view her proofs, explaining, "I see my finished print in the proof . . . but I cannot expect others to see anything beyond what the proof presents. I avoid retouching, but prints always require spotting before they are ready. I believe that I become better acquainted with my sitter while working at the pictures, because the various steps provide ample time for the most minute inspection and contemplation." The relationships Ulmann forged with her subjects through "the pictures" point to a theme in her otherwise solitary existence as a portrait photographer. The camera aided Ulmann with her shyness. She got to know those with whom she was most intrigued by studying them thoroughly through the lens. For one portrait sitting of a single individual, Ulmann would expose a tremendous number of glass plates. She recognized the complexity of human existence and felt that too few shots would simplify and ultimately distort a life. She believed one photograph could not define a person and so offered her sitters "twenty or thirty finished prints instead of the scant dozen or so proofs submitted by the craftiest of commercial photographers."[15] In the process Ulmann gained a deeper understanding of people she admired by looking at two-dimensional renderings on glass plate negatives, not unlike the nineteenth-century phrenologists who determined character traits by measuring physical attributes prominent in visual depictions of famous politicians and military leaders. Systematic examination of Ulmann's photography reveals that she embraced several long-term projects that would allow her to revel in her infatuation with particular individuals.

A fascination with the literary mind turned Ulmann's attention to writers, editors, and poets in the 1920s; it was an attraction that she fostered throughout her life, never turning completely away from the wordsmiths who so impressed her. She once told an interviewer, "The faces of the men and women in the street are probably just as interesting as literary faces, but my particular human angle leads me to the men and women who write." Her desire to capture penetrating and revealing images of literary figures points to her own love of literature and the word culture in which she had grown up. She reveled in language, read the German classics aloud, and, according to John Jacob Niles, spoke "flawless" English. From Ulmann's childhood to her young adult days, the American population's reliance upon words, not only for information but also for entertainment, had gradually waned. Visual images grew to dominate the messages promoted and delivered by both the public and the private sectors. During the Great War, posters urging support for the cause and "100% Americanism" employed compelling signs and symbols; and by the 1920s, advertising had reached a new height in its sophisticated use of pictures and graphic designs to convince consumers that buying the right products would ensure an easier or more enjoyable lifestyle. Ulmann's exposure to new educational trends and to advances in technology, including those in photography, helped her realize that the written culture was undergoing a radical transformation in the twentieth century. An avid reader, she attempted to sustain and illuminate the world of the literati by opening her apartment doors to a host of exciting American authors. In the same year that the First World War ended, she began taking professional photographic portraits, thus launching her career at a particularly crucial time for the arts and literature, so significant that it led the writer Gertrude Stein to note, "After the war we had the twentieth century." Meanwhile, Ulmann's reputation as a professional photographer grew with each passing year as she photographed many prominent writers, including H.L. Mencken, Dorothy Parker, Edna St. Vincent Millay, Carl Van Doren, Charlotte Perkins Gilman, James Weldon Johnson, Ellen Glasgow, and William Butler Yeats. Among these portraits were publishable images that prospective readers later saw on book jackets, in magazines, even in the Literary Guild of America's book club advertisements. Ulmann apparently tolerated and even enjoyed listening to a variety of opinions, given her selection of sitters who were decidedly modern and often confrontational voices of the 1920s.[16]

Ulmann developed a rudimentary understanding of her subjects, carefully observing mannerisms and gestures, long before she stood behind the tripod to study their faces. Dale Warren, himself an Ulmann subject, described the photographer's handling of a portrait session. She would serve cocktails and sweets and cigarettes, not solely for her subjects' enjoyment but to "draw [them] out." Warren intimated, "She studies your hands as you pass her a plate of cakes, observes which leg you cross over the other, notices the expression of your eyes, tells you a funny story to make you laugh, and another not so funny to see if you are easily reduced to tears." Ulmann even engaged her sitters in conversations that required them to articulate and defend their opinions. Brief responses were unacceptable to her, noted Warren.[17] After isolating certain peculiarities in each individual, Ulmann built on these in her portraits. There was no one chair or single backdrop or unique angle she preferred. Faces mattered most, with hands nearly as important. The authors Ulmann photographed could choose from a limited collection of props she kept in her Upper East Side apartment-studio. A fountain pen, pads of writing paper, or various sizes and types of books satisfied most of them. But exceptions kept Ulmann and her household staff busy—Robert Frost, who never worked at a table, requested a wooden writing board; and E.V. Lucas, the prolific British essayist, demanded an inkwell instead of the fountain pen that his portraitist offered.

Patient, gracious, and soft-spoken, Ulmann accommodated her subjects to a certain degree, hoping to create "bonds of sympathy" with them. She claimed to have allowed a few individuals to direct the day's events if they wished, but her desire to manage the portrait process is revealed in her fond recollection of a session at Sherwood Anderson's Virginia home: "I arrived at ten o'clock in the morning and did not leave until after midnight. Certainly no photographer could ask for a more interesting subject than Mr. Anderson, nor could anyone have put himself more completely in my power. He even led me to his clothes closet and asked me to look over his suits and choose the one I wanted him to wear."[18]

Ulmann preferred to direct a portrait sitting to this extent. And into the 1920s, as she grew increasingly confident in her abilities as an artist, she wielded greater control over her subjects. Her dominant hand in the positioning of heads and upper torsos is present throughout the bulk of her portraiture. Because she abhorred artificial light, with few exceptions her arrangements are determined by the available light from open windows or doors. Allowing sunlight to shine on her authors' faces, Ulmann disclosed her own reverence for those who were masters of language, moving their readers to anger or compassion or laughter. Sherwood Anderson, who bemoaned the barren nature of a society driven by industrialization, idealized rural life as simple and carefree, and therefore rich. Ulmann came to express similar ideas in her field photography in the early 1930s. The portraits she composed of Anderson show him sitting comfortably in front of a stone wall. One (reprinted on the cover of a 1958 journal issue featuring her photography) reveals a slightly rumpled Anderson. With his tie askew, a full inch off his starched shirt's placket, and his jacket gaping open over his crossed legs, he rests his right arm over the back of a straight chair. Anderson's demeanor is marked by carefully set lips and a furrowed brow, the latter probably due to the direct sunlight in which Ulmann positioned him. But his tanned skin reveals that his face has known days of sunlight, as much as any face of an ordinary Virginia farmer. Disheveled, Anderson's look suggests that he rarely dons a suit and tie and might prefer to be sitting instead in work clothes. At the very least, his outer trappings cannot shake his informality, sitting as he does slung back into the chair. Ulmann's positioning of Anderson makes him appear unassuming and nonthreatening, someone to be trusted,

believed. By the end of the day, Ulmann had so thoroughly surveyed Anderson that he said, "I feel as if you are taking a part of
me away with you."[19]

Ulmann's keen powers of observation and her hours spent
with her sitters' portraits helped her to understand her subjects
better, but in many cases she had read an author's works before
the portrait appointment and so had made her initial acquaintance
from the printed page. Never requesting money for her work,
Ulmann sought compensation, if at all, in other ways. She preferred to receive a copy of a writer's latest book or a dedication
inside her own personal edition. Her gracious letters to her sitters
show that such rewards greatly satisfied her. To South Carolina
writer and artist John Bennett, Ulmann expressed thanks for the
"precious book" he sent to her; the "beautiful page" he illustrated
and inscribed was "so delightfully done and so in harmony with
the whole book." No bills or order forms or contract agreements
accompanied an Ulmann portrait session. Independently wealthy,
she never worried about money. She considered herself an artist,
not a commissioned employee, and so refused to assume the role
of court painter who made every subject appear beautiful or brilliant. She preferred that a subject's portraits "be worthy" of him
or her, as she told John Bennett.[20]

People who came to Ulmann to be photographed frequently
did so at her request, rather than their own. Her friend Olive Dame
Campbell noted that Ulmann "rarely took a photograph unless
interested in the sitter." She created images of writers so that she
might forge relationships with those whom she admired; she often asked them to sign their portraits for her, next to her own
signature, in essence sealing the relationship between herself and
another artist and making it a matter of both personal satisfaction and public record.[21] One such image Ulmann created of Elizabeth Madox Roberts, a Kentucky poet and novelist whose stories
described customs and traditions of the Kentucky mountain people.
In Ulmann's most stunning portrait of Roberts (fig. 3), she highlighted the author's hands. Set against a darkly draped background
and a dark, nondescript dress, Roberts's left hand emerges at the
end of her long arm to set the horizontal plane and thus the compositional stability of the photograph. Ulmann achieved the perpendicular balance by highlighting Roberts's graceful right hand.

Fig. 3. Doris Ulmann. *Elizabeth Madox Roberts. 1928.* Audio-Visual Archives, Special Collections and Archives, University of Kentucky Libraries.

The long, slender fingers create the image's vertical line and, because they are slightly bent, hint at gentleness. Although she had done so with other writers, Ulmann chose not to put a pen or a book or manuscript pages in Roberts's hands, instead making the right hand itself the prominent feature in the portrait. Secondary in importance is the author's face, half-obscured by shadow on one side and adorned with a contemplative, yet comfortable expression. Ulmann relished studying masters of words like Roberts, those who were simultaneously reflecting and molding the American cultural landscape. And perhaps on a more personal level, Ulmann could appreciate the author's observations on romance. In Roberts's best-known work, *My Heart and My Flesh* (1927), a young woman attempts to find happiness in several different love affairs. Given her own discontent with her personal life and intimate liaisons in the early 1930s, Ulmann may easily have identified with the protagonist, appreciating Roberts's skills on an even deeper level than a mere literary observer would have.[22]

Perhaps nowhere is Ulmann's concern about the influence of writers more evident than in her 1925 publication *A Portrait Gallery of American Editors*. The proliferation of new magazines in the 1920s meant increased circulation of a variety of editorial opinions. Though an admirer, Ulmann also remained skeptical about some editors and the periodicals they produced. She prefaced her portrait collection with a sharp yet diplomatic statement about journals and those who controlled them, contending, "Magazines are so great a part of our daily life that almost unbeknown to us they mould our opinions and colour our views on most of the great problems of the day. Insidiously they have become a part of us and often times the views we hold as our own have in truth been formed by the editors of our favorite magazines. It is but natural that we should care to know what manner of men are these, who have thus formulated our ideas, coloured our thoughts and directed our perception of humour."[23] She does not discuss the public's curiosity, or her own for that matter, to see merely what famous people look like. She drives at something deeper by pinpointing a desire to know "what manner of men are these." In this collection of forty-three images, Ulmann showed that her interests in psychology and portraiture were inextricably bound. Given her confidence that she could, as she told Dale Warren, "draw

[individuals] out," Ulmann trusted that her psychological studies—the portraits—would reveal the layered complexities that made
up each individual's personality and enable the American public
to examine more closely the sources of their thinking. She warned
viewers against being fooled by other visual images: "Personality
and character are often so illusive, so intangible that they defy
and escape the most seductive efforts of reproduction and instead
of rendering a living likeness, little more than an anatomical copy
is made." Ulmann intended for her portraits of these selected men
and women to exceed such limited dimensions, by providing insight and lending definition to their influential lives. In her book,
each editor's portrait carried alongside it an essay written by the
editor, describing the nature of his or her work, but these were
ancillary to the real purpose of the volume, an effort "to portray
[the editor's] personality and something of their character by means
of photographic portraits."[24]

Among those featured in *A Portrait Gallery* were Ellery
Sedgwick of the *Atlantic Monthly,* Carl Van Doren of the *Century,* and Lawrence F. Abbott from the *Outlook.* Ulmann used
four general poses for the collection. The most common one had
the subject seated in a chair, holding an object in his or her hands,
most often a cigarette, a book, or a sheaf of papers; a second pose
or position had the subject seated behind a desk table and either
attending to business or looking up from work. Twenty-seven of
the forty-three portraits feature these positions, with attention given
to the hands as well as the head of the editor. Richard Walsh, editor
of *Collier's,* is viewed from his upper right side, so that the camera eye captures not only his face and upper torso, but also the
page surface where he has previously directed his attention. The
paper in his hand shows several lines of text that have been vigorously marked out by a well-sharpened pencil. The dark and light
contrasts, charcoal scratches against white paper, provide Ulmann
the aesthetic qualities she sought but also the story she wished to
tell about Walsh's professional tasks. Her emphasis is on the aspect of manual labor required for this editor to fulfill his responsibilities. As in most of her survey projects on specific subjects,
Ulmann connected individuals with the objects they touched and
the work they accomplished with their hands. Whether a doctor
with test tubes, a farmer with a scythe, a quilter with a T-square,

or an artist with brushes, the subjects' hands play a prominent role in determining the quality of their work and thus in defining them.[25]

In the same year that *A Portrait Gallery of American Editors* appeared, Ulmann's career took an evolutionary turn. The change coincided with two events in her personal life—the legal dissolution of her decade-long marriage to Charles Jaeger and the sudden death of her mentor, Clarence White. The shift in Ulmann's photographic vision led her on a search that would require her to approach the men and women she most wished to photograph rather than summoning them to her Park Avenue studio, as she had become accustomed to doing. Ironically, this experimental phase with new subject matter that required greater travel manifested itself about the time Ulmann suffered a fall that temporarily immobilized her. The injury, a shattered knee cap, left Ulmann dependent upon a cane from 1926 on and, according to one friend, "colored the remainder of her life." As a result, she employed traveling companions and servants to escort her and help her move camera equipment from one shooting location to another.[26] In spite of her losses, but more likely because of these significant changes in her personal life, Ulmann's career entered a new phase in the second half of the decade. The confining walls of the studio, treks to the New England coast to find picturesque landscapes, and strolls around Columbia University's campus gave way to the wide open spaces of the Carolina lowlands, hikes to southern Appalachian homesteads to find mountain handicrafts, and walks around the grounds of Berea College, the John C. Campbell Folk School, and several rural settlement schools.

This second stage in Ulmann's photography career, marked by her physical impairment and by the absence of her mentor, allowed room for new mentors and companions to enter Ulmann's life. In cultivating close personal relationships with four of them, in particular, Ulmann merged her private life and personal needs with her professional goals. She met John Jacob Niles, a singer-actor who exchanged his self-proclaimed expertise as a "Kentucky backwoods-man" for Ulmann's financial support of his fledgling musical career. Ulmann also met Julia Peterkin, a Pulitzer Prize–winning author who lived in Fort Motte, South Carolina; Lyle Saxon, a Louisiana writer of romantic southern tales; and Olive

Dame Campbell, a New Englander who spent most of her life running a folk school in Brasstown, North Carolina. Each new acquaintance helped Ulmann circulate through a specific American culture where she was able to gaze upon "the folk" there. Niles, who had collected ballads of mountaineers in the eastern parts of Kentucky and Tennessee, accompanied Ulmann on several Appalachian trips. Julia Peterkin invited Ulmann to her family's own Lang Syne Plantation for extended periods, guiding her through the African American Gullah village where workers were only sixty-five years out of slavery. Lyle Saxon hosted her at a northern Louisiana plantation named Melrose and also in New Orleans, where Ulmann took her cameras into the religious communities, the cemeteries, and among vendors in the lively Vieux Carre. Olive Dame Campbell, in western North Carolina, acquainted Ulmann with an educational method directed toward building "an enlivened, enlightened rural population."[27]

Using these contacts, Ulmann pushed her work in new directions. Had she never created more than her architectural studies, pictorial landscapes, and portraits of famous people, she still would have secured a place for herself in the annals of fine photography.[28] But she developed her photographic eye further, carving a unique niche that bridged pictorialism and documentary. She combined what was considered an old-fashioned tonalistic photographic style in the late 1920s with a documentarian's sense of reform. She arranged scenes and people and objects in rural America in order to show them to audiences with urban sensibilities, not just people who lived in cities but transplanted reformers who wished to celebrate and manipulate agrarian traditions and symbols. Her photographs took their place on the walls of rural schools and country hotels in equal measure to the space and influence they wielded in New York City galleries and at the White House. And as the objects of her camera eye slowly changed, Ulmann ultimately came to realize what photography could achieve. In 1930 she told Allen Eaton, with whom she later collaborated on a book, "I am of course glad to have people interested in my pictures as examples of the art of photography, but my great wish is that these human records shall serve some social purpose."[29] Ulmann's articulation of this desire places her alongside other "documentary" photographers who used their pictures in the 1930s to evoke social change.[30]

Since the term *documentary,* amorphous at best, is understood differently by its various interpreters, the task of labeling Ulmann's work remains difficult. One biographer maintains that her "need to complete a group of pictures for a specific purpose . . . suggests the documentary intent of her work." This "need" applies to a significant portion of Ulmann's photography, even the portraits of physicians made between 1918 and 1922 that she claimed to have published as a favor to her husband's colleagues. Yet those portraits have not been employed by photography scholars as examples of documentary. Beyond Ulmann's thematic approach to her subjects is her style, described by Jonathan Williams as "earthy, yet savory." Williams places Ulmann in the "great realist tradition," likening her photography to that of Julia Margaret Cameron, Alvin Langdon Coburn, and Eugène Atget. The wide range of her subject matter certainly justifies placing her alongside these three. But when considered a distinct genre, documentary photography frequently hosts subjects that are considered ordinary, common, often anonymous. William Stott has identified documentary's subjects as "individuals belonging to a group generally of low economic and social standing in the society (lower than the audience for whom the report is made)." And James Guimond, in *American Photography and the American Dream,* notes that the great documentarians of the twentieth century focused upon "'ordinary' people, like black students, child workers, sharecroppers, office clerks, and factory workers, rather than the famous people in history books, on the front pages of newspapers and the covers of magazines."[31]

Indeed, Ulmann provided a vast body of scenes and faces never before captured on film. Her subjects, especially those tucked away in remote valleys far from modern conveniences and mass culture, seemed appropriate ones to document. But to define documentary photography solely on the basis of who shows up in the pictures confines its purpose. Documentary, like all discourses, comprises message-laden records. The records require scrutiny, but so does the record-maker. What exists within the four corners of a photographic print is a combination of personal cultural background, technical or artistic training (or both), and expectations of one's subjects. These factors produce a photograph as much as do compositional arrangement, the sophistication of mechanical equipment, and the subjects themselves. All contributed to the camera "eye" that Ulmann had developed and was fine-tuning in her last decade as a photographer. Together, these characteristics made her a documentarian.

In the mid-1920s, her observations of several religious communities, including Dunkards, Mennonites, and Shakers, prepared her to enter similarly small, distinctive enclaves in the American South, where she chose to invest her principal energies in the late 1920s and the 1930s. Although Ulmann never ceased making portraits of important writers in New York, her attention grew increasingly more focused on groups in rural America. The philosophical link among these groups was that each represented a category of people that she believed would eventually disappear. Louis Evan Shipman pointed out in his introduction to Ulmann's 1925 portfolio of American editors that Ulmann had "collected a distinguished and tragic group—of a fast-disappearing species. Tragic, because they are the last of a line of notable progenitors, who vitalized and adorned a notable profession." This statement could have applied to any of the groups on which Ulmann focused her cameras after 1925—the utopian religious communities whose numbers were dwindling, the Kentucky mountaineers whose land and population suffered under the weight of large mining companies, or the craftspeople who strived to preserve the skills of their ancestors while corporate production lines turned out chairs, brooms, baskets, quilts, and other items by the thousands each year. With her camera Ulmann found what she wanted to find among these people—idealized worlds that she and others had imagined; worlds created in order to challenge the pace of change in America that was rapidly homogenizing its citizens with radio, movies, and mass production.[32]

One of Ulmann's direct challenges to fast-paced technological change in the twenties was her refusal to embrace it in her own work. Despite the ease newer photographic technology could have contributed on her sojourns into southern Appalachia, Ulmann remained devoted to her old equipment. She could personally afford any kind of new camera or accessory, but she most frequently used a 6½ × 8½ inch whole-plate camera positioned on a tripod. Even in her travels on foot across creek beds and to out-of-the-way home-steads, Ulmann took along the bulky camera, the tripod, a lens box, and scores of glass plate negatives. Since she hated artificial light and never used a flash, she "always carried some white sheets along . . . for reflecting," remembered Allen Eaton. She did not use a shutter or a meter but made a practice of sliding the lens cap off in order to admit light. She once declared, "I am the light-meter." Because composing a single photograph required signifi-cant mechanical preparation, Ulmann took no action shots or candid photographs. John Jacob Niles noted that "moving objects were never effective as subjects for her photography."[33]

Nor did Ulmann wish to freeze a moment of action. She much preferred to capture facial expressions or a pair of hands, some-one sitting quietly rather than in motion. In most of Ulmann's portraits, time itself is arrested. Not captured, but arrested. She allows very few clues about the era to enter into the frame, so that the photograph possesses a timeless quality. The latest gad-getry, current fashion, or other marks that would situate the pho-tograph chronologically are absent. In figure 4, a typical Ulmann portrait of rural America, the absence of motion is achieved by the stillness of the figure, who, sitting bent over in a straight-backed chair, has folded her hands carefully over her rough, thread-bare apron. No prominent prop or clue situates the image in the twen-tieth century. Rather, worn wooden planks on the cabin porch reveal the dwelling's age and contribute to the premodern spirit of the scene. Given its place in the photographer's light, the featured subject of this photograph is the spinning wheel. The pristine rope of thread balanced on the end of the wheel glistens, the spokes shine, and the rest of the object stands in sharp focus while the woman, herself a timeless character, provides a softly focused contrast to the principal icon. Together the human subject and the inanimate object create an atmosphere that transcends time.[34]

When Ulmann's method is examined, her perception of time takes yet another fascinating shape. Ulmann believed a better, truer image would emerge from a technical process that required min-utes, even hours, than from one that took only seconds. She never worried that the sharpness of one moment would escape her. In her opinion the photographers who made quick pictures distorted reality by failing to study their subjects thoroughly and from many

Fig. 4. Doris Ulmann. *Nancy Greer. Trade, Tennessee, ca. 1933-34.* Audio-Visual Archives, Special Collections and Archives, University of Kentucky Libraries.

angles. Ulmann even rejected the newer camera models designed for professional photographers, such as the Rolleiflex, the Leica, and the Speed Graphic. Niles recalled, "There was no hurry-up, no snapshot business. Snapshot photography was the end of vulgarities so far as she was concerned. When I demanded a Roliflex [sic] and got it and everything that went with it, Doris immediately looked upon me as a complete faker." Ulmann was less interested in grabbing a piece of time than in grasping the essence of human character. And she believed it could be accomplished through a collection of images of the same subject over time, as her many proof prints of single individuals demonstrate. Although her choice of photographic subjects changed, from the urban intellectual set to a rural, semiliterate class, her artistic philosophy remained the same. In her opinion a single picture could not adequately portray a person, regardless of his or her background, appearance, or socioeconomic status. She observed a Kentucky knife maker as intensely as she had studied a famous fiction writer or a stage actor.[35] This is evident from the number of shots she composed of each subject; the proof books of her Appalachian work contain not lone portraits, but serial studies. Ulmann often took several exposures of closely similar poses. In one series an unidentified man uncomfortably holds a Bible stamped "Placed in this motel by the Gideons." The fellow's tense facial expression changes just slightly from portrait proof to portrait proof. Perhaps he was an unwilling subject or the prop seemed alien to him or he became too warm to sit patiently in the summer sun. Whatever the reason, it is the *series* that reveals his prolonged discomfort. One frame could have been an aberration, capturing a momentary letdown of defenses; four portraits in sequence reveal to the viewer the subject's awkwardness before the camera. The sitting may have required fifteen minutes or two hours, depending upon the changes in natural light and how swiftly Ulmann removed the exposed plate, covered it, and inserted a new one into her camera. The session's duration, perhaps too long for the sitter, abolished any hope for spontaneity. Through this complex and extended process, then, we see that Ulmann's intentions went beyond reproducing the semblance of one moment's reality. Her old equipment and her methodology prohibited it. And she did not apologize for either.[36]

In 1930 Ulmann stated with conviction, "I have been more deeply moved by some of my mountaineers than by any literary person, distinguished as he may be." The photographer's reference to "my" mountaineers implies her ownership of them, yet her photographs do not project an air of condescension. Rather, the subjects and their surroundings are glorified, cast in what Jonathan Williams described as a "wistful tone of reverie." That Ulmann did not consider her relationship with her subjects reciprocal is obvious; but her tone of possessiveness more likely suggests her personal pride at having mined out some valuable raw material that would engage artistic and literary circles in the late 1920s. Ulmann, with Niles's help, had surveyed small pockets of people who remained untouched by the forces of industrialization and mechanization. Or so her photographs would suggest. Even though she traversed rich coal mining territory in Kentucky, she deliberately avoided it and other similar topics that would reveal the encroachment of corporate America and materialism upon otherwise unique areas. The messages Ulmann recorded on glass plates were nostalgia-laden reminders of an older, simpler America. As a native New Yorker and an urban dweller, Ulmann had witnessed the growth of the city and the vice and deterioration that accompanied industrial progress there. Her training at the Ethical Culture School had only heightened her awareness of these issues.[37]

Her expanding interest in rural America in the late 1920s situated Ulmann in an intellectual context peopled by men and women striving to highlight the virtues of technologically unfettered enclaves. Among the strongest proponents of these virtues were the Agrarians, a group of southern poets, historians, and literary critics committed to what they viewed as the superior side of the dichotomy "Agrarian *versus* Industrial." They laid out their arguments in a collection of essays entitled *I'll Take My Stand* (1930), a provocative book that created tremendous controversy, inviting a wave of criticism in the early Depression years. Seen by some commentators as "utopians" and "nostalgic eccentrics," the Agrarians clung to "the supremacy of tradition, provincialism, and a life close to the soil." Like the Agrarians' impassioned sentiments, Ulmann's photographs fed the imaginations of viewers who, in just a few years, had found much to revere in the hills of the southern highlands. Her images served as an antidote to the wild-eyed fanatics caricatured in H.L. Mencken's newspaper reports of the 1925 Dayton, Tennessee, Scopes "Monkey Trial." Ulmann's images of mountaineers were described by one critic as "strong reminders of the richness and depth and variety of dramatic treasure to be found in the mountains and valleys of country districts."[38]

The mountain "types" Ulmann produced also fed the nativists' imaginations in twenties America, although she did not align herself with the nativist movement, nor did she intend for her photographs to serve that purpose. Her enlightened upbringing in the Ulmann household and her training at Ethical Culture would have made that an unpalatable option. Throughout her youth, hundreds of thousands of immigrants had wandered through Ellis Island, the large majority packing themselves into New York City tenements. As a child of financially successful immigrants, Ulmann stood apart from many who could not find a place for themselves and their families in the crowded city or who were taken advantage of in the urban, industrial whirlwind. The pressure on eastern and central European immigrants to assimilate into mainstream American culture increased during the world war, reaching a frenzied height in the years immediately following the armistice. Urban riots, local police crackdowns on labor union activity, U.S. Attorney General Palmer's raids on suspected subversives (many of them immigrants), and eventually, the passage of immigration quota laws in 1924 marked an environment where ethnic difference proved dangerous. The culmination of anti-immigrant hysteria came in 1927 with the execution of Nicola Sacco and Bartolomeo Vanzetti, two Italian immigrants who lost their appeals after being convicted of a 1920 murder on circumstantial evidence. Theirs was a cause célèbre for many writers and artists, including some Ulmann subjects, such as Edna St. Vincent Millay. As a participant in literary and intellectual circles, Ulmann no doubt questioned the impact of such drastic solutions to fast-paced change. And she seemed particularly afraid that the forces of conformity and assimilation would be so successful that distinctiveness would be erased and lines blurred between groups of people. Of her focus on Appalachian mountaineers, Niles observed, "These were the people she really wanted to get down on paper for posterity. She thought they would finally disappear, and there would be no more of them."[39]

Ulmann realized that the unique world the mountaineers had preserved could slip easily away, especially as elements of mass culture, bureaucratic government, and corporate business bled into all Americans' lives. The traditions, rituals, and mores of southern Appalachia would not be overtly attacked as had those of urban ethnic groups, but they would vanish slowly. To counter such erasure, Ulmann squared her views, added her own creative touches, and produced hundreds of images preserving the actual faces and gestures of an American pastoral ideal. Highly praised, her work was featured in the June 1928 issue of *Scribner's*. But a *New York Times* writer imposed his nativist interpretation (in an election year) upon Ulmann's subjects when he identified them as "the mountain breed" and stated, "These Republican mountaineers are of British and North Irish stock without intermixture. There are no more clearly pedigreed native Americans than they." Another observer of the Appalachia work described a Kentucky subject as a "strong youth out of the purest English stock of the nation" who hears sung in her home ballads dating from "the beginnings of English history."[40] Ulmann's photographs spurred the nativistic impulse in the United States, even though this had not been her objective.

To accomplish her task in the mountains, Ulmann looked to individuals as her best subjects. Her cordial, gentle manner endeared her to people who might otherwise have balked at the sight of strangers carrying cameras, boxes, and other heavy equipment. Allen Eaton, a collaborator of hers, once remarked that she had a personality that was "very attractive to most anyone. They knew she was earnest and not pretentious. . . . [and had] a way of getting along with people." William J. Hutchins, president of Berea College, witnessed Ulmann's relationships with her camera's subjects and saw in her "a singular gentleness and grace, a self-abnegation joined with an amazing human interest." She would ask men and women about their work, and Niles warmed them up by singing to them or with them, encouraging any inhibitions to dissolve. Olive Dame Campbell noted that Ulmann's "gentle and generous personality disarmed criticism and suspicion."[41]

She arranged individuals in chairs or in doorways, giving her attention to one person at a time. Nearly every photograph has a single individual as its main focus, not a couple, a family, or a group. Ulmann's corps of images demonstrates that she turned away from group interaction and collective efforts. The lone woodcarver or chair maker or quilter received her fullest attention.[42] As she highlighted a character, Ulmann drew from the person's face a desired "look" that became the trademark gaze of her subjects—intense thought, a faraway glance, neither smile nor frown. Rarely does her sitter peer directly into the camera. Ulmann preferred a three-quarter view or a profile angle (see figs. 5 and 8). In her study of Lydia Ramsey, the sitter changes her facial expression just slightly for each of the five plates. In one photograph she has switched hats, exchanging a flower-topped bonnet for a casual straw. In a series of a different subject, the woman stares contemplatively as she holds up her printed apron for Ulmann to admire. There are few smiles on these faces. In fact, rarely did a smile appear on any of the ten thousand plates that Ulmann exposed.[43]

The mountain people who most intrigued Ulmann were the elderly. She believed their years of experience made them perfect character studies. Attempting to justify her fervent concentration on older people, she explained, "A face that has the marks of having lived intensely, that expresses some phase of life, some dominant quality or intellectual power, constitutes for me an interesting face. For this reason, the face of an older person, perhaps not beautiful in the strictest sense, is usually more appealing than the face of a younger person who has scarcely been touched by life." Through her lens Ulmann focused on long white beards, furrowed brows, bespectacled dim eyes, and parched, wrinkled faces. In an extended series marked "unknown, before 1931," she composed fourteen portraits of an older, copiously whiskered man. In one photograph he tugs slightly at his beard, holding it for her approval. Given her control over portrait sessions, Ulmann probably asked him to show his whiskers proudly for her camera. In another series she photographed a man sporting a long, oddly fashioned, unkempt mustache. Several other portraits highlight her fascination with the bearded elders of Appalachia. About those preferences, Niles astutely observed, "You had to be an individual, a character more or less, before she was interested in you even a little bit. . . . I think she loved most the white mountaineers, the old patriarch types. . . . She saw in their faces the care and the trouble of the awful effort they had made to carry on life now that they had reached the afternoon or evening of their days."[44]

In their portraits, the venerable give an impression of strength and perseverance that Ulmann seemed drawn to. Since she was physically quite frail, it may not be coincidental that she pursued individuals whose bodies had not betrayed them, whose wrinkled faces and roughened hands were made so by seventy or eighty years of living. When she first encountered them she was not yet fifty years old. The "dominant quality" that she said she sought often revealed itself in her Appalachian portraits as a subtle hardiness, a quiet energy evident behind contemplative faces and dreamy eyes. And she was prepared to look for it at great length. Niles recalled Ulmann's tenacity in seeking out her preferred subjects during their travels together, noting that she was "willing to put up with any kind of weather, any kind of heat, any kind of rain . . . for the sake of getting to some out-of-the-way, God-forsaken spot where some ancient with a long white beard and a shock of white hair was sitting in front of his little cabin." Besides healthy beards, other symbols of age or wisdom appear in Ulmann photographs to categorize further the individual subjects. One sage holds a thick cane; in another print, a toothless woman has donned a huge sunbonnet; in yet another, a man wears tiny spectacles reminiscent of an earlier time. Ulmann's method of convincing viewers that this culture was dying was to narrow her vision overwhelmingly to those who would not live too many more years. These wrinkled faces belonged to the last true pioneers, her images warned. She must have been pleased to read in a 1930 feature that she had produced "an authentic record of a phase of our national living that will soon be no more." Capitalizing on symbols of age, Ulmann challenged the intensely popular movement that celebrated youth. Her few words, combined with thousands of images, spoke to her contemporaries who declared the 1920s a decade of and for the young. Popular fiction thoroughly dissected the attitudes and habits of those in their late teens and twenties, while magazine editors filled their pages with stories concerning the new generation. As literary critic Frederick Hoffman has pointed out, "No aspect of the decade was more thoroughly burlesqued or more seriously considered than the behavior and affectations of the young generation. They lived all over Manhat-

Fig. 5. Doris Ulmann. *Christopher Lewis. Wooten, Kentucky, ca. 1933-34.* Audio-Visual Archives, Special Collections and Archives, University of Kentucky Libraries.

tan, at both ends of Fifth Avenue, and disported themselves in a manner that amused *Vanity Fair*'s humorists, impressed its book reviewers, and provoked replies and analyses from its sophisticated journalists."[45] Ulmann's artistic renderings stand in opposition to this popular social phenomenon, challenging the trendy focus on youth and vigor with a message that linked wisdom with the aged.

Another way Ulmann attempted to shape cultural knowledge about southern highlanders was to give them a spiritual quality that was characteristically Protestant. In her attempts to portray religious devotion among mountaineers, she employed the symbol most clearly connected with fundamentalist Protestantism—the Holy Bible. Whether the photographer carried one with her as a prop remains an unanswered question. She could easily have borrowed from her hotel the Gideon Bible that showed up in one series of pictures. In another series, Winnie Felther, of Hyden, Kentucky, holds a Bible in her lap. Ulmann highlights the striking contrast between the dark, leather-bound book and the light apron where it rests. The larger frame for the object is the busy print of Felther's dress. The viewer's eye is drawn to the book, illuminated and cradled in the pure white fabric of Felther's apron and made even more prominent by its place in the woman's lap. In another series of an unidentified sitter, a woman holds open a book of hymn tunes stamped "Revival Gems." The woman's facial expression, stern and serious, suggests the fire-and-brimstone pitch of emotionally charged revival meetings.[46] Ulmann's portrait of preacher Christopher Lewis, of Wooten, Kentucky, carries her message most poignantly (fig. 5). Lewis sits in a cane chair in front of a soft natural backdrop, the sun shining on his white hair. The slight downward tilt of his head marks his humility, a prized Christian virtue. His seriousness of purpose and religious devotion are unquestionable, given the solemn expression on his face. His barely raised shoulders suggest an upward movement of his arms, bringing the sacred book in toward his heart, resting it securely on one hand, while clutching it protectively, even lovingly, with his other. Clearly "the word" is with him. As a symbol the book carried Ulmann's message about this Kentucky mountaineer more clearly than other religious icons would have. A crucifix, for example, would have hinted at Roman Catholicism, a volatile subject in the United States in the late 1920s. Earlier in the decade, Ku Klux Klan terrorism against Roman Catholics had been widespread. By the late 1920s, the most public arena for debate was the 1928 presidential election, which pitted Al Smith, a Roman Catholic Democrat, against Herbert Hoover, a Protestant Republican. In tying southern highlanders to the Book, the word, Ulmann not only implied that they were literate, but she also secured the quality of spiritual devotion for them. Her message about religious faith and her attention to it echoed that of groups like the Agrarians who felt that the worship of industry and materialism had squeezed out more traditional expressions of faith.[47]

The South Carolina Gullah people gave Ulmann another rich and distinctive culture on which to build her composite studies and promote her antimodernist sentiments. Ulmann made her way into the African American settlements after having met and befriended the writer Julia Peterkin in New York. Peterkin, owner and matriarch of Lang Syne Plantation near Fort Motte, South Carolina, employed more than three hundred Gullah workers to cultivate her two-thousand-plus acres of cotton, asparagus, and wheat. She had lived next to the Gullah village since 1903, and its inhabitants became the principal subjects of her poetry and fiction. The Vanderbilt Agrarian Donald Davidson praised Peterkin's work, emphasizing that she "let the Negro speak in full character," and that she had "forgotten more about Negroes than Joel Chandler Harris (for all his greatness), Thomas Nelson Page, and that prurient modern, Carl Van Vechten, ever knew." In planning what would become her novel *Scarlet Sister Mary* (1928), Peterkin said, "I have been very close to life. It is all about me—several hundred negroes." Equating herself with the Gullah people set her apart from many of her southern literary contemporaries, who depicted African Americans in a romantic Old South haze; this led Davidson to note further that Peterkin's knowledge was "so intimate, so detailed, so exact that one is overwhelmed and asks how a white person could ever know so much (if what she knows is true) about Negro life."[48] At the MacDowell Writers' Colony in the summer of 1928, Peterkin built a story (*Scarlet Sister Mary*) around the life of a poverty-stricken Gullah woman who is expelled from her church but readmitted after her son dies and she sees a vision of

Christ's suffering. The book won Peterkin a Pulitzer Prize the next
year.

The religious theme of Peterkin's work inspired Ulmann's photography in South Carolina. Primitive foot-washing ceremonies,
river baptisms, and healing rituals in the church were among the
practices she photographed. She was privy to these rural scenes
because Peterkin paved the way for her, accompanied her to Gullah
homes, and talked with those she knew well. Ulmann needed
Peterkin, especially on her first trip to Lang Syne in 1929, when
she found it difficult to work with and understand the individuals
she called "Negro types." Despite her initial insecurity on this
venture, she created images that earned her accolades. With regard to the photographs hung in New York City's Delphic Galleries, one *New York Times* reviewer claimed that Ulmann's
pictures were "good" because her subjects were "grand." Lang
Syne's remoteness, in the central "backwater" section of the state
near the confluence of the Congaree and Wateree Rivers, made
it a suitable location for Ulmann's work. And it served as a
compelling venue that urban imaginations could cling to in their
pursuit of rural American "folk."[49]

Ulmann and Peterkin became "very close friends," traveling
together throughout the South and living at Peterkin's two homes,
the plantation near Fort Motte and the summer retreat known as
Brookgreen, located on Murrell's Inlet just south of Myrtle Beach.[50]
Ulmann's devotion to Peterkin is evident in several portrait series
she completed of the writer, including one of Peterkin in costume
for her role as Hedda Gabler in a local production of Ibsen's drama.
Peterkin sits in an elegantly trimmed formal gown, reminiscent of
a turn-of-the-century wedding trousseau. The friendship transcended the others Ulmann had developed with writers, and its
intensity may be felt in a unique series of portraits she arranged
for Peterkin and herself. In one of these compositions (fig. 6),
Ulmann has positioned herself so as to require Peterkin's support.
Peterkin appears comfortably stable, so much so that she is able
to smoke while balancing Ulmann on her right shoulder. Their
relationship to each other in the image reveals their intimacy, with
Ulmann's left arm on her companion's back and Peterkin's erect
posture offering a firmness on which Ulmann, who is sitting in a
more precarious position, is completely dependent.

Fig. 6. Set by Doris Ulmann. *Doris Ulmann and Julia Peterkin. ca. 1930.*
Photo used by permission of the South Carolina Historical Society.

Their friendship led them to collaborate on a project juxta-
posing Ulmann's Gullah records and Peterkin's narrative descriptions. In 1933 *Roll, Jordan, Roll* appeared in two editions: the
trade edition contained seventy-two Ulmann images, and a finer,
limited edition exhibited ninety hand-pulled gravures. The NAACP
gave *Roll, Jordan, Roll* tremendous approval the year it appeared,
with Walter White, an NAACP leader, calling the book a "magnificent achievement." Ulmann's South Carolina work showed her
expansive artistic range, with landscapes, group scenes, and portraits combined in one volume. However, in keeping with her
previous work in the mountains, the more outstanding, more introspective compositions are of one or two individuals, not of
groups. She pictured a father and a son sitting quietly on their front
porch, a couple of men standing beside the cotton weight scale at
the edge of the field, and a middle-aged man sitting in the doorway
of his barn. Ulmann identified several of her sitters, including a man
known as "Black Satin" and another called "Papa Chawlie."[51]

As in her Appalachian portraiture, Ulmann attempted to define character through her direction of facial expressions, arrangement of hands, and body postures, but she also used the
accoutrements of a pastoral setting to promote the notion of an
idyllic, rural landscape. In her portrait of a South Carolina fisherman (fig. 7), the still water and the small vessel support the main
character, who carefully (almost lovingly) guides with his hands
the line on which his sole attention is focused. The line he has
pulled cuts a perfect horizontal angle across Ulmann's frame,
matched in parallel by the boat's rim in the foreground. The man's
position and the heavy, hanging net provide a solid vertical line.
The angle of the boat provides the necessary depth for the photograph, and the result is an image with unwavering compositional
stability. Here artistic arrangement coupled with bucolic subject
matter transmits one of Ulmann's clearest messages about rural
America—that quiet scenes like this one should not be ruined. Her
composite examination of Gullah African Americans revealed yet
another regional culture that Ulmann perceived to be unharmed
by the forces of modernization yet vulnerable to their encroachment. This seemingly stable, pleasant rural existence hid the fact
that depressed economic conditions in the early 1930s had sub-
merged many American farmers under the weight of overproduc-

Fig. 7. Doris Ulmann. *Untitled (Man Standing in Boat Fishing). ca. 1929-30*. Collection of the University of Kentucky Art Museum.

tion, accelerated mechanization, worn-out soil, and for a decade, an unsupportive federal government. Romanticizing the country life in visual images steered viewers away from realities that faced agricultural workers at the onset of the Great Depression. Ulmann's glowing portrait of a working plantation in the twentieth century recreated the nineteenth-century ideal that antimodernists like the Agrarians had hoped to promote through local organization and public debate but which failed to get off the ground.[52] Among the group's numerous philosophical and practical disagreements was the question of what role blacks would play in the revived rural South. Although Ulmann did not vocalize her opinions on race, her photographs from those years reveal her close attention to the issues of color and ethnicity.

In *The Darkness and the Light,* a collection of Ulmann's images of African Americans, William Clift argues that Ulmann wanted to record the South Carolina Gullah culture because "she envisioned a gradual blending of the races in which these types of people would lose their particular distinctiveness." Her work at Lang Syne evokes a nostalgic tone, but more than likely this was due to Peterkin's influential presence and her own literary interpretations of the world she inhabited. Ulmann's larger body of photography from this period reveals that the results of miscegenation fascinated her. Some of her more remarkable compositions in terms of subject positioning, prop use, and overall message emerge from her work in the late 1920s and early 1930s. Her curiosity about people of mixed race and various ethnic backgrounds, those who could not be categorically defined, is evident in a number of photographs she took during this period. On one of her trips through the Carolinas, she pictured several "undefinable" individuals, including a young Melungeon woman, an elderly South Carolina woman labeled "Turk," and a South Carolina man called "Cracker." The Melungeon woman was one in a series Ulmann completed on Native Americans living in the Appalachians. The sturdy young wife bends her back over a washboard and tub while staring pointedly at Ulmann with her penetrating, deep-set eyes. In the background a lone log house fixes her situation, emphasizing her physical isolation and implying her social isolation as a native of mixed blood, likely Cherokee and Celtic. Hers is one of the many compelling faces of mixed ancestry that

Ulmann chose to photograph; others revealed African American and Native American features gracing the same face. The South Carolina "Turk," an elderly white-haired woman with a sun-parched face and a cob pipe, was described as a possible descendent of Turkish sailors who, according to legend, were shipwrecked off the eastern coast of North America. More likely she had descended from ancestors who were Native American, African, and Caucasian. In the portrait labeled "Cracker" by *Theatre Arts Monthly* editors, a dark-haired, olive-skinned, broad-shouldered man, with kind eyes and a slight smile, confounds ethnic categorization. He probably appealed to Ulmann because he crossed several defining lines or because he was, as the article description states, "a rural type of uncertain origin." In his portrait, he sits quietly among brush and tall grasses, exuding an air of gentleness, even vulnerability.[53]

Ulmann was probably influenced by the novel that Peterkin was writing during this time. Titled *Bright Skin,* the plot turns on proper skin color and features, as the demands of racial purity dictate. Two sisters have disgraced an African American family by producing children with traits unrecognizable to its members; one sage announces early in the story, "Right is right. Color is color. I would hang my head wid shame if one of Jim's chillen had bright skin." The principal relationship developed in the story is between two cousins, Blue and Cricket. Shocked when he first meets Cricket (who looks like no one he has ever seen), Blue describes his cousin as the "color of a ripe gourd." He grows to love her, even more as she is taunted by the community as an outcast. The novel's message is summarized in an outwardly pious church member's announcement that "A bright skin ain' got no place in dis world. Black people don' want em an' white people won' own em. Dey ain' nothin but *no-nation bastards.*" The concern over belonging nowhere, to no tribe or group, had engaged a number of writers in the 1920s who built on the theme of painful alienation from kinship ties. On into the 1930s, the added dimension of national economic turmoil led to further artistic probes about belonging, particularly about what constituted and defined Americanness. In 1932 Peterkin dedicated *Bright Skin* to Ulmann.[54]

The photographer's curiosity about other cultural enclaves dotting the American landscape was further touched when, with

Peterkin's assistance, she went deeper into the South. The two traveled together to Louisiana between 1929 and 1931, connecting
there with the writer Lyle Saxon. A historic preservationist, travel
writer, and raconteur, Saxon articulated many of the same concerns that Peterkin and Ulmann expressed in their respective art
forms. He appreciated local color, delved into questions about race
and mixed blood, and feared the homogenizing influences of modernization upon rural Louisiana. Although his fullest expression
of these themes did not appear until he published *Children of
Strangers* in 1937, he spent more than a decade dealing with such
issues in his conversations, his diary, and his varied other writings. As Ulmann attempted to preserve on glass plates those people
and places she believed would disappear, Saxon once wrote that
he wanted the rural Cane River country "put down upon paper
before it lost its quality and became standardized."[55] Sharing similar
passions, Saxon and Ulmann were ambivalent about some of the
same things: both thrived in their urban worlds, living among artists
and writers, but both had had their imaginations sparked by
American rural life, its inhabitants, and its characters. Neither could
throw off the romance and nostalgia of bygone days in the country, whether real or imagined. Saxon introduced Ulmann to both
of his worlds, the Cane River Country in north Louisiana and the
New Orleans French Quarter. In the latter she photographed the
faces of several characters, including an ancient bookseller and a
vegetable hawker in the French Market. A number of her regionally defined visual studies bespoke Ulmann's keen interest in spiritual and religious influences. In a predominantly Roman Catholic
city, whose mingled French and Spanish histories still shaped its
character, Ulmann found a craftsman whose work helped fill the
community's spiritual needs (fig. 8). Cast in a glow of light, the
sacred object and its caretaker mirror each other, with the artist
admiring his work, lovingly touching the "saint" he has repaired
and retouched. The balance Ulmann achieved in the image by
splitting the focus leads the viewer to believe that the icon is as
vital as its creator. The product itself seems to have life, a technique that Ulmann cultivated in the early 1930s as she began to
feature in her photographs the fruits of artisan labor. Her Appalachian handicraft images, in particular, feature the products as
prominently as their makers, so that the objects themselves come

Fig. 8. Doris Ulmann. *Man with Sacred Image. New Orleans, Louisiana, 1931.* Audio-Visual Archives, Special Collections and Archives, University of Kentucky Libraries.

to define a region or a locale. Certainly this may be argued of Ulmann's study of the New Orleans sacred image maker, whose product stands as a symbol for the city's culture and history. In pursuing her curiosity about the city's religious foundations, Ulmann entered the confines of monastic life, composing portraits of women in the Ursuline order and in the Sisters of the Holy Family. Ulmann used her opportunities with the African American Holy Family nuns to experiment with degrees of lightness and darkness. Her visual studies show the blinding, starched white of the sisters' habits framing the quiet solitude of their smooth brown faces. New Orleans provided Ulmann yet another compelling venue in which to pursue her cultural interests in American religion and race.[56]

As with her other sophisticated liaisons, including those with Niles, Peterkin, and Campbell, Ulmann found Saxon a curious study for her portraiture. In one series of him, Saxon is dressed up in a three-piece tweed suit, sporting a boutonniere. A French doll dresser figures prominently in these photographs, as does Saxon's young black servant. The dominant figure in the portrait, Saxon appears to be a protector of both the dedicated craftsman and the uncorrupted black youth. Thus Ulmann cast Saxon in the role he had fashioned for himself in his attempts to stave off change and preserve the past. Ulmann's reverence for Saxon and his talents is evident in this portrait series. That she also was infatuated with him is suggested by John Jacob Niles's jealousy-laden annotations to this series after Ulmann's death. In the posthumous proof books, Niles terms Saxon a "Writer of Sorts," even though Saxon was leading one of the most successful Federal Writers' Projects in the country. Louisiana and its characters mesmerized Ulmann, yet she never returned to accomplish her proposed projects on two unique Louisiana subcultures, the Atchafalaya Basin Acadians (Cajuns) and the urban Creoles.[57] The Appalachian mountains beckoned her again, and it was there where she engaged her most physically demanding, and arguably her most significant, theme-centered photographic project—a survey of the southern highlands craftspeople.

In the late 1920s Ulmann had enjoyed a conversation with Allen Eaton, a fellow New Yorker interested in producing a book on mountain handicrafts. Eaton, by training a sociologist, had become involved with craft exhibitions in his home state of Oregon fifteen years earlier. He had been named the first field secretary of the American Federation of Arts in 1919 and joined the Russell Sage Foundation the next year. That foundation, established in 1907 by Mrs. Russell Sage, directed its activities toward "the improvement of social and living conditions in the United States" by supporting "programs designed to develop and demonstrate productive working relations between social scientists and other professional groups." As Eaton worked for the foundation, he felt driven to combine social work and art, a goal that he felt could be realized in the southern highlands. With the help of Olive Dame Campbell, who ran the John C. Campbell Folk School in western North Carolina, Eaton's objective came to fruition. The Southern Mountain Handicraft Guild (SMHG) was established in 1929 and four years later renamed the Southern Highland Handicraft Guild (SHHG). Impressed with Ulmann's studies of Appalachian people, Eaton asked if she would be willing to illustrate his proposed book on the subject. She eagerly agreed and, according to Eaton, "offered to undertake the project at her own expense." In this instance Ulmann typified the photographic artist described by Stieglitz in his 1899 essay "Pictorial Photography," where he maintained that "nearly all the greatest work is being, and has always been done, by those who are following photography for the love of it, and not merely for financial reasons." Stieglitz's turn-of-the-century pictorialists, many of them independently wealthy, could afford to concentrate their efforts in such a way. Ulmann, following her principles as a pictorialist, embraced Eaton's project "for the love of it." She told Eaton, "If you will make the contacts for me . . . I will be able to get the subjects that I want most and I will photograph the people working with handicrafts that you want." Eaton and Ulmann had the same ideas about this unique group of people tied to the land and dependent upon their hands for survival; both feared that self-sufficient existence and the stability it represented could vanish. The sociologist and the photographer knew that the highlanders' creative endeavors, the "living craft, the moving tradition" had to be recorded. In presenting fifty Appalachian portraits to the SMHG at its 1932 fall meeting, Ulmann took a significant step in committing herself to the arts and crafts enterprise in the region. Her collaboration with Eaton shifted into high gear in the following months as she entered the third and last phase of her career as a photographer.[58]

Ulmann's focus shifted once again as she embraced the social and educational functions that her photographs could serve. The public's thirst for traditional American craft design had been built up and sustained by the turn-of-the-century Arts and Crafts Movement, but in the early 1930s the deepening economic depression increased the focus on American-made and American-inspired designs. When the editor of the *Annual of American Design 1931* called for complete "emancipation" from European design so that a thorough cultural independence could be enjoyed in the United States, he was referring specifically to freedom for industrial design. Ulmann's exposure of traditional southern highlands handicrafts through her photographs, then, could both fulfill and challenge the call for "Americanness" in design. On the one hand, her focus would be upon processes and products that had been around for decades and thus were firmly situated in the nation's regional histories and cultures; on the other, the processes and products themselves challenged the burgeoning machine-age aesthetic that was inspired by industrial lines and shapes and mass production. Ulmann's visual images, to be successful, had to illuminate the appealing qualities of hand-made, carefully created items. She had to picture beauty in tradition if the SMHG's twenty-five craft-producing centers, many of them settlement schools, were to continue operating during the Great Depression. In the Southern Highland Handicraft Guild meeting reports, Eaton noted that "there had been many calls for inexpensive articles" at the 1933 Century of Progress Exposition in Chicago. He, Ulmann, and other craft supporters could take solace in the fact that although the exposition's nationwide quilt contest, sponsored by Sears, had fielded many entries with innovative themes focused on industrial power in a modern nation, the winning quilt featured a traditional nineteenth-century design made by eastern Kentucky needlers.[59]

Ulmann chose to picture people as well as their products. For years, the constituent elements of character in a person had intrigued Ulmann, who often tied those qualities to the work people did. Faces were important for her portraiture, but so equally were hands. Though she had always striven to entwine a person's occupation with his or her individual character, this aspect of her photography took on greater weight in her Appalachian handicraft portraits. As one Ulmann biographer has noted of these photographs, "the sitters' hands and the objects they are holding are often as important as the faces themselves." Her portraits included specific materials and tools of trade. Emery MacIntosh, a tombstone maker in Breathitt County, Kentucky, holds a chisel and a hammer in his portrait. Oscar Cantrell, a North Carolina blacksmith, works at his equipment with a wrench, and Aunt Lou Kitchen sits at a low spinning wheel. Attracted to a "person who was doing something," Ulmann often pictured her industrious subjects engaged in the creative process or surrounded by their finished products (see figs. 9 and 10). Her photographs reveal the beauty and simplicity she sought to reveal in Samuel Clark's handmade looms, Cord Ritchie's woven baskets, Ethel May Stiles's tufted bedspreads, and Enos Hardin's chairs. Ulmann situated the objects at the forefront, casting them in the brightest available light, to emphasize their pristine, geometric lines, while their makers occupied the background of the frame. Ulmann's focus on the relationship between creator and product revealed how craftspeople were defined by what they made with their hands. Thus she could picture Enos Hardin in a softly screened light with one of his chairs in the foreground and write about the "appreciation of beauty in the man" and at the same time express astonishment that he could make "such fine chairs when the home is so dirty and slovenly."[60]

Although Ulmann hoped to connect these people with their respective vocations, she rarely had her sitter observe his or her task. The sitters stared off into the distance somewhere while their hands attended to the job. In one portrait (fig. 10), Cord Ritchie protectively secures the raw materials for her task of basketmaking. In another study, Hayden Hensley, an accomplished young carver, holds a knife and a piece of wood but has turned away from his job. In the large majority of these vocation portraits, Ulmann revealed busy minds and active hands without necessarily connecting the two. She seems to have tried to picture craftspeople thinking grand thoughts as they engaged in their work or held their finished products, surrounded by the richness of their talents. The men and women appear to be contemplators and artists, not mere tradespeople. Compositional stability in these portraits is achieved through a kind of motionlessness, or arrested action, on the part

Fig. 9. Doris Ulmann. *Hands of Lucy Lakes. Berea, Kentucky, ca. 1933-34.* Audio-Visual Archives, Special Collections and Archives, University of Kentucky Libraries.

Fig. 10. Doris Ulmann. *Aunt Cord Ritchie, Basketmaker. Knott County, Kentucky, ca. 1933-34.* Audio-Visual Archives, Special Collections and Archives, University of Kentucky Libraries.

of the featured individual. This technique Ulmann had employed
for years in her portrait studio, and by using it in the handicraft
documentary project she elevated the status of each sitter to that
of a featured actor in a vital drama that could not be allowed to
cease. Her portrayal of the drama intimated that these artists
possessed the secrets for carrying on productive, fulfilling lives.
At the end of her first summer on Eaton's project, she wrote, "All
these people give one so much to ponder about and meditate."[61]

Ulmann's deliberate compositional style and her artistry continued to reflect the influence of Clarence White even after she
had embarked on her documentary project in the southern highlands. In a series picturing Kentucky cloth weaver Aunt Tish Hays,
Ulmann combines three complex patterns in the frame. The first
and most imposing is a background made entirely of a hanging
bed cover sporting nearly two hundred small, tightly woven wheels
of contrasting light and dark yarns. The second pattern, which
cuts the middle plane of the surface, is Hays's dress, made of a
fabric splashed with a floral design. Ulmann, who enjoyed studying the printed fabrics of mountaineers' dresses, often directed her
lens toward the lines and shapes of those patterns. The third pattern,
covering the lower plane of the frame and lying in Hays's lap, is
another bed cover, whose intricate design combines circular shapes
and rectangular squares. The play of textures provides sensory depth
in this portrait, with Hays's face giving the portrait a necessary
softness to break the busy confusion of three distinctive patterns.
Compositional stability is achieved by the presence of a large spinning wheel that separates Hays from the constructed background
but echoes the smaller wheel patterns in the coverlets. In many of
Ulmann's portraits of Appalachian women, their softly screened
faces allow exquisite contrast to sharp, geometric designs.[62] In them
we see the aesthetic mix Clarence White impressed upon his students—a combination of textures and pointed details that would
yield both harmony and multi-dimensionality.[63]

A similar study of Mrs. C.O. Wood, of Dalton, Georgia (fig.
11), reveals Ulmann's attraction to geometrical design. In this image
the continuity of patterns is obvious in the backdrop composed
of large diamonds on a quilt set against Wood's diamond-patterned
dress. The photograph is a study of angles. The table edge provides perceptual depth, running from the lower right plane of

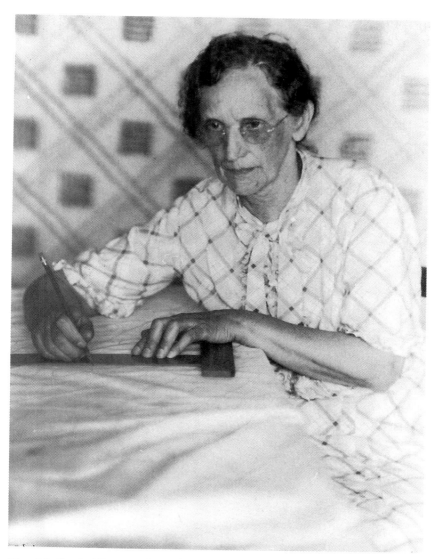

Fig. 11. Doris Ulmann. *Mrs. C. O. Wood. Dalton, Georgia, ca. 1933-34.* Audio-Visual Archives, Special Collections and Archives, University of Kentucky Libraries.

the photo to the back center. Cut perpendicular to this angle is Wood's left arm, which continues into the T-square, forming a secure horizontal line in the image. Beyond the artistic use of geometrics, Ulmann insists in the image that here is craft at its best, not accomplished by chance but by careful calculations (much like her own work). Although Wood looks away from her task and from the camera, intensity shapes her face. Her meticulously manicured nails and smooth fingers hold the T-square securely on the table, where she draws a pattern onto the cloth. Among the other tools the subject employs to accomplish her goal is a new pencil with a sharp point and a glistening metal band, deliberately arranged to articulate Ulmann's message about this craftswoman extraordinaire. White's influence carried through to Ulmann's survey of southern highlander handicrafts. His vision, as well as Stieglitz's, may also be detected in those images where Ulmann remained the traditional pictorialist, especially where she manipulated light to its fullest advantage. In one work of art, she pinpointed a lone hand on a weaving machine, the threads illumined so they appear as fine as angel hair. The cross strands, barely focused, look even more fragile and delicate. As a photographic artist, Ulmann clung to pictorialist principles on into the 1930s, while her contemporaries left behind the soft focus and glimmer.[64]

In Ulmann's attempts to preserve what she believed was vanishing, she occasionally created idyllic scenes out of imagination and mountain lore. She posed her subjects and put props in their hands. She saw nothing wrong with trying to create a certain look, and if a simple tilt of her sitter's head achieved it, Ulmann asked that it be done. Niles recalled that Ulmann wanted to pose "everything she photographed" and that her frequent instruction was "Pose but don't let it seem posed." Just as she had done with her city friends and literary acquaintances, Ulmann attempted to shape her images of Appalachian men and women. She had no intention of falsifying their existence, but she sought to portray the ideal world of a pastoral rural experience. She had people wear costumes they would not normally have worn. According to Niles, young women "would bring down spinning wheels and portions of looms and cards and other things, and show us how their ancestors carried on, and we would photograph them in their granny's old linsey-woolsey dresses." Gene Thornton, in a 1975

Fig. 12. Doris Ulmann. *Wilma Creech. Pine Mountain, Kentucky, 1934.* Used with special permission from Berea College and the Doris Ulmann Foundation.

New York Times piece, noted that "in 1933 . . . Ulmann was still dressing her Appalachian hillbillies in grandmother's dresses and posing them with spinning wheels that they no longer knew how to use."[65]

The photographer's posings were not designed to reflect realities of daily living in the mountains, but rather were set up to highlight an appreciation of the rural past that many settlement school students had willingly embraced. In Eaton's text on the Pine Mountain Settlement School, he points out that "several of the grandchildren and other members of the Creech family dressed in old costumes and were photographed at work as in pioneer days." These participants, in what has become Ulmann's most infamous series (completed the last week of May 1934), held deeply personal interest in dramatizing the history of the Pine Mountain institution—their grandfather, William Creech, had donated 136 acres in 1913 for the school to be built. In one image (fig. 12) granddaughter Wilma Creech stands next to a high wheel, which at one time in the mountains was the most commonly used model for spinning wool. Outfitted in a dress and bonnet that belonged to an earlier generation of Creech women, the young barefooted woman peers out into the distance with a deliberate and focused look, neither sheepish nor downcast. Her grip on the wheel spoke and her graceful posture (the ill-fitting dress notwithstanding) deliver a message of pride and confidence. The central place she occupies in the frame, accentuated by her well-lit head against the dark doorway, makes her portrait image stunning. Although Ulmann's purpose was to feature the spinning wheel as part of Eaton's handicrafts story, Wilma Creech's sidelong gaze, her penetrating eyes, and her easy composure dominate the image. Situated in time in her own family's history and culture, she nevertheless defies time in a documentary photograph suggesting that she may regularly operate the wheel, when in fact she was a medical school student home for the summer at the time Ulmann created this image.[66]

Of the many craft locations Ulmann surveyed, two places in particular captured her imagination and made her a believer in traditional arts education—Berea College in Berea, Kentucky, and the John C. Campbell Folk School in Brasstown, North Carolina. The latter, established by Olive Dame Campbell in 1925, was modeled on several types of folk schools Campbell had visited and studied in Scandinavia in the early 1920s. Her European venture followed the death of her husband, who had been a Russell Sage Foundation field officer in the Appalachian mountains from 1908 until 1919. After his death Olive Dame Campbell completed and published his study of the area, *The Southern Highlander and His Homeland* (1921). Convinced that the ideals of the Danish rural projects she had witnessed abroad could be transplanted in the United States, she returned to Appalachia to begin a school "to enrich the whole content of rural life" by building and sustaining community spirit. With financial support from the Brasstown community for her project, Campbell started the school under the creed "I Sing Behind the Plow," a line she had taken from a Danish folk song. Focusing heavily on cultural education and rural cooperative experiments, the school became one of the producing centers in the Southern Highland Handicraft Guild. Ulmann first visited the Campbell Folk School in July 1933. Within a week she wrote, "The more I see of Mrs. Campbell and her work, the more interested I become, and I admire her and her achievements tremendously." When Ulmann returned the following summer to continue photographing in the region, she considered Brasstown the highlight, noting that "in all our wanderings, we have not found anything or any place that can compare to Mrs. Campbell's." Ulmann grew so attached to the institution that she made plans to help fund a "boys' building" and a "little hotel" in Brasstown.[67]

Located two hundred miles north of the Campbell Folk School was Berea College, a unique institution where students helped keep alive the arts of their Kentucky ancestors. The school charged no tuition; it provided education to poorer students who worked to keep the college facilities operational. One of Berea College's goals was to uphold "the ideal of simplicity" while "contributing to the spiritual and material welfare of the mountain region of the South, affording to young people of character and promise a thorough Christian education." The college president, William J. Hutchins, had made Ulmann's acquaintance initially through a display of her photographs at the 1930 Mountain Workers' Conference in Knoxville. In reply to a note of appreciation he sent to her, Ulmann wrote, "Your words have made me feel that I have perhaps succeeded in expressing a little of the great and deep humanity of these fine and sturdy mountain people. It helps to know that one's work has been of some value." Hutchins hoped to take advantage of Ulmann's work for his own institution's cause, and his persistence eventually paid off. In the fall of 1933, as a result of the cooperation between President Hutchins and John Jacob Niles, an Ulmann photography exhibit went up on the walls at Berea College. Niles appeared to be quite devoted to what he termed the "Berea movement," which was, in his opinion, "constructive education." At one point Niles recommended that Hutchins try to make some money for the college from an Ulmann photography exhibit—"Perhaps you can do this later at some other city where the rich could be caught and impressed with your needs. . . . Don't [sic] mind about paying me, or Miss Ulmann, we carry on somehow. . . . My motto is 'OTHERS.' . . ." Perhaps it was, until he found out that someone had arranged for *him* to sing at a benefit concert for Berea College in New York City. He abandoned the altruistic tone and expressed displeasure when he learned about the proposed benefit, and the concert was immediately canceled. In the weeks following, Niles's communications centered around Ulmann's possible contributions to the college, including a set of her prints for a permanent exhibition there, an idea she had "long considered." Following Niles's suggestion that the college might like to have photographs of "old men . . . young ones . . . children . . . ancient females . . . musicians," Ulmann selected "enough pictures to cover the Mountain subject" sending them ahead to the college so they would be on display when she arrived during the fall semester of 1933.[68]

Niles and Ulmann made appearances at Berea the last week in October. Their visit was a brief one, allowing Ulmann only a small glimpse of Berea College. But she described her trip as "a very beautiful experience." In a letter to Hutchins and his wife, she said, "It is with impatience that I am looking forward to the time in the Spring when I hope to make pictures of the activity and interesting people at your college. There are many things which I saw which ought to be recorded in the best possible way." Ulmann had little time to take photographs while at Berea in October, but she did manage to take formal portraits of President and Mrs. Hutchins, then cordially offered to make any number of prints the couple might want "to use for publication or any other purpose." Limited to these few portraits, Ulmann grew anxious to return to the college in order "to do some real work." Her admission that photographing the institution's chief official did not thoroughly satisfy her indicates how far she had gone beyond cataloging famous faces. By 1933 she had pushed her work in a decidedly more challenging direction.[69]

Moved by Berea College's mission, Ulmann wrote to the Hutchins that the college had made "a deep and delightful impression" on her—"My thoughts have been busy with your very remarkable and effective institution. It is a blessing to know of a place in the world where everybody is giving out of the fullness of his heart without ever thinking of a spiritual or material return." Hutchins, however, could not ignore the possible financial returns of a wealthy New Yorker's blessings. He and Niles discussed such matters in their correspondence. Ulmann would have been appalled to learn that Niles mentioned their personal finances to Hutchins, and since she had no intention of seeking payment for their work there, she promptly returned a twenty-five-dollar check that Hutchins sent to Niles after his morning assembly performance at the college. Ulmann willingly assumed any expenses Berea College incurred regarding her portraits. She paid all costs to ship her photographs, even if the institution had requested them. Returning to New York taken with Berea, Ulmann thought words were inadequate to describe it and insisted instead that the "whole atmosphere" of the college "must be felt."[70]

In the following months, Niles's influence in Ulmann's life expanded. He often accompanied her to various kinds of winter season entertainment in New York City, including the theater and the opera. And because her intestinal problems had worsened, she finally relented and let Niles assist her as she spent long hours in her darkroom developing the faces and views of her southern summer. But she continued to hold the reins tight on her artistry. Niles remembered, "If I assisted her, she stood over me and watched everything I did, telling me forty times how to do the simplest operation, how to pick up the glass plates and how to put them in a frame to dry. She mixed all the chemicals. She did let me make prints occasionally, but I would get weary of the enterprise about midnight, and move on, and she'd continue working until the early hours of the morning."[71]

Niles held a substantial interest in the plates, since he and his musical fieldwork were featured in many of Ulmann's photographs. He found these trips to Appalachia advantageous; he collected tunes and ballads of the highland culture that he hoped to publish someday, and he searched for unusual words or phrases with possible early Saxon roots that he could include in his prospective "Dictionary of Southern Mountain English." As such, Niles's goals complemented Ulmann's objectives. They conducted their respective searches on common ground, with Niles's face and hands recorded on the plates. In one photograph, labeled "Old Timers' Day," Niles sits with a dulcimer in his lap as several elderly gentlemen listen to his performance. In another pose Niles holds a dulcimer while busily taking notes as a young black boy talks to him. The ballad singer attempted to interview as many dulcimer makers and performers as he could locate. The dulcimer, a stringed folk instrument, remained a southern Appalachian specialty, and Niles made sure Ulmann photographed the unusual ones that he found. One journey took Ulmann and Niles to a remote area where a lone cabin bore a sign announcing, "American Folk Song Society Presents the Second American Folk Song Festival." Nothing indicates how many people attended the event, but in Niles's identification of the image he wrote, "Cabin Where Festival Was Held: *Mostly Bunk.*"[72]

Niles figures prominently in Ulmann's Appalachian work, but he also sat for portraits on several occasions in her studio, appearing to be a willing subject who offered a variety of poses for his companion's camera. Three extended series of him, all taken in a studio, indicate that he was a showman, quite the grimacer. In one series Niles is dressed up in evening clothes, complete with top hat, white tie, and cane. In a couple of these portraits he wears a mysterious facial expression, halfway shielded by his coat. His penchant for music is highlighted in another series. Ulmann focused particularly on his hands—playing the piano or writing notes on a musical score. She connected him with his livelihood and his artistry, just as she had done with the weavers and carvers of Appalachia. But Ulmann also used Niles in a number of "couple" studies, where Niles served as a companion figure to young women in Appalachian scenes. In nearly all of those studies, the young woman and Niles sit close to each other, their hands touching, their gazes fixed dreamily on a musical instrument or on a point in the distance. In several of them the fresh-faced innocent girl appears to be guided by the teacher, the role performed by Niles. Among the female protagonists Ulmann chose for these series were Carol Deschamps and Virginia Howard at Brasstown, North Carolina. Ulmann posed Howard in several individual portraits, but she also paired the young woman with Niles in an extended series. As the two sit very close to each other in heavily shadowed light, Niles's hands cover Howard's hands, which rest lightly on a dulcimer. Despite their best intentions to show the seriousness of their forced relationship, their tense facial expressions provide an image more comical than romantic. Both the extensiveness of the "couple" studies and the tones of intimacy conveyed in them suggest that Ulmann may have imagined herself in these idealized scenes with Niles, revealing more of her true feelings on the glass plates than she could exhibit in her own real life.[73]

The Ulmann entourage returned to Appalachia in the spring of 1934. Niles took care of the preparatory details, announcing that "on about the 10th of April the 7th Ulmann Niles Folk Lore Photographic Expidition [sic] will set out. With cars and trailers and cameras and note books." One of the early stops took them to the White House, where they were entertained by Eleanor Roosevelt, who had seen and "enjoyed" the Ulmann photographs on display at the Library of Congress. As they moved farther south into the Carolinas and then into Kentucky, Ulmann corresponded regularly with Eaton, who made contacts with particular settlement schools and individual craftspeople for her. He provided introductions well ahead of time so that Ulmann would not startle potential portrait sitters. In one request Ulmann wrote to Eaton that it would "help a great deal if you could prepare the people for our visit so that they could have things ready for us." Eaton's role from a distance preceded that of another photographers' mentor, Roy Stryker, who paved the way for his Farm Security Administration (FSA) government photographers from 1935 until 1943. Stryker's contact with regional FSA offices, especially in the early years, prepared the locals for an invading photographer who might ask too many probing questions of government aid recipients. In Stryker's case his organization depended upon his scheduling and assignments; his photographers expected him to grease the skids for them. Eaton's preparation for Ulmann amazed her to the extent that she thanked him repeatedly for communicating with her subjects, reiterating to him that it would be "so difficult to work without your introduction."[74]

Eager to please Eaton, Ulmann asked him to send her lists of subjects he wanted to have in his book. Seeing her work as a sort of mission, she desperately wanted to provide publishable illustrations and was willing to accommodate Eaton's wishes, noting that she hoped he would "find all the important and interesting things here." She acquainted herself with individuals and schools before she arrived to see them. She studied maps, attempting to use her time on the road as efficiently as possible. And she judged her productivity according to the willingness of craftspeople to cooperate with her and to demonstrate the steps in their various enterprises. Describing her study of tufting specialist Ethel May Stiles of Ringgold, Georgia, Ulmann wrote, "I worked with her and in my pictures tried to show the *process* of making a tufted spread." Tufting yarn, a coarsely spun cotton of many strands in various colors, was used to decorate bed covers, curtains, and other decorative articles made of heavy cloth. In the 1930s tufters in northern Georgia fed a national market hungry for their products. Some handiwork they sold on the roadsides in Georgia, but most went through a dealer in Dalton who sent the merchandise on to department stores. It was Ulmann's careful attention to processes like tufting that revealed the documentary spirit she harbored and nurtured.[75]

In four weeks at Berea in 1934, Ulmann provided an exhaustive treatment of the college's departments and operations. As soon as she arrived, President Hutchins requested permission to make "all manner of suggestions to you as to pictures of men, women, and things, always with the understanding that you will do exactly as you please, without feeling the remotest constraint from me." Whether Hutchins directed Ulmann's work is unknown, but the photographs she completed in her few weeks at the college must have pleased him. She observed Berea's various departments and provided extensive coverage of its operations. She took photographs of drama students parading in costume and of woodworking students using hammers and saws. Her documentation of the students' sheepshearing process began with an unsheared animal. She followed the complete procedure, recorded each step, and ended her series with a telling photograph—a sack stuffed with wool sitting on the market scale. Berea College, since it made no money from student tuition payments, perpetuated itself through such agricultural endeavors. Ulmann hoped to reveal the unusual nature of the school, along with the contributions made by its hardworking young men and women. She recorded the dairy and creamery operations, again showing various stages in the production process. She took pictures in the candy kitchen, the bakery, and other food service areas. Many of the products made in these kitchens served guests in the college's downtown hotel, Boone Tavern. As with all her Berea work, Ulmann wanted to highlight the "process." She made photographic studies of girls kneading flour, stirring pastry mix, and decorating cakes. Several still-life photographs show the intricate work done on tiny sugar candies. After her thorough documentation of the college's operations, she more than likely was unable to fill President Hutchins's self-described "*impossible* order" of choosing one photograph that would "incarnate the ideals of Berea."[76]

The Berea images by Ulmann stand closely in line with traditional documentary photography as it came to be measured in the 1930s. The visual records Ulmann made of procedures anticipated the similar subject matter that emerged a few years later in U.S. Government–sponsored photography. WPA photographers' assignments included recording various stages of building projects, photographing audiences at Federal Music Project concerts, and shooting Federal Art Project shows as they went up on gallery walls. Other government employees with cameras recorded living conditions in poverty-stricken areas, then later returned to those regions to evaluate the effectiveness of New Deal aid programs by examining how the money was spent. Certainly the candid expressions Ulmann captured on some Berea College students' faces bear out the documentary nature of her work there and show how far she had reached beyond conventional portraiture. In one series Ulmann framed a classroom scene where the dean of the college was surrounded by students, all men. The boys attempted, with obvious difficulty, to get rid of their smiles, but the serious business of education was lost on them. They managed to look contemplative in only one photograph in the series. This series, among others at Berea, sets Ulmann's survey apart from a similar project carried out by photographer Frances Benjamin Johnston at Virginia's Hampton Institute in 1899. The *Hampton Album* shows the seriousness of purpose of all the students and the graduates pictured. The Victorian expectations of balance and conformity dominate, resulting in highly stylized scenes in the classroom and laboratory settings. James Guimond notes of the students in Johnston's photographs, "None of them is crowded, inattentive, bored, or distracted by anything—least of all the photographer and her camera." Ulmann's collective portrait of Berea College students and their work shows the extent to which she developed a documentary style that could allow spontaneity and ease in front of the lens.[77]

But Ulmann's need to control her artistic vision reveals the limits of her willingness to operate as a pure documentarian. Despite Eaton's initial direction, Ulmann did impose her aesthetic tastes on him and the handicraft book project. When he discussed using other photographers' illustrations alongside Ulmann's work, she forcefully replied that he should either use her images exclusively or omit them completely. In a heated letter, she wrote, "I would not care to have some of my pictures used and some done by somebody else. After all the rather intensive work of last year and this year I do think that you will find enough subjects from which to choose." Ulmann, on advice from culture workers in the region, had begun making decisions about which states, towns, and craft workers to visit, despite Eaton's appeals for her to travel elsewhere. She avoided a few schools that did not exhibit the "real mountain spirit." Her conflict with Eaton revealed her steadfast belief in achieving depth over breadth. The portrait photographer who spent hours with each woodcarver wanted to cover those individuals thoroughly rather than snapping a plate of every whittler within a one-hundred-mile radius. Ulmann's vision had been molded by her training as an art photographer, which she never relinquished. Her disputes with her colleague in 1934 bear similar characteristics to the clashes a few years later between FSA mentor Roy Stryker and his "art" photographers, especially Walker Evans and Dorothea Lange. Stryker struggled with Walker Evans, who tended to disappear on the road, failing to communicate with the Washington office for days at a time. Lange, whom Stryker once labeled his "least cooperative" photographer, refused to send her film back to the Washington lab, not trusting the technicians to develop it properly. And although Lewis Hine sought employment on the FSA team, Stryker would not hire him because he feared the well-established photographer was too inflexible to produce the kind of documentary work the agency needed. This generation of artists who worked with cameras declared their independence by refusing to have their individual visions shaped by organizational demands.[78]

Since Ulmann was funding the handicrafts tour almost entirely, she could afford to preserve her artistic integrity, and Eaton could not afford to lose her. The determination Ulmann exerted here underscores her status as a financially independent woman whose money gave her a freedom that other photographers in the Depression years did not enjoy. Being wealthy also allowed her to sidestep some of the challenges other female documentarians faced while on the road. As no organization's employee, Ulmann could choose the causes and the specific subjects she wished to photograph. She traveled comfortably in a luxury sedan with a driver and set her own shooting schedules. Any restrictions Ulmann faced were self-imposed, based on her dependency upon companionship and, more specifically, her need to forge relationships with writers, artists, and educators. Her emotional needs coupled with her ability to pay assistants kept her from having to travel or work alone or meet the challenges of attempting to do so, unlike several of her documentary counterparts in the 1930s. WPA photographer Berenice Abbott suffered the pranks of construction workers on New York City streets; FSA photographer Marion Post not only aroused suspicion but faced hostile inquiries in several states arising from her status as a woman alone on the road. Ulmann managed to avoid such treatment, in part because she did not fit the "modern career girl" stereotype or face its challenges, as did Abbott, Post, and Margaret Bourke-White. Ulmann's primary conflicts remained more personal than public, enlarged by her willingness to reveal her physical and psychological vulnerabilities.

Shaped by her experiences in the summer of 1934, Ulmann's truest devotion was to Olive Dame Campbell and the Campbell Folk School, where she spent the final weeks of her last Appalachian trip. Earlier she had told Eaton, "How I do love Mrs.

Campbell and all her people!" By the end of the summer Ulmann
had come to depend on Campbell's advice and direction for her
photographic studies. When Ulmann and Eaton suffered their
breach over the book illustrations, Campbell mediated the alter-
cation. Although Eaton was directing the book project from New
York, it was Campbell who advised Ulmann on which craft centers to omit from her travel schedule, suggesting that some places
"would probably not have any new or any exceptionally interesting material." As Ulmann gained more confidence with each successive visit to new schools, she increasingly questioned Eaton's
lists, replying that she felt "doubtful about Hot Springs" and did
not "know anything about Higgins." Some of her doubts about
traveling to particular places were driven by her own fatigue and
deteriorating physical condition. She argued that she wanted to
cover individual subjects thoroughly and in one place, rather than
fanning out to come away with repetitious images of the same
crafts in many places. She claimed it would be a poor use of her
time to drive to South Carolina, Alabama, and Virginia to photograph the same processes and products she had surveyed in Kentucky, North Carolina, Tennessee, and Georgia. Ulmann used the
power of Campbell's clout and expertise in the realm of southern
handicrafts to back her up, implying that Campbell was a completely objective observer who had sided with her in the dispute.[79]
In one of the last letters she wrote, Ulmann said, "It is always
hard to leave Mrs. Campbell,—the more I see her the more deeply
I appreciate her. I think that she is the wisest and finest woman
whom I have ever met. . . . Many times I wonder whether all these
young mountain girls and boys will ever appreciate the rare opportunity which they are having in being so greatly influenced in their
lives by a person like Mrs. Campbell. There is not a school or a
university where they could have this beautiful, personal attention."[80]
    In August, a few days after she expressed these poignant sen-
timents, Ulmann's physical limitations, the exhausting schedule,
and the summer heat overtook her. She was near Asheville, North
Carolina, making photographs for the Farmers' Cooperative Union
of Asheville.[81] Following the session at Top of Turkey in Buncombe
County, Niles and the rest of the Ulmann entourage rushed the
photographer to Pennsylvania for medical treatment, then on to
New York City.

In the days before she died Ulmann composed a will, generously leaving the bulk of her life's work and wealth to the John C. Campbell Folk School, Berea College, and John Jacob Niles. She was quite explicit in her wish that no one person profit from her photography. She indicated that "in no event shall any one except a public charitable, scientific, art or educational institution or other public institution or the public in general acquire any beneficial interest in the money or property held hereunder." She also requested that any printing, exhibiting, or distribution of her work be done solely to "further the development of photography as an art and further the understanding of American life, especially of the South." The South's multitude of subcultures had drawn Ulmann to the region, their complexities enticing her out of her urban environment. William Clift argues that Ulmann's photographs of Appalachian residents and Gullah African Americans prove that she was not interested in "social commentary" since they recorded no poverty, misery, or racial conflict. Perhaps that is so, but certainly she wished her photographs to provide cultural commentary. Focusing on ethnic distinctiveness, personal talents, regional rituals, religious devotion, and rich local traditions, Ulmann made a significant contribution to an "understanding of American life" that she hoped would continue after her death.[82]

Reared and trained as a turn-of-the-century educator at the Ethical Culture School, the rich urban dweller combined her vision and her camera work in order to present a composite picture of a fleeting rural America. She attempted to record specifically defined groups that did not yet bear the marks of an ever-encroaching mass culture destined to blur the lines of ethnic and geographical distinctiveness. She sought to arrest time through the vehicle of compositional stability, creating portraits of individuals situated in natural stillness, mental solitude, or fruitful work. Ulmann hoped to illuminate the nobility and dignity of the men and women she encountered, an idea carried on through the decade of the 1930s by better-known documentary photographers such as Walker Evans and Dorothea Lange. Like her successors, Ulmann found herself working for an organization dedicated to improving living conditions of lower-class groups in the United States, those who would come to be celebrated as "the people" and would receive support from federal government programs in the second half of the decade. But Ulmann's vision transcended that of the U.S. Government employees with cameras who succeeded her. She delivered the prints back to the schools and the subjects she had photographed; her fellow documentarians did not. Knowing that three thousand undeveloped negatives remained, Ulmann made provisions in her August 1934 will to have distributed to Berea College and to the Campbell Folk School the prints concerning their respective activities. She further instructed "that each person who was the subject of any photograph taken in the Southern Mountains . . . be given a copy of the print." It is this specific request that made Ulmann more than an uninvited observer or a voyeur. The ultimate students of the highlanders' lives and work became the mountaineers themselves. Ulmann, who referred to her portrait sitters as "patients," always kept in mind that her subjects themselves were their own most vital observers, whether they were physicians at Columbia University, writers in New York City, or craftspeople in the Appalachian mountains.[83]

Although Ulmann produced an impressive body of photographic work for Allen Eaton, she never saw full fruition of the Appalachian project. *Handicrafts of the Southern Highlands* appeared in 1937, three years after she died. Eaton later said that Ulmann "didn't realize that she had made the most definitive collection of rural characters, certainly in the field of handicrafts, that's been done any place in the world." Loyal Jones, who directed the Appalachian Center at Berea College for a number of years, noted that Ulmann easily bridged the gap between "her world and theirs," offering up thousands of images bearing "no condescension for the harshness of her subjects' lives, but rather an appreciation of mountain people for the uniqueness of their faces." But Jones also pointed out that "those who look at Southern Appalachia usually find whatever they are conditioned to find." Ulmann had been educated to seek beauty in each individual, had been convinced of the bucolic pleasures of rural existence, and had embraced the traditional craft movement as an antidote to mass production. Her tastes and methods kept her from recording some of the harsher aspects of mountain life, particularly in Kentucky's coal mining counties. People in western North Carolina in the 1930s much more easily recognized themselves in the snapshots taken by their neighbor, the "Picture Man" Paul Buchanan, than in Ulmann's portraits or the photographs made a few years later by Bayard Wootten for Muriel Sheppard's text *Cabins in the Laurel* (1935).[84]

Using her finely tuned observation skills as a student of psychology and photography, Ulmann first studied individuals and then created portraits, scores of them. Her subjects were appropriate for documentary work, but her equipment was not. Since she never used a Rolleiflex, a Speed Graphic, or a roll of 35mm film, she never snapped a quick picture. Although she captured a few smiles on the faces of students, her subjects were rarely caught off guard in the split second of a shutter snap. Ulmann's subjects' faces were recorded over minutes and hours on heavy glass plates, most of which were destroyed after her death. President Hutchins of Berea College declared, speaking of Ulmann, "It is fortunate indeed that before her passing she was able to capture and hold forever before the American public these portraits of types which may pass all too soon from the American scene." In Ulmann's search for types, she found individuals, many of them living on land and in homesteads that their great-grandparents had occupied. She helped to lead the cultural resurrection of rural life in the late 1920s and early 1930s by picturing rural existence in all "its color and picturesqueness and excitement." Yet she never considered one picture's worth. Instead, she believed in the magic a thousand pictures could wield over the power of mere words. As a result she helped set the stage for documentary to become the principal vehicle of photographic expression in the 1930s.[85]

2

Portraitist as Documentarian

DOROTHEA LANGE'S DEPICTION OF AMERICAN INDIVIDUALISM

> The secret places of the heart
> are the real mainsprings of one's action.
> —Dorothea Lange

DOROTHEA LANGE SPENT MORE THAN SIXTY YEARS living what she called a "visual life." She admitted that such an existence was "an enormous undertaking, practically unattainable," yet she claimed to have pursued it from a very early age. Lange's most vivid childhood memories revolved around the skills of observation she cultivated and finely tuned on New York City streets in the first decades of the twentieth century. Her experiences there as a young girl affected her approach to the world and to the lives she studied throughout her career as a photographer. Like her uptown New York contemporary Doris Ulmann, Lange chose to focus her camera on individuals rather than groups or their external surroundings. The primary carrier of Lange's personal messages was the human face, as it was for Ulmann. But Ulmann's approach to a hobby assumed midway through her life stands in significant contrast to Lange's full-fledged visual life. Lange claimed to have made no conscious decision to become a photographer, but she felt that she had always been, at some stage in the development of her craft, either "getting to be a photographer, or wanting to be a photographer, or beginning to be a photographer."[1] Things visual piqued an insatiable curiosity in Lange, one to which she devoted a lifetime of care and attention.

Although she found it difficult to explain, Lange felt that she had been led to photography by instinct. After spending many years in her chosen career, she learned that her great-uncles who had come to America from southern Germany had all been lithographers. The sons had followed their fathers in the trade, making visual images with grease, water, and ink. Lange believed that her love for pictures came from those ancestors, whom she never knew. She once said, "I've sometimes wondered whether these things that we do on our own, the directions that we take and the choices of work, are not determined by something in the blood. . . . I think that there is some kind of memory that the blood carries. . . . There are certain drives that we have." Despite the instinctive impulses that her forebears may have passed along, Lange grew resentful of her Teutonic heritage at a young age, consciously rejecting a trait that she believed to be inherent in all Germans— an unwavering respect for authority. Exhibited by her mother and other relatives, such reverence bothered Lange, who complained that Germans were constantly concerned about "what other people would think of them."[2] Throughout her career, Lange enjoyed a personal independence that she had nurtured as a young girl. A profession in photography suited her, since it freed her from the control of supervisors and superiors and enabled her to preserve "a very great instinct for freedom." Lange, born in 1895, matured in an age when women were challenging the boundaries that had been set for them earlier in the nineteenth century. Many found that professional lives in social work, political organizations, or radical labor movements offered alternatives to the constricted ground their mothers and grandmothers had occupied. Living in New York City as a young adult, Lange was poised to view some of the outlets that provided independence to a new generation of women in the new century.[3]

The fierce sense of independence that buttressed Lange's disdain for authority also set her apart from others. As a child Lange developed a serious outlook toward the world, and she coupled it with a strong will. Both developed out of a yearning for self-protection. At the onset of Lange's pubescence, her father abandoned the family. The incident was terrible enough that Lange uttered not a word about it, even to two husbands and her children, for the rest of her life. In the wake of her father's disappearance, Lange's mother, forced to become the sole breadwinner, sought employment across the river from their home in Hoboken, New Jersey. She took a job in a library on New York's Lower East Side and enrolled young Dorothea in a nearby public school. Lange saw firsthand the steady stream of immigrants who packed themselves into the bustling Jewish neighborhoods. And she competed daily with ambitious children who strove diligently to succeed in their new homeland. Lange was a German Protestant, the only non-Jew in a school of three thousand Jewish students. It was here that she learned what it was like to be in the minority, but as she later pointed out, "I was a minority group of one." A loner, Lange shielded herself from her classmates' world, closely observing it but never becoming part of it. After school she spent the late afternoons staring out the windows of the library where her mother worked, studying the frenetic activity of "the sweatshop, push-cart, solid Jewish, honeycomb tenement district." The window panes separating Lange from the subjects she viewed served much the same purpose as the camera lenses she used in later years. Lange hid behind them, appeared unobtrusive, and studied tirelessly with her eyes.[4]

Her desire to become part of the environment without drawing attention to herself developed as a response to her physical disability. A victim of polio as a child, Lange walked with a limp, and at her mother's constant urging she attempted to conceal it and appear normal. She tried to wrap herself in a "cloak of invisibility," hoping to avoid notice by those around her. Lange would later attempt to do the same when she began photographing, so that her subjects would not be distracted by her presence. In downplaying, even hiding her appearance, Lange challenged the cultural prescriptions of the teens and twenties, when advertisements directed at female appearance and health multiplied by the hundreds. These ads were frequently accompanied by visual images of the physical ideal, depicted as a rosy-cheeked and athletic young woman, a bicycler in the 1910s and a dancer in the 1920s.

Ever conscious of her disability, Lange perfected her method of invisibility as a teenager, hiding from the inherent dangers in a poor, urban environment. Walks home through the threatening Bowery section taught her to mask her emotions so she would attract no attention and therefore no trouble. The effort helped her to dispel any fears, further creating a shield between her and the strangers she encountered. She remembered consciously altering her facial expressions so that "eyes [would] go off me." In essence her manipulations of face gave her power. Using the face to effect certain reactions became a common tactic of Lange's. She began with her own face as a teenager; as a photographer she worked with other faces to achieve the looks, and thus the images, that she desired. In drawing out certain expressions on her subjects' faces, Lange attempted to control the reactions of those who viewed her photographs. It seems that she deliberately brought the human face into focus so that the rest of the body would be overshadowed by the complexities found in eyes, mouths, and other compelling features.[5]

Though Lange attempted to keep others from watching her as a teenager, she continued to watch them. Her short stint in an all-female high school in uptown Manhattan further encouraged the solitary individual in Lange to grow. She had few close friends there and chose to spend more and more of her time outside the institutional walls. Lange recalled wandering up and down the streets of the city, watching the people, looking at pictures, sharpening her observational skills. Because she felt that she was developing her artistic sensibilities, Lange defended her frequent absences from school, claiming that such activity was *not* "unproductive truancy." She managed to graduate from Wadleigh High School and at her mother's insistence pursued a teaching career by enrolling at the New York Training School for Teachers. Her senses dulled by the educational routine, Lange decided to indulge her visual curiosity. A visit to Arnold Genthe's Fifth Avenue studio proved fruitful. He gave her a job and a camera, but more importantly, he revealed to her his tremendous love for the human individual. Lange learned little about the technique of photography but much about the art of engaging people before the camera.[6]

Genthe possessed a sense of beauty that his young assistant came to appreciate, even though she was already composing a

definition of beauty that would set her apart from her male mentors. Lange observed Genthe's interactions with his subjects, most
of whom were women, and later described him as "a man who
really loved women and understood them." Genthe had one particular client, the modern dancer Isadora Duncan, who fascinated
Lange. Duncan's departure from traditional movements and stric-
tures of classical ballet secured a place for her in dance history.
But so, too, did her challenge to conventional social expectations
regarding the female body. Duncan's veil-like costumes, some similar
to Greek chitons, revealed bare legs and bare feet. Off stage, Duncan
dressed in classical Greek-style clothing that draped loosely over
her body. Lange observed that Duncan was "rather sloppy-look-
ing, rather fat, with very heavy upper legs" but that the dancer's
movement in space was beautiful. In a penetrating examination
of Lange's obsession with the body as a defining characteristic of
identity, Sally Stein argues that Lange's own lameness impelled
her to express difference "in terms of the well and ailing body as
it intersected and could be aligned with more conventional forms
of difference like class, race, and gender." Lange's scrutiny of the
body and her unrelenting attention to its component parts became
an underlying theme in her work. And it was her own physical
difference, her alienation from the acceptable ideal of human health
and perfection, that informed her artistic vision very early in her
professional apprenticeship. That element joined with the other
overriding impulse from her youth, the desire for personal freedom, as Lange continued her informal study of photography.[7]

Alongside her aesthetic ponderings, Lange gained expertise
in the photographers' trade through various part-time associations.
In one job, for Armenian portrait photographer Aram Kazanjian,
she solicited business by telephone. For another employer, Mrs.
A. Spencer-Beatty, Lange became a camera operator. She was responsible for exposing the plates; others in the studio took care of
spotting, printing, and mounting the photographs.[8] Finally, in 1917,
armed with plenty of applicable knowledge, Lange turned her back
on assorted part-time jobs and enrolled in a photography seminar
given by Clarence H. White at Columbia. Upon Arnold Genthe's
advice, she sought to absorb everything White could teach her about
the art. White's influence had already touched a generation of young
photographers. Doris Ulmann Jaeger, among others, had listened

to his meditations on style and the direction of American photography. Lange found the teacher's methods unusual, but she learned a great deal from him. She appreciated his subtle way of getting his message across without criticizing his students' work. She remembered that he "always saw the print in relation to the person." Though she clung to his words, Lange never completed White's seminar assignments. She drew more from his personal and professional example than she learned in his classroom. Lange considered White an outstanding teacher despite being "vague, indefinite, [and] non-didactic."[9]

Perhaps she recognized in White one who shared her love for the visual life. Rarely had Lange found anyone with whom she could claim a common interest. White was "a man who lived a kind of unconscious, instinctive, photographic life." Her respect for his pursuit of photographic excellence led her to remark, "I never saw any photo of his that had a taint of vulgarity." A few months into White's seminar, Lange knew she wanted to devote her life to pictures. But as she made quite clear in later years to interviewer Richard Doud, "[i]t was more a sense of personal commitment . . . [than] a conscious career." She was convinced that her mind "had made itself up" about photography. Lange's reflective look and personal narration of her path to photography places her squarely within the tradition in which she grew up, where careers and causes "found" women and swept them away. We do not know whether Lange read published autobiographies of the nation's more accomplished women in the early twentieth century, but if their public personae or written testimonies impressed her, then she followed their paths and told her own story as they had articulated theirs. Literature scholar Carolyn Heilbrun, in *Writing a Woman's Life*, describes a study conducted by Jill Conway on independent, successful, cause-driven American women in the early twentieth century. Each woman in her autobiography downplayed her career choice, following the only acceptable "script" for the female life which "insisted that work discover and pursue her, like the conventional romantic lover." For example, the controversial subject of the Standard Oil Company just "happened to be there" for Ida Tarbell. Beyond their lives designed for public consumption, though, these women's personal letters revealed aggressive, authoritative, and deliberative individuals. Even though Lange never admitted having consciously chosen photography as a career, she carefully molded her artistic style and staunchly defended her independence as an artist; her attitudes at times alienated some of her friends and family and at least one employer.[10]

In 1918 Lange left New York City on a journey to see the world. She and her traveling companion, Florence Ahlstrom, made their way to New Orleans by boat and from there to the California coast by train. On her first day in San Francisco, Lange was robbed of everything but the change in her pocket. Forced to find employment, she immediately landed a job as a photo finisher in a store on Market Street. What had first appeared to be an unlucky turn of events proved fortuitous; across the counter Lange met a number of well-known photographers, including Imogen Cunningham, Roi Partridge, and photojournalist Consuela Kanaga. Photography curator Joyce Minick later remarked that Lange's employment there was "one of the first instances in her career when time and events were clearly in her favor." Not long after she took the Market Street job, a friend of a friend supplied three thousand dollars so that Lange could set up her own studio, which she opened on Sutter Street.[11] Achieving instant success, she marked her accomplishment as an up-and-coming photographer with a symbolic gesture—she dropped her father's family name, Nutzhorn, and took her mother's maiden name, Lange. This expression of independence, spurred by financial success, separated her from a ghostlike figure whose memory she had battled for over a decade. That she abandoned the patrilineal name in favor of the matrilineal name instead of choosing a completely new name suggested that her true freedom lay in separating from the principal male figure in her life and that the connection to her mother, however remote and filled with resentment, remained intact. These gender identifications would press Lange later when she married and had two sons, embracing burdens that would manifest themselves in the commissioned family portraits she made for her patrons.

Artistic and financial success allowed Lange to challenge her painful childhood status as a loner, as she joined a circle of artists who also happened to be friends. Roger Sturtevant, Lange's printing assistant, remembered the close-knit group that gathered daily in her studio. He described a gag picture in which photographers Edward Weston and Anne Brigman, posing as parents, held Lange's camera (wrapped in a black photographer's cloth) in their arms. Imogen Cunningham and Lange, as the other children, rounded out the "family" portrait. Such evidence reveals not only Lange's pleasure in her work but also her comfort among peers. Ironically, the mock family portrait also showed Lange's preference for sharing intimacy with friends rather than with family members; that issue created a taxing struggle for her for most of her life. Lange enjoyed having an open door to her studio, where people gathered day and night. Each afternoon she served tea in an effort to attract patrons, but, she recalled, "by five o'clock that place was full of all kinds of people . . . [as] everybody brought everybody." Like other famous gathering places, such as those on Paris's Left Bank or in New York's Algonquin Hotel, the Lange studio radiated a club atmosphere. Its members worked, however, as much as they talked about their work. Sturtevant, who at the time was a high school student with an interest in photography, was impressed by his friends' work habits and believed they all prospered because they "only believed in working" and because they were reluctant to let politics dominate their craft or popular art trends guide their hands.[12]

Combining her talent behind the camera with her subtle promotional technique in the salon, Lange developed a large and devoted clientele rather quickly. Her reputation as a portraitist grew as news of her craftsmanship spread; Lange depended upon her wealthy patrons to tell their friends about her photography. Many of them had had their first glimpses of her work in the Hill Tollerton Print Room that adjoined Lange's studio. The gallery attracted loyal customers who could afford original etchings and prints. Lange was fortunate to draw support from a such a group, comprised primarily of families willing to spend unlimited sums of money on portraits. Lange described her rich San Francisco patrons as "large families who knew each other, and had a very strong community sense . . . , [with] children and education and buildings and pictures, music, [and] philanthropy . . . [tying] their private personal life and their public life together." By an ironic twist, nearly all of Lange's customers were prominent members of San Francisco's Jewish community. As a young child in New York,

Lange had stood outside her Jewish classmates' world as a mere observer of what she could not understand. As a highly sought-after professional, she entered the mansions, strolled about the manicured lawns, and befriended the children of her loyal Jewish patrons.[13]

Employing her invisibility tactics, Lange sought to put her subjects at ease by blending into the surroundings. Richard Conrat, one of Lange's assistants, described her unique approach in trying to capture "the essential person without managing to have her presence *felt* in the finished photograph."[14] Rarely did an individual appear anxious or tense before Lange's camera, because she preferred natural poses over stiff, formal ones. Her method of arranging her subjects provides a striking contrast to that of her contemporary Doris Ulmann, whose bustlike figures of writers, doctors, and academics harked back to the Roman style of preserving great men. Lange's subjects appear carefree, their bodies resting comfortably across chairs or their arms draped over fellow family members. This ease of posture indicated Lange's personal desire for effortless movement and brought her into the frame alongside her subjects, despite her attempts at invisibility. Perhaps, too, Lange was able to cross an ethnic and cultural barrier that had seemed too great for her years before, when she sat at the library window watching her classmates in the streets of New York City.

Beyond the soft lighting and simple backgrounds arranged in her portrait studio in 1919, Lange discovered early in her career the benefits of working outdoors. In addition to the advantages of employing natural light, she realized it was vital to picture individuals within their own environments. Lange would continue this practice on a larger scale during the Great Depression, when she expressed a profound desire to picture "a man as he stood in his world." Doris Ulmann ventured outside her portrait studio less frequently in the early years of her career. Once she began her fieldwork in the late twenties, dealing with a myriad of cultures in the rural northeast and the South, she recreated her studio everywhere she went. Lange exercised a greater desire to enter and be welcomed into the various personal worlds of her subjects, rather than creating a world around them or for them, as Ulmann did. In the numerous photographs Lange took of the Katten children, most have a background of sunshine and wide open spaces. The subjects are not posed, nor have they combed, curled, or otherwise primped for the occasion.[15]

Lange's devotion to photographing families in the 1920s was juxtaposed with her own complex set of family relationships and responsibilities during that decade. In 1920 she married the painter Maynard Dixon, a friend of Roi Partridge and a frequent guest at the daily social gatherings in her studio. Dixon's carefree temperament and calculated eccentricities were well known in the San Francisco bohemian community. And although Lange was twenty years younger than her husband, she provided the necessary complement to a man who had an infamous reputation for irresponsible behavior. Lange contributed the serious, more introspective side of the liaison. Lange and Dixon had two sons together in the twenties. They often traveled to the Southwest so that Dixon could revive his artistic self and dwell on the themes that interested him most—the haunting, spiritual elements characteristic of Native American culture. These trips left Lange limited time to satisfy her own artistic curiosity, since she believed her purpose should be to maintain a comfortable environment for her husband, her stepdaughter, and her two young sons. She later told an interviewer, "The largest part of my energy, and my deepest allegiances, were to Maynard's work and my children." Her struggle with these choices eventually manifested itself in an unhappy marriage and alienation from her children. The situation grew worse as her desire for independence coincided with the onset of the Depression. Lange decided to live apart from Dixon in her photography studio and to board their sons, Daniel and John, elsewhere.[16]

The substantial body of family portraiture that she completed during the years she was married to Dixon bespeaks her longing to remain a separate and independent entity despite her role as wife and mother. In Lange's family portraiture (of other families, not her own), she highlights the individuals within a family. Although they may share a physical connection or a touch, the single individual's role in a photograph transcends the importance of the pair or the group. David Peeler has argued that Lange's "romantic depictions" of family during the years she was nurturing her own family stand apart from her later work. But a closer look at the example that Peeler included in *Hope Among Us Yet* (fig. 13)

Fig. 13. Dorothea Lange. *Mrs. Kahn and Child, 1928.* © Dorothea Lange Collection. The City of Oakland, Oakland Museum of California. Gift of Paul S. Taylor.

is necessary. In *Mrs. Kahn and Child* an infant is being pressed upon, nearly smothered, and definitely obscured by the figure of the mother. The position of the child's arm, an erratic wave indicated by the blurred portion within the frame, suggests self-defense, perhaps even a gasp for breath. Lange created this family scene when she herself was caring for two young sons, one an infant, the other a toddler. Lange had become a mother in a decade when standards of motherhood jumped significantly. Society required that women assume much more personal responsibility for their children's health and cleanliness than they had held before, thus creating heavy psychological burdens in the process. The pressure to rear children properly, to appease a temperamental husband, and to nourish a burgeoning career, was visible in Lange's family portraits. Rarely in Lange's work is an entire family posed in a portrait, because she preferred not to line up the whole clan. She did not believe that the "intense alliance" within the human family could be adequately recorded. And she found it especially difficult to photograph her immediate family members together. Later comments about her own children reveal a separateness from them: she referred to them in an interview as "the little boys" and "the two little boys" rather than "my" boys or "our" children or by name.[17]

Lange focused overwhelmingly on the individual stranger's life, using her images to imply that a single human was much more important than anything that surrounded him or her. Even in the 1930s, when Lange began photographing Depression victims, she concentrated upon the individual man, woman, or child. The human element surpassed mere symbols of financial disaster such as the eroded soil or an empty pocket. The images of Lange's wealthy patrons in the 1920s disclose many of the same characteristics present in her later portraits taken for the federal government. Lange's studio print books reveal hundreds of faces bearing the dignified, almost heroic, look that the photographer captured so well on film. The style Lange developed early in her studio career is clearly expressed in an undated series of sitter Erika Weber. A couple of poses were taken outdoors, though these offer few clues about the location. In one pose, Weber's head is tilted slightly to the right as she observes something in the distance, outside the frame of the photograph. Lange's purposeful omission of such evidence infuses the image with mystery. In another proof Weber wears a scarf tied around her head as she stands with hands on her hips. In such an unlikely pose for a rich urbanite, the subject could as easily have been a farmer's wife surveying the year's crop. But the viewer is not allowed to see the ground Weber occupies. Lange peers up at the subject, making her the principal focus and ignoring the ground on which she stands. This stylistic preference, the photographer's upward gaze, helped Lange achieve a certain "look," one that appeared in her early studio work and showed up again and again throughout the 1930s. Defining this quality as a certain "tonality" rather than a "style," Lange admitted that particular photographs would make her exclaim, "Well, there's a Lange for you."[18]

Although Lange was a commissioned artist who had to keep her customers' satisfaction in mind, she eschewed "pandering to their vanity" and offered what she believed were accurate representations of character and personality. Of her portraiture, Lange said, "I was a professional photographer who had a product that was more honest, more truthful. . . . there was no false front in it. I really and seriously tried, with every person I photographed, to reveal them as closely as I could. . . . No posturing, no dramatics." This approach helped Lange depict the world of the San Francisco wealthy beyond the trappings of wealth, narrowing the focus to those elements common to all human beings. She admitted that in her early years as a portraitist, she encouraged her subjects to wear "odd, simple clothes" so that "the images would be timeless and undated." The goal of achieving timelessness aligns Lange's intentions with those of Doris Ulmann. Both photographers sought to eliminate the present from their images. Ulmann encouraged her sitters to don ankle-length frocks and bonnets to achieve this objective, but Lange preferred clothing that would betray no particular era. In addition, Lange's philosophy for accurately revealing personalities through the medium of photography differed from that of Ulmann, who believed that certain props—pens for writers, costumes for actors, knives for woodcarvers, and other accoutrements of skill or interest—drew out individual characteristics and distinguished one person from another, thus making him or her unique. Whereas Ulmann depended upon physical objects and backdrops that set individuals apart from each other,

Lange cared more about the common experiences of human beings that tied them together. No doubt the reason Lange relied
less upon props than Ulmann did was that she considered them
secondary to the overwhelming marker of experience—the human
face.[19]

Nothing intrigued Lange more than the human face. Within
the body of her work, no other subject received as much attention. She believed the face spoke a "universal language." Of its
power she said, "The same expressions are readable, understand-
able all over the world. It is the only language . . . that is really
universal. It's [*sic*] sad, shades of meaning, it's [*sic*] explosions of
emotion and passion . . . I'll concentrate on just this part of the
human anatomy where a slight twinge of a few muscles . . . runs
the gamut of that person's potential."[20] She remembered well the
poignant, pained expressions in "that noble wasted face" of President Woodrow Wilson as he desperately tried to garner support
for the League of Nations outside the St. Francis Hotel in San
Francisco. From her earliest years as a portrait photographer, Lange
viewed sanguine, unhappy, angry, hopeful, and contemplative faces,
capturing on film the wide spectrum of expressions arising from
the human condition. Two notable studies composed on one of
her journeys to the Southwest reveal Lange's unique ability to
extract meaning from a face and record it. In one photograph the
surrounding darkness frames the face of a Native American woman
deep in thought. Because her head is covered, even more attention is directed toward her facial features and the contemplative
expression they form. In the second study, a series of an ancient
man wrapped in a blanket, Lange used light to illuminate the signs
of age on the Indian's face. A wrinkled jowl in profile shows the
desert sun's parching effects. In another image within the same
study, Lange pointed her camera directly at the old man, whose
eyes were turned away from her as he stared off into the distance.
Rarely capturing superficial smiles, Lange preferred to record her
subjects' more complex expressions. In her portraits of Hopi natives,
Lange places herself, serious and brooding, yet invisible, alongside them. Or in many ways, inside them. Her early images were
largely about who she was, not who the Hopi Indians were. Lange
realized that her pictures revealed as much about who stood behind the camera as about who sat in front of it. She once said, "A

photographer's files are in a sense his autobiography. . . . As fragmentary and incomplete as an archaeologist's potsherds, they can be no less telling." She expressed the same sentiment when she stated, "Every image [a photographer] sees, every photograph he takes, becomes in a sense a self-portrait."[21]

Her camera work in the southwestern desert paved the way for Lange's later work with federal relief agencies. Here she saw for the first time people who did not share in the abundance of twenties America. Although she experimented on a limited basis with southwestern landscape, Lange believed the native inhabitants were far more interesting subjects. Her documentation of Indian life covered the region around Taos, New Mexico, and northern Arizona. While her husband looked inward to cultivate his own artistic tastes, Lange spent time exploring the everyday life, rituals, and work habits of the Hopi tribe members. In keeping with her focus on the individual, Lange brought her portrait-studio mentality to the wide open spaces of the desert. When she discovered a particularly fascinating subject, she chose to take a series of photographs of the person or scene, thus creating a composite study. Lange's catalogs of negatives reveal the wide range of interests she nurtured during her Southwest adventures. She concentrated equally upon mothers, children, young men, and village elders. These images of Native Americans taken in the 1920s are similar in style and content to the pictures she took for the U.S. Government in the 1930s. The choice of subjects, the use of light, the preferred camera angles, even the dominant themes of individual independence and human dignity, remain constant throughout the twenty-year period.

A favorite stylistic technique enabled Lange to draw from her subjects expressions of pride and dignity. To achieve the "look" for which she later became famous, Lange crouched down a bit so she could point the camera up at a slight angle. This device, though subtle, infused an image with an easily readable message—the pictured individual deserved respect since the camera (and thus the viewer) peered up, rather than down, at him or her. Lange's equipment required her to look down into a viewfinder, with the camera held about abdomen level, thus creating the image of a reverential bow to her subjects. A few of Lange's earliest portraits, dated either "1920s" or "1923–1931," reveal this particular tech-

nique. In one experiment with light and dark contrasts, Lange
composed a portrait of an adolescent Indian boy shown from torso
to head. Though he looks directly into the camera, his head is
bent downward to make eye contact with the photographer. In
another similar series, Lange focuses on a Hopi man, maintaining
her lowered position so that the image appears full of pride and
hope. In a closer view of the same subject, the man, who has taken
off his shirt, exudes an air of confidence, but confidence devoid
of conceit. Lange managed early in her career to discern the difference. She was able to isolate and highlight commendable traits,
while neglecting the less appealing side of human nature. Perhaps
she attempted to find what was exemplary because she believed it
existed everywhere and in everyone, regardless of ethnic background
or social class or economic constraints. But she did not force a
formulaic vision upon her artistry. Her method resembled the
American expatriate writer Gertrude Stein's attempts to clear her
mind before writing a portrait. Lange's contemporary, Willard Van
Dyke, described his colleague's method in 1934—"For [Lange],
making a shot is an adventure that begins with no planned itinerary. She feels that setting out with a preconceived idea of what
she wants to photograph actually minimizes her chance for success. Her method is to eradicate from her mind before she starts,
all ideas which she might hold regarding the situation—her mind
like an unexposed film."[22]

All of the images that Lange created reflected her aesthetic sense.
An examination of her 1920s studio portraits, her commissioned
family pictures, her Native American studies, and her 1930s farm
relief photographs reveals that Lange altered her visual preferences
very little, if at all. The years from 1919 to 1940 were her most
active as a photographer, and during those years her focus on the
human individual remained unchanged. Because of Lange's respect
for humanity, she depicted inner stability and downplayed external
circumstances. The basic vision she had developed as a young child
in New York City endured, but it was augmented by two memorable episodes that helped solidify Lange's direction and purpose as
a photographer. The first occurred one summer when Lange was
experimenting with landscape photography in the California mountains. She was caught in the midst of a raging thunderstorm and

said that she realized during those brief, frightening moments that she should devote her attention only to human subjects, that she must "take pictures and concentrate upon people, only people, all kinds of people, people who paid me and people who didn't." After her life-changing experience in the desert thunderstorm, Lange began advertising her studio work on printed flyers that read simply "PICTURES OF PEOPLE—DOROTHEA LANGE PHOTOGRAPHER." Documentary filmmaker Pare Lorentz once said of Lange's work, "You can usually spot any of her portraits because of the terrible reality of her people; in short, she is more interested in people than in photography."[23]

The second memorable episode that helped Lange to clearly define her role as a photographer occurred three years after the thunderstorm epiphany. At the height of the Depression in 1932, Lange sat in her studio window watching the unemployed men drift by on the streets while she examined the proof prints of her wealthy sitters. She recalled that "the discrepancy between what I was working on in the printing frames and what was going on up the street was more than I could assimilate." That view through the window crystallized for her a few months later when she and Dixon went to see the Noel Coward movie, "Cavalcade," a story about nineteenth-century Victorian life. As they strolled out of the theater, newsboys walked the streets yelling "EXTRA! EXTRA!" about President Franklin Roosevelt's declaration of a bank holiday. Roger Sturtevant remembered that Lange visited him the next day, still "stunned" by her experience. She "had come out at the end of one era and the beginning of another." He noted that "from then on she became interested in photographing the poor [and] downtrodden."[24]

Lange's pivotal experiences coincided with the burgeoning documentary tradition in the United States that encouraged writers, artists, sociologists, filmmakers, photographers, and others to go out into city streets, rural highways, and fields to observe people coping with the effects of economic depression and social dislocation. The prevalence of such subject matter, hundreds of thousands of hungry and displaced Americans and their environs, compelled the creative eye and mind in the 1930s. Lange joined fellow artists in the 1930s whose close scrutiny of Depression Americans resulted in texts, written or visual or both, that were designed to shape public opinion. The persuasive nature of documentary texts produced in that decade engaged the viewer and the creator in direct dialogue with each other. Even the titles of the works suggested that readers could claim part of the examination process or, at the very least, identify with the subject. Among those published were *You Have Seen Their Faces, Let Us Now Praise Famous Men, Twelve Million Black Voices, Preface to Peasantry,* "The Plow That Broke the Plains," and "The River." As William Stott pointed out in *Documentary Expression and Thirties America,* "Thirties documentary constantly addresses 'you,' the 'you' who is we the audience, and exhorts, wheedles, begs us to identify, pity, participate." British film critic John Grierson, who in 1926 had coined the term "documentary" to describe films, argued that such films "should educate and persuade," not merely chronicle factual data.[25]

Though Lange never mentioned and probably never heard of Grierson, her pictures nonetheless satisfied the simple requirements he had articulated. She created visual images that spoke to a viewer's conscience. Lange believed that a photographer could cause a viewer to look at a picture of something commonplace and find some extraordinary quality in it. The photographer as facilitator, then, helped the viewer eradicate mental walls, rethink misconceived ideas, and sharpen dulled senses. Insisting that a photographer always keep the potential viewer in mind, Lange articulated the desired results. Of the person who saw her pictures, she said, "My hope would be that he would say to himself, 'oh yea, I know what she meant. I never thought of it, I never paid attention to it.' . . . You have added to your viewer's confidence or his understanding." Well beyond enhancing a viewer's understanding of a chosen subject was the responsibility of selecting what would be looked at. The tremendous power of documentary resided in the choice of subject matter for an audience's consideration. This selection process made documentary form a type of "biased communication" that allowed a photographer or reporter or social scientist to reveal only as much as he or she wished.[26]

Lange believed that her deeply personal experience and struggle with the Great Depression and its victims led her into documentary photography, although she later said that she was practicing a style that "at that time . . . [had] no name." Years before the term *documentary* gained popularity and well before the Depression, Lange had cultivated the photographic vision and accompanying technique necessary for shaping solid social documents. Human subjects had always been the most important to Lange, and they continued to demand her full attention in the 1930s. Alone in her Montgomery Street studio in 1933, Lange found time to contemplate her own situation while she stared out the windows at unemployed passersby. Street life seemed to divide her concentration. She neither wanted nor could afford to turn away from her wealthy patrons, but she felt the outside world pulling at her. She said of this need, "I was compelled to photograph as a direct response to what was around me. . . . I was driven by the fact that I was under personal turmoil to do something." The inner disquietude of her solitary existence without her children combined with the restlessness she witnessed in others heightened in Lange the impulse to create photographs with social significance. This motivation placed her squarely within the emerging documentary tradition, and it led her to return to the streets, to observe the people there just as she had done as a child in New York City. This time, though, she carried a camera with her.[27]

On her first trip out among the unemployed, Lange managed to capture an image that would become an icon of the worst Depression years. Neglecting her friends' warnings to stay away from the areas where desperate victims congregated, Lange ventured over to a breadline set up by a wealthy San Franciscan called the "White Angel."[28] In the hungry crowd Lange found a solitary individual (see fig. 14). Having turned away from the other men, his singular presence overwhelms the others so that they serve as a mere backdrop for the portrait. The man's eyes are hidden, but his bewhiskered mouth and jowl are set in a pose from which his misery emanates. His hands together, the left hand is spread to cover and protect his right hand, which at first glance appears to be a fist. However, the position of his right thumb, which rests on top of his left thumb, keeps him from harboring a readied fist. The gesture suggests, instead, a wringing of one's hands. The figure only loosely steadies his empty, beaten tin cup, rather than clinging tightly to it, and thus exudes a patience not demonstrated by the other men in the photograph. The other element that sets this individual apart from the others is his appearance. He has the only unshaven face and the dirtiest, most ragged hat in the crowd. He stands in a breadline among men who have enjoyed higher stations in life than he obviously has. And he represents numerous poverty-stricken Americans in the early years of the Roosevelt Administration who seemed to fall through the cracks of federal government relief. The conservative measures that constituted the early New Deal in 1933 left men like the loner in *"White Angel" Breadline* as they had always been—voiceless and without help. Those in the better hats and cleaner coats with neatly trimmed mustaches received the first nods during the Depression; here, in this scene, they push through to the front of the breadline. Lange's ultimate message is one of desperation, not defiance, and it set the stage for the way she would depict the poor throughout the 1930s.[29]

The subject of her first "street" photograph, the man in the White Angel breadline provided the substance Lange had been searching for. Lange remembered, "I knew I was looking at something. You know there are moments such as these when time stands still and all you do is hold your breath and hope it will wait for you. And you just hope you will have enough time to get it organized in a fraction of a second on that tiny piece of sensitive film."[30] What Lange found fulfilled her need to answer the Depression, to react against a horrid economy that had forced her to live in her studio and place her children elsewhere, away from her. But the *"White Angel" Breadline* print represented much more. It was Lange's reacquaintance with street life; it suggests a rekindling of her own memories of a dismal childhood spent as a loner on the Lower East Side. The *"White Angel" Breadline* image, an autobiographical document, shows a man turned against the flow of the crowd—at once unique, despondent, and alone. Perhaps Lange saw in this man's circumstances an accurate, if uneasy, parallel of her own experiences. In those few moments at the breadline, Lange's training as a portrait photographer and her inspiration as a documentarian came together. At once, her documentary style solidified. The substance would be the patient and deserving poor, with dignity intact, waiting for help rather than violently demanding it; the form would be the single individual contemplating his or her circumstances. But one question remained—Who would be her audience?

Fig. 14. Dorothea Lange. *"White Angel" Breadline, San Francisco, 1933.*
© Dorothea Lange Collection. The City of Oakland, Oakland Museum of
California. Gift of Paul S. Taylor.

The day after she discovered the White Angel breadline, Lange developed her film. She immediately hung the print in her studio, and in a significant move, she also bridged the distance to her long-neglected East Coast home by sending a copy of the print to a gallery in New York City. Edward Steichen, a highly regarded photographer and critic, was astounded when he viewed the Lange picture in an exhibit. As Roger Sturtevant remembered, "Steichen saw it and everything else in this exhibit was wiped out as far as [he] was concerned. This man with the tin cup . . . this was real photography and this was the essence of what photographers should be doing." A representative image, *"White Angel" Breadline* helped open Lange's eyes to new possibilities for her career, while encouraging her to examine the very personal reasons behind her use of the camera. She hesitated to abandon her wealthy patrons, but she yearned to explore what she called the "bigger canvas *out there.*" Despite her exciting initiation on the streets of San Francisco, Lange chose to continue studio portraiture for a few more months so that she could eat, pay her sons' board, and finance her "other" photography. But she soon realized that doing both kinds of work was "a strain."[31]

General unrest over her work and her personal life sealed the transition for Lange. Her initiation into street photography sparked a political conscience. Strikes on San Francisco's waterfront and other worker demonstrations encouraged her to take her camera into highly charged, sometimes violent situations that she could view from her Montgomery street studio. But Lange's photographs rarely highlight mayhem or destructive action. At a 1934 May Day demonstration, Lange shot scenes of the featured speakers and of listeners in the crowd. In one image (fig. 15) a woman listens intently to the message, but the sun on her face distorts her expression to one of skepticism. Lange positioned herself so as to show the subject in a dignified pose, with her chin up and her head held at a slight angle. The woman's calm demeanor implies that she will not resort to violence, despite the impassioned rhetoric that the rally speakers spout. The newspaper headlines—"The Workers Took the Wrong Path in Germany and Austria" and "What is 'Americanism?'"—promote a tone of responsible, educated citizenship. This urban worker is a reader who after listening to the speeches will peruse the written arguments and make an informed

Fig. 15. Dorothea Lange. *May Day Demonstration, San Francisco, 1934.*
© Dorothea Lange Collection. The City of Oakland, Oakland Museum of
California. Gift of Paul S. Taylor.

decision based on logic and rational thinking. Nothing in Lange's frame suggests a situation out of control. In fact, the participants appear ennobled by their controlled reactions at the rally. Lange's signal to the viewer of this image is one of reassurance about the notion of mass demonstration. She dilutes the crowd by showing the full head and expression of only one participant. In the rest of the frame are partial faces, including one person's ear, another's jawline, and yet another's profile. But we do not see their eyes, because they would have detracted from the principal subject of the portrait, the calm, informed, attentive listener. Nor does Lange show the feet of the crowd, which could have revealed frenzied activity or, at the very least, a sense of motion. Nowhere does she hint that the labor situation was volatile enough to turn into the bloody fiasco that it did a few weeks later. In July 1934 the worst strikes in San Francisco's history injured many laborers and claimed the lives of two people, whose funeral procession included an estimated fifteen thousand to twenty thousand union members and supporters. Yet Lange's portrayal of labor chaos evokes control, not madness. Photo historian Anne Tucker has described Lange's intent in her documentary work as "gentle," noting that "[s]he did not want to introduce, but to remind us of things."[32]

Like many of her contemporaries, who also grew more politically aware in the 1930s, Lange did not allow radical leftist ideology to overwhelm her. Clark Kerr, who was a Stanford graduate student writing a thesis about the unemployed in the 1930s, worked with Lange in the midthirties. He described her as "one of the most nonideological persons I have ever known. . . . I never heard one word or saw one gesture that in any way indicated that she was a committed ideologue." David Peeler has cast her with a group of well-known thirties documentarians who he says were more interested in their personal "emotional quandaries" than in ideology, so that there "was a certain intellectual softness, an unwillingness to employ ideology, among American thinkers of the 1930s." Such an unwillingness led Lange to create images that exuded quiet desperation, controlled action, and individual, not collective, contemplation. As a result the subjects in her thirties work take on a characteristic that was common among figures used to appeal to viewers' emotions. They were "sentimentalized," argues William Stott.[33]

Lange's primary emotional quandary during this transitional phase in her career was a widening philosophical gap between her and Dixon. He argued against her newfound interests and discouraged her from getting involved in political activities. Lange later said of the left-wing groups that sought her support, "I'm not sure that it wasn't the right thing to do . . . participating in groups of people who were ready to take action," but she resisted the temptation. Dixon looked skeptically upon political loyalties, and he dissuaded his wife from attending party gatherings. But armed with a fresh sense of social responsibility, Lange continued to pursue subjects outside her studio. The distance separating Lange and Dixon increased as their respective artistic philosophies took divergent paths. Dixon clung tenaciously to an aesthetic standard that praised art for its own sake, whereas Lange felt that her photography could serve a larger social purpose for the cause of humanity. She sensed a deeper connection to her surroundings than her husband did to his. Though she was not speaking of Dixon specifically, Lange once noted that an "artist" photographer maintained only a "very slight . . . alliance with the world." This explains in part why she felt more comfortable working on the West Coast, away from her native New York City with its artists' colonies and photography groups. Lange's friend Willard Van Dyke offered similar sentiments, explaining that there was "a feeling that the Eastern photographic establishment was too theoretical . . . too parochial—they couldn't see anything except their own little worlds." Lange, it seems, began to realize the same about her husband's frame of reference, contrasting the contributions her photography could make to society with the limitations of his paintings. Therese Heyman, curator at the Oakland Museum, in discussing the differences between Lange's and Dixon's work in those crucial months, notes, "In her 1934 photographs, Lange was an articulate witness to the most stubborn and intractable truths of her time—the possibility of civil insurrection. She made memorable images, partly eulogistic, partly despairing as in "White Angel Breadline," but unlike Dixon, the rebellion is never out of control. Dixon's visions are suffused with fear; Lange's, on the other hand, are confident and human."[34] Even if her photographs exuded confidence, Lange herself remained unsure about forging ahead in a new direction until she had received reassurances. These came from two sources, Oakland gallery owner Willard Van Dyke and University of California economist Paul Taylor.

Van Dyke put up a Lange exhibition in his Oakland studio, Brockhurst, in 1934. Comparing Lange's talents and pictures to those of Civil War photographer Mathew Brady, Van Dyke said in his commentary, "Both Lange and Brady share the passionate desire to show posterity the mixture of futility and hope, of heroism and stupidity, greatness and banality that are the concomitants of man's struggle forward." Attempting to prove his point, he showed the public Lange's images of hungry bread-seekers, May Day demonstrators, impassioned union leaders, and determined strikers. One viewer who was impressed by what he saw was social scientist Paul Taylor of the University of California. He had been gathering information on labor by carrying out field research among striking agricultural workers in the San Joaquin Valley. An assistant praised Taylor's approach, noting that "Paul was always out there finding out what was really happening while others played around with their theoretical models and ran their regression analyses." Part of Taylor's hands-on methodology included the use of photographs. He recognized immediately the breadth of Lange's talent and her vision. Within a few months, the two were working side by side for the Federal Emergency Relief Administration in California.[35]

Taylor helped Lange gain the confidence she needed to continue photographing in the outside world among the unemployed poor and dispossessed. Although she had always energetically pursued her visual interests, she lost a fraction of her verve when Dixon failed to confirm her socially inspired work. His opposition created doubts in Lange's mind, forcing her to seek approval for her pictures elsewhere. Recognizing Lange's apprehensions, Van Dyke noticed the impact of Paul Taylor's support of the government's newest employee. Van Dyke said that Lange was seeking "validation for an approach that she was struggling with as far as this photography was concerned. And Paul gave it." The professional collaboration soon expanded to include a personal relationship between the two, as they undertook field trips throughout the state. In spite of their respective marriages, Lange and Taylor found their intimate liaison a necessary complement to the work they were doing. Lange reciprocated the support Taylor offered and justified her request to Dixon for a divorce in 1935 by saying, "I want to marry Paul, he needs me." In the same year that Lange forfeited her life with the wild-natured, often unpredictable artist, she closed her San Francisco portrait studio for good, casting her future with a man who was the personification of solid, respectable citizenship. In a professional sense, both the photographer and the economist felt that their complete devotion to each other would strengthen their collaborative efforts.[36]

The realization that she was serving an expanded audience, beyond the wealthy families who had supported her portraiture business, prompted Lange to reconsider the function of her photographs. If her pictures were to be as useful as she hoped, she thought they needed to be accompanied by words. She believed one was incomplete without the other—words had to help explain pictures, and the visual had to illuminate the verbal. Lange saw the inherent weakness of photography, one person's act of selecting a small piece of the present. She noted, "When you take it out and isolate it, a good deal falls through the slot." To characterize her concern, Lange described the life of one particular woman depicted in a photograph, pointing out that the viewer might see the woman in the picture but would never see all the things that surrounded her and influenced her life. As the photographer, Lange experienced what she termed a "visual flood," but knowing that her viewer would not have the same opportunity, she asked disconcertingly, "How can you put that . . . so somebody else will understand it?"[37]

Lange felt that Taylor's words would help bridge those gaps and would complete her pictures; she found a voice for her images as Ulmann had discovered several wordsmiths for hers. Novelist Julia Peterkin penned stories around Ulmann's photographs; Allen Eaton provided sociological data; and John Jacob Niles was the one who engaged Ulmann's subjects, seducing them into having their pictures made. Thus for both Lange and Ulmann the most important "talkers" were the men with whom they traveled in the field, the voices of authority who asked questions of their subjects, recorded or wrote down the answers, and lent a certain amount of credibility to the photographers' visual images. Both women felt it necessary to include their research partners alongside their subjects inside the photographic frame. Taylor appears with pen in hand in several of Lange's photographs (although he was usually cropped out by editors), and Niles is a subject in several Ulmann series, including an extensive one in which he engages a young Appalachian woman holding a symbol of the mountains, a dulcimer. After Ulmann died, Niles's captioning of her proof books left an indelible mark on her work, expressing opinions that reflect his tastes and prejudices much more than hers. Taylor's influence on Lange's photography was similarly significant; it was characteristic of the word-and-picture marriage that social documentarians of the 1930s embraced so readily.

The combination of visual and written sources gained popularity among socially conscious artists, writers, and scholars during the Depression. Many pooled their diverse talents to produce pamphlets, books, exhibits, films, and reports that employed both visual and written evidence. A photograph had the capacity to reach millions through the popular press, but if the picture had no caption, viewers could interpret it for themselves in any number of ways. With the right message attached, however, a viewer's opinion could be molded, crystallized, even changed completely. Lange believed that words were necessary to sharpen the visual acumen of most viewers and help them "see"—she once said that all photographs could be "fortified by words." The words Taylor wrote to accompany Lange's photographs came directly from the notes he took during field research. His training as a social scientist, with its requisite demands for objectivity, played a vital role in his choice of evidence. Both Taylor and Lange abhorred "poetic captions," favoring instead comprehensive, fact-driven, descriptions of the circumstances they had studied. Contemporary critic John Rogers Puckett has said that Taylor's captions "have the virtues of being interesting in their diversity and being factual descriptions or transcriptions rather than fabrications." Lange's 1930s photography was distinguished from the earlier work of Jacob Riis and Lewis Hine precisely by being "buttressed by written material and by all manner of things which keep it unified and solid." In Lange's opinion Hine and Riis had not produced true documentary, since both men considered words subordinate to the visual image. Lange agreed with her friend Willard Van Dyke, who later declared, "Words at that point were so important . . . there was so much to be said. Our feelings were so strong about the Depression that just to show a picture just wasn't point enough." Despite their reflections years later on the role of words in thirties documentary, the image makers, the photographers themselves, wielded tremendous power through their construction of particular messages to be delivered to American audiences. A number of those photographers, including Dorothea Lange, gained access to a very large American audience by signing on with the U.S. Government, one of the most significant producers and users of the visual image–written word combination in the 1930s.[38]

In 1935, in the wake of heavy criticism from radical protesters and desperate pleading from displaced Americans, the Roosevelt administration turned its attention to those groups that were either left untouched or adversely affected by early New Deal programs.[39] Among the neglected groups of Americans were tenant farmers, migrant laborers, and victims of drought in the southern plains. A new federal agency named the Resettlement Administration (RA) was created by executive order in April 1935 to aid those hit hardest during the Depression years—rural Americans. Among the RA's proposed plans were to remove poor farmers from unproductive land, to provide housing and instruction about better land use, and to reorganize migrant labor camps, which to some conservative Americans and their congressional representatives seemed too radical, even socialistic. To counter expected opposition, RA administrator and Roosevelt brain truster Rexford Tugwell immediately organized a division within the agency called the Historical Section to sell the RA's objectives to the public.[40] Tugwell brought in Roy Stryker, his former graduate student assistant at Columbia University, to head this information division. Stryker's credentials included a brief tenure at the Agricultural Adjustment Administration in 1934, but he also had, more importantly, a commitment to disseminating information through the use of visual aids, including pictures, graphs, and other imagery. His previous work as illustrations editor for the textbook *American Economic Life and the Means of Its Improvement* (1924) and the place of photographs in his classroom presentations years earlier demonstrated Stryker's long-nurtured interest in visual images as educational material.[41]

That photography was chosen as the most appropriate and expedient vehicle for increasing and sustaining support for the RA, then, is not surprising. Glossy "picture" magazines, such as *Fortune, Look,* and *Camera,* were flourishing, and Stryker knew he could get the RA's photographs published nationwide, especially if he offered them free to any magazine, journal, or newspaper that would print them. Supporting the visual images produced by RA photographers would be words the agency deemed appropriate. And the overwhelming message in the RA's early months was that taxpayer dollars would be spent on people who deserved government assistance. Potential rebels and rioters, drifters, and derelicts stood outside the RA photographic frame; those who had been struck by natural forces or circumstances beyond their control—dust storms, flooding, erosion, mechanization—became the preferred subjects.[42]

Having worked briefly with the Federal Emergency Relief Administration in California, Lange understood the role of a government propagandist well enough that Stryker wanted to draw upon her experience. He hired her, Ben Shahn (a muralist), Walker Evans (an art photographer), and one or two others to complete assignments for the RA in the early months. Given the artistic backgrounds of this cadre of employees, it is evident that Stryker wanted to call upon not only their collective experiences but also their respective aesthetic senses. He believed that these three individuals would provide a balance for the agency's otherwise scientific survey of economic conditions. A close student of John Dewey's philosophies, Stryker sought to combine the products of art with the strengths of social science in order to appeal to the American public. The RA's objectives went beyond pure reportage and statistical tables to include actual people, whose facial expressions and bodily postures revealed a myriad of economic, social, and physical hardships. Distinguishing the Historical Section's work from other informational sources, Stryker interpreted RA photography as "the adjective and adverb. The newspicture is a single frame; ours a subject viewed in a series. The newspicture is dramatic, all subject and action. Ours shows what's in back of the action. It is a broader statement—frequently a mood, an accent, but more frequently a sketch and not infrequently a story."[43]

Though propaganda, the RA sketches and stories possessed

human interest value, a requisite quality for softening the role of the messenger, the federal government, without harming its credibility. In the 1930s the aim of social publicity was, as historian Maren Stange has pointed out, to "portray social and economic management as a matter of smooth, humane, bureaucratic administration." Lange understood completely her role in promoting the New Deal government's newest objectives because they had become her own passions a year earlier. She explained her devotion: "The harder and more deeply you believe in anything, the more in a sense you're a propagandist. Conviction, propaganda, faith." Lange's "innocent" victims of the Depression bore the appropriate characteristics of thirties propaganda, which according to William Stott arranged subjects so they appeared "simplified and ennobled." More than one analyst of Lange's RA pictures has said that the photographer knew how to create an image that made viewers turn their heads again and again to absorb her messages. When scholarly interest in RA photography began to emerge in the 1960s, one of the earliest critics to examine the RA file, Robert Doherty, noted that Lange "probed deep into humanity to come up with fragments of life that pulled the heart strings of all that could open their eyes to see." This is exactly what the RA hoped to accomplish—to make viewers look again, a second, a third, even a fourth time. Seeing their work "invested with spiritual significance," Allan Sekula suggests that certain documentarians are more than mere witnesses to situations and appropriately describes them as "seers." Indeed, Lange's desire for social change, combined with an unaltered camera style that highlighted human dignity and independence, made her that well-suited *seer* for the *RA*.[44]

From the desperate yet proud individuals who were her subjects, Lange created an inextricable combination that proclaimed a powerful message—she recorded personal character wrapped in adversity in order to argue that government funds could promote stability in the nation if they were distributed to people whose personal initiative and self-respect remained intact. Whereas Doris Ulmann had discovered that work promoted dignity, Lange seemed to intimate that this intangible quality had been sustained, perhaps enhanced, by a lack of resources. And that the government could serve to reward those whose fortitude had brought them through several years of economic devastation. In a series on drought refugees who had made their way from Oklahoma to California, Lange produced an image whose female character exudes personal strength (fig. 16). She sits unflinchingly, with arms draped in a graceful, relaxed pose around an aggressively nursing child who has misshapened his mother's exposed breast with a turn of his head. He appears as determined as she is, drawing on her life nourishment and also internalizing the courage that has enabled her to travel a thousand miles west in search of a better existence. She appears not unlike her probable ancestors, whose facial characteristics she wears proudly. Her wide, flat-cheeked face with a sharp-angled jaw and close-set eyes reveal a tie to the Native American inhabitants of her home state; their one-thousand-mile trek one hundred years earlier had placed them in an environment as foreign to them as the one she now inhabits. She makes no effort to hide the tears along the side seam of her dress as she sits erectly in a makeshift shelter. Of the photographs Lange made in this series, this frame's iconography sends the most powerful message. Others, which show the woman in a less confident posture alongside her much more desperate, reclining husband (who takes up half the frame in the foreground), do not offer the simultaneous poignancy and complexity that the mother and child alone offer. Without the father figure in the photograph, Lange promoted within the RA her preferred view of the family, a fragmented unit comprised of staunch individualists.

That Lange's personal vision and the RA's initial objectives dovetailed reveals the weight of nineteenth- and early-twentieth-century standards in the prevailing cultural values of the 1930s. Lange had matured in an America that revered the strength of the individual making his or her own way regardless of the costs. Dependence on others was an admission of cowardice or laziness. From her painful childhood days Lange had developed a psychological stamina, an inner source of strength. One of her cousins pointed out that Lange was "tougher than any of the ones in our family. She had a much better sense of where she was going and nothing was going to stop her."[45] Lange believed that the potential for personal courage resided within everyone and that it came to the front when circumstances demanded it. A large portion of her work illuminates her ability to isolate and record this intangible quality on film. In a 1937 study of South Carolina share-

Fig. 16. Dorothea Lange. *Oklahoma Drought Refugees, 1936.* U.S. Farm Security Administration Collection, Prints and Photographs Division, Library of Congress.

cropping, Lange's values are emitted by a farmer who has paused in his field to strike the portrait pose that Lange so often preferred (fig. 17). With a firm grip on his manual plow, the man looks confidently into the sunlight, his spectacles reflecting it, appearing to ponder some serious question. Holding his chin and head up in a position of assurance, he exhibits the 150-year-old American sentiment that hard work in the soil can bring prosperity and lift one above other laborers who will never enjoy such independence. Although the viewer is allowed to see the farmer's neat furrows, Lange does not provide a glimpse of the immediate ground on which her subject stands. She omitted from the frame the actual object to which the sharecropper is bound by a wrist rope, allowing an element of mystery in the image. Any contemporary viewer familiar with farm equipment could have filled in this gap, but the uninformed viewer, especially the urban dweller, had to internalize the larger symbol—a man tied to the land but ennobled rather than trapped by his connection to it. This American farmer, as Lange depicts him, is the perfect icon for the RA audiences, who desired reassurance that RA money was being spent on the "right" kind of people. Photography historian James Curtis has argued that the principal viewing public for RA photographs was an urban middle-class audience who gazed upon images that had been carefully selected for them by Stryker's photographers.[46] The South Carolina sharecropper was an ideal subject for such audiences. His field suit shows no signs of carelessness, no rips or tears, and his chest pocket reveals the outline of a watch attached by safety pins to his suit. The secondary message promoted, then, is that the farmer's careful attention to his appearance denotes personal responsibility. Would this man not be a safe bet for RA funding, a good investment for the government aid? What better insurance for the nation's future than this proud, astute, and clean tiller of the soil? Like this lone farmer, many people depicted in Lange's thousands of RA photographs run parallel to the RA's fundamental beliefs about the role of individual fortitude.

The price of visually depicting personal independence was forgoing a focus on human intimacy. As adept as she was at getting people to cooperate with her, Lange seldom photographed them interacting with each other. More often than not, members

Fig. 17. Dorothea Lange. *South Carolina Sharecropper, 1937*. U.S. Farm Security Administration Collection, Prints and Photographs Division, Library of Congress.

of a group or a family had little or no contact in Lange's pictures. No image better illuminates this point than Lange's most famous photograph, *Migrant Mother* (fig. 18). Even though the mother's children surround her, touching her on all sides, she is separated from them by her thoughts and her faraway gaze. The children are merely appendages to the primary subject, the woman. Photo historian John Rogers Puckett, who believes the image is indicative of the entire Lange oeuvre, has said, "Like many Lange photographs . . . [Migrant Mother] depicts a moment of withdrawal into self—of isolation and alienation from the world. The theme of human estrangement runs through all of Lange's work." A very personal reflection of her own life, Lange revealed this isolation as a common thread in human existence, focusing upon it in her fieldwork as often as she could. Critic George Elliott identified *Migrant Mother* as "a sort of anti-Madonna and Child. . . . The mother, who, we feel without reservation, wants to love and cherish her children, even as they lean on her, is severed from them by her anxiety." Thus, the most widely published photograph that Lange ever took can be interpreted to reflect the most difficult decision she had made in her life up to that point—to board her children at school when the family finances would not allow her to keep them with her. James Curtis, who has conducted the most thorough study on the *Migrant Mother* series of photographs, discusses Lange's conscious arrangement of the composition. In her notebook Lange recorded that the woman had seven children, yet only four are present in the pictures. Curtis suggests that Lange did not want several more people in the photograph, since "five figures posed enough of an obstacle." Lange even positioned the children to achieve the desired effect. Curtis explains, "She had the youngsters place their heads on their mother's shoulders but turn their backs to the camera. In this way Lange avoided any problem of competing countenances and any exchanged glances that might produce unwanted effects. She was free to concentrate exclusively on her main subject."[47] And the "main subject" for the RA employee whose training had been in portraiture was the single individual, contemplative, perhaps anxious, but more stable than rebellious.

    Lange's decision to focus on the separate individuals helped her avoid the complications of human exchange. This preference

Fig. 18. Dorothea Lange. *Migrant Mother. Nipomo, California, 1936.* U.S. Farm Security Administration Collection, Prints and Photographs Division, Library of Congress.

may be seen in numerous other photographs, including a 1938 study of displaced tenant farmers in Hardeman County, Texas (fig. 19). Though all six men are in the same predicament, unemployed and disfranchised, they remain separate from each other, neither looking at nor talking to one another. Their personalities emerge, in part, through their various stances, two men with their arms folded, another with his hands in his pockets, and yet another with his hands behind his back. Their caps and hats, ranging from wide-brimmed to deep-billed, sit at dissimilar angles, from one that obliquely covers a right eye to another that shows a man's full forehead. Although the men constitute a group, through common distinguishing characteristics of gender, geographical location, and economic circumstance, they are not a unit. All of them emit personal dignity, but each is so different from the others that Lange could have made six individual portraits and achieved exactly the same end with her message. Another frame in this series shows the same men sitting on the ground. They cast their gazes in myriad directions, with not one crossing another's visual path. Despite their similar economic condition, individual diversity transcends what they have in common. Lange made an interesting observation about her penchant for seeking out individuals among the masses. Describing a photograph she took at the Richmond ship-yards during World War II, she said, "The shipyard workers were coming down steps. . . it was a mass of humanity. . . . But what made the photograph so interesting was that they were all look-ing in different directions. There was no focus, there was no co-hesion in this group. They were not a group of people united on a job. It showed so plainly. Their eyes were all over."[48] Lange understood the importance of remaining independent while un-der pressure to conform. Forced to become self-reliant at an early age, she always felt sympathy, even empathy, for individuals strug-gling against powerful voices of authority that sought to squelch individual expression. Although Lange and Doris Ulmann both preferred to focus on the single human life, Lange brought to her socially inspired photography one vital element that distinguished her from her rich New York counterpart. Lange always sought words to provide information that a visual image could not emit, regardless of its aesthetic beauty or its penetrating portrayal of the human spirit.

Fig. 19. Dorothea Lange. *Displaced Tenant Farmers. Hardeman County, Texas, 1938.* U.S. Farm Security Administration Collection, Prints and Photographs Division, Library of Congress.

Of all the field photographers Stryker employed, Lange wrote
the longest captions. They rarely contained her personal interpretations, since she relied upon her eyes and her camera for that. In
fact, Lange's word accounts showed much less conscious creative
construction than did her visual images. She chose to record the
exact words her subjects uttered, sending to Washington "verbatim accounts to accompany her photographs." Her husband, Paul
Taylor, remembering that Lange carried a loose-leaf notebook for
such purposes, claimed that she carefully practiced "using her ear
as well as her eye." Taylor wielded a great deal of influence in this
area, since he accompanied Lange on most of her field assignments
and in some cases interviewed the people whom Lange photographed. She had learned the technique from Taylor after observing his style on their earliest trips together.[49]

The best-known Lange image that includes Taylor in the frame
conducting an interview was one she took near Clarksdale, Mississippi, in 1936 (fig. 20). A husky plantation owner, in a cocky
stance with one foot propped on his new automobile, converses
with Taylor as one black man stands and four others sit quietly in
the background. Various editors cropped Taylor out of the picture for their publication purposes, even though the original negative
shows him prominently. When Archibald MacLeish used the photo
in his 1938 publication, *Land of the Free*, he further cut the picture, leaving out the African Americans. The plantation owner's
image illustrated MacLeish's poetry, which included the words
"freedom," "American," and "pioneers." The irony of MacLeish's
verbal message placed beside the original photograph would have
been too overwhelming for the Jim Crow South, so MacLeish
purposely avoided the potentially explosive combination. Three
years later, Richard Wright and Edwin Rosskam used the photograph in *Twelve Million Black Voices* as a parallel text on the same
page with a definition of the word "Negro," which does not include the word "freedom." Wright exposes the socially constructed
weight of the term "Negro" by declaring it "a psychological island" sustained by "a fiat which artificially and arbitrarily defines, regulates, and limits in scope of meaning the vital contours
of our lives, and the lives of our children and our children's children." His words more aptly reflect the conditions represented in
the original photograph, but the departure from reality that docu-

Fig. 20. Dorothea Lange. *Plantation Owner. Clarksdale, Mississippi, 1936.*
U.S. Farm Security Administration Collection, Prints and Photographs
Division, Library of Congress.

mentary texts allowed can be seen in both books. Since Lange's picture belonged to the federal government, which had sole control over its use and distribution, her thorough captions could be changed, deleted, or ignored altogether. But by altering Lange's word stories that accompanied her photographs, one compromised the photographer's original intention, which was to create "documents that were both visual and verbal." As a thirties documentarian, Lange depended upon words and had developed such keen interviewing skills that "her conversations with the people she photographed and her understanding of their way of life were as much a part of the print she achieved as the camera or the film." When she felt that her aesthetic sense was being challenged, Lange fought back.[50]

As Lange was allowed increasingly less control over her words and her photographs, her relationship with Stryker grew more tense. The slow start the RA experienced in distributing photographs left most images in the hands of government agencies and inside government publications. But eighteen months after its creation, the RA's Historical Section was placing its photographs in widely circulated national magazines and journals; in addition they were gracing the walls of exhibit halls and galleries in several U.S. cities. A reorganization at the agency in 1937, which gave it a new name, the Farm Security Administration (FSA), and placed it under the control of the Department of Agriculture, meant even greater exposure than the RA had enjoyed. Accustomed to personally completing every step of the technical process as a portrait photographer, from selecting subjects to developing film to making prints, Lange had never been comfortable sending her undeveloped RA film to the Washington office to have it become U.S. Government property. Early on Stryker had allowed her limited control over her own film development when she was on assignment in California, but when she traveled in the South and had no studio darkroom, Lange was forced to send the film to the FSA office. As her audiences grew larger, Lange desired even more control over her negatives in order to see whether she was capturing on film what she had intended. Often several weeks passed before she was able to see the proof sheets the FSA darkroom had produced. Entire assignments were jeopardized, Lange thought, without her constant scrutiny of her creative process. Lange had never ceased to consider herself an artist, even though her subject matter had changed dramatically from her days in the portrait studio.[51]

The uneasy working relationship between Lange and Stryker was further complicated by Paul Taylor's overwhelming presence. Taylor posed a threat to Stryker, who needed to have his FSA photographers follow his direction and inspiration. Of all the traveling companions that accompanied Stryker's photographers, Taylor was the one for whom Stryker showed the least support. It is through this case, where the FSA employee was a woman and her traveling companion her husband (a university professor and respected economist, no less), that Stryker's bias toward his female employees reveals itself. He showed no such concern, for example, when Jean Lee accompanied her husband, FSA photographer Russell Lee, a few years later. It is not surprising, then, that nearly every female field worker that Stryker chose to hire in the wake of his troubles with Lange was an unknown and unattached photographer to whom he could dictate specific orders. Although Stryker has been praised for allowing his photographers to "follow their instincts," he clearly treated FSA women differently from their male counterparts, writing letters to them more frequently and admonishing them more casually.[52] Foreshadowing the fractious relationship that emerged between Lange and Stryker were the two motivations for his hiring her in the first place. First, he wanted a portrait photographer, an artist with an aesthetic sense; Lange, the independent artist, expected the same kind of control over the creative process she had always enjoyed. Second, Stryker sought experience, which Lange had gained in her early fieldwork as Taylor's photographer; Lange, as Taylor's companion, remained devoted to the person who had given her initial support as a documentarian. Drifting slightly from the agency but still on its payroll on a per diem basis, Lange maintained her truest loyalty and her hallmark style. Both found expression as she and Taylor embarked on their own word-photo project, an examination of 1930s displacement and the ensuing massive movement across the continent.

After two years of traveling, recording, interviewing, and photographing, the Lange-Taylor project appeared under the title

An American Exodus: A Record of Human Erosion (1939). The work challenged a similar and wildly popular 1937 publication entitled *You Have Seen Their Faces,* by the photographer-writer team Margaret Bourke-White and Erskine Caldwell. But the Lange-Taylor collaboration bore the marks of social science that were missing in their counterparts' earlier study; their extensive field work penetrated contemporary economic, social, and political realms of activity and was reinforced by statistics and the actual utterances of the people Lange had photographed. In addition, *An American Exodus* played upon the vital dialectical relationship between the past and the present in American rural life, examining the effects of mechanization, drought, sharecropping, and the South's severest tyrants, cotton and the boll weevil. Taylor even compared the displacement of landless farmers in the United States to the devastation caused by sixteenth-century British enclosure policies. In composing this portrait of a substantial segment of the populace, Lange found full expression in what became the second distinguishing theme of her camera eye: movement. In addition to her focus on the American "individual," she highlighted Americans in transit. In a larger sense, this recurring theme in Lange's work could be defined as the culture's obsession with mobility. Lange captured figures walking along the highways, relaxing in overstuffed jalopies, and dwelling in makeshift shelters. Many of her FSA pictures were taken before government efforts had effectively addressed the small-scale farmers' problems, when thousands were moving—away from tired soil or dust-filled houses and toward fresh land and new prosperity.

The final chapter of *An American Exodus,* entitled "Last West," opens with a statement Taylor composed while on U.S. Highway 99 in the San Joaquin Valley: "For three centuries an ever-receding western frontier has drawn white men like a magnet. This tradition still draws distressed, dislodged, determined Americans to our last West, hard against the waters of the Pacific." The first image in the chapter shows a loaded automobile sitting in the hot desert sun (fig. 21). The driver has stepped out of the car, leaving the door ajar, while a young girl peers out of a side window at the camera. The shadows on the ground, as well as the man's slightly turned head, indicate that he is talking to someone on his immediate left. The cropped photograph omits

Fig. 21. Dorothea Lange. *Stalled in the Desert. California, 1937. An American Exodus* (1939). U.S. Farm Security Administration Collection, Prints and Photographs Division, Library of Congress.

the interviewer, allowing a much more mythic image to emerge. The sharp shadows on barren land suggest a waiting place, appropriately the desert, that must be crossed before the promised land is reached. The angle Lange chose allowed the highway to cut across the entire horizontal line of the frame at a roughly thirty-degree angle, thus situating the travelers on a slight incline. The incline may be read as a gradual path toward socioeconomic betterment, but it seems to represent more clearly the upward, perhaps difficult, climb that the transients have before them. If the photo were superimposed upon a map of the United States, this Oklahoma family would be headed in the right direction; both the male guide and the front of his vehicle focalize left, suggesting their westward movement. The active portion of the frame is the left center third, from the back wheel of the vehicle to the left edge of the image. Were the young girl not distracted by the photographer, all of the photograph's action would be centered in one-sixth of the frame. What remains behind the automobile is past, forgotten, gone. Cultural historian William Jordy has connected Lange's depiction of mobility with the regional fervor that gained popularity in the 1930s. He sees the jalopy as an important symbol of the age, with westward travelers imitating "the saga of earlier pioneers." In Lange's image of the Oklahoma family, the heavily weighted automobile is not unlike nineteenth-century wagons bearing the accoutrements of life necessary for cooking, cleaning, sleeping, and storing goods. Here a milk can, buckets, and tubs in several sizes are prominently visible. The family's holdings are out in the open, for everyone to see. Privacy no longer exists for them.[53]

The hope inherent in mobility was that a better future lay ahead, an idea that had long been a principal element in the American consciousness. Freedom to move was viewed as an opportunity and a privilege in the dominant culture. As early as the 1830s, Alexis de Tocqueville had identified this cultural peculiarity. He believed that Americans possessed an anxious nature, a driving quality, that kept them constantly uprooting themselves in search of better lives. In the years just prior to the Civil War, the threat to mobility affected both northerners and southerners. Among the concerns of Americans in the Northeast and the Midwest was preservation of western lands as places to which they could move and be afforded the privilege of a familiar lifestyle—

one of "free soil" and "free labor." Southerners thought along similar lines. James Oakes, who has turned traditional perceptions about the South upside down, has shown that transience was "a normal part of existence in the Old South." The great majority of antebellum slaveholders had not sunk their roots into the soil, nor had they lived on plantations for generations with grandparents, parents, and their own children. Instead, most southerners were migrants. Women "complained that their husbands and children seemed determined to move every time they got the chance. Success seemed as much an excuse for packing up and leaving as did failure." In the 1890s one of the great contributors to the psychological crisis in the United States was Americans' fear that there was "no more frontier." When historian Frederick Jackson Turner gave his renowned 1893 address, "The Significance of the Frontier in American History," he emphasized the importance of land access to the society's evolution, success, and prosperity. Thus, the safety valve for Americans who wished to create better lives and new starts for themselves or their families had historically rested upon their ability to move.[54]

In the 1930s, when nearly every excuse for moving was tied to failure, Lange photographed Americans who were on the road in search of a prosperous, successful existence. The idea of a restless nature in Americans had appealed to Lange even in her early years as a photographer. In her pictures of the southwestern Indians, Lange frequently depicted men and women on horseback, turned away, riding out across the desert. She photographed Native American family members perched on wagons or walking alongside them. These images were strikingly similar to those she created for the FSA and others she made while working on *An American Exodus*. To broaden the theme of mobility in her work, Lange included those who had not discovered the good life they had wished to find in California. In one study (fig. 22), a reticent, contemplative, antihero is the focal point. He is the disappointed searcher returning home. We see little of the road behind him; the road is no longer important, since it failed to lead him to the promises he had hoped for. His words belie the misleading messages regarding the West, and California in particular: "[T]hey appreciate the cheap labor coming out. When there's a rush for work they're friendlier than at other times." His value rested in

Fig. 22. Dorothea Lange. *Laborer Returning Home. California, 1937. An American Exodus* (1939). U.S. Farm Security Administration Collection, Prints and Photographs Division, Library of Congress.

what he could offer as a laborer, nothing more. The good life eluded him and his family; there was no more frontier that could guarantee success and prosperity in America. Mobility, then, had lost its strength and its attraction, developing a tarnished reputation among many Americans during the Depression.[55]

For all of their appeal in the mid-1930s, these messages in Lange's photographs grew increasingly less popular at the FSA office by the late thirties. Stryker's other photographers, especially Russell Lee and Marion Post, were concentrating upon the small town, the community, and the stability offered within those comfortable congregational confines. The idea of mobility, which Lange had viewed as a thoroughly American notion indicative of courage and independence, diminished as an American value, as did the single individual or the lone hero. These cultural barometers were surpassed by newer ones—community effort, group support, and identity with place. Yet Lange chose not to embrace them. The government photographs that she took from 1935 to 1939 reveal that her primary visual interests and her aesthetic approach changed very little. Beyond her squabbles with Stryker, perhaps it was her intractability, combined with what was essentially a nineteenth-century view of the United States and its inhabitants, that cost her the job with FSA. She admitted to an interviewer, "I go over some of the things . . . done in this Sutter Street period and I see plainly that I'm exactly the same person, doing the same things in different forms, saying the same things. It's amusing sometimes . . . to look at my own early endeavors and [say], 'There she is, there she is again!' It's built-in. Some things are built in."[56]

Although Lange had embraced FSA goals in the agency's early years and helped to mold the Photography Unit, by the end of the 1930s the FSA had shifted its objectives and its vision. Stryker needed to prove that the FSA was working, so photographs of contemplative migrant mothers and exhausted travelers no longer filled his needs. In fact they tended to hurt his case with the much more conservative Congress that had swept into office in the 1938 midterm elections. Stryker had begun to see his photography group as a producer of a tremendous historical record, and in his opinion the more records they amassed, the better. Stryker wanted photographers who were willing to cover a lot of ground in a short time and send back hundreds of photographs composed in brief stints on the road. He was willing to forgo lengthy captions to get a substantial number of visual images in the file. As a result, the facets of Lange's work that had made it so compelling, so appropriate for the FSA in the midthirties, had made it obsolete by the end of the decade. Lange's two-year struggle with Stryker over her approach, her loyalties, and her handling of negatives, gave him the excuse he needed to release her for good. In 1939, when Stryker was forced by budget constraints to fire one photographer, he claimed to have chosen the "least cooperative" one. Lange's signature style and its accompanying themes, individual independence and success through mobility, were surpassed by fresher ideas coming from new directions.[57]

3

A Radical Vision on Film

MARION POST'S PORTRAYAL OF COLLECTIVE STRENGTH

Speak with your images from your heart and soul. . . .
Trust your gut reactions; Suck out the juices—the
essence of your life experiences.
—Marion Post Wolcott

WHEN MARION POST MOVED TO WASHINGTON in 1938, she carried cultural baggage heavy with radical politics and innovative
art. About her personal convictions, she remembered, "I had warm
feelings for blacks, could communicate effectively with children,
had deep sympathy for the underprivileged, resented evidence of
conspicuous consumption, [and] felt the need to contribute to a
more equitable society."[1] Nearly every image Post produced with
U.S. Government equipment and on U.S. Government time reflects
those convictions; her photographs also show that she sought out
the vitality effusing from everyday life and ordinary people and
their relationships with one another, even in the midst of the
Depression. Her previous camera work, although limited, had been
devoted to cooperative efforts, revealing her inclination toward
the strength of collective action. And her preferences translated

Seeing America — page 126

onto film a firm commitment to the group as an effective vehicle for handling economic and social needs.

Unlike her fellow documentarians Dorothea Lange and Doris Ulmann, Post could not claim a history or a reputation as a distinguished portrait photographer. She had no experience recreating a studio in the field and had spent very little time filling her frames with faces of single individuals posed to appear talented or dignified or hardworking. But she could effectively show the intensity of human interaction, the omnipresent heat emitted from people who lived together intimately or worked together closely. Twenty-eight-year-old Marion Post brought to her job as FSA documentarian a worldview colored by her upbringing in an unconventional household, an aesthetic sense molded by her training as a modern dancer, a photography education initiated in Europe amid Nazi terrorism, and sharp, recent memories of abuse heaped on her by an all-male photography staff at a Philadelphia newspaper. Her experiences in all of these venues had shaped her political and social opinions; they also defined her artistic temperament, which led directly to her developing the camera eye she turned on Americans in the late 1930s.

A vital member of Roy Stryker's team, Post came to the FSA Photography Unit at a transitional time for the agency. What had begun as a public relations project in 1935, a job to convince taxpaying Americans that New Deal resettlement programs were desperately needed, changed a couple of years later.[2] Stryker, who recognized that pictures of poverty-stricken farmers and disease-plagued families overwhelmed the FSA photo file, reconsidered his strategy and the role of the Photography Unit in 1937. An economic downturn that year had left New Deal agencies open to severe criticism, and by 1938 an increasingly skeptical and often hostile Congress demanded evidence that programs such as the FSA were effectively handling the problems of poverty. Stryker saw that the FSA would survive only if he redirected his hired cameras and expanded the agency's goals. The immediate usefulness of FSA photographs had already been demonstrated by their broad exposure in the country's best-selling picture magazines and newspaper photogravure sections. Stryker willingly shared the photographs with popular publications of the day, including *Life*, *Look*, *Camera*, and *Midweek Pictorial*. Other federal agencies, such as U.S. Public Health, took advantage of the mounting FSA collection, as the "honest photograph" proved its "inherent educational power." With an expanding FSA file on his hands, the director recognized what tremendous opportunities lay before him to show America to Americans. In Stryker's opinion, the collection would someday be of historical value, and he wanted to make it as complete and representative a picture of the nation and its people as possible. Alan Trachtenberg has pointed out that Stryker "adopted the idea of a historical record with evangelical fervor." As a result the original New Deal public relations project became a nationwide search for America, one conducted principally with cameras and only peripherally with notepads.[3]

Post's contribution to the search was in volume, breadth, and subtle social commentary. Before joining the FSA staff in 1938, she had completed a couple of "documentary" projects—a photo shoot of a farmers' cooperative for *Fortune* magazine and a job taking still photos for a film about political activism. At the time she did not articulate any personal theories about or definitions of documentary photography. Nearly fifty years later, though, she pointed out to a group of photographers and scholars that the most important quality a documentary photographer could possess was the ability "to empathize with the people directly and indirectly involved."[4] She felt that her cultivation of this ability had prompted her to shape the images she created for the FSA in her three years there. Because of the social conscience she had honed, she developed not just a rudimentary understanding of, but an empathy for her subjects and their lives. Her socially charged political education began in an extraordinary household.

Marion Post was born in June 1910 in Montclair, New Jersey. Located eight miles northwest of the county seat of Newark, Montclair sat on the edge of the most concentrated industrial area of the state. Between Montclair and Newark was Bloomfield, where Dr. Walter Post had made a home for his wife, Nan, and their older daughter, Helen. They lived just a fifteen-minute commuter train ride from bustling New York City. Two hundred miles south of Boston and ninety miles east of Philadelphia, Marion Post's hometown was located in the center of the most heavily populated urban-industrial region in America. As a result she was exposed at an early age to the region's social and economic problems. She attributed much of her awareness to her mother, an idealistic woman who supported racial equality, labor union activity, and leftist political causes. Nan Post's liberal opinions on everything from Bolshevism to modern art and her own unfettered sexual expression caused her neighbors and, eventually, her own husband to consider her too eccentric and too radical. Walter Post, a practitioner of holistic medicine, divorced his wife when their daughter Marion was thirteen years old. Older sister Helen, the "overachiever" of the two Post girls, aligned with her father; Marion chose to spend more time with her mother. Marion watched an already-scandalous situation become even more so when her mother refused the customary financial assistance from her former husband. Old friends abandoned Nan, but the staunch independence she displayed made a lasting impression on her adolescent daughter.[5]

In her capacity as a registered nurse, Nan Post took a job with Margaret Sanger at the Birth Control Research Bureau in New York City. Fighting hard to eliminate the social stigma and oppression that accompanied unwanted pregnancy, Sanger organized birth control clinics all over the country, addressing in particular the birth control problems of working-class women. Post took pride in working with Sanger, particularly as they guided the New York clinic through the growing pains of its early years. Sanger was so impressed with her colleague's work that she asked her to travel nationwide to set up new clinics. Marion remembered, "Mother became a field worker and a pioneer, traveling alone in her car around the country to rural and urban areas, first doing the ground work (obtaining support of the clergy and local influential citizens) for the establishment of birth control clinics—for her a crusade. Then after the Sanger Research Bureau and American Birth Control League merged . . . my mother became their spokeswoman. I admired her not because she was my mother, but for her courage, dedication, strength and compassion." Perhaps unknowingly, the daughter was describing herself as well—as a young FSA photographer, she too was a "field worker" who traveled alone by car to villages in New England, cities in the South, and wide open spaces in the West, experiencing rural and urban life, talking to officials, laying the foundation before she took her cameras into the places the federal government wanted her to see.[6]

Her mother's influence loomed large. In the 1920s Nan rented an apartment in Greenwich Village, a fertile artistic environment that exposed Marion to musicians, painters, and actors. The messages she received while living with her mother encouraged her free expression of the physical and the sexual. As a result the teenager chose to develop an art form in which she used her body as the primary vehicle for self-expression—dance. After completing high school at Edgewood, a private boarding institution in Greenwich, Connecticut, Post enrolled in the New School for Social Research in 1928. The New School served as a training ground to inspire and encourage liberal thought and action among its artists, educators, and intellectuals. Sensitive to the avant-garde, the institution sponsored modern dance teachers as lecturers. By this time Post had devoted much of her time to studying dance with Ruth St. Denis, a pioneer of modern movement. The lessons she learned from St. Denis at the Denishawn School of Dance often echoed her mother's emphasis on maintaining and projecting a positive sense of self. "Miss Ruth" viewed dancing as "an elixir for the attainment of health and beauty." Her style brought dance down from the high–culture pedestal it had long occupied, allowing a wider audience to appreciate the spectacles she presented. This style helped to influence Post's own developing ideas about the close connections between culture, experience, art, and education. But Post's truest mentor in guiding the development of her aesthetic sense was Doris Humphrey, a longtime student of St. Denis and a classmate of Martha Graham. Post was so impressed with Humphrey that she abandoned her university studies in order to study dance full time.[7]

Among the lessons Humphrey taught, three in particular resonate through Post's later artistic expression. The first was that the ultimate achievement in art (in this case, dance) was balance; the second called for a stage to be precisely marked in time and space; and the third was that emotion should guide one's life. As a result of her belief in the last of these, Humphrey designed many dances offering social commentary.[8] Although Post constantly relied upon these maxims as a photographer, the venue where she employed her teacher's lessons to their fullest was in the Mississippi Delta region in 1939. So often reproduced in the last twenty years that it has become a Jim Crow South icon, one image (fig. 23)

draws upon all that Humphrey had impressed upon Post. An
African American ascends the exterior stairs to a movie theater in
Belzoni, Mississippi. Amazingly, Post managed to situate the human figure at almost the dead center of her frame. The words and
other objects are secondary to the single most vital element in the
photo, the lone black man being pulled by a heavy, deliberative
arrow that dictates "Colored Adm." He occupies center stage. Had
Post waited one moment longer, the figure's much larger shadow
would have occupied a more prominent position against the sun-
hot white bricks, dwarfing the real man. But Post did not wait,
and her result was a perfectly set stage in time and space, with
balance achieved, the principal dancer in place, and each subordinate object playing its own role. In the left corner, a restroom
door marked "White Men Only" serves to support the arrow's
directive. The sharp shadow cutting precisely through the clock
at ten and four provides an inner frame as well as a breath of
mystery about the moviegoer's ultimate destination. And the cheery
message "Good for Life!" on the Dr. Pepper advertisement offers
an ironic twist on the social situation for this black Mississippian
and all other African Americans living in the segregated South.
Whether or not she deliberately intended it, Post's emotional investment in this message on race fills the frame. In *Man Entering
Theatre* and many other images that Post produced for the FSA,
Humphrey's aesthetic precepts are detectable; but the dance teacher
provided even more to her pupil by sending her off to Europe to
study dance in a volatile political climate.

Encouraged by Humphrey to take advantage of the new
directions being explored in dance, Post crossed the Atlantic in
1930. An excited twenty-year-old, she arrived in Germany anxious to take on new challenges, but the physically and emotionally draining program quickly took its toll on her. She left the dance
troupe, returned to the United States to work briefly on an el-
ementary education degree, and taught in a small western Massachusetts town. Post's teaching experience in the mill town alerted
her to striking disparities based upon socioeconomic status. Beverly
Brannan has pointed out that Post became "increasingly uncom-
fortable with the class differences she observed between the upper middle class children she taught and the poor children in the
neighborhood where she boarded." Carrying the seeds of a grow-

Fig. 23. Marion Post Wolcott. *Man Entering Theatre. Belzoni, Mississippi, 1939.* U.S. Farm Security Administration Collection, Prints and Photographs Division, Library of Congress.

ing political conscience, the restless Post returned to Europe in 1932, joining her sister in Vienna. She chose not to pursue serious dance any further, but instead resumed the studies in education and child psychology that she had begun in New York.[9]

The University of Vienna offered new experiences and provided a stimulating education for a young, impressionable student. Living amid student activists, Post got a firsthand glimpse of the tense and often violent political scene in Austria. During her second year at the university, a general strike shut it down. As the National Socialist Party gained ground in Austria, demonstrations became more frequent. Post recalled, "I heard the heavy artillery bombardment of the Socialist workers' housing in Floridsdorf, whose occupants had demonstrated against encroaching Fascists. The complex was demolished and the uprising quelled."[10] Students in Vienna played active roles in politics, and significant changes took place in a city previously known for its political conservatism. The students employed every possible method to spread their messages but found photography to be a particularly useful medium. Because of the popularization of photography on the European continent, especially in German and Austrian metropolitan areas, the public saw a great number of photographs in the daily newspapers and in other publications. The emergence of photo clubs provided outlets for both amateur and professional work. But perhaps the most significant influence was the workers' photography movement, which emphasized the importance of recording everyday activities. The photographer and the subject were bound together through their understanding of each other. Above all, the movement preferred the documentary approach and focused its cameras upon "working environments and working conditions." Photography scholars Hanno Hardt and Karin Ohrn point to the emphasis on collective endeavor in the worker photographs, noting, "The theme is not one of individuals, but of a mass—even 'the masses'—of people engaged in a coordinated activity. The context of that activity, whether in the streets or in the work place, defines their role, while the size of the group suggests their power to define the setting as their own, one in which they, as a group, assert control."[11] The workers' movement reached its height while Post lived in central Europe, where she received not only a sound political education but, perhaps more importantly, her first camera.

Borrowing a small 4 × 4 Rolleiflex camera from her sister's friend, photographer Trude Fleischmann, Post gave the instrument a trial run. Since she knew little about cameras and had nothing to compare the borrowed camera to, Post was impressed by the Rollei's twin lens reflex, which allowed the photographer to view a scene before recording it. This feature permitted more accurate composition than older camera models allowed. A type of "candid camera," the small four by four was an extremely popular, quite versatile, model. Post remembered that the Rollei was appropriate for the documentary style of photography that was growing more fashionable at that time in Vienna and throughout Austria. A good price and a favorable exchange rate made it possible for Post to purchase her own Rollei. Having her photographs praised by Fleischmann increased Post's excitement about her newfound hobby, but she had only a limited time to capture Austrian scenes. By the summer of 1934, all political parties had supposedly been dissolved by Chancellor Engelbert Dollfuss's authoritarian regime. The strongest opposition remained the growing National Socialist Party, whose Austrian members opted to protect themselves by murdering Dollfuss in July of 1934. Though interested in the political scene in Vienna, Post sensed the oncoming defeat of any democratic movement. She had seen local fascists harass peasant families in the mountains and destroy their crops. An unforgettable experience in Berlin also left its mark on her. While on a brief visit, Post attended a Nazi rally. She found herself surrounded by hundreds of people aroused by the emotionally charged words of Adolf Hitler. The frenzied exhibition shocked and frightened her. Unable to withstand the uncertainties connected with increasing political violence, Post returned to the United States armed with plenty of fresh ideological ammunition and a new camera.[12]

Upon her return Post accepted a teaching job at Hessian Hills, a private "progressive" school in Croton-on-Hudson, New York. She was fortunate to be hired at a time when many schools were closing or the school years were being shortened because of the Depression. What made it more remarkable was that she, as a young single woman relatively new to the teaching field, received the offer even though national attitudes had grown more antagonistic toward single women who filled the places unemployed men needed so desperately. Men made considerable gains in the teaching profession at this time, whereas young single women "lost ground." Post had returned from abroad to teach in a country where "economic competition, structural changes, and public sentiment all worked to the advantage of men at the expense of female teachers." Post seized the opportunity in her teaching environment to sharpen her photographic skills; she snapped pictures of her school children busily engaged in daily activities. With help from her sister, Helen, who had been a photographer's apprentice in Vienna, Marion learned the basics of developing and printing film. When some of her students' parents offered to buy her photographs, Post saw the potential profit her hobby could bring. She decided to leave what she considered an unappealing job to spend more time with her camera, hoping to "earn at least a partial living on it."[13]

A venue more exciting to her than the classroom had further inspired Post's camera work. While teaching, she had spent her weekends in New York City, capturing scenes of Group Theatre activities—initial rehearsals, dressing-room preps, and opening-night performances. She even attended their summer workshops. The enterprise not only gave Post more camera experience, but it also exposed her to another art form inspired by radical politics. Group Theatre, which determined to bring "social consciousness to Broadway," built itself on the strength of community interaction. Modeled after the Moscow Art Theatre, the company's individual players were charged to develop their personal identities through the group's identity, which each actor and participant helped to forge. Moreover, the audience was to be brought into the stage action, making the entire artistic production an effort in which all present were actively involved. The earliest and perhaps the most successful Group Theatre production to elicit this effect was Clifford Odets's *Waiting for Lefty* (1935), which called on full participation as the characters in the drama (who are contemplating a workers' strike) direct their appeals to audience members. Witnessing the idealism on which Group Theatre perpetuated itself, Post absorbed its message that collective strength transcended individual power. The emphasis on collective identity would soon become the single most important defining element of her photography. But she still needed training.[14]

Enthusiastic about her future possibilities, Post set out in 1936 to establish herself as a freelance photographer. In the midthirties, the Workers' Film and Photo League frequently sponsored lectures and exhibits at various photo clubs. Just a few weeks before she resigned her teaching job, Post attended a photo club meeting in Manhattan, where Ralph Steiner spoke as the featured guest. After the lecture Post approached him, introduced herself, and expressed her growing interest in photography. Steiner asked her to assemble some of her photographs and bring them to his studio the following weekend. After scrutinizing Post's work and discussing it with her, Steiner invited her to join a small group of students who met with him on weekends. For several months Post attended Steiner's informal workshops, and she found them to be helpful. Each week the students examined samples from Steiner's large assortment of photographs. He guided them in analyzing other photographers' work, as well as their own. He gave weekly technical assignments to the young aspiring photographers, and when they reassembled he critiqued each student's work and offered advice when appropriate. Post recalled that Steiner's assignments were "very vague"— he wanted the novices to use their own material and to employ individual creativity to its fullest extent. Steiner felt it was necessary for them to record their "own impressions" in their "own way." He used to ask the photographers, "Which is worth more— an apple or a photograph of an apple? After all, the apple has taste, smell, nutrition, three dimensions, life, while a photograph is flat, inedible, etc. Naturally, the answer is that if the photographer has included a bit of himself in the photograph, then the photograph outweighs the apple itself, since a man outweighs an apple." Adhering to a strict policy of honest photography, Steiner believed if people made statements in pictures *or* words that were false to their own identities, they were lying.[15]

The inspiring combination of Ralph Steiner's teaching and his colleague Paul Strand's photography left an indelible impression on Marion Post. The critical eye Steiner applied to her work— the eye concerned solely with what photographs had to say—helped Post finely tune her technique. Steiner's earliest criticism addressed her approach as being "too artistic and somewhat directionless." Though not as explicitly critical of Post's photography, but certainly as influential, Strand led by example. Post revered his work, the rich tones and textures he created on photographic paper. As she strove to improve, Post reaped the benefits by marketing some of her photographs, a few to educational magazines and similar publications. Her rapid development as a photographer did not escape Steiner, who respected Post's work so much that he asked her to accompany him to Tennessee on a documentary film project undertaken by his new organization, Frontier Films. Frontier was guided by Steiner's desire to explain social problems with actual participants rather than actors and fictional texts. The Tennessee project, codirected by Group Theatre actor Elia Kazan, focused on the Highlander Folk School, an institution near Chattanooga that trained people to lead campaigns for social justice, among them rights for labor and enfranchisement of blacks. The school posed a threat to southern businessmen and appeared to some politicians to be a breeding ground for subversive ideas. Steiner and Kazan's film, entitled *People of the Cumberland,* promoted union organization by showing inspired group meetings at Highlander in addition to union celebrations. The bleakness of isolation and individual effort is set against the brightness of cooperation, played up by a narrator's voice offering carefully timed lines such as "They're not alone anymore; they've got a union." Post took still photographs for promotional purposes during the filming. The trip proved more than educational for her, as she admitted: "My social conscience was nourished." In April 1937 *Variety* labeled Frontier Films the "Group Theatre of motion pictures." Of Post's experiences with Group Theatre, Frontier Films, and the Photo League, she remembered that their concerns about fascism and about "social and economic injustices were in tune with my own." Her participation in leftist groups that blended art, politics, and culture made her as partisan a documentarian as anyone who claimed the title in the late 1930s.[16]

Post's other travels in 1937 included a *Fortune* magazine assignment that sent her into the rural Midwest. Her photographs were to accompany a feature story on consumer cooperatives, an increasingly popular means used by farmers and small community dwellers to obtain necessities at cheaper prices. In Waukegan, Illinois, Post captured scenes of a Finnish co-op where workers baked their own bread, homogenized and bottled local milk, carried a full line of groceries, and sold gasoline. With the assignment she also traveled to Noble County, Indiana, to photograph a farmers' cooperative with its own flour mill and to Dillonvale, Ohio, where one of the country's oldest cooperatives was located. The *Fortune* assignment gave Post considerable exposure as a serious photographer. Of the negatives printed, twenty-one were published in the March 1937 issue of *Fortune,* along with a thirteen-page story entitled "Consumer Cooperatives."[17]

After completing a few Associated Press projects, Post received a recommendation to work as a staff photographer at the *Philadelphia Evening Bulletin.* With no previous newspaper experience, she took the job, joining nine other staff members, all men. Being a woman on a major newspaper staff was quite a rarity; according to Penelope Dixon, Post was probably the only woman in the United States to hold such a position. The pressure and competition in a profession dominated by men kept her busy perfecting her technical skills. The *Bulletin* provided her with a large Speed Graphic, a camera common to news work but completely unfamiliar to Post. She recalled that her fellow photographers on the staff gave her all the help she requested about learning to operate the new model, but she first had to prove that she was serious about her work. After they had properly initiated Post by spitting into and putting out cigarette butts in her developer and by bombarding her with spitballs, she lost her temper: "Finally I exploded—telling them that I was there to stay; I had to earn a living, too; this was where I was going to do it, and they'd better darned well like it. I told them how and when I could be very useful to them and that I needed their help in return. . . . We reached a truce." The first few months Post did "the regular run of newspaper assignments"; then the *Bulletin* decided to step up its fashion service. Of the ten photographers on staff, Post was chosen to cover this new area of emphasis. It seemed the editors at the *Bulletin* found it convenient to expand their coverage of fashion since they had a female photographer they could rely on. Initially Post was paired with a reporter, but she ended up doing practically all the material for the fashion and society pages alone. She soon felt stifled by having to shoot too many garden parties and designer shows. Her hopes of embracing hard news stories shrunk. So she decided to take a trip to New York City and air her frustrations to her critic-mentor Ralph Steiner.[18]

Steiner listened as Post described her disgust with the mundane *Bulletin* assignments. Sensing her strong desire for more interesting tasks, he told her about Roy Stryker's team at the Farm Security Administration in Washington. Steiner showed her a few FSA photographs and explained the purpose of the Historical Section, since she was not acquainted with the project. He suggested that she contact Stryker in order to set up an interview. Steiner not only helped her make photo selections for her portfolio, but he also wrote a recommendation for her. And he provided a second note of introduction to a friend of his at the Housing Administration. Armed with two letters from a highly reputable source, Post promptly began planning a trip to Washington. While she planned, Ralph Steiner traveled to Washington carrying some of her photographs, which he personally presented to Stryker; at the same time Paul Strand wrote a glowing recommendation to Stryker, stating, "If you have any place for a conscientious and talented photographer you will do well to give her an opportunity." The following week Stryker received a simply stated request on *Philadelphia Evening Bulletin* letterhead. It read, "At Ralph Steiner's suggestion I am writing you concerning a job. He showed you some of my photographs last weekend, and I am very anxious to discuss the possibilities with you." In less than two weeks, Stryker decided to hire Marion Post as a full-time photographer with the Farm Security Administration. He made his official offer a few days after one of Post's stills from *People of the Cumberland* graced the cover of the *New York Times Magazine*. With the nationwide exposure and a new position, Post's career took a serious turn. She made a "painless and friendly" departure from the *Philadelphia Evening Bulletin*.[19]

Stryker's newest appointment to the FSA crew may have surprised his other staff members. Post was young, still in her twenties, and she had handled a camera for only three and a half years. Although some of the original FSA photographers had left the agency, nearly all had been veterans of the craft, having mastered every step of the photographic process from initial light judgment to final print development. Several, including Dorothea Lange and former employee Carl Mydans, could boast numerous nationwide photo publications in *Life, Look,* and other magazines. Painter-turned-photographer Ben Shahn was well known for his politically charged canvases before joining the FSA. Art photographer Walker Evans, who had accompanied author James Agee on an Alabama assignment in the summer of 1936, was looking for a publisher for the book the two had produced. The FSA photographers' various successes may have outnumbered those on Post's résumé, but Stryker saw freshness in her approach that the agency needed. Speaking of her FSA appointment, Post later said, "Roy E. Stryker took a chance."[20]

Twenty-three hundred dollars a year, five dollars a day for field time, four and a half cents a mile to pay for gas, oil, and car depreciation—these figures represented the total salary offered to Marion Post as a new FSA employee. In addition, she could count on the agency to provide her with film, flashbulbs, and some other equipment. Stryker informed her that the section's photographers primarily used three camera models, the Leica, the Contax, and the $3\frac{1}{4} \times 4\frac{1}{4}$ Speed Graphic. In a letter to Post explaining the terms of employment, Stryker reiterated that *all* negatives had to go into the official file. His struggles with Dorothea Lange over control of negatives had so infuriated him that he made his expectations quite clear to Post. Each time her camera shutter clicked, the negative automatically became the property of the U.S. Government, which had sole authority over its use. Public ownership did pose a problem for Stryker, though, as he fought newspaper and magazine editors in order to secure credit lines for his FSA photographers.[21]

After finally making her way to Washington the last week in August, Post received a spectacular initiation. Rather than subject his new employee to lengthy explanations and detailed instructions, Stryker turned her loose in the photograph files, the thousands of prints, among them Arthur Rothstein's Dust Bowl scenes, Dorothea Lange's hopeful faces, Walker Evans's proud sharecroppers, and Russell Lee's yearning children. The magnitude and the depth of the imagery overwhelmed Post. She recalls that she was "overcome and amazed and fascinated. . . . The impact of the photos in the file was terrific." Besides their sheer quantity, the file pictures projected a quality of honesty that motivated Post. Viewing them not as individual works but as a unique group of documents overflowing with emotional strength, Post realized, without a doubt, that she wanted to contribute to this vast collection. The first month's experience, though wonderfully stimulating and thought-provoking, also left Post feeling somewhat inadequate, since the previous year had been devoted to nothing but soft news work. She feared that she might fail to live up to the high standards set by previous FSA photographers, most of whom considered themselves artists as much as camera operators. Logistical matters also threatened Post. She wondered if she would be able to keep up the travel pace, find time to write background notes, remember to caption every negative, and get the material back to Washington not only in an orderly form but also on schedule.[22]

Nervous as she might be, Post remained ecstatic about her new position, believing she could "contribute something important, work that might inform and influence the American people and affect legislative reforms. Be a crusader!" Her first month in Washington, Post stayed close to the FSA office, reviewing procedures and reading the literature Stryker recommended. Stryker, ever the academic, usually suggested that his photographers read a variety of books, essays, and reports to prepare themselves for the regions where they would be traveling. In her early days at the agency, Post also listened to the sound advice offered by Arthur Rothstein, a twenty-three-year-old seasoned veteran on the FSA team. When Post received her first field assignment, she collected and packed supplies Rothstein had helped her acquire, including an axe he insisted she have close at hand when traveling.[23]

The initial journey Post made as a field photographer took her into the Bluefield and Welch coal mining regions of West Virginia. Her brief time in the area fascinated her, despite the continual frustrations of inclement weather. She reported to FSA office secretary Clara "Toots" Wakeham, "If it hadn't stopped raining, I'm afraid I'd have been driven to the Christian Science Reading Room or Myra Deane's Elimination Baths (both right next to the hotel), just for a change or purge, you know." Earlier, Stryker had advised her not to let weather, among other things, get her down. As the skies cleared, Post turned her attention toward the people, especially those whose critical mine-related health problems appalled her. One photograph (fig. 24) shows a thin young woman, a tuberculosis and syphilis sufferer, perched upon a tiny porch rail. A half-smile offsets her dark, tired eyes. The softness

Fig. 24. Marion Post Wolcott. *Miner's Wife. Marine, West Virginia, 1938.* U.S. Farm Security Administration Collection, Prints and Photographs Division, Library of Congress.

and ease of her manner—arms crossed to support herself, knees slightly bent, tousled hair pulled loosely away from her face— contrast sharply with her rugged surroundings, particularly the makeshift log rail on which she kneels. More striking is her confidence. Uninhibited by nearly bare breasts, she exudes a comfortable sexuality despite her illness. Post's earlier training and her own sense of self permeate this image, as the former modern dancer focuses here on the physical, heeding her mother's lessons about bodily freedom and sexual expression. Fifty years after she made the photograph, she said of her subject, "She's pretty erotic, isn't she? I probably admired that. . . . She was so beaten down by life and yet there was all this sexuality in her, not just sexuality but sensuality." Seven photographs in the series provide more details about the woman's life; in one she holds a fist up to her mouth while coughing into it; in another, two young children appear; and in a third her surroundings show torn paper and cardboard affixed to the walls of an abandoned mining company store. Forty years after the prints became part of the immense FSA collection, Witkin Gallery in New York City displayed the famous porch scene in a Marion Post Wolcott exhibition. In his review of the show, critic Ben Lifson wrote of this particular picture that he saw not a pathetic, sick woman whose miner husband had lost his job, but "an image full of rough and perplexing sexuality." In a more disturbed tone, he continued, "[W]e're supposed to see these poor people as sufferers, not as vital and occasionally alluring."[24]

Lifson's comment not only raises a pertinent issue regarding the documentary genre, but it hints at Post's willingness to mold her own documentary vision, departing from the FSA standards set by earlier photographers. William Stott, in *Documentary Expression and Thirties America,* notes that these politically driven images (even a series of the same subject) reveal very little about the actual "real life" experiences of the men, women, and children in the pictures. He points out that the subjects "come to us only in images meant to break our heart. They come helpless, guiltless as children. . . . Never are they vicious, never depraved, never responsible for their misery." As a result Great Depression images often fail to allow their subjects complete humanity. Stott's description aptly characterizes the people in photographs made by Walker Evans, Ben Shahn, and Dorothea Lange in the pre–

Marion Post years at FSA. Although Post admitted that she and her FSA colleagues had similar views about their work, that each photographer possessed a depth of social consciousness and felt genuine concerns about "the plight of human beings," she emphasized that their individual approaches remained varied and distinctive. On the West Virginia assignment, Post was beginning to show just how different her vision would be from those of her FSA predecessors. She took interest not merely in the "plight" of human beings but in the people themselves.[25]

Her views of the tubercular woman and other subjects in the coal mining world disclose Post's developing eye, which was moderate enough to please Stryker yet sufficiently radical to satisfy her own personal politics and tastes. Although Post's image of the unemployed coal miner's wife is in some ways derivative of Lange's style, it more forcefully projects Post's own outlook. She wanted to show dignified humanity, so she ennobled the figure by the angle of her shot; the woman looks over and down at the photographer, a technique Lange frequently used. But unlike the subjects in Lange photographs, here the body approaches the viewer, moves forward in the scene with her shoulders and her toothy smile; the subject is obviously comfortable with the photographer and may even have shared with her a few intimate details of her life. This picture reveals one of the most stunning differences between Lange's and Post's approaches—the treatment of sexuality. Lange had pictured bare breasts before, but usually to show mothers nursing their children. There are no children in Post's frame, therefore allowing the suggestion of sexual desirability rather than maternal responsibility. This vital photograph from her first assignment foretells the nature of Post's contributions to the FSA files.[26]

Not only would the photographer attempt to portray the full humanity of her subjects and their lives, but she would cover issues that addressed "blind spots" in the collection. For example, she produced several images in a series showing miners of varied ethnic backgrounds and their families. A challenge to Doris Ulmann's vision of the Anglo-Saxon descendants who occupied the Appalachian hills, Post photographed several Bohemians, Hungarians, Poles, and Mexicans, emphasizing the Mexicans' role in breaking a 1926 miners' strike. In her photographs she attempted to evoke a sense of easy camaraderie between those of different backgrounds, capturing on film their conversations and leisure pursuits. On her first assignment Post provided further examples to prove her departure from earlier FSA work and her enthusiasm to take risks. In one image, a man stands fully naked in a shower he had constructed. How Post arranged such a scene is uncertain; no doubt Stryker was shocked to see it surface on the contact prints. Clearly, the transition already under way in the FSA Photography Unit accelerated as soon as Post's undeveloped film arrived in the Washington offices. The final prints from Post's West Virginia negatives, a wide spectrum of provocative images, showed the FSA director the "undisputed" superior quality of his newest employee's work.[27]

Having proven her skills on the inaugural journey, Post prepared to make her first FSA trip into the rural South. A challenging venture lay ahead, for in no other region of the country had the FSA met with such vicious opposition. The agency was considered "a disturber of the peace" in the South, where "virtually every FSA program and policy seemed to touch a nerve." The FSA had never enjoyed a strong political constituency in the region, since scores of sharecroppers and migrants were "voteless or inarticulate." The natural enemies of the agency, southern landlords and operators of large farms, possessed both votes and voices and certainly wanted to maintain their supply of cheap labor. They objected to any aid the FSA might offer to tenant farmers. As a result, strong congressional representation of the FSA's opponents purposely kept the agency's appropriations as low as possible. The rural victims of 1930s politics, according to FSA historian Sidney Baldwin, were "southern tenants, sharecroppers, and laborers for whom restriction of cash crops meant eviction from their rented lands and homes or complete destitution where they were; subsistence farmers of Appalachia who were actually outside of the farm economy; and migratory farm laborers for whom the filtering-down effect of rising prices was a snare and a delusion." A 1937 recession further deepened the problems of those living on the land. This downturn in the economy, tagged "Roosevelt's Depression," also dealt a grave blow to New Dealers, proving to them that their programs had not packed the force necessary to break economic cycles. Agencies such as the FSA then were forced to prove their

effectiveness or suffer the withdrawal of congressional appropriations.[28]

In the wake of economic and political rumblings, Stryker opted
to give Post a different role from those his earlier photographers
had assumed in southern states. The major projects, such as FDR's
"lower third" of the nation and the Dust Bowl, had been completed. Post would survey FSA successes while simultaneously
building Stryker's historical record of the nation. As she described
it, Stryker "wanted to fill in, wanted pictures of lush America,
wanted things to use as contrast pictures in his exhibits, wanted
more 'canned goods' of the FSA positive remedial program that
they were doing. Partly, I think, to keep his superiors happy so
that they would continue to support his program." The agency's
new direction would require Post to meet with FSA supervisors so
that they could show her the "good side" of local programs. And,
as a result of a directive from Department of Agriculture information director John Fischer, she and her fellow photographers
would continue to take fieldwork assignments for agencies such as
the U.S. Public Health Service and the U.S. Housing Authority in
order to keep the FSA Photography Unit solvent. Stryker was very
accommodating in scheduling and rearranging his field photographers' travel plans to include such assignments. He recognized their
importance to the system as a whole, but more importantly he remained eager to please the "higher-ups" in Washington and to
keep the FSA Photography Unit alive.[29]

Barriers stood high for any outsider traveling in the rural
South, but a U.S. Government employee who happened to be a
twenty-eight-year-old unmarried white woman driving her own
car could expect to be treated with great skepticism. Post's mere
presence would pose threats in certain areas, Stryker told her. Soon
after hiring Post, Stryker warned her about the segregated culture's
expectations, noting, "[N]egro people are put in a very difficult
spot when white women attempt to interview or photograph them."
He expressed serious doubts about subjecting her to possible
mistreatment, but at the same time he was confident in her ability to take care of herself. Although the FSA photography director treated all of his field employees with a degree of paternalism,
he seemed particularly protective of Post, who later said, "From
the very beginning, being a woman, I was considered by Stryker

to be in a different category." But he had not expressed similar concerns to Dorothea Lange. Her situation was safer, in his opinion, because she almost always traveled with her husband, Paul Taylor.[30]

Even before Post made her first trip on the FSA payroll, she "felt challenged" by Stryker's attitudes and intended to confront him about his chauvinism. One of their more heated exchanges resulted from an incident on her second field trip. In South Carolina a small group of people whom Post described as "primitive" accused her of being a gypsy. Her deep tan, bright bandana scarf, and elaborate earrings intimidated them, as did the mysterious boxes of equipment in her convertible. Afraid that she might kidnap their children, they demanded that she leave immediately.[31] When Stryker learned of the ordeal, he reprimanded Post on her appearance:

> I am glad that you have now learned that you can't depend on the wiles of femininity when you are in the wilds of the South. Colorful bandanas and brightly colored dresses, etc., aren't part of our photographer's equipment. The closer you keep to what the great back-country recognizes as the normal dress for women, the better you are going to succeed as a photographer. . . . I can tell you another thing—that slacks aren't part of your attire when you are in that back-country. You are a woman, and "a woman can't never be a man!"[32]

Assuring Stryker that he had "touched on a sore subject," Post shot back, "I *didn't* use feminine wiles . . . My slacks are dark blue, old, dirty & not too tight—O.K.?" As for the trousers, she argued that female photographers might appear slightly conspicuous with too many film pack magazines and film rolls stuffed in their shirt fronts. FSA historian Jack Hurley has noted that "Stryker never knew quite how to take this disconcertingly frank young woman, but he knew he liked her and the pictures she was producing."[33]

The gender-related issues presented irritations that separated Post's experiences from those of her FSA colleagues. As a young woman on the job, Post suffered from officials' not taking her assignments seriously and from general suspicions about her intentions. She stepped out of her car alone, not with a well-known economist as Lange had or with a controversial novelist as Bourke-White did or with a lively native son as Ulmann had. Early in December 1938, in the first days of her journey south, Post expressed anxiety about her experience at Duke University, where she confronted a number of "disturbing circumstances & problems" because of the local officials' lack of preparation. They had failed to arrange an agenda of shooting sites and prescheduled meetings with FSA project clients. She found her assistants not the least bit aware of which subjects she wished to photograph. In Columbia, South Carolina, a couple of days before Christmas, Post hoped to get several shots of people celebrating the holiday season. But her slow-moving FSA guide made her wait in his office for several hours while he finished other business. Angry that she had missed good morning light for shooting, Post resolved to take advantage of the waning afternoon light. Instead, she spent the better part of the afternoon hours admiring holly trees with the FSA guide's wife, who had insisted upon going along for the ride. With sunlight fading at day's end, Post rued the precious time wasted. The next morning she awoke to pouring rain, conceded defeat in Columbia, and packed her bags.[34]

General suspicions also impeded the photographer's work, requiring her to come up with creative solutions. In particular she found the people of South Carolina's swampy lowlands extremely cautious and unfriendly. They allowed no strangers inside their homes, and they would not let her take any pictures outside the houses. Mimicking her subjects, Post explained their excuses to Stryker—they "didn't like for no strangers to come botherin' around because they mostly played 'dirty tricks' on them or brought 'bad luck.'" Post often resorted to begging and bribing to win them over and therefore kept her pockets full of coins and small pieces of food. Some individuals boldly refused the nickels she offered, asking instead for "*real* money." Another challenge Post faced in the southern states was resistance to any woman with a camera who might have the same agenda as Margaret Bourke-White, a freelance photographer whose collaborative work with Erskine Caldwell entitled *You Have Seen Their Faces* (1937) presented a scathing depiction of the South and its problems. In that popular documentary book the region's inhabitants came off as pathetic and depraved, Caldwell's objective in attempting to prove that his fictional works were based in fact.[35]

The initial southern trek presented Post with cultural differences that shocked her. Appalled by southern eating habits, she observed that vegetables for one meal usually included "fried potatoes, mashed potatoes, sweet potatoes, rice, grits, or turnips!" The treatment of time in the South also caught her off guard, as she told Stryker, "It's amazing. . . .the pace is entirely different. It takes so *much* longer to get anything done, buy anything, have extra keys made, or get something fixed, or pack the car. Their whole attitude is different—you'd think the 'Xmas rush' might stimulate them a little, but not a chance." The differences in the manner of holiday celebration also surprised Post, who was taken aback by the overt gaudiness of it all. She observed that the southern towns she visited were "decked out fit to kill, with more lights and decorations and junk than you could find in the 5th Avenue 5 & 10-cent store." But what most startled her was the constant crackle of exploding firecrackers, sounding off throughout the day and well into the night.[36]

Through January and February of the new year, Post followed the migrant camp circuit in Florida. It was the first of three trips she would make to the state in order to record the FSA's success there. The agency had attempted to provide organized, clean dwellings for migratory laborers. Before FSA intervention, living quarters in the tourist and trailer camps where migrants stayed were drastically overcrowded and unsanitary. Large families lived in one-room cabins or small tents. Privacy and respect for property were nonexistent. Frequently a man and a woman would live together as "good friends" until the season was over, and then they would go their separate ways. Among the many health problems that camp residents suffered, one of the most common was venereal disease. Post's preliminary reading, which included previous reports by FSA photographers, introduced her to migrant life, but nothing compared to her firsthand experience in Florida. And what she found convinced her that the FSA had only minimally improved the basic needs of migrant laborers. Most facilities remained unsanitary, as her photographs prove, showing dwellings without lights, running water, or toilets (indoor or outdoor). Her written records and photographs reveal afflicted, desperate communities, ravaged by natural disasters. In one of the more poignant series of images, workers in Lake Harbor stand by hopelessly as one of their homes burns to the ground. Post situated herself behind them, as a witness to their impotence to stop the swelling flames. A few items saved from the home occupy the foreground of Post's frame, signaling what poverty of possessions the occupants have been reduced to.[37]

Early encounters with fruit pickers made Post aware that the previous season had been disastrous, especially for citrus growers in central Florida. Crates of fruit sold at the lowest prices ever, leaving the pickers with even fewer resources than usual. Since Florida had experienced no rainy season that year, the added expense of irrigation drove many growers out of business. Post photographed the remnants of a bad season—houses and camps abandoned, blown down, or taken apart. The present season promised little relief, as she soon realized. Migrant pickers were idle, anxiously awaiting peak season, which had been delayed by the lack of rain. In the Tampa area, a strawberry-growing region, the crop gradually dried up. During an ordinary season, the "fruit tramps"—men, women, and children who followed the crops— would travel a route from Homestead, Florida, in January to Ponchatoula, Louisiana, in April; to Humboldt, Tennessee, in May; to Paducah, Kentucky, in June; to Shelby, Michigan, in July. By August they would start the trek back south to Florida. But Post found the 1939 season to be an exception. Her research in Belle Glade revealed that about half the usual number of transients were there and that most of those were employed in the packing houses, not in the fields. Post learned that the majority had not traveled long distances to find work but had come from nearby towns, Key West, or the Miami area.[38] Since during an ordinary season, migrant fruit harvesters spent a considerable amount of their earnings on transportation, extended traveling in a bad season was economically infeasible.

The difficulties of migratory work during a poor citrus fruit season seemed slight compared to the troubles of people engaged in vegetable field work. In the vegetable area, located in the drained swamplands of Florida, the availability of jobs depended solely on weather conditions. Tomatoes, celery, lettuce, and bean crops had to be picked at precisely the right time. The 1938–39 winter, with its frosts and its cold, windy weather, had killed almost everything, including a large crop of English peas, a substantial portion of the tomatoes, and most of the beans. Only a small amount of celery, cabbage, and sugar cane survived. A major difference between citrus work and vegetable work in the region was the necessity that individuals hired for the latter labor on their hands and knees in the fine, powdery soil dust known as "black muck." Post witnessed and recorded its effects on children, who had skin rashes and sores scarring their bodies. She complained fiercely to Stryker about the soil dust, and she struggled continuously to keep it out of her cameras and off her film. In jest, Stryker replied to her complaints, "We really don't care what the black dust does to you as long as you can work, but I hate like the devil to see it get into your camera, because it hurts your negatives." Many of Post's images in the Florida vegetable region were unusable because of the powdery soil's damage to her film.[39]

Time was a key factor in Post's Florida excursion. Because peak picking season had been delayed three to four weeks, Post was unable to capture the choice field shots she desired. She photographed a few workers combing the vegetable fields, but the action she had hoped to find was limited to an afternoon picking session during which several horrific screams and rifle shots drew her to a side road. There she eyed a ten-foot-long rattlesnake "as big around as a small stovepipe." Even though the picking season was slow, Post was willing to wait it out and stay in the Belle Glade–Okeechobee area for a few more weeks, hoping that transients would again fill the streets and the packing houses. Interested especially in packing-house activity, Post wanted to produce a photo story of life there—"the hanging around, the 'messing around,' the gambling, the fighting, the 'sanitary' conditions, the effects of the *very* long work stretch." Her desire yielded rich stories, which she took down as "General Captions," but not many photographs of group work. One woman explained to Post that the packing houses opened their doors only a few hours a week. The woman had waited outside one house all day hoping to get work; when she finally got in, she was allowed a mere nine minutes of work time, compared to her usual sixteen- or eighteen-hour day during peak season.[40]

In attempting to create her photo essay on packing houses,

Post focused on a thirty-two-year-old mother of eleven children, whose husband worked in a packing house. To connect the story with the visual image, Post photographed the woman talking (fig. 25). With clasped hands and a dependent child in a torn garment clinging to her neck, the woman looks away while relating that she had been "frozen out" when the employer got rid of the "old ones" at the packing house. He kept young girls instead of mature women in the plant for the long stretches of packing time. The woman offered Post the kind of material on gender experiences that she was seeking—in addition to releasing the woman from work, her employer told her that since she "looked pregnant" he would not give her further work even when the house needed laborers. Post hoped that the combination of her verbatim accounts in general captions and photographs of the storytellers would impress the FSA enough to step up its presence in Florida in order to improve migrants' lives and working conditions. Even though the idle laborers' stories intrigued Post, they did not offer Stryker sufficient substance for FSA photography. He wrote to her, "We mustn't spend too much time in Florida unless there is an awful lot of pay dirt." He promised her that she could return later but insisted that she concentrate on other subjects. Post was willing to move on but asserted her independence by instructing her boss, "It takes time, Roy, but you'll see that it's worth it."[41]

Turning her cameras in another direction—toward Miami Beach's rich tourists and residents who lived just a half hour's drive from the miserable migrant camps—brought the "pay dirt" that would make Post's FSA work exceptional. Among the most highly praised photographs in the collection, Post's images of the wealthy provide sharp contrasts to the plight of the underprivileged. In an indirect manner she made her case about class division in the United States. Her characterizations of the rich reflect a well-directed intention by the socially conscious young photographer to illuminate the gap between wealthy Americans and their impoverished counterparts. She later explained her objectives on the job: "To create a general awakening and concern for [the migrants'] appalling conditions, I tried to contrast them with the wealthy and with the complacent tourists in Florida." Post stayed in Miami overnight, then drove twenty-five miles to the migrant camps during the day. The disparity between the two worlds amazed her. She

Fig. 25. Marion Post Wolcott. *Wife and Child of Migrant Laborer. Belle Glade, Florida, 1939.* U.S. Farm Security Administration Collection, Prints and Photographs Division, Library of Congress.

later told an interviewer, "I thought: 'Wow, this is great. Why not use this material in exhibits and as contrast material?'" Until Post joined the staff, the FSA files lacked any pictures of those situated at the upper end of the income scale. Post noted the omission and believed photographs of the elite at play would help tell the American story. Indeed they did document socioeconomic conditions in the late 1930s, sending the clear message that some people were oblivious to the nationwide depression. Photography historian Stu Cohen contends that Post's Miami Beach series was "a perfectly reasonable addition to the FSA files; it is simply unfortunate that they stand there so much apart."[42]

To accomplish her purpose Post employed the techniques learned years before from her dance instructor Doris Humphrey—use of perfect balance and space. With the playgrounds of the wealthy as her stages, Post created images that made hotel entrance archways appear massive and long, shiny automobiles seem larger than life. Her keen eye helped her achieve a superior manipulation of space and scale. Fully reclined in lounge chairs, relaxed tourists smile carelessly as the sun beams down on their faces. Inactivity is cherished, expected, and serves as a stark contrast to the migrants' idleness, which was characterized by worry and anxiety. In one image (fig. 26) Post took her place above a trio of already-bronzed sunbathers. Her display of open space in the upper right quadrant of the frame shows one of the privileges that a private beach club could offer its patrons—plenty of room away from the hordes on a public beach. Here complete relaxation is possible, as the woman demonstrates with outstretched arms delicately crossed above her head. And between the two sun-drenched men, thoughtful conversation can transpire uninterrupted because they are far removed from the bustling activity of a crowded beach. The exhibition of so much flawless skin provides a stunning contrast to the arms and faces of migrant workers, miners, and poor children, whose bad teeth, soot-covered faces, tousled hair, and skin sores Post had recorded. Here the woman's face reveals the careful attention she has given to penciled eyebrows and colored lips and eyelids. Both men have oiled, carefully combed hair. In addition to the relaxed position of each subject—the sheer enjoyment of idleness—another indication of comfort is the healthy stomach shown off by the seated man who has his legs casually

Fig. 26. Marion Post Wolcott. *People Sunning Themselves. Miami, Florida, 1939.* U.S. Farm Security Administration Collection, Prints and Photographs Division, Library of Congress.

crossed. To experience physical hunger in this setting would be impossible, as Post notes in yet another of her series images. Analyzed by Sally Stein in her provocative commentary on Post's class messages, the photograph is a voyeur's view of a patron being served a multicourse meal by an attentive waiter on a club sundeck. Given an aerial perspective of an exquisitely set table, the viewer sees numerous plates, bowls, and utensils being used by a single individual. Among the small absurdities the photographer identifies are perfect squares of butter kept cool by miniature ice cubes, all exposed to the beaming Miami sun. Stein saw that Post's clear message was about the gap between the server and the served here, as the waiter keeps his reverent distance from the customer while providing the man's every need in a substantial brunch. Ben Lifson identified in Post's Miami Beach collection a "sense of uneasiness," noting that the photographer managed to show "just how much of her world is out of whack."[43] Post's social conscience often translated into irony in her Miami series.

Post found the leisure class as difficult to deal with as any she had previously encountered, which is why most of her scenes appear to be captured from safe distances, odd angles, or undetectable locations. These distances also meant Post rarely talked to her subjects; so the photographs had to stand alone. Most of her captions are brief, such as "A beach scene" or "A street scene" or, more frequently, "A typical scene." Use of the word *typical* on a good number of her image descriptions suggests that Post may even have grown disgusted with the opulence and display she digested on a daily basis while in Miami. Although the majority of her "idle rich" photographs were shot in resort palaces and dwellings and on beaches, Post spent time in another haunt of the wealthy—the racetrack. Besides taking crowd shots of stands full of horse-racing enthusiasts, she attempted to record some gambling scenes. The betting men, irate at her temerity, snatched her camera away. They eventually gave it back, but the gamblers kept her film and instructed her to get out and stay out. As she related the story in a personal interview years later, Post explained that the men were most annoyed that she had tried to get away with the stunt simply because she was a woman—"which was exactly what I was trying to do," she admitted.[44]

The Miami area suited Post's need for rest from lugging her

cameras and other equipment around. At the resort beaches she could safely swim and sunbathe, away from the spiders and "red bugs" that infested grasses in the migrant picking region. While taking a break, she secluded herself and began the long-overdue task of captioning some three hundred prints Stryker had mailed from the FSA lab in Washington. Post assured Stryker that she would devote full attention to the job, attempting to find "some quiet spot where I will be a completely unknown quantity & not be disturbed. I'm going to put on *very* dark glasses & hang a typhoid fever sign on my front & back & *dare* anyone to come near me." Captioning complications frustrated every FSA photographer, including Post. Since the photographers were required to send all film to the FSA lab in Washington for development, it could be weeks before they actually saw their own finished products. One group of negatives that Post sent from Florida showed watermarks when printed; another set was smudged with the "black muck"; and yet another came out blurred, revealing serious shutter mal-adjustment in the Rolleiflex camera. When her negatives turned out satisfactorily, Stryker offered his praise along with a gentle nudge for her to edit scrupulously and to complete captions quickly, so that the Washington office could "get the stuff mounted up and ready for use."[45]

Stryker gave the photographers in his unit much freedom in editing their own work. In Post's case, he believed she leaned toward taking too many exposures of a particular scene or subject, inevitably costing the lab an exorbitant amount in development. In an attempt to eliminate the problem, Stryker suggested that Post "kill" a good number of the negatives as she edited and captioned. Post respected the director's guidance and direction, but even more she appreciated the unusual liberty he allowed her in editing and cropping her own pictures. He also consulted her about lab work and exhibit layouts, causing Post to remark that he provided an opportunity for "unprecedented creative and artistic expression." She gained experience and insight unavailable to a routine staff photographer in the 1930s and said of these opportunities, "I don't know of any magazine that would give you the freedom that Roy gave you in editing your pictures. He would listen to arguments. I mean if he wanted to keep a picture that you wanted to throw out he would certainly give you your chance to say. I suppose in the end he had the last say but if your arguments were valid he would bow to them. . . . We never had any real struggle about this at all."[46]

Post found that Stryker's fairness extended even further, especially regarding her position with FSA regional advisors. As with all of the unit's photographers, Stryker gave Post the authority to handle the FSA representatives as she deemed most appropriate. Local and regional advisors, particularly in the South, often turned out to be stubborn and quite hesitant to allow a young woman to plot her own course on "their territory." Some attempted to shield her from the more unsuccessful aspects of the FSA projects and so limited her travel to the few places they personally wished to have photographed. Post recalled, "They didn't want us to photograph the seedy side of their area, their program; they wanted us to spend much more time, of course, on the progress that they had made." Having aired her frustrations to Stryker about this kind of treatment, she received encouraging replies that assured her that she was the boss in those situations. Stryker explained that she, like all the other FSA photographers, had been hired because she was a specialist. And as a specialist, she was entitled to the privilege of deciding what should be recorded on film. He further stressed the extent of her authority, telling her, "Don't take orders on what is to be photographed or how it is to be photographed from anyone. . . . Don't let those guys put anything over on you. You will have to lie and do all sorts of maneuvering, but it will have to be done—and you know I will always back you, so be sure that you are right, and you may always depend on my OK and support." Post realized that Stryker entrusted her with substantial responsibility and that he faithfully relied upon her judgment, leaving "the whole business" to her discretion. According to FSA scholar Beverly Brannan, his "willingness to let the photographers follow their instincts reflected not only a general confidence in their abilities but also the certainty that the discoveries they made in the field would lead to a diverse selection of photographs for the file." Stryker provided general shooting scripts for the photographers at the beginning of their assignments, but they served merely as guidelines.[47]

Stryker held the reins tightly, though, when it came to logistics and official details. More than once in the spring of 1939 he had to track down Post. He demanded that she correspond often with the Washington office by mail or telegram, informing the staff of her whereabouts. In mid-March Post's agenda had her scheduled to leave Florida and proceed toward Montgomery, Alabama. Several days after her expected arrival date, Stryker wrote in a distressed tone, "I don't think you realize yet the importance of keeping us informed at frequent intervals about where we can reach you. Here it is Thursday and we don't know if you got to Montgomery or not." The second major rule Post had to learn to follow involved satisfying a bureaucratic system. If she wanted a U.S. Government paycheck, her paperwork had to be completed accurately and submitted on time. The Finance Office required FSA photographers to present the following materials on a regular basis: travel logs showing miles covered, equipment sheets listing film rolls and flashbulbs used, and account forms recording all expenses incurred. At different times Post and each of her colleagues neglected the red tape, provoking Stryker's threat to "raise hell with all of you people if you don't attempt to do this." His "volatility" was what she remembered most about Stryker years later, as she observed, "I think he felt possessive of all of us. He could be so supportive, and he also could be so damn tyrannical."[48]

Stryker was forthcoming if he thought his field workers had failed to meet the agency's expectations. He expressed disappointment at their visual surveys of a well-known resettlement project in Coffee County, Alabama. Before the FSA moved in, the area had been devastated by a major flood that destroyed soil that had long been exhausted by intensive farming. Coffee County's farmers had relied solely on the cotton crop until the 1930s, when they were forced to diversify their crop production in order to survive. Some began planting peanuts and beans, while others turned to poultry farming. U.S. Government officials remained curious about the Coffee County project's successes in the wake of diversification and were anxious for Stryker's photographers to provide some visual evidence. FSA photographers Dorothea Lange and Arthur Rothstein on respective assignments had attempted to photograph the "new" Coffee County, but neither could produce the desired effects for Stryker. Like her predecessors, Post could not supply enough material to prove that conditions in this rural Alabama county had changed. After informing his friends and superiors in

Washington that Coffee County remained unphotogenic, Stryker flatly told Post, "*Three* photographers can't be wrong."[49]

Although Stryker felt Post's results left much to be desired, her Coffee County photographs *did* in fact carry the message of success that the FSA wanted to disseminate. In one image a woman stands in her smokehouse, backed by shelves full of canned vegetables and fruits and framed by substantial portions of cured meat. In another photograph Post captured a family's collaborative efforts: a mother and a father and their two sons shell peanuts in preparation for the planting season ahead. Armed with big bowls and busy hands, the four are portrayed as diligent and content to be working together in close proximity. Beyond her attempt to prove that Coffee Countians had embraced alternative crops, Post arranged the image to promote her own agenda—that collective work not only paid off in strict economic terms but also provided necessary human contact, enriching the group effort and in turn producing benefits for the entire community. In another Coffee County image, Post worked from that same theme and revealed yet another issue that she would pursue in many places over the next two years—abundance. From her angle slightly above the Peacock family dinner table, Post places the food at the center of her photograph (fig. 27). Framed on all sides by the members of the family, full bowls, plates, and glasses measure the bounty. No one has paused for the photographer; nutritious eating continues uninterrupted. In the foreground, the central act of *sharing* the food marks the image, as two individuals are connected by the representation of plenty. Light streams in on the scene, transforming the occasion from a mundane daily occurrence to an extraordinarily fulfilling, almost divine process. In her photo caption Post pointed out that FSA supervisors had helped improve the diet and health of project families; in other notes she listed the menu as having included "sausage, cabbage, carrots, rice, tomatoes, cornbread, canned figs, and bread pudding."[50] Stryker's dismissal of the series notwithstanding, Post marked her work in Coffee County and the rest of Alabama with the signs of liberal politics and communal effort—family members helping each other with housework and farmwork, home economists and vocational agriculture teachers instructing rural rehabilitation clients, and neighbors working together in a house-raising project.

Fig. 27. Marion Post Wolcott. *Family Dinner. Coffee County, Alabama, 1939.* U.S. Farm Security Administration Collection, Prints and Photographs Division, Library of Congress.

To prepare for her assignments in Alabama and Georgia, Post scanned the reading material Stryker had sent weeks earlier. One of the works she reviewed was sociologist Arthur Raper's *Preface to Peasantry* (1936), a critical study of farm tenancy in Greene County, Georgia. Raper blasted the earlier plantation system for fostering an unnatural dependency among rural laborers. Well-respected for his comprehensive research, especially his effective interviewing techniques, Raper received thoughtful recognition from FSA staffers. He was genuinely interested in the field photographers' work and never failed to offer his assistance when they came to the Atlanta area. Stryker noted that Raper possessed a keen eye for selecting key sites and angles for the FSA cameras, even though Raper himself was not a photographer. Post was scheduled to spend April and May traveling throughout the state, adhering closely to Raper's suggestions but attending to other matters as well. Post's activities during her two months in the state accurately reflect the varied responsibilities an FSA photographer faced, from documenting segregated farm projects to attending small-town celebrations to locating a decent place to get a camera fixed. Using the Henry Grady Hotel in Atlanta as a base, Post initially directed her energy toward one of the more mundane tasks—shooting a "progress report" on the Area Warehouse in Atlanta. Her principal subjects were broken-down tractors, some photographed before repair, others after repair. Besides focusing on farm equipment, Post got some inside shots of the workrooms and office space. Her comprehensive treatment of the site resulted in Stryker's glowing acknowledgments a few weeks later.[51]

In a three-week period, Post journeyed from Greensboro to Irwinville to Montezuma to Orangeburg and then back to Atlanta. A crowded schedule, horrid weather, and temperamental equipment posed numerous problems for her. Although it meant losing some time, Post decided not to drive at night since everything closed early, including the gas stations. She discovered that "the only ones who stay out are bums who are pretty drunk or tough or both." Realizing that people would not understand why she remained out alone after dark, she cautiously avoided the likelihood of confrontation. Rain kept Post from traveling too far on a couple of occasions. Unable to maneuver her car on the slippery red clay, she decided not to drive it. Stranded in a small town, she was left to endure two May Day programs, both ruined by messy weather and boring speeches. A few days later, with rain continuing, Post took her Plymouth out on the road again and got stuck twice in the thick clay of the Flint River Valley. While her car was being repaired, mechanics found a faulty rear axle, one that obviously had been broken and welded. Post then realized she had driven some twenty thousand miles across the South in a defective car that she had presumed was new, in fact right off the assembly line. She understood why it was a "lemon" after realizing that she had purchased it in the wake of the auto workers' strikes a few years before, when the demand for automobiles was high. With one breakdown repaired, yet another occurred, this time with her camera equipment. The rangefinder, which the Eastman Kodak Company in Atlanta attempted to adjust on three different occasions, had given Post constant trouble. The company seemingly bungled the job each time, leaving Post no alternative but to pack it up and ship it to Washington. Stryker's immediate response included a reprimand about her harsh handling of FSA equipment, but his admonition was accompanied by the good news that her photographs had just been published in *Collier's*.[52]

Stryker gave Post an exceedingly high overall evaluation on the pictures from her Georgia assignments, particularly the recreational scenes. His favorite shots included those taken of schoolhouse activities, of sandlot baseball games, and of sharecroppers standing in front of log cabins. Post, too, was overwhelmed by the pictures, so much so that she found editing even more difficult than usual—"It's practically impossible for me to throw any of them away because the people and their faces are so often different even tho the place or situation is similar." Additional praise came from Raper, who traveled to Washington to view the finished prints and requested that Post be assigned to the region again. Raper eventually used several of Post's photographs in his 1941 publication *Sharecroppers All,* a look at changes in sharecroppers' lives and how U.S. Government programs had affected them.[53]

After enduring sweltering heat and voracious mosquitoes in New Orleans, where she had been "lent out" to the U.S. Public Health Service, Post neared the end of her seven-month journey through the American South. Despite all the pleasure she derived from her assorted encounters with southerners and their ways of living, Post welcomed the opportunity to return to Washington. She wrote to Stryker just before heading north:

> I'll also be glad of a rest from the daily & eternal questions whether I'm Emily Post, or Margaret Bourke-White, followed by disappointed looks—or what that "thing" around my neck is, & how I ever learned to be a photographer, if I'm all alone, not frightened & if my mother doesn't worry about me, & how I find my way. In general I'm most tired of the strain of continually adjusting to new people, making conversation, getting acquainted, being polite & diplomatic when necessary.
>
> In particular I'm sick of people telling me that the cabin or room costs the same for one as it does for *two*, of listening to people, or the "call" girl make love in the adjoining room. Or of hearing everyone's bathroom habits, hangovers, & fights thru the ventilator. And even the sight of hotel bedroom furniture, the feel of clean sheets, the nuisance of digging into the bottom of a suitcase, of choosing a restaurant & food to eat.[54]

The long-awaited retreat from the field brought Post back close to the FSA headquarters for three months. For the first time she was able to see the total accumulation of her work—thousands of negatives and prints with her name in the top right corner.

A few months earlier art critic Elizabeth McCausland had written in *Photo Notes*, "Today we do not want emotion from art; we want a solid and substantial food on which to bite, something strong and hearty to get our teeth into. . . . We want the truth, not rationalization, not idealizations, not romanticizations." She believed that FSA work provided "the strongest precedent for documentary" in the 1930s. And she foreshadowed analyses of FSA imagery that situated it inside a thriving social science community in the 1930s. In his 1940 publication *Since Yesterday*, Frederick Lewis Allen discussed contemporary documentary images and described the thirties photographer as "one with the eye of an artist who was simultaneously a reporter or a sociologist." Post claimed that Stryker urged his field staff to record and document America from "the historian's, the sociologist's, and the architect's point of view." Respected art photographer Ansel Adams once told Stryker, "What you've got are not photographers. They're a bunch of sociologists with cameras." By the late 1930s they could also be considered preservationists, since they were compiling what historian David Peeler has called "a visual catalog of thirties America."[55]

Despite the various descriptive labels they were given, FSA photographers did frequently operate as practitioners of social science. Few FSA field workers were as rapidly immersed in the discipline as Post. In October 1939 she headed for Chapel Hill, North Carolina, to see firsthand how some of the nation's finest social scientists carried out their research. Her assignment there required her to work closely with a cadre of University of North Carolina sociologists at the Institute for Research in Social Science. The institute, whose projects examined farm tenancy, race issues, and southern culture, had put the university's Sociology Department on the map. Directed by Howard Odum, the institute had fostered the work of Arthur Raper, Rupert Vance, Harriet Herring, and Margaret Jarman Hagood. Hagood, having earned her Ph.D. in the department two years earlier with a study of fertility patterns of southern white women, assisted Post while she was in North Carolina. Excited about the prospect of a comprehensive photographic survey of the area, Howard Odum cordially arranged extra help for Post. Several graduate students accompanied her in the field, taking notes while she snapped pictures. Odum seemed extremely anxious to secure as much of Post's time as possible and held out as a return favor the promise of garnering support for the FSA in its attempts to produce a solid publication within the next year. Stryker took notice and was generous with Post's time in North Carolina since photo publication kept the agency and its attempts at social welfare alive.[56]

Perhaps the photographer's lengthy stint at UNC was designed not only with the Sociology Department in mind but also with an eye on the university press. If Chapel Hill could claim the boldest social scientists in the region, it could also boast the most daring and innovative southern publishing organization, the UNC Press. Chief editor W.T. Couch had drawn national attention by accepting works on controversial topics that addressed race, lynching, and labor strikes. Couch's other position, regional director of the Federal Writers' Project (FWP), kept him closely in touch with southern voices. In 1939 thirty-five of the nearly four hundred "life histories" collected by FWP writers were published under the title *These Are Our Lives*. The volume received critical acclaim, including praise from historian Charles Beard, who said, "Some of these pages are as literature more powerful than anything I have read in fiction, not excluding Zola's most vehement passages." Historians Tom Terrill and Jerrold Hirsch, who edited a volume of previously unpublished "life histories," wrote that "[Couch] wanted to give Southerners at all levels of society a chance to speak for themselves. In his opinion collecting life histories was one way of doing this. He thought project workers could be used in an effort to provide a more accurate picture of Southern life than had either [Erskine] Caldwell or the Agrarians." Couch was the kind of ally Stryker needed, especially since continued publication of FSA photos kept the public's mind on domestic problems rather than on the gradually encroaching foreign policy concerns.[57]

Post immediately recognized the possibilities the region could offer a photographer. In fact, she initially expressed a concern that the sociologists at Chapel Hill seemed unaware of the ways in which visual images could enhance their scholarly efforts. But her diplomatic skills and affability helped her build the necessary bridges to make successful collaboration possible. The FSA written records reveal that the institute's personnel found Post much easier to work with than Dorothea Lange; Harriet Herring appended a thorough outline entitled "Notes and Suggestions for Photographic Study," with the comment, "Miss Lange does not consider this a very good outline. Too much emphasis on the economic set-up, not enough on the people at the center of it." In her nota bene, Herring pointed out that Lange had suggested the institute use movies rather than still pictures to complete their visual study of the thirteen-county Piedmont area in the state. By this time Lange was working only sporadically for the FSA, but the North Carolina survey ended her tenure with the New Deal agency. A few weeks after Herring typed her warning note about Lange's opinions, Stryker permanently released the photographer, claiming she was the "least cooperative" of his employees. Post's work on a similar assignment shows, by contrast, her willingness to cooperate with the UNC professors; she adhered closely to a shooting schedule prepared by the institute's Margaret Hagood. Post documented an array of agricultural pursuits, including wheat plowing, hay baling, and soybean thrashing.[58]

The variety of crops grown in the region sparked Post's curiosity, but she devoted her most concentrated effort to tobacco—its preparation and marketing. Her 1939 series on the tobacco industry closely follows the suggestions on Hagood's outline. Although Professor Odum often looked over her shoulder as she photographed and edited, Post nonetheless cast her North Carolina images in her own mold, focusing on the spirit of cooperation and group effort. Intent on revealing the fruits produced when family members worked together, Post highlighted the prosperity enjoyed by black tobacco farmers in Orange County (fig. 28). In this image an impeccably dressed, carefully groomed husband and wife strip and grade tobacco. They not only sit close to each another; they even face each other while working. The positions of their bodies suggest their participation in a collaborative venture, pursued toward a single goal that will help them maintain their present level of comfort. The tobacco, surrounding them on all sides, serves as symbol of their prosperity; they are engulfed by it. Post also revealed in her tobacco series the demands imposed by segregation during the tobacco auctions. She photographed the warehouse "camp rooms" where farmers congregated, conversed, and slept while awaiting the chance to sell their tobacco. Post made her way into one of the "Negro camp rooms" and documented the crowded conditions—men sleeping on low benches, their bodies tightly cramped, twisted and touching each other at odd angles. The tobacco auctions attracted hundreds of people to the larger towns, such as Durham, where Post was able to capture the atmosphere of bustling activity and constant money-changing. Auction time put dollars into people's hands; knowing this, product hawkers circulated throughout the auction area selling everything from patent medicines and socks to hound dogs and farm equipment. An itinerant preacher, anxious to market his own product to the crowds, figures prominently in several pictures. An African American "corn parer" works diligently on a white man's foot, displaying the results of his other successful operations (various corns)

Fig. 28. Marion Post Wolcott. *Couple Stripping Tobacco. Orange County, North Carolina, 1939.* U.S. Farm Security Administration Collection, Prints and Photographs Division, Library of Congress.

on a towel nearby. Post recorded some faces cautiously deliberating what to buy and others simply enjoying the frenzied environment with enthusiasm. Her images reveal the relationships between tobacco buyers and sellers, marked by the looks they exchanged; some sellers wince at the prices being offered, whereas others exude their confidence in securing profitable deals. In nearly every photograph, Post aimed to portray the connections between individuals or groups of people—their looks, their touches, or their words to one other.[59]

One agricultural field trip particularly suited to Post's vision of the collective ideal was a "corn shucking" she and Hagood attended near Stem, North Carolina. The harvesttime event, described by Hagood as "the most widespread occasion for gatherings of a semisocial nature," produced paradigmatic situations for Post's camera work. As the "form of cooperative endeavor most commonly reported" in the area, the ritual brought together male family members, neighbors, and laborers to shuck the corn, while their wives prepared the obligatory meal for them. Hagood reported that one family's corn return was so large that the event required them to serve both "dinner and supper" to their shuckers. Post illuminated the enormity of the day's task at the Wilkins family farm in a shot of corn piled high in the yard; in another frame, she pictured two men sitting and one standing in a sea of corn yet to be shucked.[60]

But the photographer's most penetrating depiction of the cooperative spirit may be found in her images of the mealtime activities. Six women, including five white sisters-in-law and one black woman, toiled together to prepare food for twenty men. Post's series shows the step-by-step process from biscuit-cutting to cooking to serving to extensive dishwashing. She highlights gender-specific responsibilities as well as race-determined roles. The meal progressed through three stages: a seating for all of the white men, a similar seating for the white women and children, and finally a meal for the black male tenants. One image in the series shows where several of these constituencies situated themselves during the first stage (fig. 29). Here in the midst of the clan gathering, the photographer peers in to capture the Wilkins wives standing fairly stationary in their places, children at their sides, while their husbands enjoy a meal at its hottest and freshest. The

Fig. 29. Marion Post Wolcott. *Dinner on Corn Shucking Day. Stem, North Carolina, 1939.* U.S. Farm Security Administration Collection, Prints and Photographs Division, Library of Congress.

camaraderie between these male family members and neighbors is marked by their hearty laughter and ease, despite the presence of a U.S. Government photographer. Aside from the smiles and limited arm motions of the eaters, the most concentrated energy is detected in the upper left quadrant of the frame, where a black woman is caught walking. Where is she going? Is she the principal server headed off to retrieve an item the diners have demanded? Or does she wish to step out of the photographer's view, feeling misplaced in a candid family picture? During each stage of the prolonged meal, she stands to serve those who eat, but nowhere is her own dinner recorded. The nature of her portrayal in this series reflects the limitations of 1930s social science, even that which challenged the comfortable confines of acceptable scholarship on race. The issues facing African Americans were defined almost solely in terms of what black men needed, so that well-intentioned efforts often blurred the gender lines.[61] Even Post, whose keen social-political consciousness attuned her to problems inherent in a racially segregated society, made precious few statements on film about segregation's peculiar effects on black women. The frames she shot at the Wilkins cornshucking event in Stem, North Carolina, are among the few.[62]

If Post's personal feelings about the segregated South had previously been revealed in subtle ways, Mississippi changed her mind about her approach. In no other assignment is her commentary on the bifurcated social system more pointed than in her study of the Delta region. Taken within a few weeks of the intense North Carolina job, Post's images expose her yearning to illuminate both the complexities and the cruelties of Jim Crow. Among her staging grounds were the cotton market, a plantation store, a movie theater, a juke joint, and several other public venues. In observing a common and rather benign activity of cotton pickers getting paid, Post realized that social "separation" actually translated into something much greater. She attempted to communicate its reality in an image highlighting the distance and the physical barriers between blacks and whites. In one image, behind a pay window stands a laborer known only by the straw hat she wears, since her face is obstructed by the heavily supported window design. Her clearest identifiable feature is her right hand, reaching through a tiny opening to receive her wages. The cash transfer is indirect, not from a white hand to a black hand, but placed at a safe distance between the two parties, so they need not come into contact with each other. The white cashier's sacred space, completely partitioned from the outside world save for the small window at the edge of his desk, protects him from potential unpleasantness. How apt that Post made this laborer's hand the focal point, the center icon, of her frame; it was a cotton picker's principal asset in the plantation economy. So valuable to white landowners, yet at the same time untouchable. Initially Post watched the scene from the African American side of the partition but was frustrated by the angle's ineffectiveness at portraying the unfolding drama. She then managed to attain a better position and viewpoint (inside the sacred space, from the "white" side of the window), by being "sweet" to the cashiers, a couple of men who failed to realize that her presence "might be to their disadvantage." Existing in a well-entrenched patriarchal system that considered blacks inferior and women politically innocuous, the cashiers were not culturally prepared to accept the possibility that a polite and appealing young woman with a camera could house a radical, accusatory voice.[63]

Post's role as an FSA commentator on race can be measured by the place she commanded in Nicholas Natanson's study *The Black Image in the New Deal: The Politics of FSA Photography* (1992). The first picture he introduces to the reader is from Post's Mississippi assignment, *Negro Man Entering a Movie Theatre by "Colored" Entrance* (see fig. 23). As Natanson observes, the figure stands as a symbol in "the psychology of Jim Crow" because "the man's blackness, and not the man, is what counts." That he is unidentifiable makes the image even more powerful, because he is everyman in the southern black world. In addition to the illuminatory postures Post captured at the cotton picker's pay station and outside the Belzoni movie theater, she detected a highly charged moment between two black farmers and a prospective buyer at a Clarksdale cotton market (fig. 30). She must have reveled in the unshielded temerity displayed by the farmer at a price that displeased him. With his hand propped on his hip and his right leg comfortably crossed over his left leg, his apparent ease in challenging the meager offer sets him in noticeable contrast to the well-suited buyer, who has backed away in a defensive posture. The buyer signals with his body his surprise and confusion at the

Fig. 30. Marion Post Wolcott. *Unsatisfied Farmers with Cotton Buyer. Clarksdale, Mississippi, 1939.* U.S. Farm Security Administration Collection, Prints and Photographs Division, Library of Congress.

dispute—holding the cotton samples in nearly fisted palms, as a boxer would position himself to protect his upper body. The image stands almost alone in the FSA file as a depiction of African American challenge to the powerful white elite in the South.

Since the social realm was the milieu where Post felt most comfortable, she sought out places that would help her characterize people's lives beyond their work responsibilities. Shortly after she took the FSA photography position, she intimated her general preference to observe and record "the hanging around, the 'messing around,' the gambling." Her attempts to capture scenes of leisure activity helped strike a balance in a photography collection filled overwhelmingly with work-related scenes, especially those of the rural hoe culture taken by earlier FSA photographers such as Lange. In the Delta Post had a hard time convincing white authorities that she needed to enter the "off hours" world of black living. She begged, insisted, and eventually bribed a white male escort (a plantation owner's son who wanted to take her out) to accompany her into a "juke joint." In the Delta juke joints, from Clarksdale to Memphis, Post continued to document the intensity of human relationships in group settings. In one image made in a Memphis juke joint (fig. 31), Post frames a couple's attempts to dance the jitterbug. The young woman's cautiousness is checked by the young man's confidence. Looking down, she measures carefully with feet barely apart, while her partner reassures her with his eyes, challenging her to balance his risky swings and steps by pulling her toward him. Post's own dance studies inspired her extended observations of various dance styles and steps that she encountered while on the road. Driven by energy expressed in two dimensions, Post's dance pictures became some of the most frequently published photographs in her FSA oeuvre.[64]

Exhibits of FSA photography attracted widespread attention in the late 1930s, particularly after a successful showing at New York's Grand Central Terminal in 1938. The agency sponsored its own exhibits in public spaces, and it lent and sold photographs to other organizations that had embraced the visual image as a powerful vehicle for delivering their messages. Una Roberts Lawrence, of the Southern Baptist Convention's Home Mission Board in Atlanta, made plans with Stryker to set up exhibits at the 1940 Southern Baptist Convention in Baltimore and at the Baptist Stu-

Fig. 31. Marion Post Wolcott. *Dancing in a Juke Joint. Memphis, Tennessee, 1939.* U.S. Farm Security Administration Collection, Prints and Photographs Division, Library of Congress.

dent Union Conference in Ridgecrest, North Carolina. She made a specific request for "a complete exhibit of Negro life" so as to "strike more surely at the intelligence and conscience of our folks" by showing "conditions as they are." She also explained the importance of FSA pictures in instructing writers of Southern Baptist church literature, who she said displayed "an appalling lack of contact with farm life and especially with these problems and the solutions that are being worked out." Art photographer Ansel Adams asked for thirty-five prints to display at San Francisco's Golden Gate International Exposition, in hopes of demonstrating the "use of the camera as a tool of social research for communication and information." But the gap between art photography and documentary photography remained open, as was proved when the New York World's Fair officials abandoned the idea of an FSA exhibit. In focusing on the future, new technology, and the urban world almost exclusively, the fair found little place for agrarian traditions or regional emphases, and its organizers chose commercial photographers to do their bidding. Another group not so willing to embrace what pictures could offer were professional historians, Stryker reported. He admitted making little progress at the 1939 American Historical Association meeting, where he attempted "to convince the historians that the camera and the photograph" could be "important tools for them to use." In the meantime, he directed Post to continue producing the kinds of photographs he wanted, scenes of the good life in America.[65]

Pictures of families and community gatherings and flourishing small towns, the places where collective strength and cooperative efforts were frequently displayed—Stryker sought these for his file. Although he was a friend of sociologists Robert and Helen Lynd, Stryker disagreed with their less-than-flattering assessment of community life in their studies *Middletown* (1929) and *Middletown in Transition* (1937). With FSA photography he attempted to provide a contrasting image, one that would prove that the small town in America was alive and well and that its inhabitants remained committed to home, family, church, and community. Stryker wanted his photographers to capture the uplifting spirit of unity that bound people together in close-knit communities—he believed such a spiritual climate characterized most small towns in America. Historian James Curtis has argued that Stryker

mounted a serious campaign "to counter Lynd's portrait of declining family solidarity with fresh visual evidence. . . . [He] needed
a counterweight to assure his contemporaries that home life still
mattered, that it was a basic determinant of American values."[66]
Fortunately, the FSA director had Post on board at the height of
his campaign. No one on his photography staff was better equipped
ideologically to handle these themes. Her personal vision, molded
by her experience with cooperative efforts and group identity both
at home and abroad, helped guide the FSA into the next decade.

For Post, 1940 proved to be her best year in producing extended
photo essays on community and its requisite human relationships.
Four extended assignments illuminated her attachment to the value
of cooperative spirit—one in New England, a second in Louisiana, another in Eastern Kentucky, and a fourth in Caswell County,
North Carolina. The New England assignment took Post to a region
blanketed in fresh snow. Exactly one year earlier, she was trying
every way imaginable to protect her equipment and film from
blistering Florida heat; now she faced the freezing cold of a Vermont winter, "Everything stuck & just refused to operate at *least*
half the time. Shutters, diaphragms, range finders, tracks; film &
magazine slides become so brittle they just snap in two." With
temperatures below zero every morning, Post devised creative ways
to keep her camera parts warm. She wrapped them in sweaters
with a hot water bottle, and when driving she tried to place them
as close to the car heater as possible. She pleaded with Stryker to
understand her struggle to prevent equipment breakdown in the
event her negatives turned out "underexposed, overexposed, out
of focus, jittered, in any other way technically not too hot."[67]
    The New England assignment was a significant one since FSA
photographers had spent very limited time in that region. The office
files included only a few prints of the northeastern United States,
so Stryker wanted Post to fill the gap with a large number of
assorted scenes and activities. His specific request was for broad,
snowy landscapes. In spite of hazardous driving conditions on icy
roads, Post made her way from western Massachusetts through
Vermont and New Hampshire and on up into Maine. In one wealthy
section where an exclusive mountain lodge was located, she photographed skiers gliding down the slopes. But she was much more

interested in the permanent residents of the area; she focused in particular on their seemingly smooth adaptations to fierce weather. Scenes of rugged activity—men and women clearing snow, cutting wood, logging, and making early morning milk deliveries— are matched with those of carefree playfulness, most notably delighted children headed downhill on sleds. Cabin villages, private farms, and a local general store served as stages for Post's artistry. After spending several weeks in New England, she received word from Stryker that novelist Sherwood Anderson wanted her to emphasize particular themes. Anderson's detailed list suggested that Post find and photograph a small-town jail, a birth scene, an abandoned roadhouse, and a country lawyers' gathering.[68]

Taking advantage of the free agency Stryker allowed, Post accommodated Anderson to an extent but ignored many of his requests. Her own fascination with the wide spectrum of personalities she encountered kept her continuously occupied, encouraging her years later to name the New England trip as one of her favorite assignments. She heard one farmer explain the extensive sugaring process and how the season's excessive snow prohibited him and many others from reaching their maple groves. "Sugaring" served as a social event, bringing people of all ages together to assist in the gathering and boiling of maple tree sap. Enamored of the old mill village Berlin, New Hampshire, Post discovered that most of its townspeople came from French, Norwegian, Swedish, or Italian backgrounds. In North Conway, New Hampshire, she was asked by "a couple of cranks who weren't too pleased" not to snap many pictures of the town meeting.[69]

The rejection in New Hampshire mattered little because Post soon gained entry into a town meeting in Woodstock, Vermont, allowing her a comprehensive look at the theme of community action. Woodstock residents had received "warning" to attend the annual business meeting, where several vital issues would be discussed—among them whether the town would allow baseball or concerts to be played or movies to be shown on Sundays and whether to permit the Prosper Home Makers Club to build an addition onto the local schoolhouse. The subject that seemed to attract the most attention, though, was the selling of "malt and vinous beverages" and "spirituous liquors" in Woodstock. In one image Post recorded the interplay among three older participants, one a selectman casting his secret ballot on the liquor issue. As a woman reaches to give the voter his ballot, the lighthearted banter shared among the three is evident in their smiles and laughing eyes. Post captured the exchange not only visually, but also in a caption reading, "If you vote yes for liquor you'd better put your ballot in a box in a different town. We won't let you stay around here long." The woman's comment, jokingly uttered, reveals one of the gendered avenues toward political influence. Although her voice and presence may be minuscule in the collective gathering (proved by the absence of women voters in Post's photographs of the meeting), her personal and nonthreatening contact with a voter on this local prohibition initiative carries her message effectively.[70]

The importance and intensity of the discussion on liquor sales in Woodstock may be observed in the crowd scenes that Post captured from various angles inside the town hall (figs. 32 and 33). Although it was scheduled on a Tuesday morning, the meeting obviously attracted a substantial crowd. She indicates the size of the group with back views and overhead views of the meeting room. Each wooden chair holds an interested discussant. In the crowd were men from different classes, the more modestly dressed wearing field jackets and no neckties. Post's direct view from an aisle reveals these participants crammed closely into tight space, their looks and actions ponderous and deliberate. One man cradling his chin in forefinger and thumb listens intently, while another, with pencil in hand, takes notes on his ballot. Prominently displayed by the two men in the foreground are copies of the seventy-five-page "Annual Town Meeting Agenda and Report," distributed to all meeting attendees. Beyond their direct and active engagement with the issues, the men Post included in *Town Meeting* (fig. 32) serve to articulate her commentary on class in small-town America. They may share in this democratic experiment, but they occupy seats at the back of the room. None of these men, whose clothes and untidy long hair set them apart from voters at the front of the meeting hall, appear at the counting table (fig. 33). Those in charge of the ballots hail from a different class. They have fresh haircuts; starched, collared shirts; vested suits; and neckties. That so many wear eyeglasses implies a certain literacy level, perhaps that they work with words or numbers in jobs that require specialized, skilled training. These men represent the

Fig. 32 (above). Marion Post Wolcott. *Town Meeting. Woodstock, Vermont, 1940.* U.S. Farm Security Administration Collection, Prints and Photographs Division, Library of Congress.

Fig. 33 (below). Marion Post Wolcott. *Counting Ballots after a Town Meeting. Woodstock, Vermont, 1940.* U.S. Farm Security Administration Collection, Prints and Photographs Division, Library of Congress.

town's power elite; their willingness to work together, as Post demonstrates, ensures their ability to maintain their honored position. Apart from each other, the two images bespeak the varied contributions of Woodstock's male citizens to the civic enterprise; together they illuminate the subtle disparities existent in an otherwise bucolic setting.[71]

Some of Post's most widely circulated prints were those she generated in her six-week venture in New England. Sherwood Anderson, in collaboration with Edwin Rosskam, printed twenty-five of Post's photographs in his 1940 publication *Home Town*. Anderson admitted developing a sense of appreciation for collective endeavor during the 1930s, all the while shaping his views on community efforts and small-town values. The week *Home Town* appeared, an editor at *PM's Weekly* penned a glowing review, suggesting, "Now is a good time for intelligent persons to start learning about small towns. City planners insist that the big city has outlived its usefulness and that there will be a migration back to the small town." Praised for "its timeliness" in some quarters, *Home Town* later suffered criticism with regard to its use of FSA photographs. William Stott believes the images, although "badly presented," greatly transcend Anderson's text. Former agency photographer Walker Evans put it more bluntly in proclaiming Anderson's words "cheap use" of FSA imagery. Evans described an example in which a Post photo was accompanied by an adjoining caption that "degrade[d] the picture." Post's photographs of New England and its inhabitants proved to her contemporaries and to later critics that they were hearty enough to stand with few or no words.[72]

Late in the spring of 1940, Post crisscrossed the South covering routine government assignments, some repeat affairs for Public Health and for Surplus Commodities as well as the FSA project areas. Her Louisiana assignments on the Terrebonne Project, the Transylvania Project, the LaDelta Project, the Allen Plantation Project, and the Bayou Borbeaux Plantation Project were designed to show off the FSA's tremendous success in the state. The images Post created elicit conceptions of model environments where comfort, cleanliness, abundance, prosperity, and cooperation characterize all life and experience. She focused particular attention on an unusual FSA venture she must have appreciated—the racially integrated Allen Plantation Project in Natchitoches. Post's frames show men and women, black and white, working alongside each other in the cotton fields there. But to reveal her personal opinions about the project's limitations, she also photographed the cooperative's all-male, all-white board of directors.[73]

The FSA's strides in home management supervision at the Louisiana projects provided Post with ideal subject matter. In one image (fig. 34) a young boy stands in his home's orderly, well-stocked pantry. Post's angle emphasizes the depth of this project family's abundance, as she shows the long sweep of canned vegetables from the left edge of the frame to just past the center line. The photograph shouts "Plenty." That the rows are reflected in the window gives the illusion of unending nourishment achieved through expert planning. The boy, wearing a spotless shirt and trousers, is literally weighted down by the bounty his family has stored up for their future. But as Post's larger photo essay describes, the scene is the result of more than good harvests. The FSA's provisions for this kind of client security were made through hands-on instructional methods, which Post recorded in the same series. The most vivid image shows a home management supervisor demonstrating the use of a pressure cooker to an enthusiastic student. As one Terrebonne Project client, a former tenant farmer, said, "I think *we* are going to put it over. I think *everybody will get together and work hard and put it across.* . . . I've never been able to get anywhere. Now I'm on the project and I'm going to make a better living. I'm working now and getting paid for it, and when *we* make a crop I'll get part of the profits." Individual success depends upon the willingness of everyone in the community to contribute to the collective strength; when "we" cooperate, "I" will reap the rewards, the farmer realizes.[74]

The third of Post's profitable long-term community studies in 1940 was set in Kentucky. She was shocked by what she encountered: "I was quite amazed at the backwardness of the Kentucky mountain country when I saw that for the first time. . . . I hadn't realized that it was still this way in this country, in the U.S." As the FSA photographer responsible for the majority of Kentucky scenes, Post took more than seven thousand photographs in the state during the late summer and early fall of 1940. Aside from conducting a brief session with Public Health officials in Louis-

Fig. 34. Marion Post Wolcott. *Child of an FSA Client. LaDelta Project, Thomastown, Louisiana, 1940.* U.S. Farm Security Administration Collection, Prints and Photographs Division, Library of Congress.

ville, she spent most of her time either on major highways connecting the larger central Kentucky towns or in the Appalachian region. The pictorial record that Post produced illuminated the social, cultural, and economic lives of Kentuckians, particularly those who were tucked away in the mountains and kept few written records. Post received a limited amount of preparatory material before entering the world of Kentucky mountain natives. She had read only the recently published WPA *Guidebook to Kentucky* and Malcolm Ross's *Machine Age in the Hills* (1933), but fortunately the *Louisville Courier-Journal* staff was helpful in lending suggestions and showing her the best travel routes. Post left the newspaper office impressed by the "superior" job they did running a paper, but more than that she was elated that one employee had repaired the shutter on her Speed Graphic.[75]

While in central and eastern Kentucky, Post used her time scrupulously and never failed to carry her camera everywhere she went. She attended county fairs and horse shows, church suppers and farm auctions, an American Legion fish fry and a Primitive Baptist Church baptismal ceremony. In September she begged Stryker to allow her to stay longer than her designated time so that she could record the sorghum molasses "stir-offs," community apple peelings, and bean stringings. Included in her vast collection are scenes of children walking to school, families mourning their deceased kin, and women carrying heavy loads of weekly supplies up creek beds. Away from telephones and Western Union offices, Post unleashed her creative spirit. She traveled several days through a remote area of the region guided by Marie Turner, the Breathitt County school superintendent. Turner's assistance proved exceptionally useful; because of it Post gained entrée into places that she would not have known existed otherwise. Having a willing guide gave Post the kind of support Doris Ulmann had enjoyed with her Kentucky escort, John Jacob Niles. But the experiences of Ulmann and of Post in eastern Kentucky were quite different upon comparison. Post and Turner faced travel difficulties that Ulmann and Niles tended to avoid by staying on main thoroughfares. On their first day out in Breathitt County, Post and Turner got their borrowed car stuck in a mountain streambed and had to be pulled out by a mule. A bit farther up in the mountains they were stopped with a flat tire. Post told Stryker of their successful attempt at repairing it: "Tore down a fence post & while our driver (a young kid who is the son of the school janitor) tried to prop the car up I was down on my belly in the creek bed piling rocks under it. But we finally got it fixed. . . . We did amazingly well over the worst & most dangerous roads & creek beds I've ever seen."[76] On another trip, the two women traveled by mule on a narrow mountain trail, finding great relief when their mode of transportation was upgraded to horses. Post spoke excitedly of the tour.

> It is wonderful country & the people are so simple & direct & kind if they know you, or if you're with a friend of theirs. I got some excellent school pictures I think, & some other things as we went along. It takes a great deal of time just to *get* anywhere & make friends. . . .
>
> Got arrested the other day in a town in the next county, even after all kinds of precautions & after having gone to the F.S.A. office & walked all over town with the F.S.A. people too. But there's a feud there between F.S.A. & Triple A, & that's where it all started. They just took me before this county judge who asked questions & looked at identificaton, etc.—it was just funny because the whole town was full of people (Labor Day) & got all stirred up over it & followed me in a big procession to the court house—all crowding around the judge afterwards to see my papers & getting into arguments about spies.[77]

The case ended quickly. The judge charged Post with being a suspicious person but dismissed the case and concluded that she was probably "as good an American as anybody."[78] Her connection to the U.S. Government sparked her accusers' doubts. Ulmann had avoided such confrontation, probably because of her relationship with the Russell Sage Foundation, a private humanitarian organization.

Stryker realized that Post's work in Kentucky was invaluable, her official contacts important, and her scenes priceless. He told her that the stories she could tell of her experience in the Kentucky mountains would fill an entire chapter of the book he wanted the FSA staff to write someday, adding that "with your contacts down in them thar hills, we should be able to bring out good gold." Beyond the contacts she made, Post stepped into traditional gender roles in order to gain her subjects' trust. In her casual attire, she helped women in their kitchens by peeling potatoes, dressed small children in the family, and drove to town to pick up household necessities for them. She engaged herself in the scenes she arranged, just as the Workers' Photography Movement had instructed its adherents in the mid-1930s. Her own cooperation within the family, the neighborhood, and the community allowed her a comfortable reciprocity from the people she photographed. They returned to Post uninhibited poses, natural images for her camera.[79]

In a series on the annual memorial service held in Breathitt County, Post surveyed the day-long ritual that brought together nearly three hundred people who were unable to hold formal funeral services immediately following a death in the family. As Post reported in "General Caption No. 1" on the assignment, travel was much easier during the summer and early fall, allowing family members and friends from remote areas to reach mountain cemeteries. The collective mourning for those who had died during the year took place at an outdoor feast where people gathered to "tell about the events of the past year." The group also listened to "a battery of preachers" for several hours and often witnessed immersion baptisms in nearby creeks.[80] In one image (fig. 35), five men share the arduous task of carrying a coffin along a river fork. Post captured on film the moment of greatest difficulty, at the place where the stream is widest. The man in the foreground stands awkwardly on an exposed rock, while another on the far left steps carefully from dry ground into the water. The communication among the five is essential if they are to safely complete the job. Another photograph in the annual memorial service series (fig. 36) exhibits the sense of camaraderie that Post sought out in all her shooting locations. Here small clusters of friends and family members, of all ages, share stories and news. The seriousness of the occasion is marked by the prominent gravestone, but Post's rendering of the scene draws more attention to the lively storyteller profiled in the center. The participants carry on despite the camera's presence, with only one person in the crowd gazing at

Fig. 35 (above). Marion Post Wolcott. *Men Carrying a Coffin. Breathitt County, Kentucky, 1940.* U.S. Farm Security Administration Collection, Prints and Photographs Division, Library of Congress.

Fig. 36 (below). Marion Post Wolcott. *Visiting After a Memorial Service. Breathitt County, Kentucky, 1940.* U.S. Farm Security Administration Collection, Prints and Photographs Division, Library of Congress.

the photographer (at the far right of the frame). Holding a baby, the woman stands out as the only individual in the scene who is not engaged in a small group conversation. Post's message has little to do with the principal reason for gathering, the funeral; it is instead a comment on communication—people laughing, listening, and touching.

Even though they traversed the same area within just seven years of each other, Marion Post and Doris Ulmann produced extremely different portraits of southern Appalachia. Both in content and in style, Post's work stands apart from Ulmann's collection. Based on disparate generational and cultural backgrounds, the two photographers created what they wanted viewers to see. Ulmann's vision reflected Progressive Era reform ideology combined with the influences of the 1920s agrarian appreciation movement. Post's views were informed by thirties radical politics, which emphasized the strength of the masses, the group, the collective ideal. As a result Post highlighted social lives and connections, whereas Ulmann defined individuals by their work, especially what they made with their own hands. Post placed value on activities that filled leisure time; Ulmann searched for personal industry. Among the events that Post examined was a "pie supper," where eager teenage men bid on pies baked by young women with whom they wished to share their desserts. Energy permeates these scenes, as well as many others, such as the Fugate School playground, the Woodford County courthouse square, and the site of a neighborhood house raising. In Post's Kentucky photographs people weep, laugh, and flirt. They rarely stare off into the distance pondering their individual lives or their fates; with an occasional exception, they connect with each other by word or touch or glance. As Sally Stein has pointed out, Post "was especially attentive to informal group interactions for the way they expressed some of the bonds and boundaries within a community. . . . Hers are rather cool observations which document not only the character of people's gestures but also the general environment and specific objects that gave rise to these forms of non-verbal communication." An excellent source for comparative analysis on this issue is *A Kentucky Album,* for which editors Brannan and Horvath selected a representative sampling of FSA photography by each field staff member who visited Kentucky during the agency's eight-year life. Here Post's images stand out because they survey the wide spectrum of human relationships, dissecting the ways people interact with each other.[81]

With the Kentucky assignment finished in midfall, Post assumed a more structured shooting schedule to cover several FSA projects in North Carolina. She traveled to Chapel Hill, where she got caught in the swarming activity of a university town. Since students had taken most of the rooms in Chapel Hill and Durham, Post had trouble finding a place to live. But the enlivened atmosphere, especially on home football weekends, gave her rich subject matter for visual studies. She recorded college student frivolities, fraternity parties, and practical jokes. Her more essential camera work took her to Caswell County on the North Carolina–Virginia border. Assigned to record the success of project activity, Post thoroughly surveyed FSA borrowers' homes, farms, and work in progress.[82]

In Caswell County Post completed the fourth of the community studies that defined her FSA vision. Two images (figs. 37 and 38) based on the same theme, land use, appear similarly derived. Each shows a small group, a community land use planning committee, examining a map of the county. Both groups sit at tables, are focused on their purpose, and appear earnest in making thoughtful decisions. Their hands on the maps symbolize their direct participation in creating change in Caswell County. Below the surface of the seemingly similar messages, though, lies Post's subtle commentary on race, class, and gender. That one county requires two land use committees intimates the limits of the ability—and perhaps of the willingness—of the U.S. Government to challenge regional social standards on race. Beyond the obvious bow to Jim Crow laws are the local community-driven expectations. The all-white committee includes women; the all-black committee includes males only. The white women planners serve not merely as onlookers but as participants, with fingers and pens pointed to specific locales. Their contributions may be checked by the two men who have assumed authority by standing over the table, but their power relations are little different from those that exist between the seated men and the standing men. By comparison, the African American committee shows one individual who has taken charge, so much so that he has bored the person seated to his right. Those

Fig. 37 (above). Marion Post Wolcott. *Land Use Planning Committee. Locust Hill, North Carolina, 1940.* U.S. Farm Security Administration Collection, Prints and Photographs Division, Library of Congress.

Fig. 38 (below). Marion Post Wolcott. *Land Use Planning Committee. Yanceyville, North Carolina, 1940.* U.S. Farm Security Administration Collection, Prints and Photographs Division, Library of Congress.

at the table wear suits and ties, while several men sitting outside
the realm of decision-making wear open-collared shirts and casual jackets or overalls. These men appear physically and socially
removed from the center of power, which poses itself as solidly
middle-class. A wider view of this same scene reveals clearly that
a white observer is seated at the table. His responsibility as an
advisor to the group allows him closer contact with the men than
the only remaining "other" in the room—a woman charged with
taking notes of the meeting. Post's photographic survey yielded
pictures used in a USDA pamphlet entitled *We Take You Now to
Caswell County.* Advertising the positive effects of federal government programs such as the Civilian Conservation Corps, the
Agricultural Adjustment Act, the Soil Conservation Service, and
the FSA, the tract announced, "They're putting some real stuffing
into democracy in this North Carolina county where farmers and
representatives of Government services are meshing their resources
and energies through the Land Use Planning Program." Accompanying the uplifting rhetoric were Post's mirrors reflecting the
limits of democracy in a community where a web of sexism, racism, and classism exists.[83]

Despite her valued contributions toward putting across the FSA
message in 1940, Post had grown weary of her job, with its tight
schedules, haughty social workers, and disrespectful clients. She
had waited on several occasions and for many hours for a Caswell
County doctor who was supposed to be a subject of her survey
there. He finally refused to oblige her. In general, Post felt she had
gotten a "temporary 'belly full' of the dear old South." Furthermore, news of the European war bothered her. Recollections of
her days in Europe seemed to promote a concern Post found difficult to dispel. Several times she mentioned her uneasiness as the
international scene grew darker. In May, she wrote to Stryker, "War
news, international and national happenings, always seem to be
even worse & more terrifying to me when I'm away & don't have
anyone I know to talk to about it." Nightmares about the war
and the "human" race, as she put it, kept Post from fully enjoying
the travel. Her job was made even more difficult because people
gradually became more hysterical about the war and grew suspicious of who she was and what she was doing. In the summer of

1940, while working on the Terrebonne Project in Louisiana, Post frightened a couple of Cajun children who ran away screaming to their father that they had seen a German spy with a machine gun (her camera). On several occasions state officials and policemen stopped her to check her identification. Unable to ease authorities' suspicions, Post found herself in a number of county sheriffs' offices. She hated to waste time on the long, drawn-out security procedures, but she concluded that the officers probably had nothing else to keep them busy.[84]

At midyear Post "complained that she felt pigeon-holed as 'the project-photographer-in-chief.'" Her claim was not unfounded: She was saddled with a large number of routine assignments and was often sent to regional FSA offices where a public relations visit from an "attractive and enthusiastic" member of the Washington team meant more than any photograph she might take while there. By the fall she expressed her accumulated frustrations in an outburst, "Now tomorrow I'd like to wake up in the morning & say—today I can go where I damn please, I don't *have* to let F.S.A. & Stryker know where I am, I don't *have* to go to Western Union or the post office or the railway express, or feel guilty because I haven't done my travel report & can't find the notebook that has the mileages of the first half of the month in it." This letter more than any other that Post wrote in her tenure at the FSA highlights the double standard to which she felt Stryker held her. The overwhelming paperwork load she bore alone, unlike most FSA photographers who had enjoyed the aid of traveling companions. Yet Stryker would not have permitted her to have a "secretary" as her colleague Russell Lee did while on the road. Post had to keep herself above suspicion in the realm of intimate relations. She later told historian Jack Hurley, "I had to be so *careful* not to meet anybody or see anybody. It was so unfair. . . . I'd never have gotten away with it because everyone was so curious about a woman traveling alone in those days and any gossip they could figure out they used."[85]

Post even had to lobby for time off, reminding Stryker in late 1940 that she had not had a real vacation in three years, a fact that illuminated the close eye he kept on her and the continuous demands he made on her. She would not be allowed one for a few more months, in the interim having her frustrations heightened by rainy Florida weather, a slow New Orleans job, an entire set of blank negatives, and more frequent assignments devoted to defense efforts in the United States. Post's growing discontent with the defense assignments is seen in an extensive typewritten report found accompanying her January 1941 assignment in Louisiana. The first paragraph reads, "If you want a taste of all-out National Defense, go to Alexandria, Louisiana. See men, women, and children living in shacks, tents, trailers, and chickencoops. Try to rent a house or apartment. Stand in line to see a movie. Pay New York City prices—and more—for your dinner in a restaurant." Later in the report she complained, "Soldiers are everywhere."[86] Her enthusiasm for the FSA waned as the agency shifted its focus.

A long-awaited vacation, spent on ski slopes in Vermont, allowed her to return to Washington refreshed, not only ready for new field assignments but also prepared to begin the most challenging and long-term adventure of her life. On 6 June 1941 Marion Post married Lee Wolcott, an assistant to the secretary of agriculture in Washington. She had met him six weeks earlier at a friend's house in Virginia and years later claimed that the two "really didn't know each other" when they got married. After the wedding they spent a few weeks near Washington; then Marion Post Wolcott resumed her full-time duty with the FSA. This time Stryker chose not to arrange a schedule filled with excessive travel connections and tight deadlines. Perhaps more importantly, he decided not to send her back to the South. Earlier she had expressed a strong desire to journey out to the wide open spaces of the Great Plains, longing to see an environment unlike any she had ever experienced. She even told Stryker that she often heard a voice shouting in her ear, "Go west, young man." A trip to the West would in some ways round out her FSA career, bring her full circle, since she had worked in nearly every other region of the United States. It would be her last field assignment for the FSA.[87]

Before she departed for the West, Post Wolcott received a preparatory reading list from Stryker. The two-page bibliography included Walter Prescott Webb's *Great Plains,* Everett Dick's *Sod-House Frontier,* Mari Sandoz's *Old Jules,* O.E. Rölvaag's *Giants in the Earth,* Hamlin Garland's *Son of the Middle Border,* and books by Willa Cather, Clyde Davis, Paul DeKruif, and John G. Neihardt. Once on the trip, Post Wolcott found the variations in terrain fascinating. Overwhelmed by miles and miles of wheat in the flatlands, she wrote to Stryker, "I never imagined such expanses." While driving through the region, she followed Stryker's shooting script closely, attempting to capture the feeling of "space and distance and solitude." She noted that the trains stretching across the Plains looked like tiny toys, and she wondered aloud if the inhabitants ever got tired of such flat, boundless distances. During the time she resided at Brewster-Arnold Ranch in Birney, Montana, Post Wolcott concentrated on the elements that composed life on a dude ranch. Careful not to duplicate Rothstein's photographs of cattle ranches, she lent her own unique style to the horse and cattle photo stories she compiled. In Wyola, Wyoming, she documented an outdoor barbeque event, with community members talking, laughing, eating, and serving each other. She remained awed by the vastness of the land on which the animals roamed, and her photographs revealed her reverence. She also experimented with color photography, studying various phases of sunsets and red dawns over the mountains in Glacier National Park. An especially memorable part of the trip were the few days she spent on the road with her new husband.[88]

Making her own discoveries about the region, Post Wolcott finally satisfied her desire for a frontier adventure. The Western trek changed her. She experienced tremendous loneliness and realized through the course of the assignment that her loyalties were divided. Her correspondence to Stryker fell off considerably. She resorted to sending most of her messages by brief telegram, a surprising departure from the humorous, detailed letters she had always written. She found herself unable to concentrate as intensely on FSA duties as she had in the past. Among other things, the international situation frightened her, and she wished to be back within the safe confines of a home shared with her husband and two stepchildren. She wrote in August 1941, "[I] keep reading in the paper that we may be getting closer, very rapidly, to the kind of world system that may drastically, & perhaps tragically & seriously, change our whole lives. There seems so little time left to even try to really live, relatively normally." In addition, Post Wolcott had grown anxious about the shift in objectives at the FSA office. In 1941 she and other FSA photographers were assigned more jobs dealing with American war efforts, and they experienced the tighter restrictions placed on the agency. Subjects dealing with defense, including war preparations, aircraft assembly, and defense housing, demanded priority but were particularly unappealing to Post Wolcott. In the fall of 1941, she depicted with irony a public weapons display in Washington, D.C., revealing the wide-eyed excitement and curiosity of young Boy Scouts examining various weapons of destruction.[89]

By the year's end Post Wolcott had decided to end her service with the FSA. Two months pregnant, she realized that she could no longer maintain the pace, the long hours, the rapid travel connections, and the risks that accompanied every FSA photographer's job. Post Wolcott's scrupulous timing took her out of the FSA just months before the agency fell under control of the newly created Office of War Information. Her last letter to Stryker differed from all the others written in her three and a half years with the FSA. The familiar handwriting replaced by typescript, the casual conversation turned formal, it read:

> Dear Mr. Stryker,
> I hereby tender my resignation as Principal Photographer in the Historical Section, Division of Information, Farm Security Administration, effective at the close of the business day, February 21.
>
> Sincerely yours,
> Marion Post Wolcott

The cool tone no doubt reflected her increasing alienation from Stryker in the previous months, a rift aggravated by Lee Wolcott's insistence that every FSA photograph ever taken by his wife have her new married name recorded on the print. As a USDA administrator further up the bureaucratic ladder than Stryker, Wolcott got his way through an official memorandum reminding all regional directors that "when a female employee in government service marries, her legal surname *must* be used by her instead of her maiden name."[90]

Marion Post Wolcott's three-year tenure with the FSA had taken her from the Maine coast to Miami Beach, from the South Carolina lowlands to the Louisiana swamplands, from the Appalachians to the Rockies, and practically everyplace in between. She had stood in fields of wheat and cotton and corn, between rows of tobacco and tomatoes and peas, and under orange trees, maple trees, and palm trees. Inside these venues she had created thousands of images of Americans working together, playing together, and simply enjoying each other's company, in spite of economic devastation. Her emphasis on the rituals of corn shucking, sugaring, bean stringing, gambling, and card-playing, among other activities, points to her steadfast belief in visually defining America according to a communal ideal rather than by individual achievement. Post's pictures stunningly portray the essence of an era guided by the simple precept uttered by her contemporary Malcolm Cowley, "Not I but we; not mine or theirs but ours." Having transcended the prevailing themes of individual independence and success through mobility depicted in Lange's work and the inspiration of the past in traditional rural environments portrayed in Ulmann's work, Post came to stand with a cadre of new photographers who measured cultural standards by other criteria. Her view of group power and human relationships augmented the visions of women who were guiding documentary expression along a different set of markers. Among the most notable of these featured the vital role of modern technological power and its connection to the human element.[91]

4

Of Machines and People

MARGARET BOURKE-WHITE'S ISOLATION OF PRIMARY COMPONENTS

> When I was discovering the beauty of industrial
> shapes, people were only incidental to me. . .
> I had not much feeling for them.
> —Margaret Bourke-White, *Portrait of Myself*

AT A 1934 CONFERENCE IN NEW YORK entitled "Choosing a Career," an advertising executive told the following story: "If [our company] had let us say a brand of peanuts that weren't selling very well at ten cents a bag because people didn't think they were worth ten cents, thought they were only five cent peanuts, usually the inevitable conference would be called, and after a half hour or so of collaboration, the conclusion of the conference would always be the same, the best thing to do would be to hire Miss Margaret Bourke-White to take a picture of the peanuts and then people would think they were worth twenty-five cents a bag." Margaret Bourke-White could claim, at age twenty-nine, one of the most recognizable names in advertising photography. She had built her career in just over five years by assuming any challenge offered her and, in some cases, by creating her own challenges, even drawing attention to herself with outrageous antics and colorful clothes. She exhibited the "bluster and self-confidence" of the quintessential American "advertising man" whose role required spreading the word about exciting new products to a mass audience. Bourke-White reveled in the freedom and power her position afforded her, intimating early in her career to an old school chum, "I have the most thrilling job in America, I believe. I can go anywhere I want to go and meet anybody I want to meet." Part of the job entailed molding a remarkable image of herself, as the daring yet glamorous woman charting new and occasionally dangerous territory. Those efforts to cultivate a public persona would have meant little if Bourke-White's photographic images had been marginal. Instead, her pictures generously fed the modern imagination, seducing viewers with extreme angles, close focus, and emotional vitality. The manipulative powers Bourke-White wielded and applied in her camera work made her one of America's most penetrating modernist *seers* and one of its most controversial documentarians.[1]

A fascinated observer of industrial and technological changes, Bourke-White genuinely loved science. Like millions of other Americans in the 1920s and 1930s, she harbored a steadfast faith in technology, finding beauty in the "excitement" and "power" of the machine. She maintained her modernist sensibilities even after turning her camera in the mid-1930s toward people and their environments, particularly those who inspired a nationwide documentary impulse in writers, artists, and photographers. Her perpetual obsession with the material world was difficult for her to set aside as she ventured into the realm of neglected people and places. She continued to view progress in terms of what machines, rather than people, could achieve. Technical processes and their products were to be studied and admired; people were to be courted and persuaded. The relationship Bourke-White fostered between these two entities, the mechanical and the human, revealed illuminating results, with people dwarfed by forces they had little or no control over. While on assignment in 1930, she wired the following message to her boss: "INDUSTRIAL SUBJECTS CLOSE TO HEART OF LIFE TODAY—STOP. PHOTOGRAPHY SUITABLE PORTRAY INDUSTRY BECAUSE AN HONEST KIND OF MEDIUM—STOP. BEAUTY OF INDUSTRY LIES IN ITS TRUTH AND SIMPLICITY." Bourke-White's pursuit of what she viewed as the most essential truth and simplicity made her an apt spokesperson for the machine age. Her unabashed enthusiasm for it warranted her inclusion in a group dubbed the "apostles of modernity" by cultural historian Roland Marchand.[2]

As a child, Margaret developed a love for mechanical things under the watchful eye of her father, Joseph White, a machinery buff and devoted inventor. In addition to his work as a factory superintendent, White spent his life perfecting the printing press. He worked on color press variations, made small presses for mapmaking during the First World War, and invented the first Braille press. White, who sought out any mechanical challenge, also recognized the tremendous experimental possibilities the automobile offered. He bought one as much to try out his technical abilities as to offer exciting Sunday afternoon entertainment to his family. The vehicle provided numerous avenues for White's testing. Margaret, an astute observer of her father's interests, learned to enjoy them with him. Bourke-White biographer Vicki Goldberg relates a pivotal experience White provided for his eight-year-old daughter inside a foundry for press manufacture: "She already had a taste for adventure and the sense that everything she did with her father was adventurous, but factories came first, for he loved them best of all. . . . He steered her quietly up a metal stairway, nodding absently at the workmen. . . . he had brought her into a secret world that other girls had never seen."[3]

Recognizing the limits imposed on females, Margaret watched her mother operate under the rules that confined middle-class women in the early twentieth century. Margaret, born in 1904, would not see measurable improvements until sixteen years later, when women finally won the right to vote and middle-class society began to embrace some degree of social flexibility for women.[4] Before then, it seemed to Margaret that her mother simply marked time in the home, exerting her energies to carry out the smallest tasks to utmost perfection. Minnie Bourke White nevertheless offered indirect encouragement to her daughter by providing an example of determination and courage amid grim domestic circumstances. She chose not to erect barriers around her daughter's world. As Goldberg points out, Margaret "realized at an early age that her mother's energies were misplaced and that women were generally fated to live within what looked like narrow boundaries. While still a child, she dreamed of enlarging the territory, and no one said a word to discourage her dreams."[5] The contributions Minnie provided to Margaret's development equaled those of her husband. Later, as a young photographer, Margaret would honor both parents and their respective families by taking on the hyphenated surname Bourke-White.

The interests Margaret cultivated early had a significant impact on the subjects she later chose to photograph. She preferred science over history, geography over penmanship. In an effort to better her daughter's handwriting, Minnie White prompted Margaret to practice it during her summer vacation and even promised her a reward if it improved. Though Margaret thought little about how legible her script was, she did pay attention to instruction in other subjects. When she was thirteen, she took home an encouraging report from her teacher, who informed Joseph White that his daughter had done "excellent work in General Science." With the world of science ever changing, boasting new discoveries and various technological improvements, Margaret invested her time and her talents wisely. Knowledge in the scientific arena helped make any future appear brighter. But the sheer gathering of information would mean little if one pictured a personal destiny lived within the confines of the domestic sphere. Margaret had no such plans; she hoped to combine her knowledge with her quest for adventure. Her deepest desire was to become a biologist and "go to the jungles to do all the things women never did before, hoping . . . to be sent on an expedition."[6]

A thorough character analysis performed on Margaret in 1919 revealed facets of her personality and prospects for her future that amazingly prefigured her actual life experience. In several places the study suggested that Margaret use her adventuresome spirit toward productive ends. Phrenologist Jessie Allen Fowler recommended that Margaret teach geography since she harbored a "desire to travel, explore and see new localities." Fowler even made a prophecy based on the teenager's character, noting, "This temperament enables you to observe the details of everything you see in nature, and when you visit a new city you come away with a clear idea of what you have seen. Therefore it would pay you to travel and see the world, for you will be able to retain ideas of where you have been and what you have done. You should always take photographs of places you have visited." The personality described was well suited to the active life, because Fowler had determined that her patient was "always ready to go to any place that is suggested." But the analyst also addressed Margaret's weaknesses, about which she issued the clear warning, "Cultivate a little more reserve, tact, and diplomacy. . . . Try to moderate your approbativeness or sensitiveness of mind, your desire to excel, and your love of popularity." Even at age fifteen, Margaret White realized that others saw the enormity of her potential and the amount of energy she would need to fulfill it. She set her goals accordingly, keeping in mind her father's creativity and her mother's determination as examples to emulate.[7]

In Margaret's freshman year at Columbia University, her father suffered a massive stroke and died. His financial mismanagement forced Margaret to find another source for her college tuition. Fortunately, her uncle Lazar assumed that responsibility and allowed his niece to resume her studies, including those with Clarence H. White, whom she later described as "a great teacher and one of the most outstanding of our earlier photographers." But after Margaret's freshman year, her uncle could no longer finance her education at Columbia University. As the White family's money matters grew worse, it appeared that Margaret would not be able to continue her education. A philanthropic neighbor who knew about the family hardships offered to send Margaret to college, asking only that she do the same someday for a needy student. When Margaret expressed her interest in herpetology, she was told to enroll at the University of Michigan, where Professor Alexander Ruthven taught. She left her New York home for Ann Arbor at age eighteen, entertaining two passions, one for reptiles, the other for photographs. After one semester, Margaret's talent for taking pictures surpassed her abilities in the zoology classroom. Professor Ruthven urged his would-be protégé to move away from a career in herpetology and toward one in photography. He found employment for Margaret in the university museum, where she could combine her interests in science and photography by making negative prints of exhibit material. In her sophomore year at Michigan, Margaret began to consider herself a photographer. She took pictures for the yearbook, assisted other photographers, and even cultivated a romance with a fellow photographer, Everett

Chapman. "Chappie," as his classmates called him, reminded Margaret so much of her father that she was immediately attracted to him. Bourke-White biographer Vicki Goldberg has said that Margaret adored Chappie because he was "modeled on her dreams and her upbringing." Through his death, Joseph White had etched the outline of his daughter's future. She had gone to Michigan, where she reclaimed him in the person of Everett Chapman.[8]

Their common interest in taking and developing pictures led Margaret and Chappie into an intimate relationship, one that would revolve around photography and ultimately be dissolved by it. Although each took on various jobs, Margaret was more determined than Chappie to make something of her talent with the camera. Her boyfriend had his eyes set on a graduate degree in engineering, and he nonchalantly assumed Margaret would follow him to his chosen destination. As they became more seriously involved, their highly charged emotional lives crossed and caused each a great deal of suffering. Both confused energy, jealousy, and uneasiness for love. Believing that simultaneous elation and misery could be cured only one way, they got married. To celebrate their unconventionality, they scheduled a brief ceremony on Friday, 13 June 1924. Since none of the White family members attended, Margaret wrote to tell her mother about the less-than-spectacular event, which she described as "a miserable affair," made so by an Episcopalian minister who insisted they rehearse in order to make the ceremony "as dignified as possible."[9]

Both the bride and the groom were exhausted, having worked in the darkroom until two o'clock that morning. Immediately following the wedding ceremony, they went back to making prints, while Margaret's new mother-in-law "went home and cried two days afterward, and said that she'd never feel right about [the marriage]." Margaret's troubles with her mother-in-law began early, foretelling a conflict that would grow increasingly tense as the marriage wore on. Once when the newlyweds managed to get away to a quiet lakeside cottage, Mrs. Chapman unexpectedly showed up for a vacation of her own. In her frequent pouting she required that her son display more devotion for her than for his wife. That he frequently switched his loyalties from one woman to the other complicated an already stormy marriage. To ease the domestic tension, Margaret delved further into her work, nurturing her ambition to succeed. She continued with her photography jobs even after her husband took a secure position at Purdue University. Torn between her desire to have a child and her fear of sacrificing her individuality for a satisfying domestic life, Margaret agonized over the choice between home and career.[10]

The couple's move from Michigan to Indiana and then to Ohio left Margaret with various college credits but no degree. The single constant in her life was her photography. She always managed to find jobs, either on campus or around town. Margaret's camera work generated excitement in her life while her temperamental husband offered only silence. In 1926 Margaret finally worked up enough courage to leave Chappie and pursue her education again. She returned to New York, enrolled at Cornell, where she resumed her studies in herpetology, and continued to take pictures. Recognizing the advantages a single woman could enjoy in the 1920s, Margaret never mentioned her marriage. She wanted desperately to succeed on her own. On a trip to New York City, she spent most of her time "running around trying to make connections . . . in the photographic world." An observant aunt of Margaret's described her as "a Janus faced person," possessing inside "a little girl who won't stay down, who peeps out constantly, and comes out boldly when the adult person gets tired and retires for recuperation." Aunt Gussie, whose expectations reflected her nineteenth-century upbringing, good-naturedly offered a few helpful hints to her niece, such as "Don't wear yourself out. . . . I think your photography will be just the kind of a thing you can combine with being a housekeeper, as you will be able to a large extent to select your time for your work." The last thing Margaret wanted to do was "select" time for her photography. It came first. Although she considered a reconciliation with Chappie, she finally decided not to forfeit her passion or ambition in order to keep an unhappy marriage intact. A job offer in Cleveland persuaded Margaret to give her full attention to photography. It also prompted her to tie up loose ends.[11]

New circumstances and old problems combined to ensure Margaret a fresh start, one offering a path toward professional success and away from domestic strife. For two years Margaret had listened to Chappie's mother express hurt that her son had "deserted her," choosing "a wife instead of coming back to her."

When the Chapmans finally decided to dissolve their union, Margaret looked back on her trials with her mother-in-law as good experience that toughened her. She wrote, "I owe a peculiar debt to my mother-in-law [who] left me strong, knowing I could deal with a difficult experience, learning from it, and leaving it behind without bitterness, in a neat closed room. . . . I am grateful to her because, all unknowing, she opened the door to a more spacious life than I could ever have dreamed." With the divorce Margaret dropped her husband's name and reclaimed "White," determined to erase from memory her fiery, short-lived romance. At age twenty-three, she started over again, recreating both her identity and her image. Margaret cut her hair, bought brightly colored dresses and gloves to match her camera cloths, and set out to record the world in pictures. She began her trek in Cleveland.[12]

A thriving midwestern city, Cleveland boasted various industrial plants and plenty of river traffic. Margaret was intrigued. She walked the streets studying the architectural design of houses and buildings, always carrying along her portfolio and her camera. In a public square she took what became her first commercial photograph, a shot of a black man with his arms raised to the sky, preaching his gospel to an audience of pigeons. The backdrop for his public sermon is an imposing two-towered neo-Romanesque structure. The Cleveland Chamber of Commerce paid Bourke-White ten dollars for the scene. She worked on a commission basis for several months, taking pictures for architectural firms and a number of wealthy patrons. Her bosses quickly recognized their employee's eye for artistry and design, as she provided them with slick, high-quality images emphasizing the geometrical exterior lines of office buildings and private estates. She embraced this work in order to support herself, even though she wanted to devote time to more stimulating subjects. She decided that the massive Terminal Tower, a structure still under construction, was a perfect place for visual experimentation and personal adventure. The pictures she took there got her professional reputation off to a brilliant start, because the tower's controlling interests, the Van Sweringens, named her "official photographer" for the project. She had free reign to photograph the inside, the outside, even the top of the tower, if she wished. The Van Sweringens paid her "to feed her own excitement."[13]

The daring young photographer loved standing high above the city on steel scaffolding. Within months she had rented a studio on the twelfth floor of the new skyscraper and was able to enjoy the breathtaking heights anytime she pleased. At the front end of her career, Bourke-White wedded herself to a truly American design form, the modern skyscraper, whose towering presence dwarfed turn-of-the-century structures. The view allowed her to look down below at people, who seemed like tiny ants making their way through the streets. At eye level she could see factories sprawled out over the city. A glance up into the sky showed her industry's trademark of success, gray smoke billowing from automobile plants, paper mills, and steel mills. These signs reminded Bourke-White that industry dominated America, that it promised the best hope for the nation's economic future, and that it brought great financial success to those who took it seriously. Beyond its practical attributes were industry's strictly visceral appeals, which had been heightened by visual artists who portrayed what James Guimond has called "a kind of sooty romanticism." As early as the turn of the century, photographers such as Alfred Stieglitz, Edward Steichen, and Alvin Langdon Coburn had depicted industrial landscapes, dramatizing in particular the smoke and dust produced by locomotives and factories. The pictorialist vision cast industrial by-products in a soft haze, giving them an ephemeral, even mystical, quality. Remnants of this characteristically romantic view of industry still existed in the late 1920s, when Bourke-White began her professional career. But she would contribute to the transformation of the mechanized world into something not merely romantic, but positively seductive. What she saw in machines and their possibilities thrilled her, just as a new automobile had excited her father, providing him endless opportunities for experimentation. Of all the available manufacturing industries, steelmaking impressed Margaret most. She exclaimed, "There is something dynamic about the rush of flowing metal, the dying sparks, the clouds of smoke, the heat, the traveling cranes clanging back and forth." Steel marked the age, an age of progress and prosperity. That industry would help launch Margaret Bourke-White's career in photography.[14]

In 1928 she accepted two offers that elevated her professional status. One, from the president of the Otis Steel Mill in Cleveland, paid her one hundred dollars a photograph; the other, from Columbia University economics instructor Roy Stryker, provided public exposure but no considerable income. Bourke-White accepted both, seeing the complementary nature of the opportunities. She began exercising the clever business sense that would sell her pictures even after economic depression swept the United States. Her timing could not have been better; she was fortunate to have begun her career in advertising photography at a fortuitous time. Photographs "won increasing favor, partly because they were cheaper than drawings or paintings." Bourke-White quickly shed her cloak of inexperience and revealed a sharp eye for self-promotion. When Stryker requested a few of her photographs to use as illustrations in a revised edition of the department's economics textbook, Bourke-White immediately sent to him a full portfolio free of charge. She believed that students needed to recognize the beauty of industry as much as corporate stockholders did. She also understood that having her photographs in a book written by the highly esteemed economics faculty at Columbia University would mean wide circulation of her name and her pictures. The department's professors were so impressed with the photographs Bourke-White delivered to them that many requested prints to hang in their offices. "They are without doubt the finest set of industrial pictures I have ever seen and we all wish to commend you upon your ability to capture the artistic in the factory," Stryker wrote. He further noted that one of his students had been so inspired by the pictures that the young man wanted "to get out and work again." Bourke-White helped bring to life the ideologies promoted by Stryker and his mentor-colleague, Professor Rexford Tugwell. In a 1924 edition of their textbook, *American Economic Life and the Means of Its Improvement,* the authors had expressed a desire to provide for readers "the understanding, the control, and the improvement of the uses of industrial forces." Illustration, particularly photography, enhanced their presentation. One of the most telling images in the text showed the dominance of industry's components over the individual. In a penciled sketch, a huge crane dips molten steel as the accompanying manual laborers, who are dwarfed by the mechanism, tentatively look on. In the Otis Steel Mill, Bourke-White captured a similar scene on film, a first-prize winner at the Cleveland Museum of Art show that year.[15]

The early images Bourke-White created fed the notion that industry could simultaneously wear two mantles: it could be both an icon of power and a symbol of beauty. In affirming the processes of technological modernization, Bourke-White pointed to the attributes she believed most indicative of beauty: truth and simplicity. That her pictures of machines graced the pages of art and culture magazines showed how deeply the industrial age aesthetic had penetrated the nation's expressive soul by the end of the 1920s. When *Theatre Guild Magazine* ran a full-page print (the magazine's first such image) of Bourke-White's favorite "dynamo" shot, the caption read, "In her camera study of the dynamo, Margaret Bourke-White has evidently caught some of the power and beauty of the machine which suggested to Eugene O'Neill his Dramatic theme. But the repose in Miss Bourke-White's interpretation is quite unlike the demoniac godliness of Mr. O'Neill's." The reference to O'Neill's new play, *Dynamo* (1929), highlights the centrality of electrical power in Americans' lives by the late 1920s. The "Electrical God" figures prominently in *Dynamo*, as does one skeptical character who must be convinced by this development in modern living. Ramsey Fife grumbles early on about "Hydro-Electric Engineering" as "stuff that gives those stuck-up engineers their diplomas." By the end of the play, though, people are on their knees before the dynamo. The suggestion by *Theatre Guild Magazine*'s caption writer that peace and tranquillity could be found in industrial shapes indicates the level of innovation wrought by industrial designers and their followers. Bourke-White saw "perfect simplicity" in the dynamo, arguing that it was "as beautiful as a vase. . . . There is no decoration and not a line wasted."[16]

The fear of industry promoted largely by early-twentieth-century radicals and reformers, especially labor advocates arguing for protective legislation, had been replaced by excitement about the changes within corporate structure designed to enhance work life and its benefits. In addition, as more sophisticated machines required more skilled technicians and fewer unskilled laborers, the equipment itself demanded a new look. Facing the machines of a more complex industrial system than had existed just ten years earlier, Bourke-White stood in awe. Through the lens she recognized her own reverence, capturing it on film for others to con-

template. Because her vehicle was photography, itself a mechanical medium, Bourke-White stood as a true beneficiary of the machine-age perception that photography recorded reality. Displaying
the widespread faith in photography as an honest medium, even
industrial critic Lewis Mumford conceded that photographers made
"right use of the machine."[17]

While Bourke-White's photographs stood as artistic renderings of elements in the nation's economic structure, they also served
a more functional and perhaps immediate purpose. The images
were meant to persuade. Twelve Bourke-White photographs graced
the pages of a booklet entitled *The Story of Steel,* the Otis Steel
Company's brochure for stockholders and clients. The Otis job
took Bourke-White six months to complete, as she filled "waste
basket after waste basket . . . with discarded films."[18] She sought
artistic perfection, realizing that her task involved changing minds
and molding opinions. The dozen printed pictures proved to be
such a huge success that other corporations wanted to commission the young woman who turned assembly lines, ore piles, and
smoke stacks into works of art. Republic Steel, Lincoln Electric,
and the Chrysler Corporation hired her to photograph their mills
and projects, allowing Bourke-White one adventure after another
inside the male-defined world she longed to explore.

The photographer's actions cast her as an intermediary of
sorts—on the one hand she contributed to the "feminization" of
American consumer culture by continuing to portray the mechanical
and technological realm as a sphere closed off to and largely
misunderstood by the culture's primary consumers, women; at the
same time, she exploited certain aspects of her own femininity and
thrived on being the ultimate female consumer.[19] On her various
expeditions, she charmed laborers, floor managers, and corporate
presidents, leaving each with indelible impressions of her style,
wit, and daring. Men rushed to assist her with her heavy cameras,
tripods, and lights. Bourke-White manipulated her advantageous
position among them and once categorized her supporters as either "high hats" or "low hats."

> My high hat friends are my advertisers, and as long as
> I have the publicity counselor and the advertising man-
> ager of the Union Trust Co., the secretary of the Cham-

ber of Commerce and the Treasurer of the East Ohio Gas company, the publicity manager of the new Union Terminal and the president of the Otis Steel, talking about me at luncheon, I shall never need to buy any advertising in Cleveland.

. . . [M]y low hat friends do all my hack work. The amount I have had done for me is marvelous, and I could scarcely buy for money all the little fussy jobs that have gone into making my little apartment the BOURKE-WHITE STUDIO.[20]

The charms and conceits Bourke-White displayed would eventually have worn thin in the business world had her work not been superb. But her pictures commanded as much attention as Bourke-White herself did. If her personality turned some heads, her photographs went a step further—they received long, thoughtful stares. Bourke-White's careful cultivation of the art of persuasion began to pay dividends.

The Otis Steel Company pictures, which showed up in several newspapers' rotogravure sections, also landed on Henry Luce's desk in New York City. Luce, the publisher of *Time* magazine, found the pictures fascinating. The Otis job had required Bourke-White to experiment with various film types, exposures, and lighting methods because of "the intense heat, splashing metal, and the extremes of brilliant lights and heavy shades." Luce, impressed with the photographer's eye, her range, and her unusual interpretative abilities, wasted no time in summoning Bourke-White to his office. He wired, "HAROLD WENGLER HAS SHOWN ME YOUR PHOTOGRAPHS STOP WOULD LIKE TO SEE YOU STOP COULD YOU COME TO NEW YORK WITHIN A WEEK AT OUR EXPENSE STOP PLEASE TELEGRAPH WHEN." Seeing a great opportunity before her, the rising star decided to take "a free ride" on the corporation's budget; according to one *Time* employee, after Bourke-White arrived in New York she "went about her own business for two or three days before she bothered to look up Mr. Luce." When Bourke-White finally met the magazine owner, the two discussed his idea for a new publication devoted entirely to business and industry. Luce intended for the magazine, tentatively titled *Fortune,* to survey the diversity and excitement inside the industrial world. He planned to reach a wide audience by covering *Fortune*'s pages with striking visual images. And he wanted Bourke-White to help him succeed.[21]

Luce offered Bourke-White a full-time position with the magazine, which was to print its first issue in January 1930. Although the opportunities and benefits appealed to her, she declined the job. Unwilling to sacrifice her freelance work for a prestigious staff position, Bourke-White explained to her mother, "I would rather develop as an industrial photographer than an executive." The meeting between publisher and photographer ended in compromise, though, as Bourke-White allowed him to borrow her steel mill pictures in order to sell advertisers on the *Fortune* idea. Luce, in turn, considered Bourke-White's offer to work part-time on the magazine while continuing to take commissioned assignments. Since the editor wanted to launch "a purely business magazine with the best procurable in industrial art," he could hardly afford to be inflexible. He needed a photographer like Bourke-White, one who had taken the steel industry by storm, had impressed architects throughout the Midwest, and was gradually making her way through the ranks of advertising from Cleveland to Madison Avenue. Within a few weeks, *Fortune*'s managing editor, Parker Lloyd-Smith, wrote to Bourke-White: "This is to inform you officially of what you might conceivably have suspected. That we are glad to accept your proposition of giving us half your time from July 1. The cash consideration being $1000 a month." In the summer of 1929, Bourke-White began her affiliation with the Luce publications. The relationship would enhance her professional reputation as well as feed her adventuresome spirit.[22]

Fortune's only photographer went to work immediately. Within a week her travel agenda had been expanded to include Europe. Bourke-White excitedly dashed off a wire to her mother, exclaiming, "I cant believe it. . . . Just came from their offices this morning where they discussed sending me to photograph the Chmpaign [*sic*] Caves of France and the marble quarries of Italy along with some possible ship building in Germany." *Fortune*'s developers had grandiose objectives in mind and planned to implement them using the talents of its ambitious new photographer. Before the magazine sent its cameras abroad, though, it would establish a reputation in North America. Since the United States had been the most prosperous industrial nation throughout the 1920s, it seemed a logical place to begin the survey.[23]

In July Bourke-White began crisscrossing the continent, completing assignments from the East Coast to Chicago to Texas to Canada and in each place fashioning views that would help shape a rapidly developing machine-age aesthetic. On the trek she photographed watchmaking, glassblowing, meatpacking, salt mining, and plow blade manufacturing. In a photograph for the Oliver Chilled Plow Company (fig. 39), Bourke-White offered an artistic rendering of speed through a repeated image, a technique commonly used by industrial designers in the 1930s. Here the repeatability of the plow blades suggests not only the function but also the beauty of mass production. In this case Bourke-White employed repetition of a simple subject to create in her viewers an attraction to the product's potential movement. The image reveals some of the basic forms of the machine age aesthetic, most notably pure, clean, curved lines. Assuming a natural quality rather than a manmade one, the pieces of steel are stunningly suggestive of the human female body in profile. Unadorned, the smooth roundness of the successive figures connotes female fecundity, a quality defined by productive capabilities. Bourke-White's image is similar to Marcel Duchamp's modernist icon, *Nude Descending a Staircase, No. 2* (1912), where the figure's action effects a sweep of motion across the canvas. Although Duchamp claims to have removed any "naturalistic appearance" from the human image in the painting, his preliminary studies for the work were focused upon experimental research done in the field of photography twenty-five years earlier. Whereas Duchamp had created a human figure based upon its abstract mechanical qualities, Bourke-White had given machinery natural characteristics—in both, motion and beauty were inextricably bound.[24]

Though she often claimed to have invented the probing focus on industry's various components, she received a good deal of assistance in solidifying her style. Bourke-White had developed her visual sensibilities in a Clarence White class, years after Doris Ulmann and Dorothea Lange had studied with him. The experiences Bourke-White had with the photography master contrasted with those his pre–World War I students could remember. Since pictorialism declined after the war, visual emphases shifted from

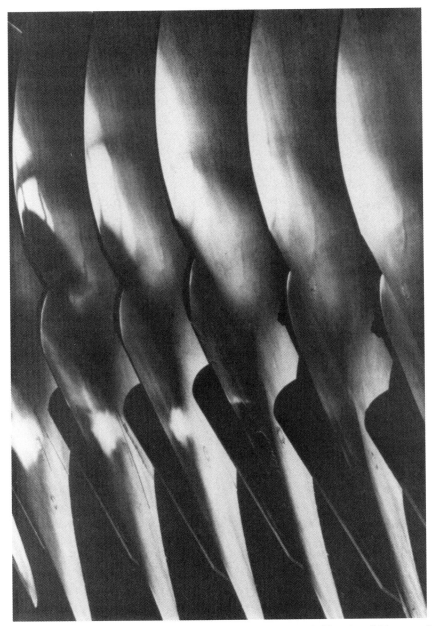

Fig. 39. Margaret Bourke-White. *Plow Blades, 1929.* Estate of Margaret Bourke-White.

shadowy scenes to more sharply defined objects. Bourke-White stepped in just as one trend in photography edged out another. Though she was on the membership rolls of the Pictorial Photographers of America for years, her work rarely, if ever, exemplified their artistic standards. The most memorable event Bourke-White recalled about the White School was an occasion when she posed nude for her classmates. One of those classmates, Ralph Steiner, reappeared in Bourke-White's life seven years later, not long after the Otis Steel job and her first taste of fame. Steiner, a skyscraper photographer, quickly became Bourke-White's most scrutinizing critic. He not only advised her on technical matters but also helped keep her ego in check. Bourke-White confided to her mother that Steiner never praised her work. The unrelenting critic remembered the night that he finally penetrated the photographer's cool, hard demeanor. He laughed at Bourke-White, who burst into tears and cried, "You're the only person in America who doesn't think I'm a great, great photographer." Steiner believed that was exactly the reason she kept returning for his advice. He taught her about different types of lenses and filters and focusing mounts, all the while improving her views and stimulating her creativity.[25]

Steiner's criticism helped Bourke-White successfully continue her mission to show industry in a new light and from different angles. His relationship to Bourke-White and her work is very different from the links he established with photographer Marion Post a few years later. Post found Steiner extremely supportive, open-minded about different styles, and hardly a tough critic. Although his steadily growing reputation in the intervening years may have mellowed Steiner, more likely the contrasts seen in his work with Bourke-White and with Post are due to the former's unwillingness to listen to anything but praise for her photography. And she seemed to hear it everywhere else she turned. Dwight Macdonald, a young editor with *Fortune* who collaborated with Bourke-White on one of the magazine's first stories, a survey of the Corning Glass factory in New York, wrote to her of the series, "It's very fine—even better than I anticipated. You certainly got a great deal out of your material; I especially liked the ones on the sand pile, the potter, and the blower. . . . Your pictures fit in well with the story. . . . It will be a great combination of pictures and text." Another *Fortune* writer, Archibald MacLeish, recognized Bourke-White's skill with the camera and begged her for prints of the glassmaking series. When she obliged him, MacLeish not only thanked her but also offered a glowing evaluation of her method, saying, "All your best things convince me that if photography is ever to become an art in the serious & rigorous & harsh sense of the term (not in the terms of the fashion magazines)—that is, if it is ever to become an art comparable to the art of painting—it will have to develop along the road you have set out upon! It will have to create its objects by isolating them in the real world, not by arranging them in a fake world." Although she never took part in the popular debate over whether photography itself was an art, Bourke-White illuminated her approach in a message to a dissatisfied client, telling him that she would rather destroy less-than-perfect prints than have them accepted with reluctance. She claimed to have followed a policy that allowed only satisfactory photographs to leave her studio; she told the patron, "I want my name signed only to pictures that I think are as artistically perfect as I can make them." In so doing, Bourke-White established herself as an art photographer as much as anyone else working in the field. Her sentiments mirror those of Doris Ulmann, who spent hours in her darkroom manipulating images to produce the desired effects after having spent hours with each subject to render the truest personality that could be captured on a glass plate.[26]

In Bourke-White's attempts to isolate objects and make them aesthetically pleasing, as MacLeish had pointed out, she often overlooked the people and the larger issues around her. With such deliberate intent, she unknowingly clouded her peripheral vision. One of the most crucial situations Bourke-White failed to recognize took place in a Boston bank the last week of October 1929. Assigned to take photographs of the bank's interior design, Bourke-White chose to work at night when the institution was closed and its customers were out. On her last night there, she became frustrated with the bank's vice presidents, who had stayed late and were frantically running about, too often in front of her camera lens. Amid Bourke-White's complaints, one official finally said, "I guess you don't know, the bottom dropped out of everything. . . . The stock market! Haven't you read the papers?" In one of Bourke-White's images in the series (fig. 40), the lines of the bank

Fig. 40. Margaret Bourke-White. *Bank Vault, 1929.* Estate of Margaret Bourke-White.

vault do not in any way intimate that disaster has just befallen the financial world, or that it ever could. Instead, the shine of the backlit steel evokes a sense of stability and security. The circular opening that protects the safe has bold concentric circles marking reinforcement, thus the solid protection of investments. The illumination coming from the vault itself makes yet another statement about the worship of material wealth in the 1920s. It would appear that heaven (or in this case, prosperity) stands just on the other side of the golden (steel) gates. In her autobiography, *Portrait of Myself*, Bourke-White spoke of the ill-fated night and admitted, "History was pushing her face into the camera, and here was I, turning my lens the other way."[27]

By the time Henry Luce published the first issue of *Fortune* in February 1930, the American business and financial realm had changed dramatically. Following the stock market crash came bank failures and factory closings. Unemployment began to rise steadily. The prosperity that had characterized the 1920s virtually disappeared. *Fortune* carried on, despite the irony of its existence. Although financial failure seemed a real possibility in any corner of the United States, Luce continued to assign projects surveying various industries and businesses. He sent Bourke-White to Hollywood to inspect moviemaking and on to Seattle to photograph the logging industry. She still gave the magazine only half her time but realized that her soaring reputation could be attributed to *Fortune*'s popularity; in turn, the magazine's success depended upon her roles as both extraordinary photographer and captivating woman. One Time Corporation executive argued that Bourke-White contributed a great deal to the early success of *Fortune*, which depended heavily on visual imagery. He wrote, "There is no denying that Bourke-White's work did most to making them [illustrations] outstanding—also, the intense little girl who took pictures for the big magazine of business added its fillip." Ralph Steiner reminded Bourke-White of the mutually beneficial relationship, pointing out, "'Fortune' is certainly skyrocketing your fame." While continuing to accept private commissions, Bourke-White also worked through the Cleveland advertising agency, Meldrum and Fewsmith. Through the firm, she got jobs that paid well; her boss, Joe Fewsmith, best remembered that his finest commercial photographer was willing to do anything, "from photographing a box of tacks to climbing up on the scaffolding on the top of a skyscraper." Bourke-White kept herself so busy that the specter of nationwide economic disaster seemed very distant. In the summer of 1930, she would forget about it entirely.[28]

On 27 June she received a bon voyage wire from her older sister that read: "SS BREMEN SAILING SATURDAY MORNING NEW YORK NY = PLEASURE ROMANCE THRILLS SUCCESS I KNOW THEY ARE ALL AWAITING YOU HERE AND ON THE OTHER SIDE." Luce had finally arranged his photographer's long-awaited trip to Europe. In the spring Bourke-White had written to a friend, "Imagine, getting paid to go to Europe! It seems too good to be true." Her itinerary included a long stay in Germany, with photography jobs scheduled at the Krupp Steel Works, the Allgemeine Elektrizitäts Gesellschaft, and the I.G. Farben chemical corporation. Upon arrival at the Farben trust, Bourke-White discovered that women were not allowed inside the plant. Bearing the double burden of being an outsider and a woman, the photographer claimed that Farben debated "six months" about whether to let her in or not. Officials eventually welcomed her, but she never knew what changed their minds. As a result of her position with a highly regarded business magazine, she managed to gain entrance into several other factories; but the business she had most looked forward to examining, the Krupp Works, was closed off to curious onlookers. The images Bourke-White took at German industrial sites show only a modest focus on the machinery, primarily because of the restrictions limiting what she could shoot. The workers themselves became her principal subjects, striking poses and expressions that made her comment that all German workers "look exactly alike." She considered her arrest at a factory near Cologne the "most exciting experience" she had while in Germany. She told her mother, "I was surrounded by police who thought I was a French spy." *Fortune* could take credit for feeding Bourke-White's daring spirit in her quest for adventure. When she finished the German assignments, she was directed to head east to the Soviet Union, a place that she discovered would teach her "a lesson in patience."[29]

Bourke-White entered the country at a particularly crucial time. The twelve-year-old Soviet experiment in government had expanded to include the huge task of mechanizing a vast, largely rural nation. Soviet leader Josef Stalin envisioned a plan that would reorganize the country within five years. A substantial part of the plan involved introducing industry to a population largely unfamiliar with its processes, rapidly mechanizing existing industries to accelerate progress, and building camaraderie among the people who participated in its execution. The stunning achievements included, for example, exponential growth in the number of heavy coal cutters and coal drills put in place; the replacement of manual labor with conveyor belts; and the creation of sophisticated machine production shops within existing factories.[30] Bourke-White knew only a little more about the Five Year Plan than she did about the political situation. Since the United States had elected not to extend official recognition to the Soviet government, Bourke-White was one of just a few Americans (and the only photographer) who had been allowed to travel inside the new Russia. Americans who had earlier garnered invitations from the Soviet bureaucracy usually came from the ranks of business and engineering; they crossed international lines much more easily than did their countrymen who were in politics. Perhaps Bourke-White's disinterest in diplomatic relations made her presence more palatable to Soviet officials. She later admitted, "No one could have known less about Russia politically than I knew—or cared less. To me, politics was colorless beside the drama of the machine." The portfolio pictures she toted everywhere grew worn as Russian bureaucrats and workers alike passed them around and marveled at the beautiful images of American industry. Bourke-White's photographs so impressed the right officials that she was presented traveling papers that stipulated few restrictions. She could move freely without fear of being searched at every turn. One Russian bureaucrat told Bourke-White she could "go to the moon" with the government papers she held. Relishing her independence, Bourke-White took full advantage of the few weeks she had in the Soviet Union to photograph its industrial progress.[31]

To her, the most appealing aspect of the Five Year Plan was its expedience. The Soviet bureaucracy was attempting to accomplish in five years what had taken one hundred years in the United States. Bourke-White realized what a rare opportunity it would be to record step by step the entire industrialization process, what she called this "great scenic drama being unrolled before the eyes of the world." Since nineteenth-century artists had shown little interest in the American Industrial Revolution, the tremendous changes that had accompanied it had gone undocumented. Bourke-White was in a position to capture similar scenes and show much more of the process because it had been greatly accelerated in the Soviet Union. Perhaps for the first time in her career, she realized that her photographs could serve a purpose beyond their immediate utility as advertising shots. A hint of the documentary impulse is evident in a vitae statement she made several months after her return: "Things are happening in Russia, and happening with staggering speed. . . . I wanted to make the pictures of this astonishing development, because, whatever the outcome, whether success or failure, the plan is so gigantic, so unprecedented in all history that I felt that these photographic records might have some historical value." While in the Soviet Union, Bourke-White had felt the contagious excitement of workers fascinated by new machines and various gadgets. Their favorite phrase, *Amerikanskoe tempo,* referred to the ultimate model in industry—the American way—one dependent upon assembly-line production. American engineers helped direct the implementation of Five Year Plan objectives, from assembly-line management to dam construction. Bourke-White remembered the Stalingrad factory workers' animated discussions about the wonders of the conveyor belt. Noticing that everywhere she traveled in Russia, people worshiped machines, Bourke-White later reflected, "They looked on the coming of the machine as their Saviour; it was the instrument of their deliverance." Given her own reverence for industrial design and machine functionalism, Bourke-White found the atmosphere in Soviet Russia congenial to her prejudices and tastes.[32]

American curiosity about the mysteries of Russian progress worked in Bourke-White's favor. She returned home to find numerous offers for her pictures of Soviet industry. Seeking advice, she informed publisher Max Schuster that she had been "bombarded with requests" from newspapers and magazines for the photographs. Although *Fortune* had first rights to the pictures, Schuster suggested that Bourke-White limit their circulation in order "to arouse anticipatory interest" in a prospective book. *Fortune* published nine of Bourke-White's numerous photographs in its February 1931 issue. A brief two-sentence introduction under the heading "Soviet Panorama" opened the photo essay. The image story included two traditional landscapes, one marked by horses in a field, the other by farm machinery. The captions for these less-than-remarkable agrarian scenes informed the American reader that 75 percent of Soviet farms were still privately owned, but that collective farms or "voluntary organizations of peasants who combine their farms" had been so successful that the Five Year Plan included a goal of 100 percent collectivization by 1933. In addition to depicting changes in rural life, the pictures focus on the "most ambitious" Soviet industrial project, the Dneiper Dam; the expanding tractor industry; and in a rather unusual undertaking for Bourke-White, a textile mill. In one image, highly lit threads coming off endless spools appear like sparks in a steel mill, sprinkling three-quarters of the frame. A stooped woman attends the machine but takes a place on the periphery as the process of manufacturing assumes center stage. Bourke-White's eye for geometric design dictates the very stylized image with its crossed, hair-thin lines connecting to their respective circular homes.[33]

Having whetted the American public's appetite for images of Soviet industrial success (while the United States floundered in the seeming demise of its own industrial economy), Bourke-White decided to accept an offer made by the Simon and Schuster publishing house. Max Schuster wanted to publish the Russia pictures, but he also wanted words to complement them. He asked Bourke-White if she would provide a few stories about her trip to the Soviet Union. She promised to lock herself away in her studio and ignore the telephone while she attempted to chronicle her experiences. Assured that Bourke-White would finish the task, Schuster drew up a contract in April that allowed the author a $250 advance and guaranteed to her "all property rights to the original photographs." Bourke-White found that showing her pictures to the American public was easy. Writing, however, was a different matter; she would have to exercise strict prudence in describing the new Soviet Union.[34]

The magnificence of Soviet industrialization had not blinded Bourke-White to the country's problems. Despite the help offered by American engineers and businessmen, industrializing the rural society proved to be a difficult task. Even with Henry Ford's assembly lines, Albert Kahn's factory designs, and Col. Hugh Cooper's personal direction of the Dnieper Dam construction, domestic problems persisted. In a letter to her mother, Bourke-White described the severity of the food situation: the peasants hoarded what little they raised, and the government exported the rest "to pay for machinery." Upon her return to the United States, Bourke-White summed up the Soviet situation to Walter Winchell in ten words: "Little food, no shoes, terrible inefficiency, steady progress, great hope." But she rarely elaborated upon the story of human suffering in public; she limited her opinions on any contentious matter to the circle of her family. Likewise, while writing *Eyes on Russia* for Simon and Schuster, she chose her words carefully. She was most afraid her critics would claim that the Soviet government had allowed her to see only what it wanted her to see. Since the government had made Bourke-White its guest and had given her "a fat roll of ruble notes," her motives could appear suspicious to the American public. She alleviated any potential controversy by concentrating on the tedious details of cement-making, steelmaking, and dam-building. She complemented her descriptive stories with charming anecdotes, scoring points with her editors. The balance she struck helped make *Eyes on Russia* a success. In her mother's opinion, Bourke-White had mastered the art of diplomacy. After seeing the book, Minnie White wired her daughter: "WISE GIRL TO STEER CLEAR OF SCYLLA OF POLITICS CHARYBDIS OF PROPHECY."[35]

Bourke-White wanted to write nothing that would compromise her skyrocketing career. At twenty-seven she had made a name for herself in the photography world, and she envisioned making even grander contributions to it. In order to be closer to the *Fortune* offices, she moved to New York City. Perhaps her most stunning embrace of the machine age was her insistence on living in one of its newest monuments. When the Chrysler Building opened in 1931, Bourke-White had to apply for a janitorial position there in order to be allowed off-hours occupancy in a city office building. But she wanted the best view in New York. She later wrote of her experience in the Chrysler Building, "I loved the view so much that I often crawled out on the gargoyles, which projected over the street 800 feet below, to take pictures of the changing moods of the city." In one image, the elements of early 1930s industrial design dominate (fig. 41). The aerodynamic shape of

the eagle's head, with its curved beak, are augmented by the concave sweep of the figure's supports, which Bourke-White highlighted
with her camera angle. The gargoyle bears several characteristics
of popular machine-age design created to connote speed and power.
The use of steel on the structure's exterior, on a decorative feature, no less, reveals the transcendence of the modern, industrial
way over the older, traditional monuments, marked by brick, stone,
and wood. Steel's smooth surface suggests its lightness, which in
turn allows for speed. The structure's aerodynamic quality sup-
plants the rough-surfaced heaviness of earlier external structures.
Bourke-White enhanced this compelling figure by carefully timing her photographic process; she scheduled the shooting early one
Sunday morning when the air was "much clearer" than on week-
days, then attempted to "catch the view at just the moment when
the sun [struck] the bright shell work of the gargoyle with the
dark city as a background." Power resides in the steel figure, whose
watchful eye stares out over the city; that the protective eye belonged to a company that made automobiles, the single most influential mass-produced item in the previous decade, marked its
unwavering status in American culture. The gargoyle image net-
ted Bourke-White four hundred dollars from Meldrum and
Fewsmith Advertising Agency.[36]

Among the photographs in the gargoyle series were several
stunning images of Bourke-White poised atop the eagle's head
looking through the viewfinder of her camera. Her fearlessness
turned heads in the photography world; it also landed her an
assignment to shoot the Chrysler Building's recent rival. Joe
Fewsmith, Republic Steel's advertising representative, tapped
Bourke-White to photograph the company's stainless steel trim
on New York City's newest skyscraper, the Empire State Building.
His specific instructions directed her to "climb out on some of the
ledges" in order to get "a very unusual view" for a full-page ad
in the *Saturday Evening Post*. Fewsmith did not hesitate to ask
Bourke-White to take the physical risks necessary to get the best
shots. His treatment of a young female employee contrasts sharply
with Roy Stryker's handling of the women on his FSA photography staff, particularly Marion Post. From the beginning Stryker
attempted to protect Post from potential problems she might
encounter and encouraged her not to take risks or conduct herself

Fig. 41. Margaret Bourke-White. *Chrysler Building, 1930.* Estate of Margaret Bourke-White. Courtesy of George Arents Research Library, Syracuse University.

in a way that might arouse suspicion. Bourke-White's deliberate construction of her public persona and her demands for individuality (with requisite guidelines for various employers) allowed her the freedom of movement that Post could not enjoy. To ensure that her reputation remained intact, Bourke-White had many portraits of herself as a photographer taken for promotional purposes. But maintaining the Margaret Bourke-White image proved to be costly.[37]

More than willing to attach her name to another eye-catching project, Bourke-White devoted her Christmas holidays to photographing America's tallest building. She hoped to enhance her reputation with the Empire State job; she also needed the money to pay her rent. Speaking engagements, advertising assignments, and movie jobs provided Bourke-White with some money, but she still struggled with her finances. She refused to give up a fashionable address and a stylish wardrobe, since she believed that these impressed her clients and drew the public's attention, making her reputation her "greatest selling factor." The appearance of success brought further success, she opined. And women, especially, had to be tireless in their pursuit of success. In a May 1933 speech, Bourke-White advised the career woman to "bring her name before the eyes of the public *in general* so as to build up a reputation, to keep her work constantly before advertising executives and industrials—who are her buyers—and to constantly keep on tip-toe turning out something new." She showed signs of harboring a feminist consciousness in her practical advice to professional women. Her own goals had kept her on site in the manmade world of machines, while she played on the new woman stereotype, telling interviewers that she had little time for a domestic life since professional demands kept her occupied; living expensively on credit; enjoying intimate relationships with several men; costuming herself in the latest fashions; and proving that she understood mechanization processes, something the advertising community had strongly suggested that women knew very little about. Constructing herself was one of Bourke-White's most stellar achievements.[38]

The Bourke-White reputation kept her employed, but it failed to satisfy her anxious creditors. She had outstanding debts throughout the city, from camera shops to Fifth Avenue department stores, where she continued to indulge her desires. Family members had even allowed her to draw on their insurance policies in order to keep her studio open during the leanest Depression years, but by 1933 she had, according to her mother, "milked the insurance cow dry." She owed hundreds of dollars to Eastman Kodak in 1934 but put off settling her account until she had equipped her new studio, a Fifth Avenue penthouse. She had been forced to find a cheaper place after the Chrysler Corporation evicted her from the skyscraper studio. Bourke-White considered her financial situation a minor inconvenience in her busy and exciting life. She was content to search endlessly in pursuit of the best scenes the industrial world had to offer. She believed that her public reputation as a professional artist would more than compensate for her mismanagement of private business affairs.[39]

LaSalle Automobile executives agreed, since they hired her to promote their new models in 1933. In her series of automobile advertisements, Bourke-White further showed herself a devotee and promoter of the machine age aesthetic. In a close-up of a LaSalle (fig. 42), Bourke-White isolated one of the more important attributes of the machine—precision. The streamlined design, with the egg-shaped headlight and the rounded front wheel cover, provides the requisite aerodynamic quality. The front panel's decorative details are simple but modern, their modernity denoted by the round decorative features. As she had done on numerous other jobs, Bourke-White isolated the circular and curved forms on the automobile. The uniformity of the design elements, all of which bear the streamlined curve, suggests that this is a carefully planned, well-functioning machine. The light color is yet another mark of the machine-age aesthetic, where the appearance of cleanness in a product became a vital sign that its operation was as flawless as its appearance. Bourke-White employed dramatic shadows to enhance the machine's true color and its brightness. With this series, her photographic work bears out historian Miles Orvell's assessment of the camera's contribution in developing industrial capitalism. He said it was not only "an instrument for analysis and record" but also "an instrument for promoting an *image* of the product that would stamp it in the public mind."[40]

In helping to foster and define this new role for the advertising camera, Bourke-White had not always met willing supporters for the modern style. Two years earlier, H.S. Bishop, an advertis-

Fig. 42. Margaret Bourke-White. *LaSalle Automobile, 1933.* Estate of
Margaret Bourke-White.

ing executive at the Pierce-Arrow Motor Car Company, complained to Bourke-White about hearing "What beautiful photography!" He wanted people to comment on the cars in the pictures rather than the pictures themselves. Having worked little with the automobile industry when offered the Pierce-Arrow job in 1931, Bourke-White had enthusiastically accepted it. She loved taking pictures in the whirlwind of city traffic and saw great "possibilities" in automobile photography, but Bishop did not share her point of view. He thought that Bourke-White's work was too abstract and dramatic to be functional and that she should show automobiles from a more "familiar angle of vision." He concluded that the photographs were "much too fine a thing for advertising. . . . These have your favored industrial flavor, but they are not pictures that we believe would help sell Pierce-Arrows to the proletariat." Within months the auto industry was moving away from such criticism, with executives growing more willing to support abstract design in advertising. Their attitudes shifted, as did automobile design, away from the black, box-shaped vehicles the industry had offered consumers in the 1920s. And Bourke-White's pictures were playing a role in the rapid transformation.[41]

By 1933, as a result of Bourke-White's experience with isolating components of the machine age, from shiny automobile hubcaps to sleek stainless steel cocktail shakers, she embraced an augmented visual form that enhanced the modernist's use of extreme angles and bright lights—photomurals. The large-scale works implied that the bigger the picture, the more influential its message, but Bourke-White saw the new form a bit differently. A ten-foot-tall picture was no better than one ten inches high, she noted, until "a large company gives over its expensive wall space to photographers. . . . that, I think, is something." Bourke-White's first photomural job took her to the RCA Building at Rockefeller Plaza, headquarters of the National Broadcasting Company. She saw the mural project as an opportunity to use photography on a much grander scale. An added benefit would be the excellent publicity the prestigious location would lend to her work. Identifying the photomural as "a new American art form," Bourke-White encouraged businesses to take advantage of these huge pictures by decorating their office buildings with them. She suggested in a speech to the *New York Times* Advertising Club that companies

choose wall coverings showing the kinds of work carried out in
the offices. She explained, "Art is too often divorced from life and,
it seems such a sound idea to me to have the subject of a mural
intimately tied up with the activities that go on in the company,
so that it will have real flesh and blood."[42]

Bourke-White believed her photomural for NBC/Radio City
exemplified the new directions and standards being developed in
the arena of public art. Since her theme was industry, the photographs brought people closer to processes they might not otherwise understand. The NBC photomural showed pictures of
microphones, switchboards, generators, towers, and transmission
tubes. Broadcasting operations could be seen in photographs covering a rotunda 160 feet around and 10 feet high. One image (fig.
43) revealed a sea of radio loudspeakers, all uniformly arranged
in closely aligned rows. Using the technique of patterned repetition that she had employed before, Bourke-White enhanced the
elements of power and beauty in sound design. Simplicity is borne
out by two unadorned features, the ringed indentations on each
speaker and the limited number of nuts and bolts holding the
speaker frames together. Whether or not her pictures helped un-
fold radio's mysteries, they stood as arguments for smooth functioning and the optimal form through which it was achieved. But
images like these also served to exacerbate the national phenomenon known in the 1930s as "cultural lag," the gap between rapid
technological growth and the culture it had outrun. During that
decade the intensive artistic search based on American sources,
regional histories, and local color could be attributed, in part, to
the attempts to close the gap by creating what Warren Susman
called "patterns of a way of life worth understanding." Bourke-
White's highly modern, stylized views of powerful technological
components emphasized their aesthetic qualities much more than
their practical applications, so that the resulting images inspired
awe in onlookers, thus limiting their ability to close the gap between industrial determinism and society's capacity to adapt to it.
Bourke-White continued to push the limits of technological progress
even further when she chose to study and photograph an experimental medium called "television" inside NBC's "secret laboratories" at the Empire State Building. Perhaps her reverence for
the machine age and its components was best seen in her photo-

Fig. 43. Margaret Bourke-White. *Radio Loudspeakers, 1934.* Estate of Margaret Bourke-White.

murals, as pictures enlarged over one hundred times revealed her truest perceptions of what she considered the guiding forces within American society, namely the "bigness and power about this industrial age."[43]

Bourke-White used the NBC job as a springboard for other photomural assignments. In her appeals to Fred Black of the Ford Motor Company, she said that a Ford mural "would cause a great deal of interest among industrialists as well as art critics." She pointed out how overwhelming the response had been at Radio City, hinting at the free publicity Ford would enjoy if it undertook a similar project. She told Black, "If the amount of newspaper space devoted to the NBC mural is any indication of the interest in this new art, I think that a mural using the wealth of spectacular material in the Ford Motor Co. would attract even more attention." Bourke-White talked her way into a contract with Ford, then moved on to deliver the same pitch to General Motors. She recommended that photomurals be used in dealer displays, since the enlarged pictures would show "how adequate your factory is and with what precision General Motors operations are carried out." Impressed with Bourke-White's earlier murals, executives at the Aluminum Company of America sent the photographer to the Midwest to take pictures for their exhibit at the Chicago "Century of Progress" Exposition. They, too, were pleased with Bourke-White's execution of the project and the attention drawn to their products as a result of her murals.[44]

While devoted clients depended upon Bourke-White's expertise, a few critics attacked her artistic approach; others thought her perception of industry's dominant role in America was overrated. Bourke-White's evaluation of the American viewing public was challenged by moviemaker David O. Selznick, who discouraged her plans to shoot industrial films. He seemed unimpressed with Bourke-White's *Fortune* contacts and told her that big industrial names were "absolutely valueless" in the filmmaking business. He informed Bourke-White that industry as a subject had no entertainment value and concluded, "Work is not the most interesting idea in America. . . . The movie goer still goes for escape. Vogue of realism is no exception; people like some other kind of realism, not their own." Bourke-White even received criticism from nonprofessional ranks. Members of a Boston women's club complained that the photographer had provided too much illustration and not enough talk at a luncheon speech. After hearing the comments, Bourke-White's booking agent cautioned her not to rely too heavily on her photographs, warning, "There is nothing better than a fine pictorial presentation to catch the eye and bring to mind what the realities actually are. But that is the end of the function of the pictures." Bourke-White adamantly disagreed with her critics. She believed photography, of all media, could best illuminate what she considered to be the overriding feature of American life—industrial capitalism.[45]

Despite the nationwide economic depression, business picked up for Bourke-White in 1934. Several of the numerous solicitations she had made the preceding year paid off. In addition to the photomurals, she completed assignments for popular magazines such as the *Saturday Evening Post* and *Vanity Fair*. The VanBeuren Corporation, a subsidiary of RKO, bought the experimental film she had taken in 1932 on her third trip to Russia. Bourke-White had hoped a bigger studio would be interested in her film's timely connection to President Roosevelt's official recognition of the Soviet Union. To entice Norman Moray of Warner Brothers, she wrote, "The news in the papers continues to be intensely interesting. . . . There is every reason to believe that when the event comes it will not be a mere trade pact but full recognition and will cause the news to be extended over a much longer period than might have been true. . . . I believe the people in the U.S. are going to want to know what the people of this new country are like." But the popular movie moguls disagreed with her prediction, remaining unconvinced that a story about Russian life and work would attract the general public to the box office. So Bourke-White accepted the only feasible offer for her film. Her campaign to sell the Russia story encouraged her increasing acknowledgment of current political news. More than ever before, she was showing an interest in matters outside the machine-age consumer culture.[46]

As she shifted her eyes to examine political issues, numerous groups solicited her support. They recognized the advantages of having Bourke-White's celebrity attached to their causes. In January 1934 she donated thirteen prints to a San Francisco exhibit and auction sponsored by the National Committee for the Defense of Political Prisoners. The committee's appeal to artists, writers, and musicians described their current effort to raise money and consciousness in order to counter "prejudice and hate" over the Scottsboro Boys case, where slight evidence had been enough to convict nine black youths of raping two white girls on a train in northern Alabama. Bourke-White also gave photographs to the New School for Social Research and lectured at the New Workers School. Eventually she joined Dorothy Day, Josephine Herbst, Mary R. Beard, and others in the League of Women Shoppers, a group that supported organized labor protests and other activities to abolish "sweat shop conditions" in factories and stores. Along with Ralph Steiner, Berenice Abbott, Reginald Marsh, and Lee Strasberg, she served as a major sponsor for the Film and Photo League, an artists' collective seeking to educate the public about the interests and concerns of the working class.[47]

But the group Bourke-White felt most strongly about was the American Artists' Congress, a coalition designed to protect cultural freedom by supporting artists in their political convictions. She wrote to several other photographers, asking for their participation in the congress and alerting them that "such reactionary tendencies as have appeared in the form of censorship and destruction of works of art, and in the suppression of civil liberties, are symptomatic of more serious conditions to come. *It can happen here.*" By directing her energies into social causes, Bourke-White developed a side of herself previously neglected. As her biographer has pointed out, "In the twenties, Margaret lost her innocence; in the thirties, her indifference." The noticeable shift in Bourke-White's attentions placed her, at least sentimentally, alongside some of the decade's great documentarians, optimists who celebrated ordinary people's lives and trials. But however deeply Bourke-White may have felt about the victims of political oppression or economic displacement or natural disaster, her photographic records lack an expression of genuine compassion.[48]

In her autobiography Bourke-White admitted, "When I was discovering the beauty of industrial shapes, people were only incidental to me . . . I had not much feeling for them." She first attempted to make the transition from photographing industrial shapes to photographing people after learning about the Roosevelt administration's New Deal programs, which supported everything from bridge building to landscape painting. Bourke-White's editor Max Schuster suggested that she look into photographing the New Deal for the Treasury Department, which was "literally team-ing [sic] with paintings and sketches by hundreds of CWA [Civil Works Administration] artists." Bourke-White, however, thought that the Tennessee Valley Authority projects offered greater op-portunities for her photographic style. Since the TVA plans in-cluded dam building, Bourke-White felt particularly qualified to document the construction. She told TVA official Arthur Morgan of her experience photographing similar works in the Soviet Union and added, "I have always thought that industrial photographs could perform a very important service in informing the public about the vital things that are happening in their country." But Bourke-White planned to go beyond merely documenting the building process; she wanted to show how a massive government project would affect its workers' families. She called the social aspects of the project an "important American development" that would "catch the public imagination." TVA executives appeared sold on hiring her until Bourke-White mentioned her rate of five hundred dollars per ten pictures. As much as they wanted Bourke-White's expertise in documenting the project, they could not commit government money to such an expense and so wired, "Regret that Photos you mention are quite beyond our means."[49]

Despite her aborted attempt with TVA, Bourke-White got an opportunity to focus more closely on humanity when *Fortune* sent her to the drought-stricken Plains in the summer of 1934. Claim-ing to have experienced a change of heart after witnessing the Dust Bowl firsthand, Bourke-White recalled its impact: "I had never seen people caught helpless like this in total tragedy. They had no defense. They had no plan. They were numbed like their own dumb animals, and many of these animals were choking and dying in drifting soil." The pictures Bourke-White took of the survivors revealed hard, shell-like faces, reminiscent of wax figures in a museum. In one sidewalk shot (fig. 44), Bourke-White peered through her lens downward on two young men who stare glumly ahead in the distance. They are diminished physically by the photographer's angle. Their eyes obscured, one pair by aviator glasses, the other by dark shades, the men project a certain vacu-ity. With their lips tightly closed, they appear resigned to the fate nature has dealt them. The visual image projects the message that nothing they might say could change their present condition. In photographing these subjects, Bourke-White neglected to pursue the single most vital facial feature in measuring human emotion: eyes. The only hint of an eye is seen through the reflected lens of the aviator specs, and even then, the eye seems out of place, set askew without its mate. Bourke-White makes the tiny circle the central focus of her image, the most highly-lit object in the frame. It stands alone, not unlike the mechanical television eye she had isolated and photographed at Radio City a couple of years earlier. In the Dust Bowl series, Bourke-White failed to capture the residual strength of individuals ravaged by environmental disaster. Whereas Dorothea Lange had illuminated what was left of human dignity amid horrible circumstances, Bourke-White focused on what had been stripped away. In Lange's photographs, the spirit survived; in Bourke-White's, it had disintegrated, much like the eroded soil (see, e.g., fig. 16). Historian David Peeler explains that Bourke-White failed to articulate the human toll taken by dust storms because she turned her cameras overwhelmingly toward buildings rather than people. In the end her "habits subverted her intentions," Peeler concludes.[50]

Although she would spend the next few years attempting to depict people and their lives, Bourke-White rarely managed to accept what they offered to the camera. The tactics of persuasion she had honed as an advertising photographer remained deeply ingrained; exaggerated angles and isolated elements yielded a picture of fragmented humanity. Even her contemporaries recognized Bourke-White's limitations when it came to portraying people. Eleanor Tracy, of *Fortune,* passed along general criticism of the images Bourke-White had made in a "family" assignment for the magazine, noting that they appeared "too stiff and posed-looking" and that they lacked "spontaneity." Tracy offered blunt advice to Bourke-White in an attempt to ensure that the photographer's next "family job" would produce better results: "After all, the only reason for candid photography is the ability to catch people unawares and to get natural expressions and actions which cannot be caught in any other way. . . . you ought to concentrate on taking lots of pictures without the subjects' knowledge if possible."[51] Tracy's suggested method was the complete antithesis to Bourke-White's carefully calculated, highly stylized work. The photogra-

Fig. 44. Margaret Bourke-White. *Dust Bowl Faces, 1935.* Estate of Margaret Bourke-White.

pher had forged a stellar career out of her ability to arrange and control scenes. Just as she had dramatized sunlit steel on the Chrysler Building gargoyles, she captured faces blinded by harsh, artificial light. She stood or crouched at extreme angles in order to record the human condition from a different, and sometimes disturbing, perspective. The resulting images often showed people at their basest level, similar to the farm animals to which she had compared the Dust Bowl survivors. The photographer's shift from machines to humans proved an arduous task.

While struggling to redefine her place in American photography in the mid-1930s, Bourke-White wrote an essay in which she explained (perhaps to herself as much as to the reading public) the significance of a photographer's "point of view." She claimed that this aspect was paramount, transcending "the illumination, modelling, choice of boundaries, focal length of lenses" and other necessary elements in the image-making process. The principal questions Bourke-White posed in the essay reveal a litmus test of sorts in judging a photographer's point of view—"How alive is he? Does he know what is happening in the world? How sensitive has he become during the course of his own photographic development to the world-shaking changes in the social scene about him?" Here the model photographer proves his or her worthiness in the profession by having developed a social consciousness along the way; the extent to which he or she may be taken seriously as a professional rides on a level of sensitivity to social issues. But how could such sensitivity be measured?[52]

    If Bourke-White came to documentary through a desire to bring her work closer to the "realities of life," as she wrote in 1936, she probably recognized the advantages that words could offer her images. At the same time that Bourke-White's pictures of people needed supportive text, southern novelist Erskine Caldwell's words about people needed pictures. In 1936 Caldwell found himself in search of "the best photographer available." He intended to make a thorough evaluation of the American South in order to prove that the scenes portrayed in the long-running Broadway play based on his best-selling novel *Tobacco Road* (1932) were authentic. Critics and censors had railed against Caldwell's fiction, stories that depicted the South in terms of its most chilling trademarks—illiteracy, racism, and poverty. Caldwell hoped to change their minds with a new piece of nonfiction that would be filled with telling photographs. His show of faith in the camera as a recorder of truth and photography as an objective medium placed Caldwell squarely within a mainstream intellectual milieu whose wholehearted embrace of technically produced visual images gave them credibility as powerful articles of truth.[53]

Early in 1936 the novelist contacted Bourke-White, asking her to meet him in Georgia six months later. She accepted his offer with enthusiasm, exclaiming, "I am happier about this than I can say! If I had a chance to choose from every living writer in America I would choose you first as the person I would like to do such a book with. . . . Just when I have decided that I want to take pictures that are close to life—seems almost too good to be true." From spring until summer Bourke-White stayed quite busy with advertising jobs, including a stint in South America for the American Can Company. Business affairs kept her in New York just long enough to delay her arrival in Augusta, Georgia, by a few days. When Caldwell threatened to find someone else for the project, Bourke-White calmed him with a letter explaining her need "to carry the overhead while doing a really creative and socially important job like the book with you." Caldwell softened, and the two started their journey together the next day. On the trek the photographer would get many opportunities to prove the expanse of her sensitivity to the "world-shaking changes in the social scene."[54]

The duo traveled through Louisiana, Arkansas, Alabama, Mississippi, Georgia, Tennessee, and South Carolina to find substance for their enterprise. In two trips, one in the summer of 1936, the other in March of 1937, Bourke-White and Caldwell gathered enough evidence to publish a shocking and sensational book entitled *You Have Seen Their Faces* (1937). Revealing human despair at its worst, the work received immediate attention. In delivering the southern portrait they thought Americans needed to see, Caldwell and Bourke-White helped to create what Alan Trachtenberg has identified as "a new cultural mode, the choreographed rapport of word and text." The viewing public had seen recent photographs of poverty-stricken farmers, displaced sharecroppers, and Dust Bowl survivors, but none of the pictures produced by the federal government or published in popular photo magazines and newspaper rotogravure sections carried the punch of Bourke-White's images. The advertising veteran used her camera to manipulate and highlight certain features of her subjects, just as she had done with LaSalle headlights, Chesterfield cigarettes, and Goodyear tires. The human features she found most appealing were oddities: a goiter distorting the face of an elderly Arkansas woman, the blind eye of a young Georgia mother (who also, from Bourke-White's angle, appears to have only one leg), and the shriveled limbs of a disabled Florida child, the kinds of choices that led historian David Peeler to comment that the photographer "willingly rendered her subjects as a set of grotesques riddled with poverty and disease."[55]

In a rare image where Bourke-White recorded a smile, a young boy stands face-to-face with the camera (fig. 45). Since a straw hat guards his head from the daytime sun, more than half of his face is cast in shadow. With his eyes and nose darkened by the shade, the most conspicuous feature he offers the lens is a profoundly maligned set of teeth. Played upon by the white-hot southern sun, these four teeth occupy the center square of the frame, prominently transcending all other elements in the composition. A fictionalized caption by Bourke-White and Caldwell attached to the boy's portrait reads, "School is out for spring plowing, and Pa said I could go fishing." Seemingly innocent, the combined word-image message points up a weak educational system subject to agrarian demands, while also hinting at the stereotype of the lazy southerner who chooses leisure activity over field work. The child's awkward mouth suggests genetic frailty, thus reinforcing early-twentieth-century literature that emphasized the natural weaknesses inherent in certain groups. One recent critic of Bourke-White and Caldwell's collaborative effort argues that they chose "only the material that sustained certain images of the South, and set the rest aside."[56]

Indeed Caldwell had taken his place among a few southern writers in the early 1930s who challenged the previous generation's construction of a magnolia-laden, peaceful region. Caldwell's writing reflected a charge made to southern writers in 1924 by playwright Paul Green. Green challenged his contemporaries to embrace the "crude, unshaped material" of lower-class southern

Fig. 45. Margaret Bourke-White. *Ozark, Alabama, 1936. You Have Seen Their Faces* (1937). Estate of Margaret Bourke-White.

life in order to find "'dynamism of emotion terrible enough in its intensity for the greatest art.'" In his reporting Caldwell depicted the tenant farming situation as a horrid, sorrow-filled existence that survived "only by feeding upon itself, like an animal in a trap eating its own flesh and bone." His prose described an exhausted South, filled with tyrannical landlords, lazy white farmers, and hopeless blacks. According to one of Caldwell's contemporary critics, Vanderbilt "fugitive" and Agrarian leader Donald Davidson, the novelist had overemphasized the negative aspects of southern life and ignored the rest, much as H.L. Mencken and other New South critics had done a few years earlier.[57]

Likewise, Bourke-White produced such an unflattering portrayal of her southern subjects that James Agee and Walker Evans reprinted a Bourke-White newspaper interview in their own documentary treatment of southern sharecropping, *Let Us Now Praise Famous Men,* which appeared in 1941. Agee and Evans clearly presented the piece for its ironic effect. In the interview Bourke-White's condescending comments about the ways she and Caldwell had to bribe and cajole their subjects in order to get from them what they wanted are juxtaposed with a concluding statement in which she argues that any *group* of photographs on a given subject offers "truth." In defense she offers, "Whatever facts a person writes have to be colored by his prejudice and bias. With a camera, the shutter opens and it closes and the only rays that come in to be registered come directly from the object in front of you."[58] After cultivating the art of persuasion for nearly a decade, Bourke-White proved she could sustain the public interest in herself and her work regardless of the subject matter. Even though she had turned her cameras away from machines and toward people for a large-scale project, she did not shed her modernist eye or sensibilities, and they in turn made her a very different kind of documentarian than her thirties contemporaries Doris Ulmann, Dorothea Lange, and Marion Post. Because of her modernist tastes and prejudices, she was an amazingly appropriate candidate to document "life" in the expanding new field of photojournalism.

While still working with Caldwell on the southern sharecropping project, Bourke-White joined the staff of Henry Luce's newest magazine venture, *Life.* In her contract she agreed to "work exclusively for TIME Inc." on the condition that she receive two months' leave of absence each year to pursue independent projects. *Life* offered Bourke-White an excitement different from that of advertising; it was a weekly publication that kept its photographers on the road, looking for the next newsbreaking story. For Bourke-White, the thrilling pace equaled her earlier adventures atop city skyscrapers. Her photo essay on the New Deal work relief project at Fort Peck, Montana, helped launch the magazine in November 1936. Curious readers purchased all newsstand copies of the inaugural issue in one day; twelve weeks later *Life* could boast a weekly circulation of one million. Bourke-White's Fort Peck Dam image (fig. 46) introduced Americans to *Life*. The icon and the photographer's reputation were merged on the magazine's cover. As a symbol of the persistence of American technological ingenuity in the midst of the Depression, the massive dam brought Bourke-White back to her first love—industrial power. To augment the grandiose structure and highlight its scale, the photographer included in her frame two tiny human figures. They stand nearly insignificant in the huge presence of the federal government on one hand and the western environment on the other. The wide open spaces of the American West allow such a structure to exist; it could not have been erected in an urban area, nor would it have been an appropriate fixture in the Midwest.[59]

But there is a cultural thread in the Fort Peck Dam photo essay suggesting that westerners could embrace the bigness of the idea and certainly could manage the construction. If, in Bourke-White's eyes, southerners had appeared pathetic, their western counterparts were tough. Her photographs and the accompanying text in the Fort Peck piece play up the popular nineteenth-century "frontier" stereotype, according to which rough men and raucous women built towns with grit and determination and washed their lives in plenty of alcohol. In one scene Bourke-White pictures relief workers dancing cheek-to-cheek in a saloon. On top of a table, one woman sits on a man's lap while he and a younger man embrace her. From her abdomen to her neck she is covered with construction laborers' hands. The women in the picture, known as "taxi-dancers," require each partner (or "fare") to buy them a five-cent beer for a dime. The caption explains the system: "She drinks the beer and the management refunds the nickel. If she can hold sixty beers she makes three dollars—and she frequently

Fig. 46. Margaret Bourke-White. Fort Peck Dam, used on *Life's* first cover. Reproduced by special permission from *Life*, Nov. 23, 1936, © TIME, Inc.

does." The last phrase casts these entrepreneurs in the stereotypical role of the "comfort" women who had populated the western frontier for nearly a century. In Bourke-White's photo essay, nearly all of the women pictured provide some kind of comfort, either as laundresses, waitresses, bar owners, or taxi-dancers. In one photograph juxtaposed with a large two-page heading that reads "THE COW TOWNS THAT GET THEIR MILK FROM KEGS," a smartly-dressed young woman turns up a full shot glass underneath a sign warning, "No Beer Sold to Indians" and another hailing President Franklin Roosevelt as "A Gallant Leader." The ways in which Bourke-White pictured women in her Fort Peck photo essay support the arguments put forth by cultural historian Wendy Kozol, who has shown that *Life* attempted to reinforce certain ideas about representative families in the United States. She notes that in the pre–World War II years the magazine "typically used the camera as a class tool enabling middle-class audiences to gaze at the family in crisis." The Bourke-White images that most clearly reflect Kozol's argument feature a young waitress forced to bring her child to work since there is no one at home to care for the little girl. The wide-eyed child, who may be three or four years old, sits atop the bar while the men around her swill beer, smoke, and flirt with the women there.[60]

Life's composite portrait of the Fort Peck Dam milieu does, however, reinforce traditional notions about the western frontier as a truly American environment, considered so because it offered a wide range of economic opportunities to both men and women and because it promoted rugged individualism. The living conditions around the Fort Peck shanty towns left much to be desired, with limited sanitation and frequent fires, and the towns offered little encouragement for traditional family life, but such drawbacks seemed less important than the larger issue—that the federal government's sponsorship of projects like these maintained the strength of American workers rather than softening them with government checks. Thus *Life* could label the venture "Mr. Roosevelt's New Wild West," connoting the ambivalence of excitement and adventure mixed and weighed against risk.[61]

Varied assignments, full itineraries, and cross-country travel kept Bourke-White busy in the late 1930s. She wrote to a friend, "I'm hardly back in New York before a new assignment takes me away again." She had to turn down numerous speaking engagements because she remained on call for the magazine. *Life*'s popularity brought attention to her photographs, just as the pictures brought buyers to the newsstands. Contemporary critic Halla Beloff has argued that *Life* made its photographer a "recording angel," one who showed America what it wanted to see of itself. In Beloff's opinion Bourke-White, as a "visual politician," reinforced the dominant culture by emphasizing "the themes of opportunity, equality, the protestant ethic, and amelioration." *Life* did attempt to show the public what ordinary Americans were doing on a day-to-day basis, but its stories and pictures covered a wide variety of subjects, from papermaking to slum clearance to Mrs. Theodore Roosevelt's needlework. The magazine, which rarely covered controversial topics, received its most heated criticism for a pictorial essay on human birth. A friend wrote to Bourke-White, "You know by now that LIFE was banned from half a dozen cities and states for telling the world where babies came from." Otherwise, *Life* and its star photographer heard praise.[62]

From the East Coast high art establishment to the Hollywood movie studios, critics and clients and interested onlookers applauded Bourke-White's work. In one intraoffice exchange, Bourke-White's photos were described as "breath-taking" with the added punch, "Boy, this will show the studios how pictures should be taken. . . . Let her take anything she sees which she thinks good, regardless of the stories we may outline for her." Such reviews allowed her to enlarge further the scope of her artistic freedom. Beaumont Newhall, of the Museum of Modern Art, congratulated the photographer for her "superb series" of "Middletown," identifying it as "the finest piece of documentary photography" he had seen.[63] One of the most arresting images in Bourke-White's series, which had been prompted by Robert and Helen Lynd's follow-up study of Muncie, Indiana, showed three members of the town's Conversation Club (fig. 47). The women appear rather solemn at a club meeting lecture on astronomy, their behavior displaying a posture antithetical to what one might expect from a "conversation" group. Bourke-White's framing of the image situates the women on the periphery, as figures who seemingly hold up the world in the center, a comfortable middle-class existence marked by its requisite accoutrements—books for intellectual stimulation, a Victorian-style

Fig. 47. Margaret Bourke-White. *Conversation Club, 1937.* Reproduced by special permission from *Life,* May 10, 1937, © TIME, Inc.

table and lamp to show the homeowners' traditional tastes, and a baby's portrait to prove the central place of family in the household. But the world these women have created and continue to support seems only to have promoted boredom. Not one of the three seems particularly excited about the way she has chosen to spend her leisure time. Bourke-White's exaggeration of the constituents' staidness, their corpse-like existence, in part reflects the photographer's own fear of succumbing to a similar middle-class life. She hoped that *Life* would keep her from it.

In spite of continued accolades for her work at the magazine, not everyone close to Bourke-White thought it was beneficial. Her long-time friend Ralph Steiner considered the magazine assignments too soft, warning, "Maybe, in time those LIFE boys (not LIFEBUOYS) will find out how to use you so that you function and so that this job is not only a financial haven but also a place where you can grow. . . . If they are going to keep you on the sort of thing they have had you do rather than on the kind of thing you have made your reputation on then something must be done—I should think and soon." Nor did Erskine Caldwell share Bourke-White's enthusiasm for *Life*. He had fallen in love with the photographer on their southern journey and was disappointed that the magazine assignments demanded so much of her time. Bourke-White later recalled that he "had a very difficult attitude toward my magazine, a kind of jealousy." When Caldwell realized that his competition with *Life* for his lover's attention was growing more intense, he pressured her to marry him. After many months, she relented, challenging the advice of a friend who said, "I have serious doubts regarding the alleged worth of monogamy to those who lead nomadic lives."[64]

Marriage failed to diminish Caldwell's insecurities; they continued to mount as Bourke-White built her reputation as a photojournalist. When *Life* sent Bourke-White to Europe to take pictures of citizens coping with war, Caldwell remained at home. Although he had accompanied her on assignment to Czechoslovakia in 1938, he stayed behind in the fall of 1939. War had changed everything. Although she was well behind the front lines, Caldwell's wife was considered a valuable asset, one who was "counted on to cut much red tape." Because of the nature of her work, she had to be prepared to move on a moment's notice, which meant traveling alone; Caldwell's presence might pose problems or cause delays. During her six months overseas, Bourke-White received numerous letters from her husband begging her to return to the states. He appealed for the sake of his reputation, her reputation, even the prospect of a future child. In December 1939 Caldwell sent Bourke-White a newspaper clipping about Marion Post, a U.S. Government photographer making her way through the South. He warned, "The young lady in the enclosed clipping is kicking up a lot of dust around the country, and I dislike seeing her getting so much glory. . . . I can't help reading the handwriting on the wall! To wit—to your everlasting glory you should make haste to get back here & do those books." Caldwell regretted Bourke-White's insatiable appetite for grabbing the latest story because he wanted her fame to rest on something more permanent. Assuming the tone of a patronizing man of letters, he instructed her, "Books, like objects of art, are not thrown away like the morning paper."[65]

But the opportunities granted a foreign news photographer were too alluring for Bourke-White to pass up. If industry had offered the greatest excitement in the late twenties, war offered the same in the late thirties. *Life* sparked Bourke-White's development as a photojournalist, and she was determined not only to participate but to help shape the new field. In assuming the mantle of photojournalist, Bourke-White took steps to shed the social conscience that had made her a documentarian. She chose not to support any political causes or group efforts; she requested that her name be taken off committee letterheads; and she sought complete independence from creeds and slogans. Her biographer, Vicki Goldberg, argues that the process helped solidify Bourke-White's position as a photojournalist, with "impartiality" being "part of her equipment." "Photography came before her causes. Photography, in fact, *was* her cause."[66] The reality of Bourke-White's transformation from documentarian to photojournalist may be best measured by the failure of her last documentary effort with Caldwell.

To placate her husband, Bourke-White worked with him on another book about the United States. *Say, Is This the U.S.A.* (1941) proved an artistic failure in that it revealed stilted communication rather than true collaboration between the writer and the photographer. The lack of public interest in the book showed the diminished place of domestic concerns in the American imagination. The documentary impulse as informed by social realities and realism inside the nation's borders had given way to photojournalism, with its smaller cameras, faster films, and wire services, whose effects were to bring the world to the United States. Making international news available quickly was the newest game in photography. In 1941 Bourke-White crossed the Atlantic as a war correspondent for the U.S. Air Force; Caldwell stayed behind. Unwilling to carry on a long-distance relationship, he asked for a divorce. Bourke-White wrote to her lawyer that she was relieved: "I have felt for so long that such a disproportionate amount of my attention went into worrying about whether someone was in a good mood or not, whether he would be courteous or forbidding to others, whether day to day life would be livable at all on normal terms, that I am delighted to drop all such problems for the more productive one of photography. . . . In a world like this I simply cannot bear being away from things that happen. I think it would be a mistake for me not to be recording the march of events with my camera." Bourke-White stayed in Europe to photograph the war. Of her decision to maintain an international focus, Bourke-White biographer Goldberg wrote, "Work was the only safeguard of her independence and integrity." A look back at her career proves that this had always been the case.[67]

Life magazine photographer Alfred Eisenstaedt once said that Bourke-White was "married to her cameras. No assignment was too small for her. If she had been asked to photograph a bread crumb at five-thirty in the morning, she would have arrived for the session at five."[68] Besides this commentary on his colleague's work ethic, Eisenstaedt isolates the material out of which Bourke-White shaped her reputation in the 1930s. She took ordinary objects and with her camera eye made them into extraordinary images that not only turned viewers' heads but often made them stare and wonder just what they were looking at. By showing up early Bourke-White assured herself the comfort of time in order to arrange her lights, pose her models, and manipulate the background so as to create dramatic effects. And having taken an otherwise absurd assignment so seriously, she then could make *herself* (as much as the exquisitely arranged bread crumb) the center of the story. Bourke-White's perceptions of the forces shaping the contemporary world, including the camera itself, made her the quintessential advertising photographer in the early 1930s. Her path from advertising to documentary photography set her apart from her female contemporaries such as Doris Ulmann, Dorothea Lange, and Marion Post, who harbored in varying degrees a sense of social reform. Although Bourke-White did recognize the desperate conditions of some Americans during the Depression, she never possessed a strong desire to lift up a specific American subculture or portray individual human dignity or expose class-based and race-based inequities. Using the camera as a tool of social reform seemed foreign to her. Yet her contribution toward shaping the age of documentary is indisputable, particularly since she collaborated on one of the first and most significant documentary books to appear in the 1930s. Her decision to leave behind the popular genre of the age seemed a natural step in her professional development, certainly in keeping with her desire to stand at forefront of new trends in photography; the advent of photojournalism provided Bourke-White a decidedly fresh challenge. As a modernist who embraced the subjects and issues of the documentary genre for a brief time, Bourke-White secured her place among visionaries who realized what sophisticated technology could mean in a complex, machine-driven environment. By isolating the elements of modernity—a radio cable, a steel plow blade, a shiny hubcap, a gargantuan dam—she cultivated an aesthetic that made mechanical parts come alive. Bourke-White's vision helped to break new ground in the photography world, and as a modernist, she was eclipsed only by a woman who addressed the cultural "lag" by arranging elements of change in their cultural and environmental contexts. The decade saw the most thorough and comfortable integration of the documentary impulse and modernist sensibility in the work of Berenice Abbott, who managed to narrow the gap between contemporary symbols of power and the existing American culture.

5

Modernism Ascendant

BERENICE ABBOTT'S PERCEPTION OF THE EVOLVING CITYSCAPE

> If I had never left America, I would never have
> wanted to photograph New York. But when I saw it
> with fresh eyes, I knew it was *my* country, something
> I had to set down in photographs.
> —Berenice Abbott

WHEN A NEW DEAL PROJECT OFFICIAL first saw Berenice Abbott's Bowery pictures, he warned her that nice girls should stay out of certain New York City neighborhoods. Abbott retorted, "I'm not a nice girl. I'm a photographer." That such a warning could be given, even in jest, revealed the widely accepted limitations still placed on female photographers, especially those who stepped outside the confines of a studio to ply their trade in the 1930s urban world. Abbott's succinct reply clearly reflected her no-nonsense view of herself, including a valued ordering of professional responsibility over certain gendered restrictions. Since Abbott had consciously chosen to put on film as much of the preeminent American city as she possibly could, she planned not to focus only on its orderly neighborhoods, beautiful facades, or attractive residents. Weather-beaten tenements and crowded markets had roles to play in Abbott's project, which was grand in scope, an attempt to capture the city in all its energy and fluidity. Early in her tenure as a Works Progress Administration photographer, she told a reporter, "This is not a 'pretty picture' project at all—it is to get the real character of New York."[1]

Abbott hoped to comb Manhattan from Harlem to the Brooklyn Bridge, and she wanted her pictures of the city to stand apart from some of those her popular predecessors in the American photography world had created. Her methods and her independence kept her from joining the Alfred Stieglitz followers, a group she evaluated as a "powerful cult" guided by one man's "stupendous ego." She avoided the soft-focus style of pictorialists like Doris Ulmann, because she thought they made "pleasant, pretty, artificial pictures in the superficial spirit of certain minor painters." Since Abbott made no pretenses about the documentary tone of her images, New York's art photography circles chose not to include her. She succeeded in spite of them, pointing to her dependence upon "a relentless fidelity to fact [and] a deep love of the subject for its own sake." Neither trait, however, illuminates her modernist sensibilities. Claiming "fidelity to fact" put her in the same camp as other 1930s documentarians, who worked in a cultural context dependent upon the public's belief in the ability of the camera to record truth and to show circumstances as they actually existed. Choosing the city itself for a large-scale documentary project placed Abbott in a distinguished line of observers who had for nearly one hundred years pondered the city's possibilities and then created and disseminated visual messages to audiences hungry for definitions of the urban landscape and its inhabitants. But, as Peter Hales has shown, these purveyors of an ideal metropolitan environment had been transplanted by the 1920s as other forms of mass culture, notably motion pictures and radio, better explained the urban experience. Unwilling to "order" the city through her lens, as other cityscape photographers had done, Abbott showed instead a world in flux; by so doing she inadvertently denied that anything could be "captured" on film. Her attempts to frame only particular moments in the evolutionary processes she viewed marked her photography as modern in the truest sense of the word.[2]

Abbott first experienced New York as a nineteen-year-old journalism student. Like many other midwesterners in the first decades of the twentieth century, she left her Ohio home in search of the opportunities the urban life offered. She intended to cast aside memories of an unhappy childhood, a disintegrated family, and a dull freshman routine at Ohio State University. Her parents had divorced soon after her birth in July 1898, and Abbott grew up having little contact with her three older siblings or her father. Feeling somewhat alienated from her family members, she assumed enough personal independence to leave them behind and seek a different life for herself. In the winter of 1918, she borrowed twenty dollars to buy a train ticket to New York. The friend who lent Abbott the money also gave her a place to sleep in a spacious Greenwich Village apartment. While living in the Village, Abbott found herself surrounded by artists, actors, and playwrights, many of whom started their workdays at dusk. Their late nights often ended in raucous neighborhood parties, where Abbott's shyness left her in the minority. But after convincing herself that the life of an artist offered greater satisfaction than the career in journalism to which she had aspired, Abbott dropped out of Columbia University, a place she later described as "a hell of a sausage factory" despite her sporadic class attendance. Admitting that she would have made "a very poor journalist," Abbott attributed her choice of the field to youth's uneasiness, the inescapable grappling for personal identity.[3]

To support herself, Abbott took a variety of jobs, from secretary to waitress to yarn dyer. She dabbled in acting and landed parts with the famed Provincetown Players, including a small role in Eugene O'Neill's drama *The Moon of the Caribbees*. Struggling to find her niche in the art world, Abbott decided to cultivate her interest in sculpture. She moved away from her acquaintances and the night life in Greenwich Village in order to live alone as an artist. Though making little money at her craft, she noticed that other artists in the city focused on how much profit their works would bring. Abbott pondered the kinds of inspiration that commissions provided, believing that such patronage might restrict her artistic freedom. Not until several years later would she be able to reconcile the need to take commissioned work in order to buttress her more experimental projects. By 1921 Abbott craved an atmosphere more conducive to creativity, and with encouragement from her friend the baroness Elsa von Freytag-Loringhoven, she decided to study in Paris. She explained her dissatisfaction with both the immediate community and the larger nation that surrounded her: "I was scared of New York, scared of America. . . . I wasn't commercial. I never dreamed about how much money I could make—it never even occurred to me. And America was so commercial, that's why I left it." As a young, carefree, self-styled sculptor, Abbott had little to lose by starting over in another place. Of the move to Paris, she remembered thinking, "I may as well be poor there as poor here."[4] Although she failed to realize it at the time, the environments Abbott chose to work in throughout her career would become either allies or enemies, inspirational havens or desolate grounds. Her surroundings were vital in stimulating her imagination.

Veritably unambitious, Abbott bought a one-way ticket to Paris. She recalled, "In March 1921 I set sail on the great big sea, like Ulysses, and I said to myself, 'Whatever happens, happens.'" For two years, as Abbott struggled to develop her sculpting talents, she held down a variety of jobs to support herself. She often posed for other sculptors and painters in order to make money. Even if her artwork failed to furnish the income she had hoped for, Paris offered the working atmosphere she had craved. Abbott explained that "Paris had a quality in those days that you can't have in an overcrowded place. People were more *people*. No one was rushed. . . . We had the illusion that we could go ahead and do our work, and that nothing would ever come along to stop us. We were completely liberated." Abbott's friends belonged to the American and British expatriate circles who huddled on Paris's Left Bank. Most were writers or painters or patrons of the arts. Leslie George Katz described them as a "community of congenial individuals" who discovered their most meaningful expression through the vehicle of the arts. He explained that "in the practice of the arts a person acting as an individual, a loner exploring a commitment to a craft, could hope to express thereby the vernacular total of human experience." Unfortunately, Abbott failed to reach a satisfying level of expression with her sculpture. Although she had escaped the commercialism of the United States, she was discontented with the European art world. Her lack of commitment combined with her personal dissatisfaction led Abbott to abandon sculpture. After nearly three years of study and practice, she willingly gave up her dream and went to work full-time as an assistant to the experimental visual image creator, artist and photographer Man Ray.[5]

When Man Ray discovered that Abbott knew nothing about photography, he immediately hired her to print his negatives. He wanted to train a novice to develop negatives using *his* vision. Man Ray realized that he could mold the inexperienced young sculptor and pay her very little while she learned the trade. Abbott surprised herself and her employer by readily mastering darkroom techniques. She employed her artistic skills in order to sculpt the faces in Man Ray's portraits. Abbott's keen ability to see "white, gray, and black" tones helped her to print sharply defined likenesses of the portrait sitters. Abbott biographer Hank O'Neal has pointed out that Abbott produced a stunning "three-dimensional effect" on paper, a pleasant consequence that attracted even more patrons to Man Ray's already-popular portrait studio. For three years Abbott worked closely with Man Ray's negatives, but she rarely observed his methods with sitters. All she witnessed emerged on film emulsion in the darkroom. Perhaps the long hours she spent alone developing, printing, and critically studying Man Ray's work encouraged Abbott to try her own hand with a camera. He showed her how to operate the machine, and soon afterward she began taking photographs of her friends during her lunch break. As Abbott took more and more photographs, her satisfied customers helped to build her reputation by word of mouth. When one of Man Ray's appointments asked for Abbott to take her portrait, the tension in the studio mounted. Abbott decided to resign rather than exist in an inhospitable working environment. Although Man Ray had been the only photographer Abbott had known and the two had worked together closely, their respective portraiture styles displayed relatively few similarities. Producing photographic work that best fit within the surrealist tradition, Man Ray was once described as "the wittiest juggler the camera has ever known." Abbott experimented with light, angle, and shadow but produced images of individuals that bore more marks of traditional portraiture than did Man Ray's. In 1926 Abbott got the opportunity to develop her ideas about portrait-making even further when two loyal patrons, Peggy Guggenheim and Robert McAlmon, loaned her enough money to set up her own studio.[6]

Out of the studio came pictures of countless faces, young and old, suspicious and carefree, affected and quite natural. Abbott took her time with her sitters and allowed them to "be themselves." She considered each subject so important that she took only one person a day. Ample time for interaction between the photographer and the sitter was vital because, in Abbott's opinion, only a comfortable session would reveal true character. Abbott explained, "I wasn't trying to make a still life of them . . . but a person. It's a kind of exchange between people—it has to be—and I enjoyed it." One critic has said of Abbott's Paris portraits, "The sitters are her peers, her accomplices, her friends, her heroes. They sit for her as her equals, one to one." And they participated. Most of the time, Abbott elected not to pose her subjects; instead, she allowed them to strike their own poses. She consciously tried not to flatter them. Some of them dressed up, whereas others appeared in casual clothes. Bookseller Sylvia Beach donned a full-length dark rain slicker and a white neck scarf for her portrait. The striking contrasts, from Beach's shiny coat to her smooth, pale-skinned profile, reveal the ways in which Abbott experimented with various textures in order to create multidimensional figures that appealed to the senses. Some of her sitters looked at the camera, some away from it; some had joyous smiles, and a few had introspective gazes. Abbott treated each sitter differently and claimed to have started every session as if she "had never taken a photograph before." She once told a group of students to avoid the "photographed expression" when making pictures of people, implying that such a conscious effort would yield more honest results in the darkroom. A rapport between photographer and subject is evident in Abbott's portraits.[7]

The relationships Abbott forged with her sitters resulted in some of the most lively, most poignant, and most haunting photographs to come out of 1920s Paris. In her series of James Joyce, a contemplative, troubled face dominates each frame. His bodily shifts from scene to scene suggest a restlessness as he looks away from the camera eye, leaning back comfortably on a sofa in one scene, then in another scene leaning forward toward the photographer, and in yet another having turned away again. Joyce's movements, which may have been orchestrated by Abbott, reveal the portraitist's attempts to record the depth of her subject's discontent. Seemingly captured in anguish, Joyce has his head propped on his hand and his chin set with lips pursed; his mood is accentuated by his furrowed brow (fig. 48). Despite the energy emanating from an Asian-inspired sofa pattern, Joyce's striped shirt, and his dotted bow tie, all potential sources of competition, the subject's emotional state is laid bare in the image, thus dominating the peripheral elements. Much in contrast to the writers' portraits by Doris Ulmann, where studio props such as ink wells and paper connected the sitters to their craft, Abbott's creation of Joyce's image shows the tormented expatriate soul, worried and disillusioned. Through the faces and bodies of her sitters, Abbott communicated the mood of European modernism, all movement, fragmentation, and fluidity. One critic commented about the Paris portraits, "There is no irony or facile cleverness in them. They are friendly and admiring, yet they never strike a false note."[8]

Among the expressive faces are the all-knowing and unshakable Princess Eugene Murat, who stares directly at the camera; the dramatic yet pensive Solita Solano, who rests her chin on her shoulder; and the straight-backed, serious editor Jane Heap, posed as the classic androgyne. Coco Chanel, Leo Stein, André Gide, Janet Flanner, and many others offered Abbott their faces and their souls for a day. They admired her work and helped make her a successful artist in a city that had not wholeheartedly brought photographers up to the level of respect that painters, sculptors, and writers enjoyed. One of her sitters and biggest promoters, Jean Cocteau, even attempted to shape Abbott's image by altering the spelling of her first name for her first one-woman exhibition in Paris in 1926. He changed her name to Bérénice, a reference to a virgin who would not marry a king, suggesting that Abbott refused to depend upon a male companion for financial support. Abbott later explained that she did not like her given name, "*Burrnees,*" so welcomed the addition of another letter in order to make it "sound better."[9] Cocteau's influence on Abbott cannot be underestimated, especially given his love of America and its indigenous sources and original products, such as jazz. Abbott had escaped what she considered stifling mainstream values in the

Fig. 48. Berenice Abbott. *Portrait of James Joyce.* Commerce Graphics Ltd, Inc.

United States, but she heard a refreshing point of view about her own country from Cocteau. The seeds he planted about the young nation's offerings were no doubt present when Abbott returned to New York City later in the 1920s.

For Abbott, Paris was defined by its people more than its landmarks, its palaces, or its history. Her profitable business was supported by clients who admired her depictions of them and reveled in her appreciation of their various artistic, literary, and social endeavors. She found comfortable, wealthy women particularly fascinating and once said, "People can be great in themselves, and not for what they do. There were women in the 1920's who knew how to live doing nothing, just pouring tea or going to the store." Of all the Parisians and exiled Americans and other individuals who inhabited the city, perhaps none interested Abbott more than the photographer Eugène Atget. Having seen a few of Atget's pictures in Man Ray's studio, Abbott sought out the photographer who provided artists with source material for their canvases. Abbott happily discovered that Atget lived just up the street from Man Ray's studio in Montparnasse. She recalled her first encounter with him as shocking, finding behind Atget's apartment door a "tired, sad, remote" man, yet one who was also oddly appealing. She befriended him, studied his photographic techniques, and bought as many of his prints as she could afford. Her curiosity about a different kind of photography marked her first step in moving away from portraiture. Several months after Atget's death in August 1927, Abbott managed to convince his closest friend, André Calmettes, to sell Atget's photograph collection to her. The fifteen hundred glass plates and thousands of prints revealed to Abbott the extent of Atget's dedication to his City of Paris project, which was later identified as "more comprehensive than anything previously attempted in European photography."[10]

For nearly thirty years, from 1898 to 1927, Atget carried his 18 × 24 cm view camera, tripod, glass plates, and covers into every corner of the city and its environs. A substantial portion of his pre–World War I photography focused on the artifices and ornamental details of Old Paris, including storefronts, balcony designs, home interiors, sculpted facades, and architectural designs. He advertised his wares as "Documents pour Artistes," peddling them to the likes of Braque, Utrillo, and Picasso, as well as other artists who might find ornate door decorations or busy storefronts interesting subjects to paint. Set designers also found Atget's images appealing for their sheer quantity of collected interior and exterior details. The cumulative body of the photographer's work impressed Abbott as a kind of "poetic epic," one that amassed numerous minuscule components in storefronts, churches, signs, fountains, and parks, among other things. Regarding his vision, she wrote, "He was obsessed by the materials of existence, the substances in which life is given form. The cobblestones of streets, the moulting limestone of houses, the bark and leaves of trees, the luster of berries, the gloss of rose petals, crumbling earth in plowed fields, marble fountains in landscaped esplanades—all these qualities and a million more controlled his imagination." With a particular focus on transitory elements, Atget was able to document a city moving out of the nineteenth century into the twentieth. The Paris he had known as a young man was fading; the fin de siècle trends crept in to modernize the city and its people. The romantic air of Delacroix's Paris and the stately manners of Manet's Parisians gradually disappeared as the sounds and sights provided by Picasso, Stravinsky, and Diaghilev demanded more of the public's attention. Atget hoped to preserve the remnants of the old world before they vanished completely. He was particularly proud of his photographic series showing neighborhoods in transition, with buildings in various stages of demolition. His theme, according to Abbott, was "society, its facades and bourgeois interiors, its incredible contrasts and paradoxes. . . . the vast scope of this world, the changing 19th century world."[11]

In 1928 Abbott began the task of promoting Atget's name and what she interpreted as his vision. She exaggerated the Parisians' neglect of Atget, who had in some quarters been thought to be a spy during the war. They considered him mysterious, this strange character toting around a huge old camera and a dark cloth. The man who focused on street signs and shopkeepers' windows seemed an oddity in 1920s Paris. Although Atget once told Abbott that the only people who appreciated his work were "young foreigners," his client records reveal that his photography interested several hundred buyers. In addition he sold more than two thousand negative plates in his "Old Paris" collection to the Monuments Historiques in 1920, a decision that allowed him, both financially and artistically, to embrace new subjects. As a close observer of Atget's development, Maria Morris Hambourg has argued that the sale marked a point of transition for the photographer, in that it allowed his 1920s imagery to become freer, more "overtly subjective, idiosyncratic, lyrical, and suggestive. . . . less concerned with facts than with life."[12]

Inspired by Atget's pictures, Abbott unknowingly began laying the groundwork for a similar project of her own, one that would take her away from Paris indefinitely. She continued to make portraits in the late 1920s, but she also took some of her first outdoor photographs during those years. If Atget's work had measured anything, it was the importance of the present, the singular moments and scenes that vanished quickly in a modern environment. In a typical scene he photographed on the Rue de la Montagne-Sainte-Geneviève, Atget peers down a narrow passageway that divides the frame. Characteristic of hundreds of his street scenes, the darkened and unseen path has a hint of mystery about it. The surrounding neighborhood seems innocuous enough, with inviting shop windows on the left side of the frame and private residences on the right half of the frame. Yet the brightness of a contemporary cheese shop window, with its baskets and milk cans neatly stacked on the sidewalk, provides a stark contrast to a darker, wrought-iron-fenced residence across the street. The shop's openness and close proximity to the street are markedly different from the home's prison-like walled closure and distance from the public space. Even the vehicles Atget used to mark the separate spheres highlight the differences; a produce wagon's wheels and a metal-fendered automobile tire stand on opposite sides of the frame yet show two worlds coexisting. One bespeaks a rural past, the other a mechanized urban future. Although Abbott had focused solely on human personalities in her studio portraiture, she came to appreciate Atget's point of view, his focus on the external world, what she called "realism unadorned." Just a few months after she bought his photograph collection, Abbott made a trip back to New York, the place she had rebelled against a decade earlier. She immediately fell in love with the city, and later she said, "I knew that I couldn't go back to Paris. Here I was, in the most complicated city in the most complicated country on earth, and I knew that that was where I had to stay." A yearning to rediscover her roots had led Abbott back to the United States, yet she never expected to feel so overwhelmed upon her return.[13]

The vitality of New York City attracted her. Her newfound appreciation for artistic realism made her aware of what tremendous possibilities the city offered as a subject. Atget's urban project may have planted the seed, but New York itself brought Abbott back. She returned to Paris just long enough to trade her furniture and pack up the Atget collection for shipping. She left behind in Paris the dadaists, the constructivists, the surrealists (whom she referred to as "microbes"), and all those who she believed focused too heavily on manipulating or distorting the human experience in their art. In her rejection of the avant-garde, Abbott carried out Atget's forbearance against highbrow innovators. He had distanced himself from them throughout his career, shunning their aesthetic tastes by drawing a line between his work and theirs. What best illustrates Atget's view of how far apart his images stood from experimental art is an anecdote describing a brief encounter between him and Man Ray. In 1926 the surrealists wanted to use on their magazine cover a 1912 Atget picture of Parisians watching an eclipse. Atget sold them the photograph but told Man Ray, "'Don't put my name on it. . . . These are simply documents I make.'" Atget circulated in a European world where documentary photographs were supposed to serve a myriad of purposes, having been defined as such at the Fifth International Congress of Photography in Brussels in 1910. It was there that images were cast as vehicles for other, more important work, such as "scientific investigation." Atget scholar Molly Nesbit has pointed out that this perception of documentary photography made a visual image simply "a study sheet. . . . beauty *was* secondary; use *did* come first." This was documentary as Atget understood it, and it is the principal reason he felt compelled to take thousands of photographs. In the end, quantity counted most.[14]

As a result of thinking and writing about Atget for years, Abbott cast him as one of the world's greatest photographers, in a way rescuing him from being remembered as *merely* a documentarian. By creating an image of Atget that would satisfy her American audience, she bolstered her own reputation as a documentarian of a large metropolitan area in the United States. Abbott helped transform Atget from a yeoman camera operator into an extraordinary *seer*, describing his work as "purely and entirely photographic." Molly Nesbit has argued that Abbott's recreation of Atget in aesthetic terms, despite his aversion to being defined in such a way while he was living, made him the kind of photographer that an American culture could accept, whereas in France he remained essentially a working-class artisan. In Abbott's interpretations of Atget, he was a wholly objective viewer, a photographic eye whose vision could be trusted. Convincing others to share her convictions and recognize Atget became one of Abbott's principal goals after returning to the United States in 1929. The other was to frame as many moments in the evolution of an urban area as she could, in the form if not the spirit of a dead French photographer whose reputation she would later shape.[15]

Abbott moved back to New York to find an array of obstacles facing her. Within months of her return, the stock market crashed, and soon thereafter a nationwide economic depression set in. The reputation Abbott had enjoyed in Europe had not accompanied her across the Atlantic, so she was forced to support herself by seeking out business rather than waiting for it to come to her. She quickly learned how to survive in New York; she took on commercial work and "anything that came up." Abbott's pictures appeared in popular magazines such as *Vanity Fair* and the *Saturday Evening Post*. She was hired by *Fortune* to make portraits of several business tycoons. While Margaret Bourke-White pictured the inner workings and tiniest elements of industrialism, Abbott photographed the men behind it all, those in charge of the machine age. Although the *Fortune* assignments kept her working with portraiture, Abbott found her new sitters much more difficult to deal with than her Parisian friends. She recalled, "I couldn't bear it. Those chairmen of the board were so different from Europeans, so unrelaxed, so uncooperative, so 'just do your stuff and get out of here.'" Yet Abbott chose not to measure New York by those men. To her, the city was defined more by its places than its people, unlike Paris. She understood and even liked the difference, the "new urgency" in her immediate environment that she found so characteristically American. "There was poetry in our crazy gadgets, our tools, our architecture," she said. Abbott planned to picture those facets that composed Americans' external world and helped them chart their progress.[16]

In between magazine assignments and other commercial jobs, Abbott reserved one day a week for roaming about New York. She spent her Wednesdays alone with the city. She felt uneasy on her first trip out, intimidated by the curious and often unfriendly stares from passersby, but she knew she would have to overcome the barriers or abandon her project altogether. Beginning at the waterfront, Abbott worked her way up Manhattan, "using a small camera as a sketch pad, noting interesting locations, storefronts, facades, and views." When she ventured out with her larger 8 × 10 Century Universal and its requisite tripod, crowds gathered. Bowery residents scoffed at the young woman hiding under the black cloth, and more than a few others played practical jokes on her. She recalled, "Men used to make fun of me all the time. They couldn't understand what I was doing. The worst time I ever had was when I wanted to photograph the George Washington Bridge, which was just going up. I wanted to shoot it from up in the crane. The construction workers put me in the pan and, once I was high in the air, they swung the crane back and forth so I couldn't take any pictures. I was *terrified*." Abbott's experiences with the men she encountered while taking pictures revealed some of the prejudices against working women that circulated during the Great Depression. Prevailing opinion viewed active, employed women as threats to men who needed work.[17]

Even though Abbott's sporadic employment could hardly characterize her as a woman who had taken a job away from a well-deserving male breadwinner, she nonetheless suffered from the public's opinion on the subject. In addition, Abbott did not enjoy the advantages some of her female contemporaries in photography could boast. Shy and unassuming, unlike the flamboyant, confident, and egocentric Margaret Bourke-White, Abbott had to force herself to contend with New Yorkers in her efforts to document their city. In the early 1930s, Abbott worked on New York's streets alone, without the benefits of a personal entourage like the one Doris Ulmann enjoyed during those years. And not until several years into her documentary project on the city did Abbott carry the credential "U.S. Government photographer," a position that garnered for Dorothea Lange and Marion Post a considerable amount of respect and clout in the locales where they carried out their agency fieldwork. Nevertheless, Abbott resolved to continue the project that had become a personal obsession, even though no institution offered to support her work in the early 1930s.

In 1931 Abbott applied for a Guggenheim Foundation fellowship but was rejected because her project failed to meet "international" character requirements. Her examination of an American city generated little interest among the referees. Several months later she appealed to the Museum of the City of New York, which also turned down the pictorial survey. The next year Abbott crystallized her objectives in a proposal sent to the New York Historical Society. She stressed the importance of preserving the city's "unique personality" in a collection of documentary photographs. Her eloquent statement included an argument supporting photography's role in historical preservation: "The camera alone can catch the swift surfaces of the cities today and speaks a language intelligible to all." She felt particularly qualified to document the American city since she had spent nearly a decade in Europe and had gained some sense of perspective about the United States from a distance during those years. Convinced she had developed an unusual appreciation for her native country, Abbott wanted the privilege to interpret what she saw before other, perhaps less objective, photographers assumed the same task. She continued taking pictures of New York, though all of her appeals for institutional support were rejected. One private museum patron sent Abbott the only contribution she received—fifty dollars.[18]

As the national economic depression worsened in the 1930s, Abbott received even fewer opportunities for commercial business. She eventually resorted to other means of income, claiming in particular to have sharpened her barbering skills. She would say to portrait sitters, "the pictures will be all wrong because of your hair," as she pulled out her scissors in order to make a little extra money. When architectural historian Henry-Russell Hitchcock asked Abbott to collaborate with him on two photo projects, she accepted his offer. They gathered enough material to mount an exhibit at Wesleyan University entitled "The Urban Vernacular of the 1830s, 40s, and 50s: American Cities before the Civil War." At the Museum of Modern Art, they exhibited pictures of buildings designed by Boston architect H.H. Richardson. The time Abbott spent working on Hitchcock's jobs prompted her to resume her New York documentation with a vengeance. Travels to several eastern seaboard cities had convinced her that America's premier city boasted an incomparable energy that could not be ignored. In Abbott's opinion, to abandon the project because it aroused no support would have meant not only artistic failure but social irresponsibility on her part. She plunged forward in spite of personal poverty.[19]

New York City offered Abbott a study in contrasts. Through close observation she saw "the past jostling the present."[20] She recognized that new skyscrapers were beginning to overshadow old brownstones and that modern technology was edging out any semblance of nature. In a 1932 series she completed on the construction of Rockefeller Center, Abbott seized on the scale of new, modern architecture. In one image (fig. 49), taken from the base of the structure looking up, Abbott used the appearance of natural elements in the foundation to highlight their place in relation to manmade structures. Although gargantuan in size, the ice formations on the limestone foundation have been surpassed by the steel framework of the building itself. To provide a sense of magnitude, Abbott included a man's head in the lower right corner of the frame. He, along with nature, is dwarfed by the architectural wonder. In another image Abbott made in the Rockefeller Center series (fig. 50), a cutaway view of the framework reveals a high level of activity and movement on the site. The workers, who stand, crouch, lift, bend, and walk, seem in command of their respective rectangular cubicles; they are important in a relatively small space, but set in the larger project they are tiny, insignificant components whose lives are overwhelmed by the structure itself. That the image, this collection of adjoining rectangles, extends beyond all sides of the photographic frame reveals Abbott's developing modernist sense. The picture represents a fragment of a moment, but even more so, a fragment of a place. There is no clear beginning or ending to the image, and there is no single definable icon on which a viewer can or should focus. What exists, instead, is a visual smorgasbord, allowing initial and subsequent fixations on any point at any place in the field.[21] A viewer's perspective in this particular Rockefeller Center image is determined by the human element;

Fig. 49. Berenice Abbott. *Rockefeller Center, 1932*. Commerce Graphics Ltd, Inc.

Fig. 50. Berenice Abbott. *Rockefeller Center, 1932.* Commerce Graphics Ltd, Inc.

no other reference points in it so clearly define scale as the persons, who appear like worker bees in a colony.

Since the face of the city changed on a daily basis, Abbott wanted to preserve the pieces of New York's past that were slowly disappearing. "I wanted to record it before it changed completely . . . before the old buildings and historic spots were destroyed," she remembered. But she also hoped to capture on film the fleeting moments known as the present. She believed that one of her primary responsibilities as a photographer was to capture as many elements of the present moment as could be comprehended. As she explained to Meryle Secrest, "[t]he present is the least understood thing about life. It's harder to gauge, to know fact from fiction, to know what's going on behind and in front of the scene. You can't tell from people as much as you can from things. You can see our villages and towns and they are really expressive of our real people." A city in flux, New York provided a spectacular continuum for Abbott's work. She saw her pictures as connectors, bridges that linked her subjects simultaneously to the past and to the future. In a 1934 series on Trinity Church, Abbott juxtaposed the neo-Gothic structure with the Wall Street buildings that had come to not only surround but also minimize the religious edifice (fig. 51). Built in the 1840s, Trinity Church embodied in stone the medieval-age principles of gallant chivalry and Christian Church domination. But here Abbott shows the mid–nineteenth century romantic revival structure taking a lesser place than the turn-of-the-century bastions of commerce and finance. The modernist observer has framed an icon of the past jostling the omnipotent symbols of the present. Realizing that the tiny pieces of the present she recorded became part of the past as soon as the shutter clicked, she maintained that a photographer's duty was to recognize "the now."[22]

A few years after she began photographing New York, she told a reporter that she considered cities important social indicators because they had "a personality. Not the people in them, but the buildings, the little odd corners. . . . And New York, especially. It's so changing. It's in the making. We're making it. There's so much movement. It gets into your blood. You feel what the past left to you and you see what you are going to leave to the future."[23] The thrill of the city's possibilities led Abbott to explore

Fig. 51. Berenice Abbott. *Trinity Church, 1934.* Commerce Graphics Ltd, Inc.

its streets, observe its statues, and study its structures. Her vision was distinct from those of earlier observers of New York as an urban space. Best known for his penetrating views of slum life and high crime areas was Jacob Riis, a Dutch immigrant whose police reporting yielded him sufficient ammunition to produce a groundbreaking publication entitled *How the Other Half Lives: Studies among the Tenements of New York*. Appearing in 1890, the book's critical acclaim was due primarily to its use of photographs "of a sort that had been seen by very few."[24] The cityscapes showed crowded, dirty streets and even more crowded sleeping rooms that doubled as living and working spaces for large families. The environment that bred violence and disease was laid bare in Riis's work. Abbott viewed similarly poverty-stricken areas, but she did so with a much larger scope; instead of going into an alley where vagrants loitered, she set a working-class tenement house against the backdrop of New York's fresh twentieth-century skyline. Or she provided a sense of overcrowding not by picturing people but by showing instead their endless lines of laundry strung crisscrossed from window to window. Abbott's vision of New York also varied from that of her contemporaries Lewis Hine and Margaret Bourke-White, both of whom made photographic studies of the Empire State Building and other skyscrapers. Bourke-White's aesthetic demands were for material textures, articulated through the isolation of polished steel components or machine-age design elements. Her highly stylized arrangement of such objects provided a sense of compositional stability, in that the eye could find one or two central points in an image on which to fix a gaze, with the realization that those points were not themselves movable (see, e.g., fig. 41). Abbott's studies of skyscraper development do not offer such clearly defined points of reference, nor do her subjects emit a sense of comfortable solidity.

Because of the surface similarities between Abbott's New York and Atget's Paris, some of her contemporaries argued that she was merely copying her mentor's work and possessed no creative impulses of her own. She shared his values and his vision, and as photographer Minor White pointed out, "a store front full of shoes, or a store window filled to the edges with hardware had a fascination that neither could resist." But Abbott saw what different and exciting possibilities the modern age offered a young nation,

whereas Atget had viewed remnants of the past in an aged culture. Unlike Atget, who had stood in the present looking back to
the past, Abbott stood in the present looking forward to the future. Many of Atget's prewar Paris scenes were marked by the
soft haze and the indefinite backgrounds that the American
pictorialists had employed; by contrast, the New York of Abbott's
pictures is alive and bustling and sharply defined on photographic
paper. Art critic Elizabeth McCausland, comparing the two photographers, found "two worlds and two ages portrayed—his the
romantic last glow of nineteenth century nostalgia, hers the modern technological triumph of man over nature." Photographer
Lisette Model agreed, arguing that Abbott's pictures revealed some
sense of the daily struggle people carried on to control the external, their immediate surroundings.[25]

Even though they showed two very different worlds from
opposing perspectives, Atget and Abbott did share a few basic
artistic principles. Perhaps of most importance to Abbott was her
recognition of the power a cumulative body of images could wield
over a select few, perfect pictures. Her New York colleagues, the
highbrow art photographers, created singular points of reference
in their symbolic images, perpetuating what Abbott later called
"the stale vogue of drowning in technique and ignoring content."
Abbott took numerous shots in the same places on various occasions; often she returned to sites in order to capture subtle changes.
In 1936 she told a reporter, "Once, way up at the end of Broadway, I saw a little church next to a row of new apartment houses.
I went back the next week to take it, and the church was gone."
This was the kind of surface alteration she believed constantly
reshaped the spirit of the city. Abbott could not accept the idea
that a few striking photographs could represent New York in all
its complexity. For that reason she continued her quest to document the whole city over the course of days, months, and years,
just as Atget had done in Paris. The step-by-step building of
Rockefeller Center, the erection of the George Washington Bridge,
the nighttime city glowing with electric lights—all of these and
more went into Abbott's amorphous portrait of a thriving New
York City. Eventually, the composite received notice. The power
of Abbott's cumulative corpus of images caught someone's eye.[26]
Curators at the Museum of the City of New York decided to

exhibit forty-one of Abbott's city pictures in the fall of 1934. The
show "proved of such great interest to the public" that the museum kept it on the walls for several months. Elizabeth McCausland
published a glowing review of the exhibit in a Springfield, Massachusetts, newspaper and repeated her praises in the next issue
of *Trend* magazine. The list of Abbott supporters continued to
grow. The New School for Social Research offered her an instruc-
torship in photography for the fall semester of 1935. Marchal
Landgren, an official on the state's Municipal Art Committee,
suggested that Abbott seek public funding for her "Changing New
York" project. He had already made appeals to private citizens in
Abbott's behalf but thought that one of the federal government
New Deal programs might be willing to sponsor her work.[27]

Abbott prepared a new proposal describing the thrust of her
documentary survey and submitted it to Audrey McMahon, the
director in New York City for the Works Progress Administration's
Federal Art Project (FAP). In the proposal she described the mul-
tifarious elements that made up "New York City in 1935," and in
summary she asserted, "It is important that they should be photographed today, not tomorrow; for tomorrow may see many of
these exciting and important mementos of eighteenth- and nineteenth-century New York swept away to make room for new
colossi. . . . The tempo of the metropolis is not of eternity, or even
time, but of the vanishing instant." Abbott laid out her objectives
specifically, making it clear that she alone intended to guide the
project, rather than have government officials direct her hand or
her camera. Even if she did need financial support, she did not
intend to forfeit her artistic integrity at the hands of bureaucrats—
"I presented them with a complete plan of what I wanted to do.
I also had five years' work to show them," she remembered. In
September, after she had begun teaching at the New School, Abbott
learned that Federal Art Project officials had accepted her proposal. They gave her a new title, "Superintendent of the Photographic Division—FAP/New York City" and allowed her to
continue building on the collection she had started six years earlier. Each new picture she took would be part of WPA/FAP project
number 265-6900-826, more familiarly known as "Changing New
York." The independent-minded photographer, who became employee number 233905, welcomed the federal government as a

benefactor. Abbott and many others like her would discover that public patronage of art and artists greatly enhanced facets of American culture that had been previously neglected.[28]

The Federal Art Project supported scores of struggling artists in the middle to late 1930s. At the inception of the project, the *New York Times* announced, "U.S. to Find Work for 3500 Artists."[29] The FAP, along with the Federal Theatre Project, the Federal Music Project, and the Federal Writers' Project, comprised the Four Arts program known as Federal One. Sponsored by and supported with funds from the Works Progress Administration, the arts projects sought to provide steady, weekly work for unemployed, creative talents, some of whom were well established before economic depression swept the United States. The WPA, established under President Franklin Roosevelt's Executive Order 7034, removed Americans from relief rolls and placed them on employment payrolls of public or private work projects. The president's endeavors caused resumption of active, productive lives for hundreds of thousands of Americans. They picked up hammers, hoes, scripts, songbooks, pens, paintbrushes, chisels, and cameras. They built highways and swimming pools, erected bridges and courthouses, cleaned streets, cleared lands, and painted post offices. Their efforts changed the face of America. What they could not clean or restore or erect fresh, they counted or recorded for the sake of historical preservation. Alan Lomax trekked through the deepest South to record Delta Blues lyrics and tunes that were inspired by conditions in the cotton fields; southern writers interviewed former African American slaves about their experiences in bondage; and throughout the states, FWP employees on the State Guide projects collected folklore, ghost stories, tall tales, and other distinctly regional narratives that revealed local color and character. The mixture of indigenous influences and ethnic variety in the United States provided a wealth of material for researchers on federal arts projects, among them Berenice Abbott.

Illuminating her own pursuit of such evidence and her passion to incorporate it into a usable visual narrative, Abbott told a *New York Herald Tribune* reporter in 1936, "For instance, Front and Pearl Streets are very Dutch, and Cherry Street is pure English. . . . And from all this melange comes something that is uniquely New York."[30] Abbott invested substantial time and en-

ergy in the streets on the Lower East Side. Along with the garment sellers who hung their wares on the sidewalks and the ped-
dlers who pushed carts through the streets were the food merchants.
Abbott's direct frontal view of a kosher chicken market (fig. 52)
provides a sense of vitality. Although the brightly sunlit birds have
their legs tied together, the fanned wings suggest flight, their feathery
lightness implying potential movement despite the fact that the
animals are no longer alive. That half the display rack is nearly
empty suggests the merchant's success; for Abbott's purposes, the
market window was a perfect example of the fleeting nature of
the present. Throughout a typical day, this window would regularly change in appearance, depending upon the availability of fowl,
the demand for chicken in the neighborhood, and each purchaser's
selection of specific birds from the rack. The curious produce seller
may have gazed out his window at another point during the
workday, seeing nothing because his vision was obscured by a full
cadre of fowl, or he might have stared out, as he does in this image,
and clearly observed those gazing at him.

As a WPA photographer, Abbott captured on film merely a
small portion of the country—its largest city—yet her work reflected and, at the same time, contributed to the growing national
search for American identity. As well as accomplishing important
physical and structural changes, the WPA encouraged an internal,
and more introspective, examination of national character. One
FAP artist noted, "It has nourished my enthusiasm to know that
I was contributing to the social order, that I was part of it." Federal One artists felt motivated to carry out the tasks of isolating
and scrutinizing regional characteristics that, when taken as a whole,
would add to the understanding of the collective cultural pulse of
the United States. One music critic noted that the Federal Music
Project (FMP) orchestras showed "the promise of a new growth
of art feeling in America—'local and vital.'" According to the same
critic, FAP painters, muralists, sculptors, and photographers elic-
ited similar responses after taking "living art into schools, hospitals and public buildings." The sole guideline for their FAP work
was to choose American subject matter, whether "naturalistic,
symbolic, legendary or historical." By exploring what was essentially American and bringing it before the public in the form of
art, FAP employees participated in a reciprocal agreement with

Fig. 52. Berenice Abbott. *Chicken Market.* Commerce Graphics Ltd, Inc.

the government. They could see their artistic visions realized, and even expect audiences to see their work, if they spoke merely from their own experiences as Americans. Project administrators would meet artists halfway by setting up exhibits in tax-supported public institutions that normally could not afford to obtain works of art, including "colleges, high schools, normal schools, trade schools, libraries, hospitals, museums, sanitoriums, prisons, and other Federal, State, and Municipal institutions and buildings." The government sought not only to raise the level of appreciation for art in general, but also to celebrate American artists and American subjects.[31]

WPA officials sometimes needed to defend the work of Federal One employees to the more cautious elements within the American public. Some skeptics believed that those on the public payroll should carry only shovels or rakes. Aubrey Williams, a WPA administrator, said, "We don't think a good musician should be asked to turn second-rate laborer in order that a sewer may be laid for relative permanency rather than a concert given for the momentary pleasure of our people." The WPA encouraged workers to continue using their own skills by giving them jobs in their chosen fields. For painters and sculptors the Federal Art Project resumed on a larger scale what the aborted Public Works of Art Project (PWAP) had abandoned in 1934 and what the Federal Emergency Relief Administration (FERA) could no longer afford. The differences in the new program resulted in more funding, better administration, and employment of all kinds of artists, photographers included. A *New York Times* article about the FAP stated in its subheading, "Nation-Wide Enterprise Is to include Photographers as Well as Painters," and it emphasized the vital roles of these men and women hired to "record the progress of the various activities and compile a record of WPA projects." A large number of FAP photographers, particularly those categorized in "Part 11, Publicity and Research," countered the cries of "boon-doggling" by providing pictures of artists and other WPA employees at work.[32]

Well over half of the New York City photographic staff devoted its time to recording citywide WPA activities. In one of her first items of correspondence to FAP national director Holger Cahill, Audrey McMahon explained that the New York photography project was "primarily a service project and only secondarily a creative one." Photographers charged to record work in progress made their medium a purely documentary one, in that their pictures served the larger purpose of fulfilling agency goals. As the vehicle through which WPA officials and the American public were informed about the agency's projects, the photographers' visual records helped to solidify the role of photography as a significant contributor to the documentary impulse working in the culture. Thus the federal government, in promoting nearly unlimited visual recordings with cameras, enforced the notion that photographs could themselves stand as clear renderings of actual conditions and real processes. In the first few months, FAP photographers snapped so many pictures of various WPA projects that Cahill was forced to limit the excessive documentation. He asked McMahon to scrutinize all photography assignments and make sure that her photographers did not step outside the FAP's jurisdiction. Although he wanted sufficient evidence to prove the value of the project, Cahill also recognized the importance of fostering artistic excellence in visual expression. In the long run, quality work would convince the American public of Federal One's worth.[33]

Cahill operated under the influence of John Dewey's aesthetic philosophy, which considered a person's (particularly an artist's) experience with the immediate environment to be important. Cahill emphasized this relationship as a measure of the national cultural awareness in a speech he gave: "The finest way of knowing our own country is to consult its works of art. . . . In these works we participate fully in the rich spiritual life of our country. In our own day it is the works of artists in all fields which concentrates [*sic*] and reveals [*sic*] contemporary American experience in its fullness. It is works of art which reveal us to ourselves. For these reasons I consider the Government art projects of the greatest importance." In the FAP reference material on procedure, suggestions were made to photographers "to work on research projects and on other projects of value to the community." No other artist than Berenice Abbott was more appropriate to be selected to create a documentary record of New York City. Even if she had been unaware of the FAP's stipulations about "American" subject matter, her pictures would have met federal government requirements. For she was not attempting to paint a glowing picture of the metropolis nor to provide a nostalgic walk down memory lane. Rather, she was trying to capture all the elements that composed a burgeoning city characterized most distinctly by its heterogeneity. Within the larger context of FAP objectives, her work fit in, since it was largely dictated by her own fascination with the city and guided by her experiences there. Abbott's photographs celebrated what one observer has called "the aleatoric quality of American culture." It was exactly what the FAP hoped to reveal.[34]

"Changing New York," as one of the FAP Photography Division's "creative" projects, stood apart from the service assignments. Since Abbott served as its director, photographer, consultant, and printer, she enjoyed a great deal of freedom on the job. When listed in reports along with other creative projects, Abbott's work was usually treated separately, as a unique undertaking.[35] Photography division supervisor Noel Vincentini made the distinction soon after Abbott's arrival at FAP. In his November 1935 report, Vincentini briefly described the activities of the general photography project:

The work of Photography, Part 6, consists of the taking of photographs requested for their records and official work by the various W.P.A. projects in the metropolitan area. . . .

In addition to this work, the Photography section is carrying out a creative project called, "Changing New York." On this project, photographs are being taken of buildings, locales and street scenes, which are meant to serve as a permanent graphic record of the transformations of the city and the changes in the habits of its life.[36]

The FAP allocated money for Abbott to have a driver, two technical assistants, and two researchers on the "Changing New York" project, but she personally assumed most of the responsibilities; she chose her own subjects, fussed over her negatives, and preferred to make her own prints. Government sponsorship neither diluted her fierce independence as a photographer nor redirected her professional goals on her urban project. Here Abbott's motivations and artistic sensibilities mirror those of Dorothea Lange, who during the same period found herself struggling with

Photography Unit director Roy Stryker over control of the negatives she was producing for her own employer, the Farm Security
Administration.

Abbott's pictures gained immediate recognition. Within six
months of her initiation with the FAP, she had photographs on
exhibit at the WPA Gallery on East Thirty-Ninth Street. The *New
York Herald Tribune* interviewed her, providing her a forum in
which to tell the city's residents why she felt the "Changing New
York" project should be important to them. Abbott's explanation
echoed the numerous appeals for support she had made in the
early 1930s, including her appreciation for its variety: "New York
is a spectacular city full of contrasts. The city is changing all the
time." Requests for Abbott's pictures and negatives and prints
poured into administrators' offices. As cosponsor of the FAP's
"Changing New York" project, the Museum of the City of New
York wanted to retain Abbott's negatives. Hardinge Scholle, the
director of the museum, went over local heads directly to the FAP's
Washington headquarters in an effort to obtain the project negatives for future print requests. FAP director Holger Cahill explained
to Scholle, "Artists like Miss Berenice Abbott are very reluctant
to have anyone else do their printing," but he reassured Scholle
that the museum would eventually receive all of the negatives to
keep permanently. The Grand Central Palace displayed several of
Abbott's photographs in a May 1936 show entitled "Women's
Exhibition of Arts and Industries." The following month, twenty
"Changing New York" project pictures were selected for inclusion in a traveling exhibit known as the "Washington Circuit."
For the circuit, called Project Number 8, the national offices chose
outstanding examples of project art and sent them to federal galleries throughout the country. Nearly all were mixed-media exhibits that included paintings, prints, pictures, designs, and
sculpture, but two traveling shows were devoted exclusively to
photography—one to pictures taken for the Federal Writers' Project
Washington, D.C., *City Guide Book* and the other to Abbott's
"Changing New York" images.[37]

Because of the increasing demand for her photographs in 1936,
Abbott spent an extremely busy year taking new pictures and
making prints. She completed extensive studies of the Brooklyn
Bridge, the Manhattan Bridge, Union Square, and Pennsylvania

Station. She focused on construction projects all over the city, but those that continued to hold special interest for her were the magnificent skyscrapers that dwarfed older, adjacent buildings. This theme would persist throughout her remaining years on the "Changing New York" project, thus proving poignantly true her intimation to a *New York World-Telegram* reporter about the inspiration for her study, "I'm sort of sensitive to cities. . . . They mean something to me."[38] Abbott's attempts to juxtapose landmarks from several generations may be seen in a 1938 view she took of Fortieth Street (fig. 53). Working from the Salmon Tower on West Forty-Second Street, Abbott depicted nearly one hundred years of New York history. The image exudes a sense of vitality because the frame is almost entirely filled with interconnected plane levels offering a variety of horizontal and vertical touch points in addition to layers of depth. The mid–nineteenth century structures occupying the lower right segment appear almost lost amid the towering modern structures, but the text stands out. The charge to "Save Regularly" at the Union Dime Savings Bank dates these old buildings, which continue to exist in the midst of a contemporary consumer culture that prizes buying over saving. The late-nineteenth century skyscrapers that initially dwarfed their counterparts of the 1800s bear the marks of a young urban landscape, complete with neoclassical and neo-Romanesque flourishes. In the lower left quadrant stands the Bryant Park Studio Building, with its revival window frames and small balconies. On the right edge of the image, the World Tower Building is distinguished by its uniformity all along the exterior, suggesting an upward sweep, until one reaches the top, where the European decorative details dominate the tower. In great contrast stand the modern skyscrapers filling up the center and top portion of the frame. Starkly simple and functional by design, the only variation in structure is the terracing at certain levels. The terraces provide layers of depth for the image, but more importantly, they send the eye up or down ladder-like steps. The climb higher to the top, whether inspired by ambition or greed or some other driving force, leads to security by contemporary standards but takes one further away from the monuments and ideas of the past. That such fluidity in movement is even possible in this relatively small area makes the feat all the more incredible.

Fig. 53. Berenice Abbott. *View of 40th Street.* Commerce Graphics Ltd, Inc.

Abbott's desire to depict the generational layers and their corresponding messages remained steeped in her belief that change occurred so rapidly that previous scenes could become unrecognizable to an eye that once knew them. In a 1938 study of an East Forty-Eighth Street block (fig. 54), Abbott framed two dwellings that had been identical to each other just five years earlier. The brownstone on the left, erected around 1860 and designed with classical-style window frames and decorative motifs, represents old New York. The newer and more prominent brick facade on the extreme left stands out just enough to obscure the brownstone's door front, but the more obvious juxtaposition is that of the brownstone's former twin, which occupies the right half of Abbott's frame. The contemporary dwelling shows no sign of its former life; the brown stone has been covered with glass brick and lightened with tones of a modern age, including concrete steps and a smooth, molded first-floor facade. What once was a stoop with open railings has been narrowed and closed off on one side, and the doorway's significant distance from the sidewalk allows the house a much more private, exclusive entrance. Wrought-iron rails with decorative grillwork have been replaced by the clean, curved lines of stainless steel, a mark of 1930s design known as streamlining. Abbott's steadfast determination to absorb as many signals as possible and then to show the fragmentary signs of deliberate and fast-paced change made her reputation as a documentarian who casts modernist eyes on the external world.

The popularity Abbott enjoyed as a WPA photographer was accompanied by controversy. When the Museum of Modern Art (MOMA) decided to put an FAP show on its walls, the museum's public relations director insisted on handling all publicity for the exhibit. Within a few weeks the *New York Times,* seemingly unaware of the upcoming MOMA exhibit, offered to present an Abbott photograph layout in its Sunday rotogravure section. Officials at MOMA threatened to cancel the exhibit, and national FAP administrators threatened to dismiss any employees who released information regarding the forthcoming show. Abbott, although one of the more prominent artists on MOMA's agenda, let FAP administrators handle the whole unpleasant affair. Her principal interest remained the same—to take pictures of New York, rather than to get mixed up in its highly volatile art community.

Fig. 54. Berenice Abbott. *East 48th Street, 1938.* Commerce Graphics Ltd, Inc.

In early 1937 the Museum of the City of New York expressed an interest in showing off the project it had cosponsored. The museum directors wanted to present a comprehensive exhibit composed entirely of Abbott's "Changing New York" pictures. When the show opened in October, 110 of Abbott's 260 project photographs revealed the multifaceted exterior of New York City to its residents. Public response was overwhelming; the museum reported increased general attendance, especially during the Christmas holidays when college students "flocked to the exhibition." Museum officials were forced to put off the exhibit's closing date in order to accommodate all of the curious patrons. Without doubt, "Changing New York" had accomplished several of the FAP's major goals, not the least of which involved generating Americans' interest in visual images created by their own, rather than European, artists.[39]

Abbott broke new ground with her "Changing New York" exhibits. Photography-only shows were rare because the value of the medium had not yet gained full recognition from other visual artists. Photography critic Beaumont Newhall, who served as an administrator on the Massachusetts FERA art project, pointed out that holding a photography exhibit in 1937, particularly at a New York museum, was a risky and controversial business practice. He explained that some critics abhorred the idea of such pictures being shown on the same walls that had previously held great paintings.[40] If any photographer could expect to be accepted among artists or even art photographers, it certainly was not Abbott, who told a newspaper reporter, "Next to golf and bridge, I loathe people who try to make photographs look like paintings or engravings."[41] Having stayed away from art photographers, Abbott aligned herself with very few others in her field. Perhaps she found more in common with her audience than with colleagues, since the former looked at her pictures with enthusiasm similar to that she felt when photographing the scenes. But what is even more likely is that Abbott had shown herself to be a modernist in tone and style, even though she had accomplished her feats with a camera rather than a paintbrush. So despite keeping her distance from other modernist photographers, such as Alfred Stieglitz and Edward Weston, she had defined herself as an innovative *seer* who embraced the principles, if not the practitioners, of modernism.

The reputation Abbott built from the popularity of "Changing New York" helped to open up other avenues for her photographic pursuits. The FAP administrators in Washington looked on with scrutinizing eyes, while Abbott received countless compliments, inviting offers, and numerous requests for pictures. Although they appreciated the tremendous attention given to an FAP project, government officials seemed uneasy with Abbott's personal success. They flexed their bureaucratic muscles in order to keep her independence in check. Working through the channels of bureaucracy grew tiresome to Abbott as the Washington FAP office insisted more emphatically on approving material released for publication. The policy, which had been in place since the inception of the WPA, hampered the New York City office because it handled such a large volume of book, magazine, and newspaper requests. The New York City division of the FAP appeared to federal administrators a bit cavalier in its approach toward publishable material. Constant reminders about necessary federal "clearance and approval" landed on Audrey McMahon's desk. Thomas Parker, Holger Cahill's assistant, addressed the subject on numerous occasions. He usually reiterated the FAP's responsibility to its public, in one instance writing, "I think it would be well for the information service to remember in the preparation of such articles that they are the employees of the Federal Government, and that the public who read the articles are the taxpayers who are providing the money for their employment." Ellen Woodward, another assistant in the Washington office, supported Parker's criticism of the New York City project and suggested that "material which goes to a nation-wide audience must be approved from the standpoint of the national program."[42]

The success of the New York City FAP, and particularly Abbott's contribution to it, no doubt prompted the national office of FAP to pay closer attention. One official from the Information Section made a special trip to New York to meet with Abbott about several photographs that might have publication potential. In the summer of 1938, Parker caused an uproar by asking Abbott to send her negatives to the Washington office. Without hesitation, McMahon replied, "Miss Abbott does not wish her negatives to be sent out and it is not procedural with us to act against the wishes of the artist." Although she was on the government payroll, Abbott enjoyed a great deal more freedom than other FAP artists and other government photographers. While Abbott was winning battles with Washington bureaucrats, Dorothea Lange was fighting to maintain some sense of artistic independence in her job at the FSA. While Abbott got to develop her own negatives and toss out the ones she disliked, Lange and her FSA colleagues had to send their undeveloped rolls of film to Washington to be printed. But Abbott's freedom was short-lived. In 1938, when E.P. Dutton offered to publish selected "Changing New York" images in a book, FAP officials assumed a great deal of control over the project, refusing Abbott permission to select all of the photographs or to make suggestions regarding the book's design. Several months later, when the FAP discontinued the "Changing New York" project, Abbott left the agency with hopes of photographing the 1939 World's Fair in New York. Under WPA guidelines, an agency employee could not be gainfully employed elsewhere and receive a paycheck from the federal government. Abbott's release was timely in that Federal One soon came under close scrutiny from many sides.[43]

Throughout the year, general criticism of Roosevelt's New Deal programs had grown more severe. The changing attitudes reflected congressional sentiments in the aftermath of the president's infamous "court-packing" attempt in 1937. His plans to change the structure of the U.S. Supreme Court appalled Washington politicians and the American public alike, thus diluting the overwhelming mandate he had fashioned from his landslide victory in the 1936 election. Roosevelt lost further party support after launching an offensive before the 1938 congressional elections to reward his loyal court-plan supporters and to eliminate the disloyal elements. In addition, the economic downturn in 1937-38, labeled "Roosevelt's Depression," had caused many skeptics to withdraw their support for the administration's bold recovery plans. In a political sense, New Dealers suffered "a discouraging year" in 1938. One of the programs most closely examined was the WPA, particularly its Federal One branch. Fear that radicals might infiltrate the Four Arts projects led Congress to cut WPA funding. All project workers who had been on the payroll for eighteen months or longer were released. New relief workers were forced to take a "loyalty oath," and the FAP, the FWP, and the FMP could con-

tinue only under the direction of local sponsors who closely su-
pervised each project. The Federal Theatre Project, which had been
perceived as the most radical of the Four Arts programs, was
abolished. Fortunately, Abbott had come in as a government
employee when the New Deal programs, Federal One in particular, enjoyed widespread support, before they were transformed by
politicos. To have carried on her work under heavy editorial eyes
would have gone against Abbott's spirit as a modernist who happened also to be a documentarian. Federal patronage had been
good for her but only up to a point—when it began to dictate
individual interpretation, the most talented artists, Abbott among
them, chose to leave rather than to compromise.[44]

Beaumont Newhall called the "Changing New York" project "the
greatest example of photography in the WPA." Considered as a
whole body, a cumulative set not to be divided, "Changing New
York" achieved what Abbott hoped it would achieve. Later in her
career, she wrote, "Photography does not stand by itself in a
vacuum; it is linked on the one side to manufacturers of materials and on the other side to the distributors of the product, that
is, to publishers, editors, business leaders, museum directors, and
to the public." "Changing New York" was not only an appropriate product of the 1930s but an extraordinary example of the
substance necessary to address the nation's "cultural lag." The
project celebrated what was American, yet what was regional; what
was contemporary, yet what was historical; where urban dwellers
had been, and at the same time where they were headed. The
photographs commanded viewers to place themselves in the scenes,
to examine the buildings, the doorways, the architecture, and the
statuary; the picture urged them to imagine that they were standing in the scene, personally judging the interplay between past and
present and past again. Abbott's pictures asked onlookers to gauge
the weight of history in relation to the fleeting present, the "vanishing instant." Abbott knew that at the moment her shutter clicked,
a piece of the present became the past, yet she continued to study
those fragments of time and place, hoping that others could participate in a specific moment just as she had. By recognizing the
power that external forces wielded over internal lives, Abbott led
viewers to evaluate their own places within their shifting environ-

ments. She argued that photography should be "connected with the world we live in," and she challenged photographers to guide their senses, contending, "The eye is no better than the philosophy behind it. The photographer creates, evolves a better, more selective, more acute seeing eye by looking ever more sharply at what is going on in the world." By surveying New York with a documentary spirit, Abbott experienced firsthand the transitory nature of an American city. By photographing it, she fit a few brief moments of time and place into the larger scheme of the past, the present, and the future—the whole gamut of human experience through the eyes of an American modernist.[45]

Conclusion

It is the spirit of his approach which determines the
value of the photographer's endeavor.
—Elizabeth McCausland

Berenice Abbott's confidante, the art critic Elizabeth
McCausland, wrote in 1939 that documentary photography's rapid
growth indicated "strong organic forces at work, strong creative
impulses seeking an outlet suitable to the serious and tense spirit
of our age." Arguing that documentary stood worlds apart from
earlier, more introspective and short-lived experimental fads in photography, McCausland characterized documentary's purpose as "the
profound and sober chronicling of the external world." That same
year, FSA Photography Section director Roy Stryker sat on a panel
addressing "Some Neglected Sources of Social History" at the
American Historical Association's annual meeting. In his presentation Stryker challenged historians to embrace photographs as
evidence when constructing social history; he noted that visual
images were "the raw material from which to compound new
histories and make old ones more vivid." Stryker's entreaty reveals his hope that the American past could be seen and not merely
read. Having observed the nationwide yearning for pictures during the Depression, Stryker tried to convince this rather skeptical
group of scholars to join other social scientists who had already
effectively used photographs to propel their arguments or secure
their conclusions. The motivations pushing McCausland and
Stryker to engage in discourses about documentary resemble those
of many documentarians in the 1930s, including Doris Ulmann,
Dorothea Lange, Marion Post, Margaret Bourke-White, and
Berenice Abbott. At the foundation of each woman's camera work

lay her desire to have her photographs contribute to a usable past or a practicable present that would help Americans shape a more promising future. All five women wanted their pictures to serve a purpose that would bring about social reform or cultural health or, at the very least, awareness if not implementation of these goals. And in varying degrees, each expressed visually the "serious and tense spirit" of the age in which she lived. Assuming the mantle of documentarian gave credibility to the wide-ranging interests of each photographer; in addition it assured each one that her pictures would be bought or circulated or filed temporarily, until some opportune moment for publication.[1]

The early-twentieth-century appeal of identifying "vanishing races" on the American landscape set the stage for Doris Ulmann's work. Her focus on marginal cultures in the United States, particularly in southern Appalachia, reflected the expectations of some Americans that isolated groups of rural inhabitants had been untouched by modernizing forces such as industrial development and urban growth. Like her counterparts who sought out cultural peculiarities in groups and ethnic enclaves that had experienced contact with modern America, Ulmann attempted to recreate their disappearing or already-dead worlds. She promoted the romanticization of people living in Appalachia, in religious communities, and in protected rural environments by situating them in the soft glow of pictorialism, a photographic style fashioned after nineteenth-century pastoral landscape painting. With noble intentions, Ulmann hoped to foster greater appreciation in the United States for traditional craftwork and the rural inhabitants who carried it out. Her willingness to pour a substantial amount of her own money into the Southern Highlands Handicrafts survey in order to promote the sale of handmade goods instead of mass-produced items is a testament to her ideological inclinations. Despite her old-fashioned camera equipment and a predictable style that had sitters pose for portraits, she exhibited the spirit of documentary by wanting her records to be used productively for educational purposes. Her generous donations to Berea College and to the John C. Campbell Folk School, in both photographs and money, secured her legacy as a documentarian and a supporter of innovative experiments in cultural education. Several years before the

Farm Security Administration photographers heightened American regionalism by showing southerners to New Englanders and Dust Bowl refugees to other Americans, Ulmann manipulated regional and racial stereotypes by celebrating differences among distinctive groups of people. That rural inhabitants captured so much of her attention shows her active participation in the popular debates that pitted modern urban life against an imagined bucolic agrarian existence.

Dorothea Lange built her reputation portraying the perseverance of human dignity in debilitating circumstances. She developed this posture early in her life when her own challenging environment toughened her reserve; as a result she fashioned and applied a fail-safe formula for enhancing her subjects' lives by casting them as American heroes. Her camera angles on individuals ennobled them, fostering an illusion of optimism and steadfastness. Combined with the enduring American theme that equated freedom to move with potential success, Lange's negatives emit a widespread sense of hope despite the grim realities of the Great Depression. The subjects in her photographs stand as pioneers and adventurers, often worn down but certainly not worn out. Her visual messages read like nineteenth-century diary pages full of struggle and persistence in the fight against nature's forces. Like Ulmann, Lange undertook photographic assignments in remote areas, often under primitive conditions. But in the starkest cabins and the driest fields, the two photographers created invisible portrait studios where the focus remained centered on the uniqueness of an individual. Lange, who was devoted to the contemporary trend of matching images with word texts, underpinned her visual messages with verbatim accounts of conversations and responses offered by her subjects while they were being photographed. Building upon the work of her husband, economist Paul Taylor, her pictures stood as mutual supports for the statistical tables, graphs, and other weighty evidence that he offered in their collaborative efforts.

Marion Post, born a generation after Ulmann and Lange, prepared fresh, revisionist messages through her efforts to reveal the potential of group power—that inspired by the family, the community, and the masses. Having picked up her first camera in the midst of National Socialist hysteria in Austria, she shaped her

vision through the lens beside workers and students who demonstrated in the streets against encroaching fascist forces. Further
experiences in leftist movements in the United States helped mark
her documentary efforts with subtle reminders of social inequities
that pervaded the nation. Her attempts to rebalance the scales meant
that her employer, the U.S. Government, kept many of her more
suggestive photographs in a file drawer, safe from the public eye.
Her countless scenes of cooperative spirit in the midst of economic
depression clearly indicate a departure from both Ulmann's and
Lange's deliberations on individual attempts to recoil from hard-
ship. Post's messages focus much more on human relationships
than on the lone individual. Stylistically, her images evoke a sense
of busyness; the smallest elements of everyday living, from spoons
and clocks to tattered hats and shiny shoes to casual conversations, fill her frames to the edges. The historical record according
to Post was full of the mundane details that made up daily existence. Through her use of the mundane, the spillage of minutiae
over the four edges of a photograph, Post's images hint at a
modernist aesthetic. In many of her photographs, she suggests that
she has captured or created only a fraction of a part of a scene.
From the time she joined the FSA in 1938, a period of transition
for the agency, she was able to experiment stylistically. Roy Stryker
was so desperate at that point to have his photographers survey
as many areas of the country as possible that Post enjoyed almost
unlimited freedom to shoot. The sheer quantity of negatives marked
with her name bears this out. Post made an initial attempt to imitate
Lange's fieldwork approach but abandoned it after completing her
first assignment. Pictures in succeeding assignments show that she
favored a more participatory role for herself than Lange had
claimed. This manifests itself most clearly in scenes where Post
engages her subjects in conversation, and they speak to her in the
camera images. As in her earliest experiences using photography
in Austria, Post developed scenes where she herself became part
of the group under examination, as a player in a political or social drama. Just as her counterpart Berenice Abbott required viewers
to place themselves in scenes, where they examined construction
processes or building facades, Post created an atmosphere where
viewers entered the sites of her images, sometimes as voyeurs but
more often as full participants.

Margaret Bourke-White attracted a large viewership with her innovative looks at industry and mechanization. Having considered herself a scientist from an early age, she reveled in visually deconstructing engines, electrical processes, and automated systems down to their tiniest components. She embraced the latest technologies, contributing her talents to the business world and the realm of profit, while the country struggled to survive economic devastation. Enamored with machine-age design elements, such as streamlined curves and light, clean, unfettered surfaces, Bourke-White arranged and isolated industrial components that exuded characteristic clarity and precision, the apparent fruits of assembly-line production. Inanimate objects so forcefully dominated her early work that people, if they had a place in her photographs at all, were peripheral, used most often to gauge the power and scale of industry. When Bourke-White chose to focus upon people in the South in her collaborative efforts with Erskine Caldwell, the results revealed a quality of brutishness pervasive and inescapable in poor Americans' lives. Bourke-White's application of her modernist sensibilities to human beings yielded disturbing depictions of magnified and distorted elements in their daily existence. Such realism, combined with the photographer's popular byline, guaranteed her images wide exposure in the 1930s. The success of her book *You Have Seen Their Faces* prompted other writer-photographer teams to pursue similar projects. In some cases these endeavors were undertaken to correct Bourke-White's sensationalism as much as anything else. That she provoked such action further enhances her reputation as a shaper of documentary expression in the 1930s.

Fascinated by shifting urban landscapes, Berenice Abbott roamed the streets of New York City in the 1930s, where she sought out evidence of the fleeting moments she called "the now." The myriad facets of a vibrant external world were her primary material. After determining that the clearest signals of cultural change could be found in juxtapositions of the historic with the contemporary, she focused on the processes of skyscraper construction and building demolition, in addition to the constant transformations inside storefront windows and along busy streets and sidewalks. Picturing the fragmentary and peripheral signs of human experience, such as clothesline laundry and daily advertisements,

Abbott created a visual framework that was indisputably modern. Virtually free of the overt distortions that characterized European modernists' views of humanity, Abbott brought her own way of seeing in line with the popular documentary impulse in the United States. In so doing, she managed to produce a comfortable marriage of modernism and documentary that addressed contemporary concern over "cultural lag" in the United States. Her images helped close the gap between fast-paced technological advances and the existing social fabric. Abbott's scenes of a Lower East Side chicken seller in his daily routine or of tenement laundry lines set against new high-rise apartment buildings sent reassuring messages about American cultural identity, especially when compared with Bourke-White's pictures of diseased sharecroppers and shackled black men, which promoted general uneasiness about what constituted Americanness in the 1930s.

Although they stood at opposite ends of the documentary spectrum in terms of aesthetic philosophy and stylistic production—and as a result are situated at opposite ends of this study—Ulmann and Abbott shared a desire to capture distinctive entities before they disappeared from the American scene. Their respective attempts to preserve specific faces and particular moments were exercises in historical preservation, as much as were the endeavors of the WPA workers who interviewed ex-slaves or gathered folktales. Ulmann achieved her goals by arresting time; consciously removing any elements within the scope of her lens that would date her images, she cast her characters in personal dramas that transcended specific dates. Abbott's continuous efforts to capture passing moments meant that she produced a piece of the past as soon as her shutter clicked. The resulting documentary images themselves became pieces of history, elements in the passage of time. Compared to Abbott's views, Ulmann's scenes appear decades older, even though both photographers' projects were conducted within a twelve-year period. The varieties and levels of experimentation displayed in their work as well as in that of Lange, Post, and Bourke-White, attest to the rich complexity of the documentary genre in the 1920s and 1930s.

All five photographers created powerful representations of pervasive, penetrating ideologies circulating in the two decades between the world wars. In the last twenty years, scholarship on photography has been heavily informed by postmodern theories that posit images as representations, entities dependent entirely upon their contexts.[2] The photographers in this study recognized the role of context and realized that the uses of their pictures would far outweigh the pictures themselves. Perhaps this is why their work and their ideas about photography seem so fresh over half a century later. Certainly another reason is that the photographers' lives and experiences figure as important texts as well. Their stories as women command attention. At a 1995 conference session, Gerda Lerner urged historians of women to strike a balance between theory and narrative. She reminded her listeners that the past is full of real women whose stories require telling and retelling. This study seeks to strike such a balance by integrating ideas about visual representation with the experiences of several women who were largely responsible for shaping the field of documentary photography. Ideally their stories will become as familiar as traditional historical narratives, so deeply internalized that we naturally see "a vision of U.S. history as women's history as much as it is men's history."

Notes

The names of manuscript and photograph collections are abbreviated as shown in the bibliography.

Introduction

1. Bourke-White, typescript draft for "Popular Photography," Sept. 1939, MBW.

2. In 1922 Walter Lippmann had stated emphatically, "Photographs have the kind of authority over imagination to-day which the printed word had yesterday, and the spoken word before that. They seem utterly real" (quoted in Orvell, *The Real Thing,* 151). In the 1930s, when the camera was thought to be able to "support" and "prove" various contentions, a "picture hunger" swept the United States, argues David P. Peeler in *Hope among Us Yet,* 75–76.

3. Remarks by Houk in Edwynn Houk Gallery, "Vintage Photographs by Women," n.p. The glaring omissions in particular works are discussed by Naomi Rosenblum in the introduction to her *History of Women Photographers,* 7–9. The scholarship gap on women photographers has begun to close; the most notable new work in the field is Davidov, *Women's Camera Work.* Kerber, Kessler-Harris, and Sklar, *U.S. History as Women's History,* 14.

4. Most helpful for their exposition of the diverse ways in which the role of gender may be examined are the essays in Alpern et al., *The Challenge of Feminist Biography.* Myra Albert Wiggins, quoted in Davie, "Women in Photography," 138. Beloff, *Camera Culture,* 61. Glazer and Slater, *Unequal Sisters;* Anne Noggle's introductory remarks in Mann and Noggle, *Women of Photography,* n.p.; Gover, *The Positive Image;* and Mary Abrams, "Women Photographers," *Graduate Woman* 25 (Sept.– Oct. 1981), 22, 24.

5. Paul Katz, introduction to Edwynn Houk Gallery, "Vintage Photographs by Women," n.p.

6. A note on use of names: recognizing that a woman's choice of name and/or title (e.g., Mrs., Miss, etc.) reflects her sense of identity, I have given these careful consideration in the text. I have noted where and when these professional photographers chose to change or alter their names. For example, Marion Post did not marry Lee Wolcott until near the end of her stint as a documentary photographer for the U.S. Government, so throughout most of the text, she is referred to as Marion Post;

at the point where she legally assumes the name Marion Post Wolcott,
I begin referring to her that way.

7. Orvell, *The Real Thing*, in which he comments, "In some ways
the modernist sensibility . . . with its commitment to photography as an
art form, albeit a uniquely modern one, was at odds with the documentary mode" (228).

8. Borchert, "Historical Photo-Analysis," 36. Schlereth, "Mirrors of
the Past: Historical Photography and American History," in Schlereth,
Artifacts and the American Past, 46. Peeler, *Hope among Us Yet*, 7.

9. Singal, *The War Within*, xiii; Sekula, "Invention of Photographic
Meaning," 37. Beloff, *Camera Culture*, 75.

1. Doris Ulmann's Vision of an Ideal America

1. Doris Ulmann to Allen Eaton, 9 July 1934, JJN.

2. Crunden, *Ministers of Reform*, 3–38.

3. Seixas, "Lewis Hine," 381–82; Hine, "Charity on a Business Basis"; Hine,"What Bad Housing Means to Pittsburgh"; Hine, "Toilers of
the Tenements." This represents a fraction of Hine's work in those years.
A substantial Hine bibliography may be found in Rosenblum and
Rosenblum, *America and Lewis Hine*.

4. Heilpern, "Vita"; Seigfried, "Invisible Women"; Cook, "Female
Support Networks"; and Whisnant, *All That Is Native and Fine*, esp.
chaps. 1 and 2.

5. Warren, "Photographer-in-Waiting," 142; on White's influence upon
his women students, see Davidov, *Women's Camera Work*, 90-94.

6. Ulmann, *Darkness and the Light*, 8–9; Warren, "Photographer-
in-Waiting," 142; I owe a great debt to Ulmann biographer Philip W.
Jacobs for describing to me Ulmann's various illnesses and physical problems.

7. A fine discussion on the relationship between pictorialism and
Photo-Secession may be found in Rosenblum, *History of Women Photographers*, 95–99; Jussim, "'Tyranny of the Pictorial,'" in which Jussim
notes, "Muckraking journalism was firmly established in the very year—
1902—that Alfred Stieglitz . . . was planning the first issue of *Camera
Work*." (54) See also Trachtenberg, *Reading American Photographs*, where
he notes that Stieglitz was "more attached to aesthetic, individualistic
alternatives than to social or political solutions" (167).

8. Seixas, "Lewis Hine," 386–89.

9. Boxes 1, 2, DU-NY.

10. Box 2, DU-NY; Sarah Greenough, "How Stieglitz Came to Photograph Clouds," 151–65. Stieglitz altered his style as the popularity of
tonalism faded, abandoning the characteristic blurred image for sharper,
more geometrical features. See Trachtenberg, *Reading American Photographs*, 180–84; and Orvell, *The Real Thing*, 198–220.

11. Lovejoy, "Photography of Doris Ulmann"; David Featherstone

declares Ulmann an "ethnographer" in *Doris Ulmann*, 31–35; Ulmann to Warren, in Warren, "Photographer-in-Waiting," 144.

12. Signed, dated prints of Ulmann's early work may be found in DU-NY, DU-OR, and DU-UK. Since Ulmann preferred to sign her prints in pencil rather than ink, the alteration process was rather simple. It is apparent that she re-signed earlier photographs after her final break with Jaeger.

13. Henry Necarsulmer to William J. Hutchins, 10 July 1935, DU-BC, emphasis mine. The discrepancy is revealed in two features on Ulmann's photography that appeared in *Theatre Arts Monthly*, one in 1930, the other in 1934; the former referred to her as "Miss Ulmann," the latter as "Mrs. Ulmann." For her family's reactions to her traveling companion, the Kentucky folk singer-actor John Jacob Niles, see the heated correspondence in DU-BC; see also "Notes by Doris Ulmann concerning Her Will," an odd dictation to Niles in the days immediately preceding her death, where Ulmann asks that her reputation and her sister's peace of mind not be dragged "into an unfortunate & undignified position" after she dies (box 51, JJN).

14. Books 61, 62, DU-OR.

15. Photographer Laura Gilpin remembered "a quiet Doris Ulmann who sat in on a [Clarence White] class she herself was attending" (quoted in Ulmann, *Darkness and the Light*, 8–9); see also Allen H. Eaton's description of Ulmann in Eaton, "Doris Ulmann Photograph Collection," 10. The quotations in the text are from Warren, "Photographer-in-Waiting," 142, 144.

16. For her breadth in studying writers, see books 21, 61, 64, 70b, 71b, 73, 74, 75, 78, 79, DU-OR; Ulmann quoted in Warren, "Photographer-in-Waiting," 139; Jargon Society, *Appalachian Photographs of Doris Ulmann*, n.p.; Stein quoted in Hoffmann, *The Twenties*, 220.

17. Warren, "Photographer-in-Waiting," 131.

18. Philip W. Jacobs, "Doris Ulmann's Search for Meaning," Ulmann Symposium, Gibbes Museum of Art, Charleston, S.C., 8 Nov. 1997; Warren, "Photographer-in-Waiting," 136.

19. Anderson's disillusionment with an industrialized society is best evident in his novels *Winesburg, Ohio* (1919) and *Perhaps Women* (1931); *Call Number* 19 (spring 1958). The quotation is in Warren, "Photographer-in-Waiting," 142.

20. Ulmann to John Bennett, 12 Oct. [?], JB.

21. Campbell, "Doris Ulmann," 11; selected prints in DU-OR, DU-UK, and DU-UKA.

22. The complications of Ulmann's private life are revealed in Williams, *Devil and a Good Woman*, 232-36.

23. Ulmann, preface to *Portrait Gallery of American Editors*, v.

24. Warren, "Photographer-in-Waiting," 131; Ulmann, preface to *Portrait Gallery of American Editors*, v.

25. DU-OR, DU-UK, and DU-BCP.

26. John Jacob Niles in Jargon Society, *Appalachian Photographs of Doris Ulmann*, n.p. The size of the Ulmann entourage is described in Bill Murphy to H.E. Taylor, 11 April 1934, DU-BC. Murphy, of Boone Tavern, Berea, Kentucky, expresses fear that the large group will tie up too many rooms at the local hotel during "commencement time" at Berea College.

27. Niles to W.J. Hutchins, 14 Sept. 1934, DU-BC; Campbell quoted in Whisnant, *All That Is Native and Fine*, 137.

28. A perfect example of Ulmann's place could be seen in the National Portrait Gallery's 1994 exhibit entitled "Art and the Camera, 1900–1940," where her work was more widely represented than that of any other photographer. Alongside her seven images were photographs by Clarence White, Alfred Stieglitz, Edward Curtis, Edward Weston, Julia Margaret Cameron, and other famous camera artists.

29. A compelling discussion of these early-twentieth-century reformers or "culture workers" may be found in Whisnant, *All That Is Native and Fine*; a fine set of primary source materials is Stoddart, *The Quare Women's Journals*; Ulmann to Allen Eaton, 25 Aug. 1933, JJN; Niles to William J. Hutchins, 16 Aug. 1933, DU-BC; the quotation is in Eaton, "Doris Ulmann Photograph Collection," 10.

30. See Lewis Hine, "Social Photography," reprinted in Trachtenberg, *Classic Essays on Photography*; Peeler, *Hope among Us Yet*; Stott, *Documentary Expression*; Fleischhauer and Brannan, *Documenting America*. For opinions from Farm Security Administration photographers on what their intentions were, see the interviews in Amarillo Art Center, *American Images*.

31. Featherstone, *Doris Ulmann*, 19; Warren, "Photographer-in-Waiting," 142; Williams quoted in Jargon Society, *Appalachian Photographs of Doris Ulmann*, n.p.; Stott, "What Documentary Treats," in Stott, *Documentary Expression*, 62 (page references are to the 1986 edition); Guimond, *American Photography*, 4.

32. Shipman, introduction to Ulmann, *Portrait Gallery of American Editors*, 1. Shipman referred to Ulmann's "gallant survivors" in a publishing world that had shifted to cover the "right shade of lip-stick . . . the merits of a cigarette or the cut of a dinner-jacket" (1); Among the scores of contemporary critiques commenting on the effects of mass culture on the nation's distinct regions was an essay entitled "A Mirror for Artists," in Twelve Southerners, *I'll Take My Stand*.

33. One of the best photographs of Ulmann standing next to her camera accompanied Olive D. Campbell's recollections of the photographer in "Doris Ulmann"; Allen Eaton interview, conducted July 1959, partial transcript, DU-BC; Jargon Society, *Appalachian Photographs of Doris Ulmann*, n.p.; Niles, "Doris Ulmann," 7.

34. Because Ulmann treated time in such a fashion, she produced hundreds of photographs that scholars have been unable to date definitively. A range of dates may be placed on particular subject matter be-

cause we know the months and years she took certain trips to specific places. However, a substantial number of the Appalachian elders remain anonymous, mere examples of a "type" that Ulmann was seeking to capture. Many of the subjects Ulmann photographed before 1931 remain unidentified and were marked accordingly in the proof books (DU-OR) after Ulmann's death. But unless Ulmann herself signed and dated her prints, doubt remains. Niles's field notebooks are somewhat helpful, as are Ulmann's occasional references to particular subjects in her letters (JJN; JJN-BC). She and Niles kept more thorough written records on the journeys they took together a few years later, especially in the summers of 1933 and 1934.

35. Niles, "Doris Ulmann," 7; Enos Hardin study (book 45), Paul Robeson study (book 72), Thornton Wilder study (book 78), DU-OR. Of the portrait session with Wilder, Ulmann recalled he was so enthusiastic that he canceled all his morning appointments and then said to her, "We might go on all the afternoon if you have nothing else to do" (quoted in Warren, "Photographer-in-Waiting," 132, 142).

36. Book 12, DU-OR. Throughout the chapter the word *series* refers to photographs made of the same subject or person in a single sitting.

37. Warren, "Photographer-in-Waiting," 139; Jonathan Williams, in Jargon Society, *Appalachian Photographs of Doris Ulmann*, n.p.

38. Conkin, *The Southern Agrarians*, 86–87; Singal, *The War Within*, 198. On the characterizations of Tennesseans and southerners in general, see Hobson, *Serpent in Eden*, 11–32, 147, 184 (page references are to the 1978 edition). The critic is quoted in Ulmann, "The Stuff of American Drama," 132.

39. Many Ulmann critics and interpreters have referred to the "types" she photographed. Although Ulmann initially sought out particular groups, she focused almost entirely on individuals. See Featherstone, *Doris Ulmann*; Niles, "Doris Ulmann," 5.

40. "The Mountain Breed," *New York Times*, 2 June 1928, 16, col. 6; Ulmann, "The Stuff of American Drama," 141.

41. Eaton interview, July 1959, DU-BC; Loyal Jones, interviews with Ulmann subjects, discussed in Banes, "Ulmann and Her Mountain Folk," 41–42; William J. Hutchins to John Jacob Niles, 1 Sept. 1934, DU-BC; Campbell, "Doris Ulmann," 11. Sarah Blanding, dean of women at the University of Kentucky, said after Ulmann's death, "The remembrance of her is like a Past bright ray of autumn sunshine—warm and glorious and meaning more than all the other myriad of rays of the preceding months" (undated letter, Sarah Gibson Blanding to John Jacob Niles, box 51, JJN).

42. Niles once wrote, "We would pull up in front of someone's house right beside a very nicely paved road, take out the camera, set it up, and I would say, 'Folks, we have come to take your picture,' and they would line up in a row and that was all there was to it" ("Doris Ulmann," 7). Niles's flippant recollection disregards all of the groundwork laid for

their arrival at certain locations; in addition, it deflates Ulmann's deliberations as an artist, ignoring the overwhelming character of her Appalachian portraiture.

43. Books 7, 15, DU-OR.

44. Ulmann quoted in Warren, "Photographer-in-Waiting," 139; books 9, 14, DU-OR; DU-BCP; Niles, "Doris Ulmann," 5.

45. Niles, "Doris Ulmann," 4–5; DU-BCP; Ulmann, "The Stuff of American Drama," 132; Fass, *The Damned and the Beautiful;* Hoffman, *The Twenties,* 110.

46. Books 12, 52, DU-OR; DU-BCP.

47. Singal, *The War Within,* 202–7.

48. Davidson, "Julia Peterkin"; Peterkin quoted in "Calhoun Times," 1927, untitled typescript of a newspaper article, JP; Davidson, "Peterkin"; at her death the *Charleston News and Courier* ran an obituary featuring a portrait of Peterkin that bore the caption "Understood Negro" (12 Aug. 1961).

49. Ulmann to Averell Broughton, 4 Oct. 1929, quoted in Featherstone, *Doris Ulmann,* 43; "Current Exhibition News in Brief," *New York Times,* 3 Nov. 1929, sec. 9, p. 12. Evidence of the burgeoning interest in rural African American communities may be gleaned from the popular dramas written in the 1920s and early 1930s by Paul Green, Eugene O'Neill, and DuBose Heyward. On Ulmann's connection to this kind of theater, see Brown and Sundell, "Stylizing the Folk," 335–46.

50. Henry Necarsulmer, Ulmann's brother-in-law and one of her three living relatives, wrote of the friendship to Roger Howson, 17 Oct. 1936, DU-BC. The intensity of the Ulmann-Peterkin relationship is wonderfully described by Williams in *Devil and a Good Woman,* 153–54, 232–34.

51. Ulmann, *Roll, Jordan, Roll.* The hand-pulling process is described in Greenough et al., *The Art of Fixing a Shadow,* 504; White quoted by Coles, in Ulmann, *Darkness and the Light,* 82; DU-NY; DU-SC.

52. On the failure, see "The Movement" in Conkin, *The Southern Agrarians,* 89–126.

53. Clift, in Ulmann, *Darkness and the Light,* 10; The Melungeon woman's portrait was first published in a 1929 issue of *Pictorial Photographers in America.* Interest in this group may be seen in Kennedy, *The Melungeons.* Both the "Turk" and the "Cracker" were featured in Ulmann, "The Stuff of American Drama," 135, 139. Ulmann did not label the photograph "Turk" (DU-UK).

54. Peterkin, *Bright Skin,* 22–23, 31, 94, emphasis mine. Two of the best builders on the earlier theme of belonging were Nella Larsen, *Quicksand* (1928) and *Passing* (1929) and Anzia Yezierska, *Breadgivers* (1925). Into the 1930s, those employed in the public arts projects made regional ties and sense of place a key element in their artistic productions.

55. Megraw, "'The Grandest Picture.'"

56. Ulmann had photographed African Americans since her earliest years as a photographer (e.g., her study of New York City laborers in 1917). In 1924 she submitted a portrait of a black man for the Kodak Park Camera Club's Fourth Annual Exhibition. Ulmann's Deep South tours allowed her to build on her already-substantial collection of African American portraits.

57. Books 43, 50, 67, 72, DU-OR; Ulmann's promise to complete future work in Louisiana was made in "The Stuff of American Drama," 141.

58. Eaton, *Handicrafts*, (1973 ed.) v; Eaton, "Doris Ulmann Photograph Collection," 10; Stieglitz, "Pictorial Photography," 117; Ulmann quoted in Eaton, "The Doris Ulmann Photographs," in Eaton, *Handicrafts*, 17; Rayna Green, introduction to the 1973 edition of Eaton, *Handicrafts*, xiv; see also Tanno, "Urban Eye on Appalachia," 30; and Featherstone, *Doris Ulmann*, 50–52.

59. Orvell, *The Real Thing*, 181; Eaton, *Handicrafts*, 249; Knoxville Museum of Art, "Patchwork Souvenirs of the 1933 Chicago World's Fair" Exhibition, curated by Merikay Waldvogel and Barbara Brackman, Traveling Exhibition, 1994–96. The process of "traditionalizing" mountain handicrafts was necessary in order to increase their marketability, argues Jane Becker in *Selling Tradition*.

60. Featherstone, *Doris Ulmann*, 55; Observation by Niles, "Doris Ulmann," 5; Ulmann to Eaton, 25 Aug. 1933, JJN, where Ulmann goes on to speculate about the kind of work Hardin could accomplish "if his wife were a different woman."

61. DU-BCP; Ulmann to Eaton, 25 Aug. 1933, JJN.

62. DU-OR.

63. Davidov discusses White's theories in *Women's Camera Work*, 91-92.

64. Thornton, "New Look at Pictorialism." Thornton says that Ulmann "persisted in her fondness for outmoded Pictorial effects."

65. Niles, "Doris Ulmann," 7; Thornton, "New Look at Pictorialism"; see also Featherstone, *Doris Ulmann*, 17, 58–60.

66. Eaton, *Handicrafts*, 74, 258; Niles to Eaton, 5 June 1934, JJN, in which he notes that the granddaughters were wearing "Aunt Sallies cloths [*sic*]"; David Whisnant, lecture at National Endowment for the Humanities Institute, "The Thirties in Interdisciplinary Perspective," University of North Carolina, Chapel Hill, July 1995, in which he used the Wilma Creech portrait as an example of Ulmann's manipulations in the field, since Creech was a university student. On Creech's role in creating an idealized ancestral past, see Becker, *Selling Tradition*.

67. Articles of incorporation quoted in Whisnant, *All That Is Native and Fine*, 139; Jan Davidson, "The People of Doris Ulmann's North Carolina Photographs," Ulmann Symposium, Gibbes Museum of Art, Charleston, S.C., 8 Nov. 1997; Davidson pointed out that there were 147 pledges made by Brasstown residents to support the project; Ulmann

to Eaton, 24 July 1933, 26 June 1934, JJN; typescript of Niles's field notebook page, 31 July 1934, JJN.

68. Berea College, *Bulletin*, 11–12; Ulmann to Hutchins, April 1930, DU-BC; Hutchins to Ulmann, 28 March 1930, DU-BC; Eaton to Hutchins, 21 July 1930, DU-BC; Niles to Hutchins, 16 Aug. 1933, 20 Sept. 1933, Oct. 1933, DU-BC. The ellipsis points are Niles's; he creatively used his own punctuation system.

69. Ulmann to President and Mrs. Hutchins, 3 Nov. 1933, DU-BC; Ulmann to Hutchins, 4 Jan. [1934], DU-BC; Ulmann to Mrs. A.N. Gould [Berea College Art Department], 20 Nov. 1933, DU-BC.

70. Ulmann to President and Mrs. Hutchins, 3 Nov. 1933, DU-BC; Hutchins to Niles, 30 Oct. 1933, DU-BC in which Hutchins speaks of the kind gesture (the returned check); Ulmann to Mrs. A.N. Gould, 20 Nov. 1933, DU-BC; Ulmann to President and Mrs. Hutchins, 3 Nov. 1933, DU-BC.

71. Niles to Hutchins, 20 Sept. 1933, DU-BC, in which he noted the results of their summer's work: "She has about 1100 plates and I have three note books full of various sentimentalia." In Jargon Society, *Appalachian Photographs of Doris Ulmann*, Niles recalls accompanying Ulmann to numerous events, including "the best of the current New York plays"; Niles, "Doris Ulmann," 4.

72. Niles to Hutchins, 16 Aug. 1933, DU-BC, in which he notes that the dictionary will explain "the origins of our strange woods"; DU-BCP; DU-OR, where an entire series features Niles with various types of dulcimers; Eaton, *Handicrafts*, 138; When the Ulmann print collection was mounted into scrapbooks (DU-OR), Niles labeled many of the photographs, identifying numerous people from the trips to Kentucky, Tennessee, and North Carolina. Cross-references may be found in his own *Ballad Book*.

73. DU-G; DU-OR; DU-UK.

74. Niles to Hutchins, 11 March 1934, DU-BC; Leicester B. Holland to Niles, 11 April 1934, JJN; Ulmann to Eaton, 8 May 1934, 24 May 1934, 6 May 1934, JJN.

75. Ulmann to Eaton, 24 April 1934, 1 July 1934 (emphasis mine), 6 May 1934, 26 June 1934, JJN.

76. Hutchins to Ulmann, 2 May 1934, DU-BC; since Ulmann had already arrived in Berea, Hutchins directed his request to Boone Tavern, where the photographer was lodging; Niles to Eaton, 5 June 1934, JJN, in which he notes that Ulmann's work in the kitchen operations, including the candy kitchen, was shot on 18 May 1934; photographic series, DU-OR.

77. Photographs by FAP photographers, New York City Photography Unit, Still Picture Branch, National Archives, Washington, D.C.; Daniel et al., *Official Images*; Davidov, *Women's Camera Work*, 157-61, 165-67; Guimond, *American Photography*, 37; Berea College series, DU-OR, DU-BCP.

78. Ulmann to Eaton, 9 July 1934, JJN; Stryker correspondence, RESP, FSAWR.

79. Ulmann to Eaton, 24 April 1934, 1 July 1934 (in which Ulmann mentions those centers Campbell suggested she omit, including Hot Springs, Crossmore, Higgins, Penland, and Tallulah Falls), 26 June 1934, 22 July 1934, 9 July 1934 (where Ulmann claimed to value Campbell's judgment "more than anybody's, as she does not allow any prejudice to influence her decisions"), JJN.

80. Ulmann to Eaton, 22 July 1934, JJN.

81. The last portraits Ulmann composed were of the Hipps family, who lived just south of Asheville; Niles's notebook, JJN-BC; here he points out that "these pictures represent the last work in Doris Ulmann's life." He made photograph notations that correspond to prints in DU-BCP. Niles's *Ballad Book* mentions dates of various visits, revealing that their 1934 summer travel schedule was extremely rigorous.

82. Deed of trust, will and bequest of Doris Ulmann, ser. 2, DU-BC; Clift notes that "her intent was to show the wealth of individual character belonging to her subjects and how they had come to possess it" (in Ulmann, *Darkness and the Light,* 9).

83. Susman, "The Culture of the Thirties," in Susman, *Culture as History,* 150–83; Deed of trust, will and bequest of Doris Ulmann, DU-BC. Correspondence from August 1936 reveals that President Hutchins was quite helpful in locating people and having prints made for the subjects and their relatives, DU-BC; Ulmann quoted in Warren, "Photographer-in-Waiting," 142.

84. Eaton interview, DU-BC; Jones quoted in Warren, *A Right Good People,* 12, 9; Hawthorne, *Photographs by Paul Buchanan.* Hawthorne states that Ulmann made her Appalachian subjects "icons," romanticized subjects from an earlier era.

85. Twenty-five years after her death, the body of Ulmann's work was transferred from Columbia University to the University of Oregon for preservation. Allen Eaton, who directed the move, sought to eliminate the tremendous cost of transporting thousands of glass plates across the country. He assumed the responsibility of choosing representative photographs, perused each portrait series and broke the "unnecessary" plates, leaving only one or two poses of each individual. The complex studies Ulmann composed of Berea College's bakery operations, of the Mount Lebanon Shaker settlement, of Virginia Howard and John Jacob Niles, and thousands of other subjects, were suddenly altered by the destruction of original plates. The surviving plates of various individuals limit their reproduction to a one-dimensional sight on Ulmann's part, far removed from the actual artistic philosophy she developed and exercised; Hutchins to Niles, 1 Sept. 1934, DU-BC; Ulmann to Eaton, 8 June 1934, JJN; Ulmann, "Stuff of American Drama," 132.

2. Dorothea Lange's Depiction
of American Individualism

1. Dorothea Lange, interviews conducted for the soundtrack of the National Education Television film on Dorothea Lange, undated, unedited transcripts, transcribed by Meg Partridge, OM-ART; Dorothea Lange, "Dorothea Lange: The Making of a Documentary Photographer," interviews by Suzanne Riess, Oct. 1960–Aug. 1961, transcript, p. 1, UC-OHC.

2. NET soundtrack interviews, 136; Lange quotations from Lange-Riess interviews, 11, 5–6.

3. NET soundtrack interviews, 137; For examples of different avenues in which women exerted their independence, see Cook, "Female Support Networks"; Cott, *The Grounding of Modern Feminism;* Jones, *Heretics and Hellraisers,* esp. chap. 1, entitled "Women Are People," 1–27; and Sklar, *Florence Kelley.*

4. Meltzer, *Dorothea Lange,* 8; Stein, "Peculiar Grace," 58–59; Lange quotations from Lange-Riess interviews, 13.

5. Paul Schuster Taylor, interview by Suzanne Riess, 1970, transcript, UC-OHC; see also Van Dyke, "Lange—A Critical Analysis"; Banta, *Imaging American Women.* Among the influential promoters of ideal womanhood was artist Howard Chandler Christy, whose *Liberty Belles* (1912) featured active, nubile American beauties. In addition, the female nude as a ripe and fertile being was becoming more popular in photography, as can be seen in the work of Clarence White, Alfred Stieglitz, and their students. Lange-Riess interviews, 16.

6. Ohrn, *Lange and the Documentary Tradition,* 3; Lange-Riess interviews, 26; Lange, *Dorothea Lange,* 105; on her work with Genthe, see Lange-Riess interviews, 27–31.

7. Herz, "Lange in Perspective," 10; Lange-Riess interviews, 60–61; Stein, "Peculiar Grace," 63.

8. Ohrn, *Lange and the Documentary Tradition,* 4–6.

9 Lange-Riess interviews, 39; quoted in Herz, "Lange in Perspective," 10.

10. Lange-Riess interviews, 42; quoted in Herz, "Lange in Perspective," 11; Dorothea Lange, interview by Richard K. Doud, 22 May 1964, transcript, AAA; Conway, "Convention versus Self-Revelation"; Heilbrun, *Writing a Woman's Life,* 24–25.

11. Minick's remarks in Heyman, *Celebrating a Collection,* 9; Lange-Riess interviews, 89.

12. Roger Sturtevant, interview by Therese Heyman, Feb. 1977, transcript, DL, 7; Lange-Riess interviews, 89–90; Sturtevant interview, DL, 5, in which he claimed that the group "didn't have any ism's," unlike the avant-garde photographers living on the East coast. He said that the tension between Lange and Ansel Adams was largely due to the fact that Adams clung to a rigid political agenda. Lange believed this distracted one from creating honest pictures.

13. Lange-Riess interviews, 118–19; list entitled "Family Photographs, 1920s & 1930s," in contact sheets, vol. 1, DL. The typewritten list includes eight pages of family names who were Lange subjects during her studio years. Among those she served for a number of years were the Clayburgh family, the Shainwald family, and the Katten family.

14. Conrat quoted in Heyman, *Celebrating a Collection,* 53.

15. Lange kept well-organized records of her portrait jobs. One cost that she frequently figured in was "transportation to the site," which suggests a good deal of work outside the studio. See contact sheets, vols. 1–18 (1920–34), vol. 11: "Studio," DL; Dixon, "Dorothea Lange," 75.

16. Willard Van Dyke, interview by Therese Heyman, May 1977, transcript, DL; Van Dyke, remembering Lange's seriousness, mentioned that she constantly centered her conversation around grand themes and visions, so much that it often became "irritating" in social situations; Lange-Reiss interviews, 97.

17. Peeler, *Hope among Us Yet,* 61; Cowan, "Two Washes in the Morning," where Cowan argues that the "problem that has no name," identified by Betty Friedan in the 1950s, had its roots in the 1920s; NET soundtrack interviews, 252–53; Lange-Riess interviews, passim.

18. Weber series, vol. 3: "Studio," DL; Lange-Doud interview, 4.

19. Lange-Riess interviews, 92; Dixon, "Dorothea Lange," 71–72, in which Lange noted that she later felt that she was mistaken to have attempted a timeless quality and that photographs need to be "dated."

20. NET soundtrack interviews, 87.

21. Lange-Riess interviews, 59; Negative catalogues, vol. 2, DL; Lange quoted in Goldsmith, "Harvest of Truth," 30; Lange and Dixon, "Photographing the Familiar," 71.

22. Negative catalogues, vol. 2, DL; Van Dyke, "Lange—A Critical Analysis," 464.

23. Dixon, "Dorothea Lange," 73; on Lange's advertising, see Ohrn, *Lange and the Documentary Tradition,* 22; also Heyman, *Celebrating a Collection,* 16; Lorentz published his observation in the 1941 *U.S. Camera Annual,* here quoted in Bennett, "Dorothea Lange," 56.

24. Quoted in Herz, "Lange in Perspective," 9; Sturtevant interview, DL, 11, 24.

25. Stott, *Documentary Expression,* 28; Tucker, "Photographic Facts and Thirties America," 41–42.

26. NET soundtrack interviews, 255; Stott, *Documentary Expression,* 66; also cf. Curtis, *Mind's Eye, Mind's Truth,* preface.

27. Lange-Riess interviews, 145–47.

28. Lange-Doud interview, 5; see also Christopher Cox, introductory essay, in *Dorothea Lange,* 8.

29. For a different reading of this photograph, see Peeler, *Hope among Us Yet,* 63.

30. Herz, "Lange in Perspective," 9.

31. Sturtevant interview, DL, 11, 24; Herz, "Lange in Perspective,"

10; Lange-Riess interviews, 152.

32. Tucker, *The Woman's Eye*, 5.

33. Clark Kerr, "Paul and Dorothea," in Partridge, *A Visual Life*, 42; Peeler, *Hope among Us Yet*, 7; Stott, *Documentary Expression*, 57.

34. Lange-Riess interviews, 152; NET soundtrack interviews, 42; Van Dyke interview, DL, 14; Heyman, *Celebrating a Collection*, 52; see also Heyman interview, 12 July 1989.

35. Van Dyke, "Lange—A Critical Analysis," 467; Clark Kerr, quoted in Taylor, *On the Ground in the Thirties*, viii; see also Kerr, "Paul and Dorothea," in Partridge, *A Visual Life*, 36–43. Kerr pointed out that although Taylor was listed as an "economist," he had also studied sociology at the University of Wisconsin and had additional interests in law and public policy.

36. Van Dyke interview, DL, 11; Lange quoted in Heyman, *Celebrating a Collection*, 16.

37. NET soundtrack interviews, 75.

38. A thoughtful discussion of the words as supplementary material to photographs is Levine, "The Historian and the Icon"; Lange-Riess interviews, 204–5; Puckett, *Five Photo-Textual Documentaries*, 91; Lange-Riess interviews, 155; Van Dyke interview, DL, 3; James Curtis discusses Lange's need for public recognition as a documentarian in *Mind's Eye, Mind's Truth*, 48.

39. The nature and variety of protests from the left are discussed in Brinkley, *Voices of Protest*; Karl, *The Uneasy State*, 104–7, 146–51; and Williams, *Huey Long*. A measure of the tone of ordinary Americans may be seen in McElvaine, ed., *Down and Out in the Great Depression*.

40. Baldwin, *Poverty and Politics*, 117. For the specific purposes of the RA's Information Division, see Curtis, *Mind's Eye, Mind's Truth*, 5–20; Hurley, *Portrait of a Decade*, 36–37; cf. Stange, "Symbols of Ideal Life," 105–7.

41. The best description of Stryker's development as an educator may be found in Hurley, *Portrait of a Decade*. On the impact of Stryker's training at Columbia University in the 1920s, see Stange, "The Management of Vision," 6; see also Stange, "Symbols of Ideal Life," 90–93.

42. Anderson, *Roy Stryker*, 4; Curtis, *Mind's Eye, Mind's Truth*, preface, 5–20; see also Kozol, "Madonnas of the Fields," 1.

43. Stryker, "The FSA Collection of Photographs," in Stryker and Wood, *In This Proud Land*, 7–8.

44. Stange, "The Management of Vision," 6; Lange-Riess interviews, 181; Stott, *Documentary Expression*, 57; in Doherty, "USA-FSA," 10; Heyman, *Celebrating a Collection*, 80; see also Heyman, interview by author; see Sekula, "Invention of Photographic Meaning," 45.

45. David Lange, interview by Therese Heyman, 15 Feb. 1978, transcript, DL.

46. Curtis, *Mind's Eye, Mind's Truth*, viii–ix.

47. Puckett, *Five Photo-Textual Documentaries*, 105; Elliott, "Pho-

tographs and Photographers," 97; Curtis, *Mind's Eye, Mind's Truth*, 52, 65; on "Migrant Mother," see also Davidov, *Women's Camera Work*, 3-6, 235-37.

48. NET soundtrack interviews, 40.

49. Ohrn, *Lange and the Documentary Tradition*, 75; Taylor-Riess interview, 132.

50. In *Five Photo-Textual Documentaries*, Puckett describes MacLeish's work as "a flabby poem . . . its language unequal to the photographs it accompanies" (48); Wright and Rosskam, *Twelve Million Black Voices*; Ohrn, *Lange and the Documentary Tradition*, 75; Mann, "Dorothea Lange," 101.

51. Stange, *"Symbols of Ideal Life,"* 108–9; Stryker-Lange correspondence, RESP; the letters written in 1937 are especially heated.

52. Fleischhauer and Brannan, *Documenting America*, 10; Even Stryker's complete frustration with Walker Evans for his lack of contact and his scant production while on the road did not match the continual bickering that Stryker carried on with Lange. For comparison, see Stryker-Evans correspondence, RESP.

53. Lange and Taylor, *An American Exodus*, 107; Jordy, "Four Approaches to Regionalism," 37.

54. Tocqueville, *Democracy in America*; Foner, *Free Soil, Free Labor, Free Men*; Oakes, *The Ruling Race*, 76–77; see also Cashin, *A Family Venture*; Higham, *History*, 174–79.

55. Negative catalogues, vol. 2, DL; contact Sheets, "Southwest 1920s & 1930s," DL; Lange and Taylor, *An American Exodus*, 60.

56. Lee and Post, the FSA's most productive photographers from 1938 to 1941, spent weeks at a time studying single communities and their activities. The most famous of these are Lee's "Pie Town, New Mexico" series and Post's town studies, e.g. lots 1641, 1459, 1599, FSAPP; Lange-Riess interviews, 215.

57. For a completely gendered reading of Lange's replacement by Post, see Fisher, *Let Us Now Praise Famous Women*. Fisher summarizes, "Where Lange had allegedly been the Mother, documentary attempted to cast Post Wolcott as its 'girl'" (144); Stryker to Jonathan Garst, 30 Nov. 1939, RESP.

3. Marion Post's Portrayal of Collective Strength

1. Wolcott, keynote address, presented at "Women in Photography" conference. I would like to thank Amy Doherty, conference director and George Arents Research Library archivist, for her personal copy of Wolcott's speech notes. A note on names: since Marion Post did not marry Lee Wolcott until near the end of her tenure as a documentary photographer in the 1930s, I refer to her in the text as Marion Post until the point where she legally takes on the name Marion Post Wolcott.

2. Known as the Historical Section of the Resettlement Administra-

tion when it was formed in 1935, the Photography Unit came under the newly named Farm Security Administration in 1937 and made its final move to the Office of War Information in 1942.

3. "Special Memorandum on Photography, John Fischer to All FSA Regional Information Advisers," 4 May 1938, Office Files—Field Correspondence, FSAWR, in which Fischer notes the federal agencies enlisting support of FSA photographers"; Stryker quoted in Anderson, *Roy Stryker*, 4; Trachtenberg, "From Image to Story," 58.

4. Wolcott, "Women in Photography" speech.

5. Wolcott, "Women in Photography" speech; see also Brannan, "American Women Documentary Photographers"; Hurley, *Wolcott: A Photographic Journey*, 3–7. The most thorough treatment of Nan Post's eccentricities and the Post divorce is in Hendrickson, *Looking for the Light*, 13–30.

6. Kennedy, *Birth Control in America*, 19–23; Boddy, "Photographing Women," 155; Wolcott, "Women in Photography" speech.

7. Hurley, *Wolcott: A Photographic Journey*, 7; Marion Post Wolcott File, RESC; the quotation is from Mazo, *Prime Movers*, 61; Brannan, telephone conversation with author, 19 March 1990, in which Brannan emphasized Post's leftist leanings in connection with her study at the New School and her interest in John Dewey's ideas; Boddy, "Photographing Women," 156. Post had decided just months before to pursue an education degree at NYU. For a discussion of the influence St. Denis and Humphrey had on American dance, see Shelton, *Divine Dancer*.

8. Mazo, *Prime Movers*, 118. Mazo entitled his chapter on Doris Humphrey "The Eloquence of Balance," 117–52.

9. Boddy, "Photographing Women," 156–57; Brannan, "American Women Documentary Photographers"; Hendrickson, *Looking for the Light*, 31.

10. Post Wolcott File, RESC; Wolcott, "Women in Photography" speech.

11. Hardt and Ohrn, "The Eyes of the Proletariat," 54, 53. Cf. Osman and Phillips, "European Visions," 74–103.

12. Marion Post Wolcott, interview conducted by Richard Doud, 18 Jan. 1965, Mill Valley, Calif., AAA; Boddy, "Photographing Women," 158; Wolcott, "Women in Photography" speech, where Wolcott recalled that Fleischmann had told her she had "an exceptionally good eye."

13. Lois Scharf, "Even Spinsters Need Not Apply," in Scharf, *To Work and To Wed*, 85; Post Wolcott–Doud interview.

14. Post Wolcott File, RESC; Pauly, *An American Odyssey*, 5–34; Peeler, *Hope among Us Yet*, 5, where he notes that *Waiting for Lefty* was tagged at the time "the birth cry of the thirties."

15. Boddy, "Photographing Women," 160; Post Wolcott–Doud interview; Steiner, *A Point of View*, 8, 31.

16. Steiner quoted in O'Neal, *A Vision Shared*, 174; Murray, "Marion Post Wolcott," 86; Glen, *Highlander*; and Horton, *The Highlander Folk School*. Film text quoted in Pauly, *An American Odyssey*, 43; Post's

comments about Frontier's influence on her in Wolcott, "Women in Photography" speech.

17. "Consumer Cooperatives."

18. O'Neal, *A Vision Shared,* 174; Dixon, *Photographers of the Farm Security Administration,* 161; Post Wolcott File, RESC; Story told in Wolcott, "Women in Photography" speech; Post Wolcott–Doud interview.

19. Post Wolcott–Doud interview; Post Wolcott File, RESC; Strand to Stryker, 20 June 1938, RESP; Post to Stryker, 26 June 1938, July 1938, RESP.

20. Correspondence between Stryker and Lange, Mydans, and Evans, respectively, reveals detailed information not only on Resettlement Administration objectives and work, but also on outside employment each of the three sought at one time or another (RESP). Mydans left the Historical Section in the summer of 1936; Lange tended to venture in and out, doing special assignments and projects, then returning to Stryker's team to complete her FSA shooting scripts. In a 28 Feb. 1938 letter to Stryker, Evans described the painful process of getting "the book" published, dealing with governmental bureaucracy and publishers' rejections. Evans's assignment with Agee, originally produced for *Fortune* magazine, which decided not to publish it, later appeared under the title *Let Us Now Praise Famous Men* (Houghton Mifflin, 1941); Post Wolcott File, RESC.

21. Stryker to Post, 14 July 1938, RESP.

22. Post Wolcott–Doud interview; Amarillo Art Center, *American Images.* The program, which later aired on PBS, was taped at an FSA reunion-symposium that brought together after fifty years Wolcott, Rothstein, Lee, Jack Delano, John Collier, Jr., and Ed Rosskam.

23. Post quoted in Murray, "Marion Post Wolcott," 86; Stryker's reading assignments, RESP and FSAWR; O'Neal, *A Vision Shared,* 175.

24. Post to Clara Dean Wakeham, Sept. 1940 [?], RESP. The letter is apparently misdated (Post had failed to date it herself) because Post's tone marks her as a relatively new employee, still a bit insecure; she speaks of Roy Stryker very formally: "The last thing Mr. Stryker said to me was *that I should take my time.*" By 1940 Post addressed Stryker as "Roy" or "Papa" or any number of other humorous titles. A second clue is that in September 1940 Post was immersed in a project in Eastern Kentucky, then had to move hurriedly to Chapel Hill, North Carolina, to confer with Professor Howard Odum on his decision to use FSA pictures in an upcoming publication. A final point to be made is that Post was assigned to the coal mining areas of West Virginia only once, in fall 1938. I believe the letter in question, from which the quote was taken, was written the last week in September 1938. Marion Post Wolcott to Paul Hendrickson, quoted in Hendrickson, *Looking for the Light,* 261; lot 1730, FSAPP; Lifson, "Not a Vintage Show," 70.

25. Stott, *Documentary Expression,* 58; Post Wolcott–Doud interview.

26. Post admitted trying to reproduce the style of earlier FSA photographers on her first FSA assignment, but she abandoned it midway through the trip (Post Wolcott–Doud interview); see also Snyder, "Marion Post Wolcott," 301.

27. Snyder, "Post and the Farm Security Administration," 459; lots 1723, 1724, 1727, FSAPP, where Post wrote in a caption, "These 'foreigners' are generally thrifty and their houses are cleaner than most"; Dixon, *Photographers of the Farm Security Administration,* 162.

28. Baldwin, *Poverty and Politics,* 279; Leuchtenburg, *Franklin D. Roosevelt,* 141; Baldwin, *Poverty and Politics,* 158.

29. Post Wolcott–Doud interview; see also Snyder, "Post and the Farm Security Administration," 458–59; "Special Memorandum on Photography, John Fischer to All FSA Regional Information Advisers," 4 May 1938, Office Files—Field Correspondence, FSAWR. For further explanation of services offered by FSA photographers to other U.S. agencies and departments, see White, "Faces and Places," 415–29. See also Howe, "You Have Seen Their Pictures."

30. Stryker to Post, 14 July 1938, RESP; Wolcott, "Women in Photography" speech.

31. Wolcott, "Women in Photography" speech; On the suspected kidnapping, see Post to Stryker, 23 Jan. 1939, RESP; Post Wolcott–Doud interview; and Brownell, "Girl Photographer for FSA."

32. Stryker to Post, 13 Jan. 1939, RESP.

33. Post to Stryker, 13 Jan. 1939, RESP; Hurley, *Wolcott: A Photographic Journey,* 40; Robert Snyder gives important attention to gender issues in "Marion Post Wolcott" and in "Post and the Farm Security Administration."

34. Post to Stryker, 23 Dec. 1938, 12 Jan. 1939, RESP.

35. Post quotations in Post to Stryker, 12 January 1939, RESP; Post to Stryker, 5 July 1939, RESP; see also Snyder, "Post and the Farm Security Administration," 476–77.

36. Post to Stryker, 23 Dec. 1938, RESP.

37. Stryker to Arthur Rothstein, [1936], RESP, where he informs Rothstein of the conditions he could expect to face as he dealt with the migrant laborers; See esp. lot 1586 (Belle Glade) and lot 1590 (Canal Point), FSAPP, where Post seemingly goes out of her way to show the filth in which migrants live and work; lot 1591, FSAPP.

38. Post to Stryker, 13 Jan. 1939, RESP.

39. Post to Stryker, Jan. 1939, RESP; Stryker to Post, 1 Feb. 1939, 13 Feb. 1939, RESP.

40. Post to Stryker, 13 Jan. 1939, 3 Feb. 1939, Jan. 1939, RESP.

41. Stryker to Post, 28 Jan. 1939, RESP; Post to Stryker, Jan. 1939, RESP.

42. Wolcott, excerpt taken from the biographical essay she submitted to *Contemporary Photographers,* ed. Walsh, et al., 606; Raedeke, "Introduction and Interview," 14; Cohen in *Contemporary Photographers,*

ed. Walsh et al., 607; Carl Fleischhauer and Beverly Brannan chose pictures from Post's Miami series for *Documenting America,* 174–87.

43. Stein, introduction to Wolcott, *Wolcott: FSA Photographs;* Lifson, "Not a Vintage Show," 70.

44. Fleischhauer and Brannan, "Marion Post Wolcott: Beach Resort," in *Documenting America,* 174–87, shows clearly the safe distances at which the photographer kept herself; captions for lot 1602, FSAPP; Post Wolcott–Doud interview.

45. Post to Stryker, 24 Feb. 1939, RESP; Stryker to Post, 13 Feb. 1939, RESP, where he tells her, "Either the Rolleiflex is out, or you are at fault; something is sure as the devil wrong . . . I am most certain the fault isn't in the laboratory"; Stryker to Post, 21 Feb. 1939, 16 Feb. 1939, RESP.

46. Stryker to Post, 13 Feb. 1939, 1 Feb. 1939, RESP, where he suggests Post "cut pretty ruthlessly"; Stryker to Post, 21 Feb. 1939, RESP; the quantity of Post's early work may be determined by examining Stryker's correspondence: every seven to ten days he sent back to her approximately one hundred rolls of negatives; Post Wolcott File, RESC; Post Wolcott–Doud interview.

47. Post Wolcott–Doud interview; Stryker to Post, 1 April 1939, 11 May 1939, RESP, where he sympathized, "I know that regional people have been making life miserable for you. . . . As far as chasing around for all these regional people—forget it!" Fleischhauer and Brannan, *Documenting America,* 10; Post Wolcott–Doud interview.

48. Stryker to Post, 16 March 1939, 8 March 1939, 4 Feb. 1939, RESP; Post Wolcott quoted in Hendrickson, *Looking for the Light,* 58.

49. Stryker to Post, 6 April 1939, RESP.

50. Lot 1621, FSAWR. In a series of another Coffee County family, Post focused in closely on the family's table, a dinner consisting of roast beef, turnip greens, potato salad, stuffed eggs, lima beans, rice, pickled pears, biscuits, cornbread, milk, peaches, and cake.

51. An insightful essay on Raper's work and personal background is found in Singal, *The War Within,* 328–38; Stryker to Post, 1 April 1939, RESP; Post to Stryker, 7 April 1939, RESP; Stryker to Post, 27 April 1939, RESP.

52. Post to Stryker, 8 May 1939, 21 May 1939, RESP; Stryker to Post, 25 May 1939, RESP.

53. Stryker to Post, 20 May 1939, RESP; Post to Stryker, 1 June 1939, RESP; Raper and Reid, *Sharecroppers All.*

54. Post to Stryker, 5 July 1939, RESP.

55. McCausland, "Documentary Photography," 2. Cf. Stott, *Documentary Expression,* on the "documentary impulse" operating in American art and literature; and Peeler, *Hope among Us Yet,* where he describes the "soft" ideological bent of thirties documentary photographers. Allen, *Since Yesterday,* 212–13; Post Wolcott File, RESC; Ansel Adams quoted in Stryker, "The FSA Collection of Photographs," 352; Peeler, *Hope among Us Yet,* 100.

56. For an excellent description of Odum's background and his relationship with the institute's sociologists and their work, see Singal, *The War Within*, 115–52, 302–38, passim; Hagood built on her work in succeeding months and then published it under the title *Mothers of the South: Portraiture of the White Tenant Farm Woman;* Post to Stryker, 2 Oct. 1939, RESP; Stryker to Post, 17 Oct. 1939, RESP.

57. Beard quoted in Swados, *The American Writer*, 48; Terrill and Hirsch, *Such as Us*, xxi; In his earlier correspondence with Stryker, Couch mentioned the idea of combining FSA photography with Federal Writers' Project Guidebooks (Stryker to Dorothea Lange, 22 Dec. 1938, RESP; Stryker to George Mitchell, regional director in Raleigh, 3 Dec. 1938, Office Files, Field Correspondence, FSAWR, in which he urges Mitchell to have Post complete some work similar to that which has made possible book-length publications such as *Forty Acres and Steel Mules*). For a description of Couch's impact on the UNC Press, see Singal, *The War Within*, 265–301, passim.

58. Post to Stryker, 2 Oct. 1939, RESP; Harriet L. Herring, "Notes and Suggestions for Photographic Study of the 13 County Sub-regional Area," Supplementary Reference Files on lot 1503, FSAWR; Lange and Post actually worked on this project at the same time, but it was the last assignment Lange would take as an FSA photographer; Stryker to Jonathan Garst, 30 Nov. 1939, RESP.

59. Lots 1502, 1503, FSAPP; Hagood's typed notes/general captions for the photographs, Supplementary Reference Files, FSAWR.

60. Hagood, *Mothers of the South*, rev. ed., 132ff, 175, 177; photographs, lot 1518, FSAPP.

61. This was especially the case in rural areas and had been fostered by U.S. Government policies since the Progressive Era; see, for example, Hilton, "'Both in the Field,'" 114–33.

62. Post's other North Carolina work in 1939 expresses her personal statement about segregation in general, including an often-analyzed image of a sidewalk confrontation between two black men and a white woman carrying a baby. Different readings of its race and gender implications may be found in Fisher, *Let Us Now Praise Famous Women*, 153; Stein, in Wolcott, *Wolcott: FSA Photographs*, 6–8; and Natanson, *The Black Image*, 8–9; In her scenes of the segregated Granville County courtroom, Post downplays gender questions in favor of her more pressing messages about race in the South. See lots 1506, 1507, FSAPP and FSAWR.

63. Post's recollections about the scene told in Hendrickson, *Looking for the Light*, 91.

64. Post to Stryker, 3 Feb. 1939, RESP; Although the specific reference here was to migrant laborers, Post made efforts on nearly every assignment to capture leisure activities, which became a defining feature of her FSA work; lots 1641, 1479, FSAPP.

65. Lawrence to Stryker, 2 Nov. 1939, FSAWR; Ansel Adams to W.W. Alexander (FSA administrator), 23 April 1940, FSAWR; Stryker to E.W.

Cobb, 20 April 1939, FSAWR; Stryker to Margaret Hagood, 23 Jan. 1940, FSAWR. After amassing a formidable collection of visual data for the UNC Institute for Research in the Social Sciences, Post spent time in December 1939 editing her work for a UNC photo exhibition (Stryker to Katharine Jocher [assistant director at UNC IRSS], 19 Dec. 1939, FSAWR).

66. Curtis, *Mind's Eye, Mind's Truth*, 103. Cf. Hurley, *Portrait of a Decade*, 96–102, and Stange, *"Symbols of Ideal Life,"* 89–131.

67. Post to Stryker, 2 March 1940, RESP.

68. Stryker to Post, 27 Feb. 1940, RESP; Anderson to Stryker, 28 March 1940, RESP.

69. Post Wolcott–Doud interview; Post to Stryker, 14 March 1940, RESP.

70. "Annual Town Meeting Warning," Supplementary Reference Files, FSAWR; lot 1239, FSAPP.

71. "Agenda and Report" (sent by Post to the Washington office), FSAWR.

72. Anderson, "Elizabethton, Tennessee," in Anderson, *Puzzled America*, 145–53, reprinted in Swados, *The American Writer*, 40–47; [Steiner], "The Small Town," 46; Stott, *Documentary Expression*, 232–33; Perhaps the most personal account of the influence of Post's New England imagery is found in the prologue of Paul Hendrickson's *Looking for the Light*. He saw one of Post's Vermont scenes in a Library of Congress gift shop, was mesmerized by it, and then began pursuing her other work and her life with a missionary-like fervor.

73. Lots 1705, 1706, 1707, 1708, 1709, 1710, 1713, FSAPP.

74. Lot 1706, FSAPP; Terrebonne Project Report, 10 May 1940, FSAWR, emphasis mine.

75. Post Wolcott–Doud interview; "Things As They Were: FSA Photographers in Kentucky, 1935–1943," a guidebook to the photograph exhibit, 6 Sept. to 9 Nov. 1985, UL. Fifty-five of the seventy-seven photographs in the exhibit were taken by Marion Post; see also Brannan and Horvath, *A Kentucky Album;* Post to Stryker, 29 July 1940, RESP.

76. Post to Stryker, 9 Sept. 1940, RESP.

77. Brownell, "Girl Photographer for FSA."

78. Ibid.

79. Stryker to Post, 21 Sept. 1940, RESP; Wolcott, comments in Amarillo Art Center, *American Images;* see also O'Neal, *A Vision Shared*, 175.

80. Post, General Caption no. 1, Aug. 1940, Supplementary Reference Files, lot 1465, FSAWR.

81. For a fuller comparison of the two photographers, see McEuen, "Doris Ulmann and Marion Post Wolcott"; Stein, in Wolcott, *Wolcott: FSA Photographs*, 6.

82. Post, lots 1509, 1510, 1511, 1512, 1513, 1514, 1515, FSAPP.

83. Supplementary Reference Files, FSAWR. For an excellent discus-

sion of the intricate relationships between gender, race, and class, see Lerner, *Why History Matters*, 146-98.

84. Post to Stryker, 2 Oct. 1940, Nov. 1940, RESP; On Post's "distressing" experiences with southern welfare workers, see Snyder, "Marion Post Wolcott," 305–7; Post to Stryker, 15 May 1940, 29 July 1940, RESP.

85. Stein, in Wolcott, *Wolcott: FSA Photographs*, 9; see also Snyder, "Post and the Farm Security Administration," 477; Hurley contends that Stryker considered Post his best "troubleshooter" (*Wolcott: A Photographic Journey*, 42–43); Post to Stryker, 2 Oct. 1940, RESP; interview quotation in Hurley, *Wolcott: A Photographic Journey*, 57–58.

86. Post, untitled report, Jan. 1941, Supplementary Reference Files, FSAWR.

87. Post to Stryker, 14 March 1941, 8 April 1941, RESP; Wolcott quoted in Hendrickson, "Double Exposure," 18; Post to Stryker, 2 Oct. 1940, RESP.

88. Ganzel, *Dust Bowl Descent*, 8; Post Wolcott to Stryker, 22 Aug. 1941, RESP; lot 101, FSAPP.

89. Post Wolcott to Stryker, 22 Aug. 1941, RESP.

90. Post Wolcott to Stryker, 20 Feb. 1942, RESP; The "married name" struggle is recounted in Hendrickson, *Looking for the Light*, 202.

91. Malcolm Cowley, "The 1930s: Faith and Works," in Cowley, *I Worked at the Writer's Trade*, 101.

4. Margaret Bourke-White's Isolation of Primary Components

1. Story told in Silverman, *For the World to See*, 14; Orvell, *The Real Thing*, 223; Marchand, *Advertising the American Dream*, 1; Marchand uses the phrase *advertising man* inclusively despite its gender specificity, noting that it does accurately reflect the presumptions of a male-dominated profession that served a masculine realm, that of scientific and technological growth; Margaret Bourke-White to Chris A. Addison, 3 March 1930, MBW.

2. Bourke-White to Addison, 30 March 1930, MBW; Miles Orvell singles out *excitement* and *power* as the two overriding characteristics of the machine played upon by industrial designers in the 1920s and 1930s (*The Real Thing*, 182); see also Meikle, *Twentieth Century Limited*, 19–67; telegram, Bourke-White to Roberts Everett Associates, 17 Oct. 1930, MBW; Marchand, *Advertising the American Dream*, 1–13.

3. Goldberg, *Bourke-White Biography*, 13–14.

4. Evans, *Born for Liberty*, 173–76.

5. Goldberg, *Bourke-White Biography*, 7–8.

6. Minnie White to Margaret Bourke-White, 14 June 1915, MBW; Lindsey Best to Joseph White, 22 Dec. 1917, MBW; Vitae of Margaret Bourke-White, Simon and Schuster Review Department, 1931, MBW.

7. Character Analysis of Margaret White, conducted by Jessie Allen Fowler, 27 May 1919, MBW.

8. Description of Clarence White in Speech Notes, Bourke-White on "Careers for Women," May 1933, MBW; Goldberg, *Bourke-White Biography*, 22–32, 35.

9. Bourke-White to Minnie White, 22 June 1924, MBW.

10. Bourke-White to Minnie White, 24 June 1924, MBW; Goldberg, *Bourke-White Biography*, 51–56, in which she discusses Margaret's ambivalence about having a family. Though Margaret wanted a baby and hoped it would strengthen her marriage, she recognized the danger of bringing a child into such an unstable household. There is some evidence to suggest that she once became pregnant but carefully orchestrated a miscarriage.

11. Minnie White to F.A. Gilfillan, 28 Aug. 1928, MBW; Mrs. White mentions Margaret's difficulty in completing a degree because of the various moves the Chapmans had made. On the image of the single "working girl" in the 1920s, see Evans, *Born for Liberty*, 182–84; Aunt Gussie to Bourke-White, 17 May 1927, MBW.

12. Minnie White to F.A. Gilfillan, 12 Aug. 1928, MBW; Bourke-White, *Portrait of Myself*, 28–29; Margaret Chapman, Divorce Papers, 3 Jan. 1928, MBW. In the divorce agreement, Margaret's family name was restored.

13. Bourke-White, *Portrait of Myself*, 36. Margaret hyphenated her last name sometime in the late 1920s as she built her new image. She makes reference to it in a letter, noting, "[E]verybody calls me Miss Bourke-White as tho' I were a personage" (Bourke-White to Minnie White, 4 May 1929, MBW). I introduce it at this point in the text for the purpose of consistency and also because her move to Cleveland marked a new stage in her life, the beginning of her career as a professional photographer. On Bourke-White's relationship with the Van Sweringens, see Goldberg, *Bourke-White Biography*, 77.

14. For a discussion of American influence on modern architecture, and skyscraper design in particular, see Meikle, *Twentieth Century Limited*, 29–38; see also Orvell, *The Real Thing*, 174; Guimond, *American Photography*, 86–87; Bourke-White quoted in Silverman, *For the World to See*, 8.

15. Marchand, *Advertising the American Dream*, 149, where he points out that the magazine *Advertising and Selling* "reflected the changing sensibilities. Photographs appeared on fewer than 20 percent of the journal's covers in 1926, but on over 80 percent in 1928"; Roy Stryker–Margaret Bourke-White correspondence, 26 Oct. to 31 Oct. 1928, MBW; Stryker to Bourke-White, 16 Nov. 1928, MBW; Tugwell, Munro, and Stryker, *American Economic Life,* ix; Maren Stange explains the intricacies of the textbook's philosophy, including John Dewey's impact, in "The Management of Vision," 6, 8; Stange also argues that the production process overrides the individuals involved, a vital point in *American*

Economic Life, since it suggested that numerous ethnic industrial workers lost their cultural distinctiveness in the "melting pot" of U.S. factory
work. Such work thoroughly "Americanized" people and thus created a
more desirable working class.

   16. Tearsheet, *Theatre Guild Magazine* (March 1929), MBW; Bourke-
White received twenty-five dollars for the photograph; Eugene O'Neill,
Dynamo, in O'Neill, *The Plays of Eugene O'Neill,* 429; "Mr. O'Neill"
New York Times, 3 March 1929, 4; Bourke-White, Speech Transcript,
"Creative Staff Meeting," 1 Feb. 1933, J. Walter Thompson Advertising
Agency, MBW.

   17. Meikle discusses this transition and transformation of industry in
Twentieth Century Limited, 134–36, noting that the fullest negative
expression of industrial exploitation was in Lewis Mumford's *Technics
and Civilization* (1934); Mumford quoted in Orvell, *The Real Thing,*
202.

   18. Vitae of Margaret Bourke-White, Simon and Schuster Review
Department, 1931, MBW.

   19. Marchand, *Advertising the American Dream,* describes the feminization of the American consumer public; Daniel Pope, comments at
the Ninetieth Annual Meeting of the American Historical Association,
Pacific Coast Branch, Portland, August 1997.

   20. Bourke-White diary entry, Dec. 1927, quoted in Goldberg, *Bourke-
White Biography,* 92.

   21. Typewritten biography, dated fall 1931, MBW; Henry R. Luce to
Bourke-White, 8 May 1929, MBW; undated office memorandum, Calkins
to Hodgins, Time, Inc., MBW.

   22. Bourke-White to Minnie White, 16 May 1929, MBW; Lloyd-Smith
quoted in Silverman, *For the World to See,* 11.

   23. Bourke-White to Minnie White, 24 May 1929, MBW.

   24. Vitae of Margaret Bourke-White, Simon and Schuster Review
Department, 1931, MBW; On the idea of repetition in mass production,
see Meikle, *Twentieth Century Limited,* 24.

   25. Bourke-White membership cards, Pictorial Photographers of
America, Cleveland chapter and New York City chapter, MBW; Steiner
to Bourke-White, 24 April [1930], MBW; Bourke-White to Minnie White,
19 May 1929, MBW, where she mentioned that Steiner did manage to
slip one compliment into his stream of critical comments. He noted that
her "viewpoint was becoming more direct and creative." The story of
the tearful outburst told in Goldberg, *Bourke-White Biography,* 100;
Steiner to Bourke-White, 2 Aug. [1929], MBW, in which he suggests that
she extend the insurance on a particular lens to include European travel.
Steiner goes into a detailed discussion on the DeBrie camera, an f1.5
Meyer Plasmat lens, a six-inch f4.5 Zeiss Tessar mount, and gelatine
filters. He closes, "I'll get you the rest of the equipment with myself as
instructor."

   26. Dwight Macdonald to Bourke-White, 26 July 1929, MBW;

Archibald MacLeish to Bourke-White, undated letter [fall 1929], MBW;
Bourke-White to William H. Albers, 23 Oct. 1928, MBW.

27. Bourke-White, *Portrait of Myself*, 72. Years later, a friend of Bourke-
White's said to her, "You must have been the only photographer in the
whole United States who was *inside* a bank that night." Unaware of the
national news, Bourke-White had instead been reading a football manual
in preparation for an upcoming date with a Harvard fan.

28. Undated office memorandum, Calkins to Hodgins, Time, Inc.,
MBW; Within three weeks after Luce had distributed Bourke-White's
steel pictures to potential magazine advertisers, he "had sold enough ad
pages to fill several *Fortune* issues" (Silverman, *For the World to See*,
11). Steiner to Bourke-White, 24 April [1930], MBW; Joe Fewsmith to
Raymond Rubican, 15 Oct. 1930, MBW.

29. Telegram, Ruth White to Bourke-White, 27 June 1930, MBW;
Bourke-White to Chris A. Addison, 3 March 1930, MBW; transcript,
"Creative Staff Meeting," 1 Feb. 1933, J. Walter Thompson Advertising
Agency, MBW; Bourke-White to Minnie White, 14 July 1930, MBW;
Bourke-White, *Portrait of Myself*, 93. When she was forced to wait in
Berlin for five and a half weeks while the Soviet embassy approved her
visa, Bourke-White realized she would have to overcome numerous
obstacles posed by the Soviet bureaucracy.

30. For specific achievements, see Hiroaki Kuromiya, "The Commander
and the Rank and File: Managing the Soviet Coal-Mining Industry, 1928–
33," and David Shearer, "Factories within Factories: Changes in the
Structure of Work and Management in Soviet Machine-Building Factories, 1926–34," both in Rosenberg and Siegelbaum, *Social Dimensions
of Soviet Industrialization*, 150–53, 202–4.

31. Bourke-White, *Portrait of Myself*, 91–92; Bourke-White to Minnie
White, 10 Sept. 1930, MBW.

32. Vitae of Margaret Bourke-White, Simon and Schuster Review
Department, 1931, MBW; see also Bourke-White, speech outline for J.
Walter Thompson Advertising, 31 Jan. 1933, MBW. One of the points
Bourke-White wished to emphasize in her talk was that "[i]n Russia things
are happening now." Bourke-White, *Portrait of Myself*, 95.

33. Bourke-White to M. Lincoln Schuster, 22 Jan. 1931, MBW; Schuster
to Bourke-White, 26 Jan. 1931, MBW; "Soviet Panorama," *Fortune* 3
(Feb. 1931): 60–68.

34. Schuster to Bourke-White, 11 March 1931, MBW; original contract agreement between Bourke-White and Simon and Schuster, 24 April
1931, MBW. With only minor changes, the final contract agreement was
signed 21 Aug. 1931. Bourke-White would receive 10 percent of the retail
price on the first five thousand copies of her book.

35. Bourke-White explains the extent of American business involvement in the Soviet Union in *Portrait of Myself*, 92. She notes that these
men were there "strictly for business reasons" and that they made tremendous contributions to the expanding Soviet economy. Bourke-White

to Minnie White, 10 Sept. 1930, MBW; newspaper tearsheet, Bourke-White to Walter Winchell, Dec. 1930, MBW; see also Bourke-White, *Eyes on Russia*, 19. Bourke-White to Schuster, 8 Sept. 1931, MBW; Clifton Fadiman to Bourke-White, 19 Aug. 1931, MBW, where he congratulates her on "a splendid job," one requiring "very few changes"; Minnie White to Bourke-White, Dec. 1931, MBW.

36. Bourke-White, *Portrait of Myself*, 80. For a discussion of the machine-age aesthetic as implemented in modern architecture, see Meikle, *Twentieth Century Limited*, 29–38; Bourke-White, speech transcript, "Creative Staff Meeting," 1 Feb. 1933, J. Walter Thompson Advertising Agency, MBW; Joe Fewsmith to Bourke-White, 21 May 1931, MBW.

37. The photographs of Bourke-White on the gargoyle were taken by her darkroom technician, Oscar Graubner. Joe Fewsmith to Bourke-White, 26 Dec. 1930, MBW.

38. Typescript, "Careers for Women," May 1933, MBW, emphasis mine; Brown, *Setting a Course*, 29–47, in which she discusses characteristics of the new woman; Marchand, "Sizing Up the Constituency: The Feminine Masses," in Marchand, *Advertising the American Dream*, 66–69. He argues that the American public was reduced to advertisers' underestimation of female consumers, which led to the "feminization" of the American buying public.

39. Minnie White to Bourke-White, 12 Feb. 1933, MBW; see also general business correspondence, 1932–34, MBW; correspondence, Bourke-White Studio and Eastman Kodak Company, 1933–34, MBW. The family loans caused quite a rift in the White extended family. In business, Bourke-White's secretaries continually asked for extensions and loans. Peggy Sergent, who took over the studio books in 1935, recalled that Bourke-White "was in debt to *every*body" (quoted in Goldberg, *Bourke-White Biography*, 141).

40. On "cleanlining," see Meikle, *Twentieth Century Limited*, 101–9; Orvell, *The Real Thing*, 223.

41. H.S. Bishop to Bourke-White, 27 Aug. 1931, MBW; Bourke-White to Bishop, 24 Aug. 1931, MBW. In his letter, Bishop admitted having a "Babbit [*sic*] point-of-view."

42. Quotations from Bourke-White, speech text, "Color Photography in Advertising," Advertising Club of the *New York Times*, 16 Feb. 1934, MBW; Bourke-White to Minnie White, 30 Oct. 1933, MBW; Bourke-White to O.B. Hanson, 21 Nov. 1933, MBW; Bourke-White to Frank Altschul, 11 Dec. 1933, MBW; Bourke-White to Sanford Griffith, 5 Jan. 1934, MBW; Personal recognition from the NBC project did not come as easily as Bourke-White had hoped. She competed fiercely with the mural executor, Drix Duryea, whom she referred to as "an ordinary hack commercial photographer."

43. Warren Susman, "The Culture of the Thirties," in Susman, *Culture as History*, 156–57; Bourke-White, typescript, "Careers for Women," May 1933, MBW.

44. Bourke-White to Fred L. Black, 26 Feb. 1934, MBW; Bourke-White to C.P. Fiskin, 17 July 1934, MBW; Ethel Fratkin to Felicia White, 19 April 1934, MBW; Ethel Fratkin to George R. Gibbons, 11 July 1934, MBW; Safford K. Colby to Bourke-White, 9 Aug. 1934, MBW; Fratkin, Bourke-White's secretary, had assumed nearly all of the correspondence duties by 1934; she was even writing letters to the White family with Margaret's apologies that she was too busy to correspond.

45. "Selznick Notes," typewritten summary of Bourke-White–Selznick meeting, 11 July 1933, MBW; Elbert A. Wickes to Bourke-White, 29 May 1933, MBW.

46. On Bourke-White's 1934 business, see, for example, Bourke-White to Helen Resor [J. Walter Thompson Co.], 6 Feb. 1933, MBW; Bourke-White to Frank J. Reynolds, 7 July 1933, MBW; Bourke-White to David O. Selznick, 7 July 1933, MBW; Bourke-White to R.C. Treseda [Coca-Cola Co.], 15 July 1933, MBW; Fred C. Quimby [Metro-Goldwyn-Mayer], 25 Oct. 1933, MBW. Bourke-White enclosed portfolios with her letters; in some cases, she included as many as fifty photographs; Bourke-White set her print prices according to a magazine's circulation: $10 (up to 25,000), $15 (up to 50,000), $25 (up to 100,000), $35 (up to 250,000), $50 (up to 500,000), and $75 (over 1,000,000); Bourke-White Studio to J. Quigney [Colliers'], 14 Aug. 1934, MBW; Ethel Fratkin to Minnie White, 16 Feb. 1934, MBW; Fratkin to Felicia White, 19 April 1934, MBW; Bourke-White to Norman Moray, 13 Nov. 1933, MBW; Ralph Steiner to Bourke-White, 19 March 1933, MBW. Steiner asks, "How can cutting the Russian film take so long? . . . Russia changes fast you know—your film may be historical rather than contemporary soon. Hurry HURRY!"

47. Letter signed by Langston Hughes, Ella Winter, Lincoln Steffens, and Noel Sullivan to Bourke-White, 23 Dec. 1933, MBW; Bourke-White to Lincoln Steffens, 26 Jan. 1934, MBW; lists of donated photographs to the New School, April-May 1934, MBW; Ray Michael to Bourke-White, 7 May 1934, MBW; correspondence, Bourke-White and the League of Women Shoppers, 1935–39, MBW; program, "First Annual Motion Picture and Costume Ball by the Film and Photo League," 27 April 1934, MBW; Albert Carroll to Bourke-White, 10 Dec. 1934, MBW.

48. Bourke-White to Ralph Steiner, Paul Strand, Alfred Steiglitz, Anton Bruehl, Edward Steichen, 7 Nov. 1935, MBW; Bourke-White to John Dewey, 19 Sept. 1935, MBW, in which she asks him to make a public speech against fascism and war. Goldberg, *Bourke-White Biography,* 157; Peeler, *Hope among Us Yet,* 65–69.

49. Bourke-White, *Portrait of Myself,* 110; Schuster to Bourke-White, 3 April 1934, MBW; Bourke-White to Arthur R. Morgan, 8 March 1934, MBW; correspondence, Bourke-White Studio (Ethel Fratkin) and W.L. Sturdevant, 6 April to 13 April, 1934, MBW.

50. Bourke-White, *Portrait of Myself,* 110; Peeler, *Hope among Us Yet,* 67.

51. Eleanor Tracy to Bourke-White, 21 Oct. 1935, MBW.

52. Bourke-White, "Photographing This World"; typescript, "Photographing the World" by Margaret Bourke-White, 29 Jan. 1936, MBW.

53. Snyder, "Caldwell and Bourke-White: You Have Seen Their Faces," 396; Stott, *Documentary Expression*, 211–37; Orvell, *The Real Thing*, 198, 227, 239.

54. Bourke-White to Caldwell, 9 March 1936, EC; Bourke-White to Caldwell, undated letter, MBW. She discusses the incident and the letter in *Portrait of Myself*, 119–21.

55. Alan Trachtenberg, foreword to Caldwell and Bourke-White, *You Have Seen Their Faces*, 1995 ed., viii; see Stott's discussion of the "Documentary Book," in *Documentary Expression*, 211–37; Noggle, "With Pen and Camera"; Peeler, *Hope among Us Yet*, 69; Orvell, *The Real Thing*, 277–78. An interesting point about the criticism levied on Caldwell and Bourke-White is that the most frequently used term to describe their portrayal of southern life has been "grotesque[s]."

56. In *Portrait of Myself*, Bourke-White explained how she and Caldwell developed captions for the photographs in *You Have Seen Their Faces*. They did not use actual quotes from people in the pictures; instead, they composed the captions themselves. Bourke-White recalled, "Many times the final caption was a combination of the two—the thought mine and the words Erskine's, or vice versa. . . . I was proud indeed when either my thought or my way of expressing the subject stood up in the final test" (137); Snyder, "Caldwell and Bourke-White," 398; see also Guimond, *American Photography*, in which he notes that Bourke-White "photographed every cliché that was popular in the newspapers, movies, and popular fiction of the era" (117).

57. Paul Green, "A Plain Statement about Southern Literature," quoted in Singal, *The War Within*, 109; Caldwell and Bourke-White, *You Have Seen Their Faces*, 1937 ed., 75; Davidson, "Erskine Caldwell's Picture Book."

58. Agee and Evans, *Let Us Now Praise Famous Men*, appendix 3, 453.

59. Contract agreement between Bourke-White and Time, Inc., 4 Sept. 1936, MBW; circulation figures in Kozol, *LIFE's America*, 35.

60. "10,000 Montana Relief Workers Make Whoopee on Saturday Night," *Life*, 23 Nov. 1936, 9; On stereotypes, see Moynihan, Armitage, and Dichamp, *So Much to Be Done*, xi–xxii, 167–68; on Montana women in particular, see Murphy, "The Private Lives of Public Women," 193–205; and Murphy, *Mining Cultures*; Kozol, *LIFE's America*, 34.

61. For a different reading of the photo essay, see Smith, *Making the Modern*, in which Smith argues that the federal government weakened a potentially dangerous group by employing them "far from the centers of power to make-work projects where they struggle to survive" (347); "10,000 Montana Relief Workers," 11; For a general commentary on the photo essay as form, see Callahan, *Photographs of Bourke-White*, 17–19.

62. Bourke-White, quoted in Silverman, *For the World to See*, 81; Bourke-White to Louis Lozowick, 22 May 1937, MBW, where she explains, "Since I've gone into this work with *Life* I find it almost impossible to schedule discussions ahead of time." Bourke-White's office correspondence reveals that she agreed to speak or lecture only if an organization also scheduled a replacement for her. Beloff, *Camera Culture*, 40; listing of Bourke-White's *Life* assignments, Index Notebook, MBW; Paul Peters to Bourke-White, 17 April 1938, MBW.

63. Office Memo, Joe Thorndike to Alan Brown, 18 Oct. 1937, MBW; Newhall to Bourke-White, 10 May 1937, MBW.

64. Steiner to Bourke-White, 19 Jan. 1937, MBW; Bourke-White, *Portrait of Myself*, 169; Guy E. Rhoades to Bourke-White, 30 Nov. [1937], MBW.

65. Margaret Smith to Ruth White, 19 Oct. 1939, MBW; Caldwell to Bourke-White, 2 Dec. 1939, EC; see also Howard, "Dear Kit, Dear Skinny." Bourke-White and Caldwell had produced a second book together entitled *North of the Danube* (1939); the reviews were good because of the international subject matter, but Bourke-White was hesitant to work on yet another project with her emotionally distraught husband.

66. Goldberg, *Bourke-White Biography*, 159.

67. Ibid., 255, 235.

68. Eisenstaedt quoted in Silverman, *For the World to See*, 7.

5. Berenice Abbott's Perception of the Evolving Cityscape

1. Abbott, quoted in O'Neal, *Berenice Abbott*, 18; "Changing City Caught in Flight by Photographs," *New York Herald Tribune*, 19 March 1936. One headline describing Abbott's photography read "Woman with Camera Snaps Revealing History of New York Life in Its Homeliest of Garb," *New York World-Telegram*, 11 Nov. 1938.

2. Steinbach, "Berenice Abbott's Point of View" (an interview with Berenice Abbott), 78; see also Berman, "The Unflinching Eye." Abbott told Berman, "He [Stieglitz] took about five good pictures in his whole life, and that was only when he ventured out of himself" (88). On pictorialists, Abbott is quoted in Berman, "The Unflinching Eye," 88. Abbott wrote, "The greatest influence obscuring the entire field of photography has, in my opinion, been pictorialism" ("It Has to Walk Alone," 15); Abbott, "Photographer as Artist," 7; Hales, *Silver Cities*, 277–90.

3. Mitchell, *Recollections* (published in conjunction with the International Center of Photography's exhibition of the same name), 12. The exhibit surveyed only living photographers, with each woman providing her own biographical sketch for the publication. Steinbach, "Berenice Abbott's Point of View," 78.

4. Zwingle, "Life of Her Own," 57; Mitchell, *Recollections*, 12.

5. Abbott quoted in Russell, "A Still Life," 70; O'Neal, *Berenice*

Abbott, 9; Abbott, quoted in Russell, "A Still Life," 70; Katz, keynote address, presentation to Berenice Abbott of the Association of International Photography Art Dealers' Annual Award for Significant Contributions to the Field of Photography, 6 Nov. 1981; "Woman with Camera Snaps Revealing History of New York Life in Its Homeliest of Garb," *New York World-Telegram,* 11 Nov. 1938.

6. Arnold, "The Way Berenice Abbott Feels about Cities"; "From a Student's Notebook," 56, 174, in which Abbott stressed the importance of training the eye and noted that "tones must be viewed instead of colors"; O'Neal, *Berenice Abbott,* 10; Man Ray described by John Canaday, introduction to O'Neal, *Berenice Abbott,* 7; see also Penrose, *Man Ray,* 75–95.

7. Mitchell, *Recollections,* 12; Steinbach, "Berenice Abbott's Point of View," 79; see also Abbott, "The 20s and the 30s," n.p.; Katz, keynote address; Abbott, quoted in O'Neal, *Berenice Abbott,* 46; "From a Student's Notebook," 56, 174; see also Abbott, *Photographs,* 13-14 (page references are to the 1970 edition).

8. Kramer, "Vanished City Life."

9. Zwingle, "A Life of Her Own," 57, in which Abbott claims personal responsibility for changing her own name; Peter Barr, correspondence with author, on Cocteau's role in altering Abbott's name for artistic effect.

10. Abbott quoted in Russell, "A Still Life," 70; Abbott, undated typewritten transcript, "Eugène Atget," ML; Lifson, essay in *Eugène Atget,* 7.

11. Abbott, "Atget" transcript, ML; Nesbit, *Atget's Seven Albums,* 20, 42–44; Abbott, "Eugène Atget," 337; Lifson refers to Atget's work as "a reverie about a dying era" (*Eugène Atget,* 10); Abbott, "Photographer as Artist," 6.

12. Abbott, "Atget" transcript, ML; see also Abbott, "Eugène Atget," 336; Nesbit on Atget's "Repertoire" in *Atget's Seven Albums,* 20; Szarkowski and Hambourg, *The Work of Atget,* vol. 1, 30.

13. Abbott, *The World of Atget;* Abbott quoted in Russell, "A Still Life," 70; Mitchell, *Recollections,* 13.

14. Man Ray to Paul Hill and Tom Cooper, "Interview: Man Ray," *Camera* 74 (Feb. 1975), 39–40, retold in Nesbit, *Atget's Seven Albums,* 1; Nesbit, *Atget's Seven Albums,* 16.

15. Abbott, "It Has to Walk Alone," 16; Nesbit draws on the work of Pierre MacOrlan, who called Atget "a man of the street, an artisan poet" (*Atget: Photographe de Paris* [Paris : Jonquiéres, 1930], in *Atget's Seven Albums,* 6).

16. Mitchell, *Recollections,* 13; Russell, "A Still Life," 70; O'Neal, *Berenice Abbott,* 14.

17. Abbott, in *Berenice Abbott,* 6; Berman, "The Unflinching Eye," 92; see also Arnold, "The Way Berenice Abbott Feels about Cities"; Lois Scharf, "Even Spinsters Need Not Apply," in *To Work and To Wed,* 85.

18. Berenice Abbott, project proposal to New York Historical Society, excerpted in O'Neal, *Berenice Abbott,* 16–17; Mitchell, *Recollections,* 13; see also Sundell, "Berenice Abbott's Work in the 1930s," 269–70. Sundell, through his extensive archival work in the Museum of the City of New York, determined that Abbott sent a form letter to two hundred of the museum's contributors but received no favorable response. Abbott, "Changing New York," 158.

19. Abbott quoted in Tom Zito, "Glimpses of the Artiste," sec. E; Kramer, "Vanished City Life," 18; for a comprehensive discussion of the impact of the architectural projects on Abbott's artistic development, see Barr, "Becoming Documentary."

20. Abbott, "Changing New York"; and Zubryn, "Saving New York for Posterity," 12. Other references have misquoted Abbott's original plan to read "the present jostling the past." See, for example, Berman, "The Unflinching Eye," 92, and Merry A. Foresta, "Art and Document," 151.

21. Peter Barr provides a penetrating interpretation of Abbott's creation of continuous fields in "Documentary Photography as Poetry" and "Becoming Documentary."

22. Smith College Museum of Art, "Berenice Abbott," exhibition dates, 17 Jan.– 24 Feb. 1974; Secrest, "An Attic Studio," sec. H, p. 2; Abbott, *The World of Atget,* xxvi. In Mitchell, *Recollections,* Abbott concluded her essay, "Photography can only represent the present. Once photographed, the subject becomes part of the past" (13).

23. Arnold, "The Way Berenice Abbott Feels about Cities."

24. Hales, *Silver Cities,* 179.

25. Secrest, "An Attic Studio," sec. H, p. 2; see also Raeburn, "'Culture Morphology' and Cultural History," 255–92; White, "Eugene Atget," 80; McCausland, "Berenice Abbott . . . Realist," 50; see also Sundell, "Berenice Abbott's Work in the 1930s," 273; and Coleman, "Latent Image," 18–19; Model, quoted in McCausland, "Berenice Abbott . . . Realist," 47. In describing Abbott's photographs, Model said, "Everything is alive. Everything breathes. Everything is rooted in life" (47).

26. Abbott, "Photography at the Crossroads," 20; "Changing City Caught in Flight by Photographs," *New York Herald Tribune,* 19 March 1936.

27. Abbott, "Changing New York," 160; McCausland, "The Photography of Berenice Abbott," 17; Landgren to Ann Kelly, 27 June 1934, ML; Landgren tried to interest the Rockefeller family in Abbott's documentation of the Radio City construction.

28. Abbott, FAP proposal, excerpted in O'Neal, *Berenice Abbott,* 17; Berman, "The Unflinching Eye," 92; "Supervising Employees on Project Unit Payroll—FAP (65-1699)," 1 Aug. 1936, RG69, ser. 651.315, box 2115. Abbott was classified as a project supervisor and received $145 per month, as did all other project supervisors regardless of their assigned division (teaching, design, murals, etc.)

29. "U.S. to Find Work for 3500 Artists," *New York Times,* 4 Oct. 1935.

30. "Changing City Caught in Flight by Photographs," *New York Herald Tribune*, 19 March 1936.

31. "Portraits of the United States: The Art Project and the Writers Project of the Works Progress Administration," undated typewritten manuscript, no. 13283, RG69, Division of Information, Primary File, box 77. Unidentified critic quoted in "Government Aid during the Depression to Professional, Technical and Other Service Workers," WPA Publication 1936, RG69, ser. 0001, General Records, FAP; "Portraits of the United States"; WPA/FAP, "Uniform Procedure for Allocation," RG69, ser. 0001, FAP General Records.

32. Williams quoted in Meltzer, *Violins and Shovels*, 19; Thomas E. Maulsby (FAP research supervisor), "The Story of WPA in American Art," 1936, typewritten transcript, RG69, Division of Information, Primary File; Foresta, "Art and Document," 148–56, in which she briefly discusses PWAP as a precursor to the FAP, pointing out both the criticism and support directed at the two programs; "U.S. to Find Work for 3500 Artists," *New York Times*, 4 Oct. 1935; Audrey McMahon to Bruce McClure, 25 June 1935, RG69, ser. 651.315. McMahon informs McClure (of the FERA) that one hundred new employees may be added to the current payroll. She specifically mentions hiring photographers.

33. See copies of the "WPA Monthly Statistical Bulletin" from 1936 to 1940, RG69, Division of Information, Primary File, and copies of "New York City Monthly Report," 1936 to 1940, RG69, ser. 0004; the FAP photographs located in RG69-ANP and RG69-AN (Still Pictures Branch), show the documentary nature of the camera work in New York City: the overwhelming majority are images of works (buildings, murals, paintings, art classes) in progress; McMahon to Cahill, 26 Sept. 1935, RG69, ser. 0005; Cahill to McMahon, 17 Jan. 1936, RG69, ser. 211.5, box 443. Reiterating that the FAP's major concern was its "artists," Cahill explained that the high costs of producing unnecessary photographs might hinder the FAP's ability to take care of its creative talents.

34. See Foresta's discussion of Dewey's impact on Cahill in "Art and Document," 150; Holger Cahill, speech delivered at the meeting of Regional Supervisors, Women's Division—WPA, 2 July 1936, transcript, RG69, Division of Information, Primary File, box 77; Holger Cahill, FAP director, "Federal Art Project Manual," Oct. 1935, RG69, ser. 0001, FAP, General Records; Raeburn, "'Culture Morphology' and Cultural History," 258.

35. "Art Work in Non-Federal Buildings (65-21-3755) and the Federal Art Project (65-1699) of the City of New York under the United States Works Progress Administration," Supplement A (Dec. 1935) and Supplement C (Jan. 1936), RG69, ser. 651.315, box 2114. In these two reports, statistics for the photo projects have been compiled, but "Changing New York" is considered individually.

36. "Art Work in Non-Federal Buildings (65-21-3755) and The Fed-

eral Art Project (65-1699) of the City of New York under the United States Works Progress Administration: A Descriptive Guide," Nov. 1935, RG69, ser. 651.315, box 2114.

37. "Changing City Caught in Flight by Photographs," *New York Herald Tribune*, 19 March 1936; correspondence between Hardinge Scholle and Holger Cahill, Feb. 1936, RG69, ser. 0005, box 28; FAP/NYC Exhibition Department, Weekly Report, 20 May 1936, RG69, ser. 0004, box 20; FAP/NYC, Exhibition Department, Weekly Reports for months June to December 1936, RG69, ser. 0004. See esp. 27 Aug. 1936 report.

38. Arnold, "The Way Berenice Abbott Feels about Cities."

39. Sarah Newmeyer to Holger Cahill, 12 June 1936, RG69, ser. 0001; Elizabeth McCausland to T.E. Maulsby, 27 Aug. 1936, RG69, ser. 0001; Thomas Parker to Audrey McMahon, 13 Aug. 1936, 21 Aug. 1936, RG69, ser. 0005; McNulty to Audrey McMahon, 25 Aug. 1936, RG69, ser. 0005; T.E. Maulsby to Thomas Parker, [Aug. 1936], RG69, ser. 0005; Audrey McMahon to Mildred Holzhauer, 8 March 1937, RG69, ser. 651.315, box 2116; Exhibition Project, Weekly Reports, 8 March 1937 and 5 April 1937, RG69, ser. 0004; comment on the exhibit's popularity quoted in Department of Information, Special Release, 31 Dec. 1937, RG69, ser. 002-A; "WPA Federal Art Project Exhibitions in Full Swing," 15 Oct. 1937, RG69, ser. 002-A; Exhibition Project, Weekly Report, 25 Oct. 1937, RG69, ser. 0004; "Notes on the Exhibition Program," n.d., RG69, ser. 0001; and Holger Cahill to Edward Steichen, 4 April 1936, RG69, ser. 651.315, box 2114, where Cahill suggests to Steichen that many developments within the FAP "may furnish the basis for a new and more vital American art."

40. Beaumont Newhall, interview by Joseph Trovato, 23 Jan. 1965, Archives of American Art, Smithsonian Institution, Washington, D.C., transcript.

41. "Woman with Camera Snaps Revealing History of New York Life in Its Homeliest of Garb," *New York World-Telegram*, 11 Nov. 1938.

42. Parker to McMahon, 15 Sept. 1937, RG69, ser. 211.5, box 443; Woodward to Paul Edwards, 13 May 1938, RG69, ser. 651.315, box 2117.

43. Thomas Parker to Audrey McMahon, 18 April 1938, RG69, ser. 651.315, box 2117; Parker to McMahon, 12 May 1938, RG69, ser. 651.315, box 2117; McMahon to Parker, 23 Sept. 1938, RG69, ser. 211.5, box 444; For more on the heated controversy, see correspondence between McMahon and Parker, 13 Sept.–12 Oct. 1938, RG69, ser. 211.5, box 444; Abbott to Parker, 11 Oct. 1938, RG69, ser. 651.315, box 2118.

44. Leuchtenburg, *Franklin D. Roosevelt*, 266; Meltzer, *Violins and Shovels*, 140.

45. Newhall-Trovato interview, 12; Abbott, "It Has to Walk Alone," 17; Raeburn, "'Culture Morphology' and Cultural History," 261–62, in which he compares Abbott's work to Faulkner's *Light in August* (1932)

and *Absalom! Absalom!* (1936), suggesting that both the photographer and the novelist require their audience to participate rather than merely observe. Abbott, "Photography at the Crossroads," 21.

Conclusion

1. McCausland, "Documentary Photography," 1; American Historical Association, *Annual Report for 1939;* Stryker and Johnstone, "Documentary Photography."

2. An excellent essay examining the most influential postmodern theorists on photography is Geoffrey Batchen's chapter entitled "Identity" in his *Burning with Desire.*

Bibliography

MANUSCRIPT AND PHOTOGRAPH COLLECTIONS

AAA   Photographer Interview Transcripts. Archives of American
            Art, Smithsonian Institution, Washington, D.C.

AFA   American Federation of Art Papers. Archives of American
            Art, Smithsonian Institution, Washington, D.C.

BA   Berenice Abbott Collection. Museum of the City of New
            York.

DL   Dorothea Lange Collection. Oakland Museum, Oakland,
            California.

DU-BC   Doris Ulmann Manuscript Collection. Southern
            Appalachian Archives, Hutchins Library, Berea
            College, Berea, Kentucky.

DU-BCP   Doris Ulmann Photograph Collection. Art Department
            Library, Berea College, Berea, Kentucky.

DU-G   Doris Ulmann Photograph Collection. J. Paul Getty
            Museum, Los Angeles, California.

DU-NY   Doris Ulmann Photograph Collection. New York Historical
            Society, New York City.

DU-OR   Doris Ulmann Photographs and Proof Books. Special
            Collections. University of Oregon, Eugene.

DU-SC   Doris Ulmann Photograph Collection. South Carolina
            Historical Society, Charleston.

DU-UK   Doris Ulmann Photograph Collections. Special Collections.
            Margaret I. King Library, University of Kentucky,
            Lexington.

DU-UKA   Doris Ulmann Prints. University of Kentucky Art Museum.

EC   Erskine Caldwell Selected Papers. George Arents Research
            Library, Syracuse University, Syracuse, New York.

FAP   Federal Art Project Photography Unit. Still Pictures Branch,
            National Archives.

FSAPP   Farm Security Administration—Office of War Information
            Photograph Collection. Library of Congress.

FSAWR   Farm Security Administration—Office of War Information.
            Written Records. Overseas Picture Division, Lot
            12024. Library of Congress.

JB   John Bennett Papers. South Carolina Historical Society,
          Charleston.
JJN   John Jacob Niles Manuscript Collection. Special
          Collections. Margaret I. King Library, University
          of Kentucky, Lexington.
JJN-BC   John Jacob Niles Field Notebooks. Art Department, Berea
          College, Berea, Kentucky.
JP   Julia Peterkin Manuscript Collection. South Carolina
          Historical Society, Charleston.
MBW   Margaret Bourke-White Papers. George Arents Research
          Library, Syracuse University, Syracuse, New York.
MBWP   Margaret Bourke-White Photographic Prints. George
          Arents Research Library, Syracuse University,
          Syracuse, New York.
ML   Marchal Landgren Papers. Archives of American Art,
          Smithsonian Institution, Washington, D.C.
OM-ART   Art Department Collections. Oakland Museum, Oakland,
          California.
RESC   Roy Emerson Stryker Collection. University of Louisville
          Photographic Archives.
RESP   Roy Emerson Stryker Papers. Microfilm Edition. Edited by
          David Horvath. University of Louisville
          Photographic Archives.
RG69   Records of the Works Progress Administration. Record
          Group 69. National Archives.
UC-OHC   Oral History Collection. Bancroft Library, University of
          California, Berkeley.
UL   Photographic Archives, University of Louisville.

OTHER SOURCES

Aaron, Daniel. "An Approach to the Thirties." In *The Study of American Culture: Contemporary Conflicts,* edited by Luther S. Luedtke, 1–17. Deland, Fla.: Everett/Edwards, 1977.

Abbott, Berenice. "Berenice Abbott: The 20s and the 30s." With an introduction by Barbara Shissler Nosanow. Exhibition Catalogue. Washington, D.C.: Smithsonian Institution, 1982.

———. "Changing New York." In *Art for the Millions: Essays from the 1930s by Artists and Administrators of the WPA Federal Art Project,* edited by Francis O'Connor, 158-62. Greenwich, Conn.: New York Graphic Society, 1973.

———. "Eugène Atget." In *The Encyclopedia of Photography* 2 (1963): 335-39. Originally published in *Complete Photographer* 6 (1941).

———. "It Has to Walk Alone." In Lyons, *Photographers on Photography,* 15-17. Originally published in *Infinity* 7 (1951): 6–7, 14.

———. "Photographer as Artist." *Art Front* 16 (1936): 4–7.

———. *Photographs.* With a foreword by Muriel Rukeyser and an introduction by David Vestal. New York: Horizon Press, 1970; Washington, D.C.: Smithsonian Institution Press, 1990.

———. "Photography at the Crossroads." In Lyons, *Photographers on Photography,* 17-22.

———. "What the Camera and I See." *ARTnews* 50 (Sept. 1951): 36–37, 52.

———. *The World of Atget.* New York: Horizon Press, 1964.

Abrams, Mary. "Women Photographers." *Graduate Woman* 25 (September/October 1981): 22-24.

Agee, James, and Walker Evans. *Let Us Now Praise Famous Men.* Boston: Houghton Mifflin, 1941.

Allen, Frederick Lewis. *The Big Change: America Transforms Itself, 1900–1950.* New York: Harper and Row, 1952.

———. *Since Yesterday: The Nineteen-Thirties in America.* New York: Harper and Brothers, 1940.

Alpern, Sara, Joyce Antler, Elisabeth Israels Perry, and Ingrid Winther Scobie, eds. *The Challenge of Feminist Biography: Writing the Lives of Modern American Women.* Urbana: Univ. of Illinois Press, 1992.

Amarillo Art Center. *American Images: Photographs and Photographers from the Farm Security Administration, 1935–1942.* Amarillo, Tex.: Amarillo Art Center, 1979. Videotape.

American Historical Association. *Annual Report of the American Historical Association for the Year 1939.* Washington, D.C.: Government Printing Office, 1941.

Anderson, James C., ed. *Roy Stryker: The Humane Propagandist.* Louisville, Ky.: University of Louisville Photographic Archives, 1977.

Anderson, Sherwood. *Home Town.* The Face of America series, edited by Edwin Rosskam. New York: Alliance Book Corporation, 1940.

———. *Puzzled America.* New York: Scribner's, 1935.

———. *Winesburg, Ohio.* New York: B.W. Huebsch, 1919; New York: Viking Press, 1958.

Arnold, Elliott. "The Way Berenice Abbott Feels about Cities and Photography, Her Exhibit Is like an Artist Painting Portraits of His Beloved." *New York World-Telegram,* 21 Oct. 1937.

Baldwin, Sidney. *Poverty and Politics: The Rise and Decline of the Farm Security Administration.* Chapel Hill: Univ. of North Carolina Press, 1968.

Banes, Ruth. "Doris Ulmann and Her Mountain Folk." *Journal of American Culture* 8 (spring 1985): 29-42.

Banta, Martha. *Imaging American Women: Idea and Ideals in Cultural History.* New York: Columbia Univ. Press, 1987.

Barr, Peter. "Becoming Documentary: Berenice Abbott's Photographs, 1925–1939." Ph.D. diss., Boston Univ., 1997.

———. "Documentary Photography as Poetry: Berenice Abbott's *Chang-*

ing New York," Paper presented at Frick Symposium, 3 April 1993.
Batchen, Geoffrey. *Burning with Desire: The Conception of Photography.* Cambridge: Massachusetts Institute of Technology Press, 1997.
Becker, Jane S. *Selling Tradition: Appalachia and the Construction of an American Folk.* Chapel Hill: Univ. of North Carolina Press, 1998.
Beloff, Halla. *Camera Culture.* New York: Basil Blackwell, 1985.
Bennett, Edna. "Dorothea Lange: A Timely Tribute." *Photographic Product News* 2 (Jan.–Feb. 1966): 54–56, 64–65.
Berea College. *Bulletin of Berea College and Allied Schools.* General Catalog, 1933–34. Berea, Ky.: Berea College Press, 1933.
Berenice Abbott. With an essay by Julia Van Haaften. Aperture Masters of Photography series, no. 9. New York: Aperture, 1988.
Berman, Avis. "The Unflinching Eye of Berenice Abbott." *ARTnews* 80 (Jan. 1981): 87–93.
Bledstein, Burton J. *The Culture of Professionalism: The Middle Class and the Development of Higher Education in America.* New York: Norton, 1978.
Boddy, Julie. "Photographing Women: The Farm Security Administration Work of Marion Post Wolcott." In *Decades of Discontent: The Women's Movement, 1920–1940,* edited by Lois Scharf and Joan M. Jensen, 153-66. Westport, Conn.: Greenwood Press, 1983.
Borchert, James. "Historical Photo-Analysis: A Research Method." *Historical Methods* 15 (spring 1982): 35–44.
Bourke-White, Margaret. *Eyes on Russia.* New York: Simon and Schuster, 1931.
———. "Photographing This World." *Nation,* 19 Feb. 1936, 217–18.
———. *Portrait of Myself.* New York: Simon and Schuster, 1963.
Brannan, Beverly W. "American Women Documentary Photographers, 19th and 20th Centuries." Keynote address delivered at conference, "Woman: A Different Voice," Western Kentucky Univ., Bowling Green, 26–28 Sept. 1990.
———. Telephone conversation with the author, 19 March 1990.
Brannan, Beverly W., and David Horvath, eds. *A Kentucky Album: Farm Security Administration Photographs, 1935–1943.* Lexington: Univ. Press of Kentucky, 1986.
Brinkley, Alan. *Voices of Protest: Huey Long, Father Coughlin, and the Great Depression.* New York: Vintage, 1982.
Brown, Dorothy M. *Setting a Course: American Women in the 1920s.* Boston: Twayne, 1987.
Brown, Lorraine, and Michael G. Sundell. "Stylizing the Folk: Hall Johnson's *Run, Little Chillun* Photographed by Doris Ulmann." *Prospects* 7 (1982): 335–46.
Browne, Turner. *Macmillan Biographical Encyclopedia of Photographic Artists and Innovators.* New York: Macmillan, 1983.
Brownell, Jean. "Girl Photographer for FSA Travels 50,000 Miles in Search for Pictures." *Washington Post,* 19 Nov. 1940.

Byers, Paul. "Cameras Don't Take Pictures." *Columbia Univ. Forum* 9 (1966): 27–31.

Caldwell, Erskine, and Margaret Bourke-White. *You Have Seen Their Faces.* 1937. Reprint, Athens: Univ. of Georgia Press, 1995.

Callahan, Sean, ed. *The Photographs of Margaret Bourke-White.* With an introduction by Theodore M. Brown. New York: New York Graphic Society, 1972.

Campbell, Olive D. "Doris Ulmann." *Mountain Life and Work* (Oct. 1934): 11.

Cashin, Joan E. *A Family Venture: Men and Women on the Southern Frontier.* Baltimore: Johns Hopkins Univ. Press, 1994.

"Changing City Caught in Flight by Photographs." *New York Herald Tribune,* 19 March 1936.

Chase, Stuart. "The Tragedy of Waste: The Wastes of Advertising." *New Republic,* 19 Aug. 1925, 342–45.

Coben, Stanley. "The Assault on Victorianism in the Twentieth Century." *American Quarterly* 23 (Dec. 1975): 604–25.

Coleman, A.D. "Latent Image: Further Thoughts on Berenice Abbott." *Village Voice,* 28 Jan. 1971: 18-19.

Conkin, Paul K. *The Southern Agrarians.* Knoxville: Univ. of Tennessee Press, 1988.

"Consumer Cooperatives." *Fortune* 15 (March 1937): 133–46.

Conway, Jill. "Convention versus Self-Revelation: Five Types of Autobiography by Women of the Progressive Era." Paper presented at conference, "Project on Women and Social Change," Smith College, Northampton, Mass., 13 June 1983.

Cook, Blanche Wiesen. "Female Support Networks and Political Activism: Lillian Wald, Crystal Eastman, Emma Goldman." *Chrysalis* 3 (1977): 43–61.

Cott, Nancy. *The Grounding of Modern Feminism.* New Haven, Conn.: Yale Univ. Press, 1987.

Cowan, Ruth Schwartz. "Two Washes in the Morning and a Bridge Party at Night: The American Housewife between the Wars." In *Decades of Discontent: The Women's Movement, 1920–1940,* edited by Lois Scharf and Joan M. Jensen, 177-96. Westport, Conn.: Greenwood Press, 1983.

Cowley, Malcolm. —*And I Worked at the Writer's Trade: Chapters of Literary History, 1918–1978.* New York: Viking Press, 1978.

Crunden, Robert M. *Ministers of Reform: The Progressives' Achievement in American Civilization, 1889–1920.* Urbana: Univ. of Illinois Press, 1982.

"Current Exhibition News in Brief," *New York Times,* 3 Nov. 1929, 12.

Curtis, James. *Mind's Eye, Mind's Truth: FSA Photography Reconsidered.* Philadelphia: Temple Univ. Press, 1989.

Daniel, Pete, Merry A. Foresta, Maren Stange, and Sally Stein. *Official*

Images: New Deal Photography. Washington, D.C.: Smithsonian Institution Press, 1987.

Davidov, Judith Fryer. *Women's Camera Work: Self/Body/Other in American Visual Culture.* Durham: Duke Univ. Press, 1998.

Davidson, Donald. "Erskine Caldwell's Picture Book." *Southern Review* 4 (1938–39): 15–25.

———. "Julia Peterkin." *Spyglass,* 3 April 1927.

Davidson, Jan. "The People of Doris Ulmann's North Carolina Photographs." Paper presented at the Ulmann Symposium, Gibbes Museum of Art, 8 November 1997.

Davie, Helen L. "Women in Photography." *Camera Craft* 5 (Aug. 1902): 130-38.

Dewey, John. *Art as Experience.* New York: Minton, Balch, 1934.

Dixon, Daniel. "Dorothea Lange." *Modern Photography* 16 (Dec. 1952): 68–77, 138–41.

Dixon, Penelope. *Photographers of the Farm Security Administration: An Annotated Bibliography, 1930–1980.* New York: Garland, 1983.

Doherty, Robert J. "USA-FSA: Farm Security Administration Photographs of the Depression Era." *Camera* 41 (Oct. 1962): 9–13.

Dorothea Lange. Aperture Masters in Photography series, no. 5. New York: Aperture, 1987.

Dos Passos, John. *U.S.A.: Nineteen Nineteen.* New York: Random House, 1930.

Dreiser, Theodore. *Sister Carrie.* New York: Harper and Brothers, 1900; New York: Random House, 1932.

Eagles, Charles W. "Urban-Rural Conflict in the 1920s: A Historiographical Assessment." *Historian* 49 (Nov. 1986): 26–48.

Eaton, Allen H. "The Doris Ulmann Photograph Collection." *Call Number* 19 (spring 1958): 10–11.

———. *Handicrafts of the Southern Highlands.* New York: Russell Sage Foundation, 1937. Reprint, New York: Dover, 1973.

Edwynn Houk Gallery. "Vintage Photographs by Women of the 20s and 30s." With an introduction by Paul Katz. Exhibition Catalog. Chicago: Edwynn Houk Gallery, 1988.

Elliott, George P. "Photographs and Photographers." In *A Piece of Lettuce,* 90-103. New York: Random House, 1964.

Eugène Atget. New York: Aperture, 1980.

Evans, Sara M. *Born for Liberty: A History of Women in America.* New York: Free Press, 1989.

Fass, Paula S. *The Damned and the Beautiful: American Youth in the 1920s.* Oxford: Oxford Univ. Press, 1977.

Featherstone, David. *Doris Ulmann: American Portraits.* Albuquerque: Univ. of New Mexico Press, 1985.

Fisher, Andrea. *Let Us Now Praise Famous Women: Women Photographers for the U.S. Government, 1935 to 1944.* London: Pandora, 1987.

Fleischhauer, Carl, and Beverly W. Brannan, eds. *Documenting America, 1935–1943*. Berkeley: Univ. of California Press, 1988.

Foner, Eric. *Free Soil, Free Labor, Free Men: The Ideology of the Republican Party before the Civil War*. London: Oxford Univ. Press, 1970.

Foresta, Merry A. "Art and Document: Photography of the Works Progress Administration's Federal Art Project." In *Official Images: New Deal Photography*, by Pete Daniel, Merry A. Foresta, Maren Stange, and Sally Stein. Washington, D.C.: Smithsonian Institution Press, 1987.

"From a Student's Notebook." *Popular Photography* 21 (1947): 53-56, 174.

Ganzel, Bill. *Dust Bowl Descent*. Lincoln: Univ. of Nebraska Press, 1984.

Ginger, Ray. *Six Days or Forever? Tennessee v. John Thomas Scopes*. New York: Oxford Univ. Press, 1958.

Glazer, Penina Migdal, and Miriam Slater. *Unequal Sisters: The Entrance of Women into the Professions, 1890–1940*. New Brunswick, N.J.: Rutgers Univ. Press, 1987.

Glen, John M. *Highlander: No Ordinary School, 1932–1962*. Lexington: Univ. Press of Kentucky, 1988.

Goldberg, Vicki. *Margaret Bourke-White: A Biography*. New York: Harper and Row, 1986.

Goldsmith, Arthur. "A Harvest of Truth: The Dorothea Lange Retrospective Exhibition." *Infinity* (March 1966): 23–30.

Gover, Jane. *The Positive Image: Women Photographers in Turn of the Century America*. Albany, N.Y.: SUNY Press, 1988.

Greenough, Sarah. "How Stieglitz Came to Photograph Clouds." In *Perspectives on Photography: Essays in Honor of Beaumont Newhall*, edited by Peter Walch and Thomas Barrow, 151–65. Albuquerque: Univ. of New Mexico Press, 1986.

Greenough, Sarah, Joel Snyder, David Travis, and Colin Westerbeck. *On the Art of Fixing a Shadow: One Hundred and Fifty Years of Photography*. Boston: Little, Brown, 1989.

Guimond, James. *American Photography and the American Dream*. Chapel Hill: Univ. of North Carolina Press, 1991.

Gutman, Herbert G. "Work, Culture, and Society in Industrializing America, 1815–1919." *American Historical Review* 78 (June 1973): 531–88.

Hagood, Margaret Jarman. *Mothers of the South: Portraiture of the White Tenant Farm Woman*. Chapel Hill: Univ. of North Carolina Press, 1939; rev. ed., Charlottesville: University of Virginia Press, 1996.

Hales, Peter. *Silver Cities: The Photography of American Urbanization, 1839–1915*. Philadelphia: Temple Univ. Press, 1984.

Hardt, Hanno, and Karin B. Ohrn. "The Eyes of the Proletariat: The Worker-Photography Movement in Weimar Germany." *Studies in Visual Communication* 7 (summer 1981): 46–57.

Haskell, Thomas L. *The Emergence of Professional Social Science: The*

American Social Science Association and the Nineteenth-Century Crisis of Authority. Urbana: Univ. of Illinois Press, 1977.

Hawthorne, Ann, ed. The Picture Man: Photographs by Paul Buchanan. Chapel Hill: Univ. of North Carolina Press, 1993.

Heilbrun, Carolyn. Writing a Woman's Life. New York: Norton, 1988.

Heilpern, Alfred. "Vita" [Doris Ulmann]. Call Number 19 (spring 1958): 12.

Hendrickson, Paul. "Double Exposure." Washington Post Magazine, 31 Jan. 1988.

————. Looking for the Light: The Hidden Life and Art of Marion Post Wolcott. New York: Knopf, 1992.

Herz, Nat. "Dorothea Lange in Perspective: A Reappraisal of the FSA and an Interview." Infinity (April 1963): 5–12.

Heyman, Therese Thau. Celebrating a Collection: The Work of Dorothea Lange. Oakland, Calif.: Oakland Museum, 1978.

————. Interview by author. Oakland, California. 12 July 1989.

Higham, John. History: Professional Scholarship in America. Baltimore: Johns Hopkins Univ. Press, 1983.

Hill, Paul, Angela Kelly, and John Tagg. Three Perspectives on Photography. London: Arts Council of Great Britain, 1979.

Hilton, Kathleen C. "'Both in the Field, Each with a Plow': Race and Gender in USDA Policy, 1907–1929." In Hidden Histories of Women in the New South, edited by Virginia Bernhard, Betty Brandon, Elizabeth Fox-Genovese, Theda Perdue, and Elizabeth H. Turner, 114-33. Columbia: Univ. of Missouri Press, 1994.

Hine, Lewis. "Charity on a Business Basis." World Today 13 (Dec. 1907): 1254–60.

————. "Our Untrained Citizens—Photographs by Lewis W. Hine for the NCLC." Survey 23 (2 Oct. 1909): 21–35.

————. "Toilers of the Tenements." McClure's, July 1910, 231–40.

————. "What Bad Housing Means to Pittsburgh." Charities and the Commons 19 (7 March 1908): 1683-98.

Hobson, Fred C., Jr. Serpent in Eden: H. L. Mencken and the South. Chapel Hill: Univ. of North Carolina Press, 1974; Baton Rouge: Louisiana State Univ. Press, 1978.

Hoffmann, Frederick J. The Twenties: American Writing in the Postwar Decade. Rev. ed. New York: Free Press, 1962.

Horton, Aimee Isgrig. The Highlander Folk School: A History of Its Major Programs, 1932–1961. New York: Carlson, 1989.

Howard, William L. "Dear Kit, Dear Skinny: The Letters of Erskine Caldwell and Margaret Bourke-White." Syracuse University Library Associates Courier 23 (fall 1988): 23–44.

Howe, Hartley E. "You Have Seen Their Pictures." Survey Graphic 29 (April 1940): 236–41.

Hurley, F. Jack. Marion Post Wolcott: A Photographic Journey. Albuquerque: Univ. of New Mexico Press, 1989.

————. *Portrait of a Decade: Roy Stryker and the Development of Documentary Photography in the Thirties.* Baton Rouge: Louisiana State Univ. Press, 1972.

————. "Shooting for the File: The Farm Security Administration Photographers in the South, 1935–1943." Paper presented at the fiftieth annual meeting of the Southern Historical Association, Louisville, October 1985.

Jacobs, Philip W. "Doris Ulmann's Search for Meaning." Paper presented at the Ulmann Symposium, Gibbes Museum of Art, 8 November 1997.

The Jargon Society. *The Appalachian Photographs of Doris Ulmann.* With essays by John Jacob Niles and Jonathan Williams. Highlands, N.C.: Jargon Society, 1971.

Jones, Margaret. *Heretics and Hellraisers: Women Contributors to The Masses, 1911–1917.* Austin: Univ. of Texas Press, 1993.

Jordy, William H. "Four Approaches to Regionalism in the Visual Arts in the 1930s." In *The Study of American Culture: Contemporary Conflicts,* edited by Luther S. Luedtke, 19–48. Deland, Fla.: Everett/Edwards, 1977.

Jussim, Estelle. "'The Tyranny of the Pictorial': American Photojournalism from 1880 to 1920." In *Eyes of Time: Photojournalism in America,* edited by Marianne Fulton, 36–73. Boston: Little, Brown, 1988.

Karl, Barry D. *The Uneasy State: The United States from 1915 to 1945.* Chicago: Univ. of Chicago Press, 1983.

Katz, Leslie George. Keynote address. Award Ceremony at the Annual Meeting of the Association of International Photography Art Dealers, 6 Nov. 1981. National Museum of American Art, Smithsonian Institution, Washington, D.C. Transcript.

Kazin, Alfred. *On Native Grounds: An Interpretation of Modern American Prose Literature.* New York: Harcourt, Brace, 1942; Garden City, N.Y.: Doubleday, 1956.

Kennedy, David M. *Birth Control in America: The Career of Margaret Sanger.* New Haven, Conn.: Yale Univ. Press, 1970.

————. *Over Here: The First World War and American Society.* New York: Oxford Univ. Press, 1980.

Kennedy, N. Brent. *The Melungeons: The Resurrection of a Proud People.* Macon, Ga.: Mercer Univ. Press, 1994.

Kentucky Federal Writers' Project. *Kentucky: A Guide to the Bluegrass State.* New York: Harcourt, Brace, 1939.

Kerber, Linda K., Alice Kessler–Harris, and Kathryn Kish Sklar, eds. *U.S. History as Women's History: New Feminist Essays.* Chapel Hill: Univ. of North Carolina Press, 1995.

Kerr, Clark. "Paul and Dorothea." In Partridge, *Dorothea Lange: A Visual Life,* 36-43.

Kozol, Wendy. *LIFE's America: Family and Nation in Postwar Photojournalism.* Philadelphia: Temple Univ. Press, 1994.

———. "Madonnas of the Fields: Photography, Gender, and 1930s Farm Relief." *Genders* 2 (summer 1988): 1-23.

Kramer, Hilton. "Vanished City Life by Berenice Abbott on View." *New York Times,* 27 Nov. 1981.

Lange, Dorothea. "The American Farm Woman." *Harvester World* (Nov. 1960): 2–9.

———. *Dorothea Lange.* New York: Museum of Modern Art, 1966.

Lange, Dorothea, and Daniel Dixon. "Photographing the Familiar." In Lyons, *Photographers on Photography,* 68–72. Originally published in *Aperture* 1 (1952): 4–15.

Lange, Dorothea, and Paul Schuster Taylor. *An American Exodus: A Record of Human Erosion.* New York: Reynal and Hitchcock, 1939.

Lerner, Gerda. *Why History Matters: Life and Thought.* Oxford: Oxford Univ. Press, 1998.

Leuchtenberg, William E. *Franklin D. Roosevelt and the New Deal.* New York: Harper and Row, 1963.

———. *The Perils of Prosperity, 1914–32.* Chicago: Univ. of Chicago Press, 1958.

Levine, Lawrence W. "American Culture and the Great Depression." *Yale Review* 74 (winter 1985): 196–223.

———. "The Historian and the Icon: Photography and the History of the American People in the 1930s and 1940s." In Fleischhauer and Brannan, *Documenting America,* 15-42.

Lifson, Ben. "Post Wolcott: Not a Vintage Show." *Village Voice,* 23 July 1979.

Lippmann, Walter. *Public Opinion.* 1922. Reprint, New York: Free Press, 1965.

Longstreet, Stephen. *We All Went to Paris: Americans in the City of Light, 1776–1971.* New York: Macmillan, 1972.

Louisiana Federal Writers' Project. *Louisiana: A Guide to the State.* New York: Hastings House, 1941.

Lovejoy, Barbara. "The Oil Pigment Photography of Doris Ulmann." Master's thesis, Univ. of Kentucky, 1993.

Lyons, Nathan, ed. *Photographers on Photography.* Englewood Cliffs, N.J.: Prentice-Hall, 1966.

Maddow, Ben. *Faces: A Narrative History of Portrait Photography.* Boston: New York Graphic Society, 1977.

Mann, Margery. "Dorothea Lange." *Popular Photography,* March 1970, 84-85, 99-101.

Mann, Margery, and Anne Noggle, eds. *Women of Photography: An Historical Survey.* San Francisco: San Francisco Museum of Art, 1975.

Marchand, Roland. *Advertising the American Dream: Making Way for Modernity, 1920–1940.* Berkeley: Univ. of California Press, 1985.

Mazo, Joseph H. *Prime Movers: The Makers of Modern Dance in America.* New York: William Morrow, 1977.

McCausland, Elizabeth. "Berenice Abbott . . . Realist." *Photo Arts* 2 (spring 1948): 47–50.

———. "Documentary Photography." *Photo Notes* (Jan. 1939): 1–4.

———. "The Photography of Berenice Abbott." *Trend* 3 (March–April 1935): 17.

McElvaine, Robert, ed. *Down and Out in the Great Depression: Letters from the "Forgotten Man."* Chapel Hill: Univ. of North Carolina Press, 1983.

McEuen, Melissa A. "Doris Ulmann and Marion Post Wolcott: The Appalachian South." *History of Photography* 19 (spring 1995): 4–12.

Megraw, Richard B. "'The Grandest Picture': Lyle Saxon and the Problem of Aesthetic Localism in 30s America." Paper presented at the Southern American Studies Association meeting, Little Rock, Ark., Nov. 1996.

Meikle, Jeffrey. *Twentieth Century Limited: Industrial Design in America, 1925–1939.* Philadelphia: Temple Univ. Press, 1979.

Meltzer, Milton. *Dorothea Lange: A Photographer's Life.* New York: Farrar, Straus, and Giroux, 1978.

———. *Violins and Shovels: The WPA Arts Projects.* New York: Delacorte Press, 1976.

Melville, Annette. *Farm Security Administration, Historical Section: A Guide to Textual Records in the Library of Congress.* Washington, D.C.: Library of Congress, 1985.

"Mr. O'Neill." *New York Times,* 3 March 1929, 4.

Mitchell, Margaretta K., ed. *Recollections: Ten Women of Photography.* New York: Viking Press, 1979.

"The Mountain Breed." *New York Times,* 2 June 1928, 16.

Moynihan, Ruth, Susan Armitage, and Christiane Fischer Dichamp, eds. *So Much to Be Done: Women Settlers on the Mining and Ranching Frontier.* Lincoln: Univ. of Nebraska Press, 1990.

Murphy, Mary. *Mining Cultures: Men, Women, and Leisure in Butte, 1914–41.* Urbana: Univ. of Illinois Press, 1997.

———. "The Private Lives of Public Women: Prostitution in Butte, Montana, 1878–1917." In *The Women's West,* edited by Susan Armitage and Elizabeth Jameson, 193-205. Norman: Univ. of Oklahoma Press, 1987.

Murray, Joan. "Marion Post Wolcott: A Forgotten Photographer." *American Photographer* (March 1980): 86–93.

Naef, Weston, ed. *Doris Ulmann: Photographs from the J. Paul Getty Museum.* Focus series. Malibu, Calif.: J. Paul Getty Museum, 1996.

Nash, Roderick. *The Nervous Generation: American Thought, 1917–1930.* Chicago: Rand McNally, 1970.

Natanson, Nicholas. *The Black Image in the New Deal: The Politics of FSA Photography.* Knoxville: Univ. of Tennessee Press, 1992.

Nesbit, Molly. *Atget's Seven Albums.* New Haven, Conn.: Yale Univ. Press, 1992.

Niles, John Jacob. *The Ballad Book of John Jacob Niles*. Boston: Houghton Mifflin, 1961.
———. "Doris Ulmann: Preface and Recollections." *Call Number* 19 (spring 1958): 4–9.
Noble, David F. *America by Design: Technology and the Rise of Corporate Capitalism*. New York: Oxford Univ. Press, 1979.
Noggle, Burl. "With Pen and Camera: In Quest of the American South in the 1930s." In *The South Is Another Land: Essays on the Twentieth-Century South*, edited by Bruce Clayton and John A. Salmond, 187–206. Westport, Conn.: Greenwood Press, 1987.
Oakes, James. *The Ruling Race: A History of American Slaveholders*. New York: Random House, 1982.
Ohrn, Karin Becker. *Dorothea Lange and the Documentary Tradition*. Baton Rouge: Louisiana State Univ. Press, 1980.
O'Neal, Hank. *Berenice Abbott: American Photographer*. New York: McGraw-Hill, 1982.
———. *A Vision Shared: A Classic Portrait of America and Its People, 1935–1943*. New York: St. Martin's Press, 1976.
O'Neill, Eugene. *The Plays of Eugene O'Neill*. New York: Random House, 1929.
Orvell, Miles. *The Real Thing: Imitation and Authenticity in American Culture, 1880–1940*. Chapel Hill: Univ. of North Carolina Press, 1989.
Osman, Colin, and Sandra S. Phillips. "European Visions: Magazine Photography in Europe between the Wars." In *Eyes of Time: Photojournalism in America*, edited by Marianne Fulton, 74–103. Boston: Little, Brown, 1988.
Partridge, Elizabeth, ed. *Dorothea Lange: A Visual Life*. Washington, D.C.: Smithsonian Institution Press, 1994.
Pauly, Thomas H. *An American Odyssey: Elia Kazan and American Culture*. Philadelphia: Temple Univ. Press, 1983.
Peeler, David P. *Hope among Us Yet: Social Criticism and Social Solace in Depression America*. Athens: Univ. of Georgia Press, 1987.
Penrose, Roland. *Man Ray*. Boston: New York Graphic Society, 1975.
Peterkin, Julia. *Bright Skin*. Indianapolis: Bobbs-Merrill, 1932.
Peters, Marsha, and Bernard Mergen. "'Doing the Rest': The Uses of Photographs in American Studies." *American Quarterly* 29 (1977): 280–303.
Puckett, John Rogers. *Five Photo-Textual Documentaries from the Great Depression*. Ann Arbor: UMI Research Press, 1984.
Raeburn, John. "'Culture Morphology' and Cultural History in Berenice Abbott's *Changing New York*." *Prospects* 9 (1984): 255–92.
Raedeke, Paul. "Introduction and Interview." *Photo-Metro*, Feb. 1986, 3-17.
Raper, Arthur F., and Ira De A. Reid. *Sharecroppers All*. Chapel Hill: Univ. of North Carolina Press, 1941.

Rosenberg, William G., and Lewis H. Siegelbaum, eds. *Social Dimensions of Soviet Industrialization*. Bloomington: Indiana Univ. Press, 1993.

Rosenblum, Naomi. *A History of Women Photographers*. New York: Abbeville Press, 1994.

Rosenblum, Walter, and Naomi Rosenblum. *America and Lewis Hine: Photographs 1904-1940*. With an essay by Alan Trachtenberg. Millerton, N.Y.: Aperture, 1977.

Rourke, Constance. "The Significance of Sections." *New Republic*, 20 Sept. 1933, 148-51.

Russell, John. "A Still Life in Maine." *New York Times*, 16 Nov. 1980.

Sandeen, Eric J. *Picturing an Exhibition: "The Family of Man" and 1950s America*. Albuquerque: Univ. of New Mexico Press, 1995.

Scharf, Lois. *To Work and to Wed: Female Employment, Feminism, and the Great Depression*. Westport, Conn.: Greenwood Press, 1980.

Schlereth, Thomas J. *Artifacts and the American Past*. Nashville, Tenn.: American Association for State and Local History, 1980.

Secrest, Meryle. "An Attic Studio Hides a Woman's Vision." *Washington Post*, 17 Aug. 1969.

Seigfried, Charlene Haddock. "Classical American Philosophy's Invisible Women." Paper presented at the annual meeting of the Organization of American Historians, Atlanta, April 1994.

Seixas, Peter. "Lewis Hine: From 'Social' to 'Interpretive' Photographer," *American Quarterly* 39 (fall 1987): 381-409.

Sekula, Allan. "On the Invention of Photographic Meaning." *Artforum* 13 (Jan. 1975): 36-45.

Shelton, Suzanne. *Divine Dancer: A Biography of Ruth St. Denis*. Garden City, N.Y.: Doubleday, 1981.

Shipman, Louis Evan. Introduction to *A Portrait Gallery of American Editors, Being a Group of XLIII Likenesses*, by Doris Ulmann, 1-2. New York: William Edwin Rudge, 1925.

Silverman, Jonathan. *For the World to See: The Life of Margaret Bourke-White*. New York: Viking Press, 1983.

Singal, Daniel J. "Towards a Definition of American Modernism." *American Quarterly* 39 (spring 1987): 7-26.

———. *The War Within: From Victorian to Modernist Thought in the South, 1919-1945*. Chapel Hill: Univ. of North Carolina Press, 1982.

Sklar, Katherine Kish. *Florence Kelley and the Nation's Work*. New Haven, Conn.: Yale Univ. Press, 1995.

Smith, Terry. *Making the Modern: Industry, Art, and Design in America*. Chicago: Univ. of Chicago Press, 1993.

Smith College Museum of Art. "Berenice Abbott, Helen Frankenthaler, Tatyana Grosman, Louise Nevelson." Exhibition Catalogue. Northampton, Mass.: Smith College, 1974.

Snyder, Robert E. "Erskine Caldwell and Margaret Bourke-White: You Have Seen Their Faces." *Prospects* 11 (1987): 343-405.

———. "Marion Post and the Farm Security Administration in Florida." *Florida Historical Quarterly* 65 (April 1987): 457-79.

———. "Marion Post Wolcott: Photographing FSA Cheesecake." In *Developing Dixie: Modernization in a Traditional Society*, edited by Winfred B. Moore, Jr., Joseph F. Tripp, and Lyon G. Tyler, Jr., 299-309. Westport, Conn.: Greenwood Press, 1988.

Sontag, Susan. *On Photography*. New York: Farrar, Straus, and Giroux, 1977.

Stange, Maren. "The Management of Vision: Rexford Tugwell and Roy Stryker in the 1920s." *Afterimage* 15 (March 1988): 6-10.

———. *"Symbols of Ideal Life": Social Documentary Photography in America, 1890-1950*. New York: Cambridge Univ. Press, 1989.

Stein, Gertrude. *Look at Me Now and Here I Am: Writings and Lectures, 1909-1945*. Ed. Patricia Meyerowitz. Harmondsworth: Penguin, 1971.

Stein, Sally. "Peculiar Grace: Dorothea Lange and the Testimony of the Body." In Partridge, *Dorothea Lange: A Visual Life*, 58-89.

Steinbach, Alice C. "Berenice Abbott's Point of View." *Art in America* (Nov.–Dec. 1976): 77-81.

Steinbeck, John. *The Grapes of Wrath*. New York: Viking Press, 1939.

Steiner, Ralph. *A Point of View*. With an introduction by Willard Van Dyke. Middletown, Conn.: Wesleyan Univ. Press, 1978.

[Steiner, Ralph]. "The Small Town." *PM's Weekly*, 13 Oct. 1940, 46-49.

Steiner, Wendy. "Gertrude Stein's Portrait Form." Ph.D. diss., Yale Univ., 1974.

Stieglitz, Alfred. "Pictorial Photography." In Trachtenberg, *Classic Essays on Photography*, 115-23. Originally published in *Scribner's Magazine*, Nov. 1899.

Stoddart, Jess, ed. *The Quare Women's Journals: May Stone and Katherine Pettit's Summers in the Kentucky Mountains and the Founding of the Hindman Settlement School*. Ashland, Ky.: Jesse Stuart Foundation, 1997.

Stott, William. *Documentary Expression and Thirties America*. New York: Oxford Univ. Press, 1973; reprint, Chicago: University of Chicago Press, 1986.

Stryker, Roy E. "The FSA Collection of Photographs." In *Photography in Print*, edited by Vicki Goldberg, 349-54. Albuquerque: University of New Mexico Press, 1981. Originally published in Stryker and Wood, *In This Proud Land*.

Stryker, Roy E., and Paul Johnstone. "Documentary Photography." In *The Cultural Approach to History*, edited by Caroline F. Ware, 324-30. New York: Columbia Univ. Press, 1940.

Stryker, Roy E., and Nancy Wood. *In This Proud Land*. Greenwich, Conn.: New York Graphic Society, 1973.

Sundell, Michael G. "Berenice Abbott's Work in the 1930s." *Prospects* 5 (1980): 269–92.

Susman, Warren. *Culture as History: The Transformation of American Society in the Twentieth Century.* New York: Pantheon Books, 1984.

Swados, Harvey, ed. *The American Writer and the Great Depression.* Indianapolis: Bobbs-Merrill, 1966.

Szarkowski, John. *The Photographer's Eye.* New York: Museum of Modern Art, 1966.

Szarkowski, John, and Maria Morris Hambourg. *The Work of Atget.* 4 vols. Boston: New York Graphic Society, 1982.

Tagg, John. *The Burden of Representation: Essays on Photographies and Histories.* Amherst: Univ. of Massachusetts Press, 1988.

Tanno, Jessica. "Urban Eye on Appalachia." *Americana* (July–Aug. 1982): 29-31.

Taylor, Paul. *On the Ground in the Thirties.* Salt Lake City: Gibbs M. Smith, 1983.

Terrill, Tom E., and Jerrold Hirsch, eds. *Such as Us: Southern Voices of the Thirties.* Chapel Hill: Univ. of North Carolina Press, 1978.

These Are Our Lives: As Told by the People and Written by Members of the Federal Writers' Project of the Works Progress Administration in North Carolina, Tennessee, and Georgia. Chapel Hill: Univ. of North Carolina Press, 1939.

Thornton, Gene. "Ulmann Forces a New Look at Pictorialism." *New York Times,* 12 Jan. 1975, 23.

Tocqueville, Alexis de. *Democracy in America.* 2 vols. London: Saunders and Otley, 1835, 1840. Rev. ed., New York: Harper and Row, 1985.

Trachtenberg, Alan. "From Image to Story: Reading the File." In Fleischhauer and Brannan, *Documenting America,* 43-73.

———. *The Incorporation of America: Culture and Society in the Gilded Age.* New York: Hill and Wang, 1982.

———. *Reading American Photographs: Images as History, Mathew Brady to Walker Evans.* New York: Hill and Wang, 1989.

———, ed. *Classic Essays on Photography.* New Haven, Conn.: Leete's Island, 1980.

Tucker, Anne Wilkes. "Photographic Facts and Thirties America." In *Observations: Essays on Documentary Photography,* edited by David Featherstone, 40-55. Carmel, Calif.: Friends of Photography, 1984.

———, ed. *The Woman's Eye.* New York: Alfred A. Knopf, 1973.

Tugwell, Rexford Guy, Thomas Munro, and Roy E. Stryker. *American Economic Life and the Means of its Improvement.* New York: Harcourt, Brace, 1924.

Twelve Southerners. *I'll Take My Stand: The South and the Agrarian Tradition.* New York: Harper and Brothers, 1930; reprint, Baton Rouge: Louisiana State Univ. Press, 1977.

Ulmann, Doris. *The Darkness and the Light: Photographs by Doris Ulmann.* With essays by William Clift and Robert Coles. Millerton, N.Y.: Aperture, 1974.

———. *A Portrait Gallery of American Editors, Being a Group of XLIII*

Likenesses. With an introduction by Louis Evan Shipman. New York: William Edwin Rudge, 1925.

———. *Roll, Jordan, Roll.* New York: Robert O. Ballou, 1933.

———. "The Stuff of American Drama: In Photographs by Doris Ulmann." *Theatre Arts Monthly* 14 (Feb. 1930): 132-41.

University of Louisville Photographic Archives. "Things As They Were: FSA Photographers in Kentucky, 1935-1943." Exhibition Catalogue. Exhibit, 6 Sept.–9 Nov. 1985.

"U.S. to Find Work for 3500 Artists." *New York Times,* 4 Oct. 1935.

Van Dyke, Willard. "The Photographs of Dorothea Lange—A Critical Analysis." *Camera Craft* (Oct. 1934): 464–65.

Walsh, George, Colin Naylor, and Michael Held, eds. *Contemporary Photographers.* New York: St. Martin's Press, 1982.

Warren, Dale. "Doris Ulmann: Photographer-in-Waiting." *Bookman* 72 (Oct. 1930): 129–44.

Warren, Harold F. . . . *A Right Good People.* Boone, N.C.: Appalachian Consortium Press, 1974.

Weaver, Kay, and Martha Wheelock. *Berenice Abbott: A View of the 20th Century.* Sherman Oaks, Calif.: Ishtar Films, 1992.

Wertheim, Arthur F. "Constance Rourke and the Discovery of American Culture in the 1930s." In *The Study of American Culture: Contemporary Conflicts,* edited by Luther S. Luedtke, 49–61. Deland, Fla.: Everett/Edwards, 1977.

Whisnant, David. *All That Is Native and Fine: The Politics of Culture in an American Region.* Chapel Hill: Univ. of North Carolina Press, 1983.

White, Minor. "Eugène Atget." *Journal of Photography and Motion Pictures of the George Eastman House* 5 (April 1956): 76–83.

White, Robert. "Faces and Places in the South in the 1930s: A Portfolio." *Prospects* 1 (1975): 415–29.

Wiebe, Robert. *The Search for Order: 1877–1920.* New York: Hill and Wang, 1967.

Williams, Susan Millar. *A Devil and a Good Woman, Too: The Lives of Julia Peterkin.* Athens: Univ. of Georgia Press, 1997.

Williams, T. Harry. *Huey Long.* New York: Vintage, 1969.

Wolcott, Marion Post. Keynote address. Presented at "Women in Photography" conference, Syracuse Univ., 10–12 Oct. 1986. Speech notes.

———. *Marion Post Wolcott: FSA Photographs.* Introduction by Sally Stein. Carmel, Calif.: Friends of Photography, 1983.

"Woman with Camera Snaps Revealing History of New York Life in Its Homeliest of Garb." *New York World-Telegram,* 11 Nov. 1938.

Woodward, C. Vann. *The Strange Career of Jim Crow.* 3d rev. ed. New York: Oxford Univ. Press, 1974.

Wright, Richard, and Edwin Rosskam. *Twelve Million Black Voices: A Folk History of the Negro in the United States.* New York: Viking, 1941.

Zito, Tom. "Abbott: Glimpses of the Artiste." *Washington Post*, 3 April 1976.

Zubryn, Emil. "Saving New York for Posterity." *Photography: A British Journal of Commercial, Advertising & Portrait Photography* (Jan. 1938): 12.

Zwingle, Erla. "A Life of Her Own." *American Photographer* (April 1986): 54–67.

Index

Page numbers in italics refer to illustrations.

Abbott, Berenice, 1, 4, 5, 6, 8, 69, 232, 249, 291; and Eugène Atget, 259–63, 271–72; camera methods/styles of, 251, 252, 255, 256–57, 259, 261, 264, 266–73, 274–75, 279–81, 283, 286, 288–89, 294, 295–96, 326 n 6; education of, 253–54, 255; family background of, 253; portrait business of, 255–59, 261, 265; personal vision of, 251–52, 254–55, 256, 262, 265–66, 271–73, 278–80, 281, 283, 288–89, 294, 295–96; and Man Ray, 255–56, 259. Works: *Sylvia Beach portrait*, 256; *Changing New York* (book), 287; *Chicken Market*, 275, 276; *East Forty-Eighth Street*, 283, 284; *James Joyce portrait*, 256–57, 258; *Jane Heap portrait*, 257; *Princess Eugene Murat portrait*, 257; *Rockefeller Center (1)*, 266, 267; *Rockefeller Center (2)*, 266, 268, 269; *Solita Solano portrait*, 257; *Trinity Church*, 269, 270; *View of Fortieth Street*, 281, 282
Abbott, Lawrence, 26
Adams, Ansel, 164, 175
Adler, Felix, 10–11
advertising, 20, 197–98, 210, 216–17, 224, 318 n 1; photography in, 4, 6, 175, 206, 208, 209–10, 222–23, 225–30, 234, 238, 241, 249, 319 n 15
African-Americans, 21, 28, 41–42, 44–47, 45, 50, 64, 68, 71, 114, 115, 125, 129–30, 131, 165, 166–73, 167, 172, 174, 175, 180–81, 182, 188–90, 189, 237,

240, 274, 296, 304 nn 48, 49; 305 n 56, 316 n 62
Agee, James, 139, 240. Work: *Let Us Now Praise Famous Men*, 240
Agrarians, 34, 41, 46, 165, 187, 240
Agricultural Adjustment Administration (AAA), 103, 184, 190
agriculture, 44–46, 66–67, 100, 103, 104, 106, 108, 112, 113, 115, 118, 120–21, 122, 136–37, 144–45, 158–61, 166–70, 169, 172, 175, 177, 182, 195, 219–20, 237–40
Ahlstrom, Florence, 81
Allen, Frederick Lewis, 163
Allen Plantation Project (FSA), 180–81
American Artists' Congress, 232
American Birth Control League, 128
American Federation of Arts, 51
American frontier, 118–23, 119, 192–93, 241–43
American Historical Association, 175, 291
American individualism, 6, 11, 17, 53–56, 75, 80–81, 105–13, 109, 111, 113, 114, 118, 121, 123, 195, 241, 243, 293, 294
Anderson, Sherwood, 22–23, 177, 180, 301 n 19. Work: *Home Town*, 180
Annual of American Design (1931), 52
Appalachia: coal mining in, 30, 140–44, 141; craftspeople in, 6, 30, 51–58, 54, 55, 57, 66, 72, 292, 305 n 59; Melungeons in, 46–47; music in, 63–65, 102; John Jacob Niles and, 27–28, 64–65, 306 n 72; Marion Post and,

140–44, 181, 183–88; role of
religion in, 40–41, 185–87, *186*;
Doris Ulmann and, 34, 35–41,
50–72, 102, 187, 292–93, 306
nn 71, 72
architecture, 6, 204, 205, 210, 213,
221–22, 259, 263–64, 265–71,
319 n 14
art photography, 5, 13, 17, 51, 58,
68–69, 99, 175, 206, 208, 214,
227–28, 252, 272, 285, 291,
302 n 28
Arts and Crafts Movement, 52, 72.
See also Appalachia,
craftspeople in
Associated Press, 137
Atget, Eugène, 29, 259–63, 271–72,
326 n 15. Work: *Rue de la
Montagne-Sainte-Geneviève
scene*, 261
automobiles, 69, 118–20, *119*, 121–
23, *122*, 199, 205; Chrysler
Corporation, 208, 221–22, 225,
236; Ford Motor Company,
230; General Motors, 230;
LaSalle Corporation, 225, *226*,
238; Pierce-Arrow Motor Car
Company, 225, 227

Baldwin, Sidney, 144
Bayou Borbeaux Plantation Project
(FSA), 180–81
Beard, Charles, 165
Becker, Jane, 305 nn 59, 66
Beloff, Halla, 8, 244
Bennett, John, 23
Berea College, 36, 60, 61–63, 66–68,
71, 72, 292, 302 n 26; Appala-
chian Center at, 72
Birth Control Research Bureau. *See*
American Birth Control League
Bishop, H.S., 225, 227
Blanding, Sarah Gibson, 303 n 41
Boone Tavern, 67, 302 n 26
Borchert, James, 6
Bourke-White, Margaret, 1, 4, 5, 6,
8, 69, 163, 224, 264, 291, 295;
and advertising photography, 6,
197–98, 204, 206, 208–10,
216–17, 222–24, 225–27; and

Erskine Caldwell, 236–40, 246–
48, 295, 325 n 65; camera
methods/styles of, 147–48, 204–
7, 208–9, 214, 216–17, 220,
221–24, 225–30, 233–36, 238–
40, 241–46, 248–49, 263, 271,
295, 296; and Everett Chapman,
201–4, 319 n 10; education of,
200–203; family background of,
199–202, 224–25; and Henry
Luce, 209–10, 216, 217, 240,
321 n 28; personal vision of,
198–99, 203, 204, 207–10,
227–30, 231, 236–37, 241–49,
295; phrenological study of,
200–201; and Ralph Steiner,
213, 216, 246; and Roy Stryker,
206; and Clarence White, 12,
201, 211, 213. Works: *Bank
Vault*, 214–16, *215*; *Chrysler
Building*, 221–22, *223*; *Conversation Club*, 244–46, *245*; *Dust
Bowl Faces*, 233–34, *235*; *Eyes
on Russia*, 220–21; *Fort Peck
Dam, Montana*, 241, *242*;
LaSalle Automobile, 225, *226*;
Middletown series (*Life*), 244–
46; *Radio Loudspeakers*, 228,
229; *Ozark, Alabama*, 238, *239*;
Plow Blades, 211, *212*; *Portrait
of Myself* (autobiography), 216;
Say, Is This the U.S.A., 247–48;
Soviet Panorama, 219–20; *You
Have Seen Their Faces*, 118,
147–48, 295, 324 n 55
Brady, Mathew, 100
Brannan, Beverly, 130, 157, 187
Brigman, Anne, 82
Buchanan, Paul, 73

Cabins in the Laurel (Muriel
Sheppard and Bayard Wootten),
73
Cahill, Holger, 277–78, 286, 328 n
34
Caldwell, Erskine, 118, 147–48,
165, 236–40, 246–48, 295, 325
n 65. Works: *Say, Is This the
U.S.A.*, 247–48; *Tobacco Road*,
236–37; *You Have Seen Their*

Faces, 118, 237–40, 295, 324 n 55
Camera, 104, 126
Cameron, Julia Margaret, 29
Campbell, John C., 61. Work: *The Southern Highlander and His Homeland,* 61
Campbell, Olive Dame, 23, 27–28, 36, 50, 51, 60–61, 69–70, 307 n 79
Century of Progress Exposition (Chicago, 1933), 52, 230
Changing New York, 6, 273–87, 288–89
Chapman, Everett, 201–4
Chrysler Corporation. *See* automobiles
Civilian Conservation Corps, 190
class issues, 5, 10–11, 12, 13–15, 42, 53, 69, 71–73, 82–83, 92–95, 96–98, 105–6, 108, 114, 117–18, 125, 127–28, 130, 132, 136, 151–56, 178–80, 188–90, 232, 238–40, 243, 244–46, *245,* 249, 271
Cleveland: Chamber of Commerce, 204, 208–9; Museum of Art, 206
Clift, William, 46, 71
coal mining, 30, 34, 72, 140–42, 143, 218
Coburn, Alvin Langdon, 29, 205
Cocteau, Jean, 257, 259
collective action, 70, 125, 132, 134–37, 159, 168, 170, 175–81, *179, 186,* 187, *189,* 293–94
Collier's, 162
Columbia University, 11, 12, 13, 72, 103, 201, 206, 253
Conrat, Richard, 83
Conway, Jill, 80
Cooper, Hugh, 220
Corning Glass, 213
Couch, W.T., 164, 316 n 57
Creech, William, 60
cultural forces: in Europe, 130, 132–33, 254–62, 271–72, 296; in the U.S., 5–6, 7, 20–21, 25, 34–36, 41–52, 60–62, 67–68, 71–73, 77–78, 80–81, 84, 91–98, 100–

106, 108–12, 114–16, 118–23, 129, 177–78, 180, 185, 208, 216, 228, 237–38, 240, 241–43, 244–46, 248, 253–55, 257, 259, 263–65, 274–79, 288–89, 292–97, 322 n 38
cultural lag, 228–29, 249, 274–75, 288, 296
Cunningham, Imogen, 5, 81, 82
Curtis, James, 8, 108, 110, 175–76, 310 n 38

Davidov, Judith Fryer, 299 n 3, 305 n 63
Davidson, Donald, 41, 240
Delphic Galleries, 42
Dewey, John, 104, 278, 312 n 7, 319 n 15, 323 n 48, 328 n 34
Dixon, Maynard, 84, 99, 100–101
Dixon, Penelope, 137
documentary films, 91–92, 127, 136, 230, 231
documentary photography, 1, 67, 92, 163, 247, 262, 291–97; development as a medium of expression, 5, 73, 91, 132, 156–57, 163, 195, 262, 263, 291–97; expansiveness of, 5, 29–30, 60, 68, 91–93, 163–64, 244, 262, 272, 278, 288–89, 291; relationship of modernism to, 6, 195–98, 211–13, 222, 237–40, 248–49, 252, 257, 259–60, 262, 266, 269–72, 283, 288, 294–96, 300 n 7; relationship of pictorialism to, 27, 58, 205, 252, 272; relationship of words to, 91–92, 101–3, 112, 114, 116, 117–18, 156, 180, 236–40, 293, 310 n 38, 311 n 50, 324 n 56; role of "fieldwork" in, 5, 83, 94–96, 114, 118–20, 128, 140, 147–53, 161–62, 194, 243; role of "the people" in, 6, 7, 29, 71–72, 91–92, 104, 125, 132, 142, 198, 232, 244–46; women's contributions to, 5, 69, 123–24, 125, 194–95, 244, 249, 252, 283, 291–97, 299 n 3; women's opportunities in, 3–4,

132, 139, 180, 185, 224, 291–
    97
—, functions of: educational, 71–72,
    103, 105, 127, 164–65, 218–21,
    237, 240, 262, 275–79, 288,
    291, 292; political, 1, 4–6, 91–
    92, 114–15, 118–23, 125, 132,
    163, 277–79; social, 4–6, 28–
    29, 60, 68, 91–92, 99, 114–15,
    118–23, 130, 139–40, 142, 153,
    155, 163, 198–99, 244–46, 266,
    274–79, 288, 294
Doherty, Robert, 105
Doud, Richard, 80
Duchamp, Marcel, 211. Work: *Nude
    Descending A Staircase, No. 2*,
    211
Duncan, Isadora, 79
Dust Bowl, 104, 106, *107*, 117–23,
    119, 139, 145, 233–36, *235*,
    237, 292

Earhart, Amelia, 4
Eastman Kodak, 162, 225
Eaton, Allen, 28, 31, 36, 50–52, 60,
    65, 66, 68, 69, 70, 72, 101, 307
    n 85. Work: *Handicrafts of the
    Southern Highlands*, 72
Ederle, Gertrude, 4
Eisenstaedt, Alfred, 248
Empire State Building, 222, 224,
    228, 271
E.P. Dutton, 287
Ethical Culture School, 10–11, 34,
    35, 71
Evans, Walker, 68, 71, 104, 139,
    142, 180, 240, 311 n 52, 313 n
    20. Work: *Let Us Now Praise
    Famous Men*, 240

farm cooperatives, 127, 136–37
Farm Security Administration (FSA),
    103, 116, 144–45, 157, 159,
    180–82, 184, 188, 190, 191,
    311–12 n 2; Photography Unit
    of, 65, 68–69, 104, 105, 106,
    108, 116–17, 121, 123–24, 126,
    127, 128, 138–40, 142–47, 148,
    153, 161, 163, 165, 171, 173,
    175, 176, 180, 193–94, 237–38,

280, 287, 291, 292, 294, 311 n
    56, 313 n 20, 315 n 47
Farmers' Cooperative Union, 70
Featherstone, David, 17, 300 n 11,
    301 n 11
Federal Art Project (FAP), 67, 273–
    74, 275, 276–88, 328 n 33, 329
    n 39
Federal Emergency Relief Adminis-
    tration (FERA), 100, 104, 277,
    285
Federal Music Project (FMP), 67,
    274, 275, 287–88
Federal Theatre Project (FTP), 274,
    287–88
Federal Writers' Project (FWP), 50,
    165, 274, 280, 287–88, 316 n
    57. Works: *These Are Our
    Lives*, 165; *WPA City Guide
    Book to Washington, D.C.*, 280;
    WPA Guidebook to Kentucky,
    183
feminist theory, 2–3, 80–81, 296–97,
    299 nn 3, 4; 311 n 57
Fifth International Congress of
    Photography (Brussels), 262
Film and Photo League. *See* Work-
    ers' Film and Photo League
Fischer, John, 145
Flanner, Janet, 257
Fleischmann, Trude, 133, 312 n 12
Ford, Henry, 220
Fort Peck Dam, 241–43
Fortune, 104, 127, 136–37, 209–10,
    213, 216, 217, 219, 221, 230,
    233–36, 263, 313 n 20, 321 n
    28
Freytag-Loringhoven, Elsa von, 254
Frontier Films, 136, 313 n 16

Garland, Hamlin, 192
Genthe, Arnold, 78–79
Gide, André, 257
Gilman, Charlotte Perkins, 21
Gilpin, Laura, 5, 12, 301 n 15
Glasgow, Ellen, 21
Goldberg, Vicki, 199, 202, 232, 247,
    248
Golden Gate International Exposi-
    tion (San Francisco), 175

Graham, Martha. *See* modern dance
Grand Central Station exhibition, 173
Great Depression, 4, 7, 44–46, 52, 83, 84, 86, 91–94, 95, 102–3, 105, 123–24, 125, 126, 133–34, 195, 206, 216, 217, 241–43, 249, 263, 264–65, 293, 294
Green, Paul, 238, 240, 304 n 49
Greenwich Village, 129, 253
Grierson, John, 92
Group Theatre, 134, 136
Guggenheim, Peggy, 256
Guggenheim Foundation, 265
Guimond, James, 29, 68, 205, 324 n 56
Gullah culture, 28, 41–42, 44–47, 71

Hagood, Margaret Jarman, 164, 166, 168, 316 n 56
Hales, Peter, 252
Hambourg, Maria Morris, 261
Hampton Institute, 68
handicrafts: Arts and Crafts Movement, 52; quilting, 52, 56–58; Southern Highland Handicraft Guild (SHHG), 51, 52, 292; Southern Mountain Handicraft Guild (SMHG), 51, 52; Doris Ulmann's photographic survey of, 10, 30, 48, 50–60, 65–66, 69–70, 292
Handicrafts of the Southern Highlands (Allen Eaton), 72
Hardt, Hanno, 132
Harris, Joel Chandler, 41
Heilbrun, Carolyn, 80
Herring, Harriet, 164, 165
Heyman Therese, 99
Highlander Folk School, 136
Hine, Lewis: 11, 102, 271; and Roy Stryker, 69; and Doris Ulmann, 11, 13, 14
Hitchcock, Henry-Russell, 265
Hitler, Adolf, 133
Hollywood, 216, 230, 231, 244
Hoover, Herbert, 41
Horvath, David, 187
Houk, Edwynn, 2

Humphrey, Doris. *See* modern dance
Hurley, F. Jack, 146, 191
Hutchins, William J., 36, 61, 62–63, 66, 67, 73, 307 n 83

I'll Take My Stand (Twelve Southerners), 34–35, 302 n 32. *See also* Agrarians
immigrants, 10–11, 143–44, 177, 314 n 27, 320 n 15; backlash against, 35; urban settlement of, 35, 77, 271, 274, 275, 276
industrial design, 52, 197, 198, 204, 205–9, 211–12, *212*, 220–21, 227–30, *229*, 241–42, 263, 295, 318 n 2
industrialism: in America, 10–11, 22, 34–36, 41–42, 198–99, 205–10, 218–19, 227–31, 241, 242, 272, 292–93, 295; in the Soviet Union, 217–21, 321 n 35
Institute for Research in Social Science (University of North Carolina, Chapel Hill), 164, 165–66, 317 n 65

Jaeger, Charles, 12, 13, 18, 27, 301 n 12
John C. Campbell Folk School, 51, 60, 61, 69, 70, 71, 72, 292, 305 n 67
Johnston, Frances Benjamin, 68. Work: *Hampton Album*, 68
Jones, Loyal, 72
Joyce, James, 256–57, *258*

Kanaga, Consuela, 81
Katz, Paul, 4
Kazan, Elia, 136
Kerber, Linda, 2
Kerr, Clark, 98, 310 n 35
Kessler-Harris, Alice, 2
Kozol, Wendy, 243
Ku Klux Klan, 41

labor: in Central Europe in the 1930s, 132–33; as depicted by Berenice Abbott, 266; as depicted by Lewis Hine, 11; as depicted by Dorothea Lange,

97, *109*, 112, *115*, 120–23, *122*, 173; as depicted by Marion Post, 143–44, 148–53, 166–71, *167*, 294; as depicted by Doris Ulmann, 45, 48–49, *49*, 51–60, *54*, *55*, *57*, 66–67, 71; in coal mines, 143–44; on government projects, 233, 241–43; in migrant camps, 103, *111*, 144, 148–53, *152*, 314 n 37; in rural South Carolina, 41, 105–6, *109*; union activity in the U.S., 96–98, *97*, 128, 136, 143, 162, 165; in urban areas, 91, 96–98, 207, 266, 320 n 15
LaDelta Project (FSA), 180–81, *182*
Landgren, Marchal, 273, 327 n 27
Lange, Dorothea, 1, 4, *5*, 6, 8, 71, 138, 158, 249, 264, 291; camera methods/styles of, *5*, 75, 77–78, 79, 83–90, 92, 96–98, 108–9, 110–17, 126, 142, 143, 165, 173, 240, 279–80, 293, 294, 296; children of, 84, 86, 93, 94, 110; and Maynard Dixon, 84, 99, 100–101; education of, 77, 78, 79–80; family background of, 75, 76–78, 79, 81; and Arnold Genthe, 78–79; personal vision of, 75–76, 83–84, 90, 93, 98–99, 105–6, 116–17, 139, 195, 287, 293, 294, 309 n 16; physical health of, 77–78; portrait business of, 82–87, 89, 93, 96, 101, 309 n 15; and Roy Stryker, 68–69, 114, 116–17, 123–24, 139, 146, 165, 279–80, 311 n 52; and Paul Taylor, 100–102, 114, 117–18, 121–23, 147; and Clarence White, 12, 79–80, 211. Works: *An American Exodus* (book), 117–23; *Displaced Tenant Farmers*, 112, *113*; *Erika Weber series*, 86–87; *Laborer Returning Home*, 121–23, *122*; *May Day Demonstration*, 96–98, *97*; *Migrant Mother*, 110, *111*; *Mrs. Kahn and Child*, 85, 86; *Oklahoma Drought Refu-gees*, 105–6, *107*; *Plantation Owner*, 114, *115*; *South Carolina Sharecropper*, 106, 108, *109*; *Stalled in the Desert*, 118–20, *119*; *Untitled (adolescent boy)*, 90; *Untitled (Hopi man)*, 90; *Untitled (Native American woman)*, 88; *Untitled (Profile of Hopi man)*, 88; *"White Angel" Breadline*, 93–96, *95*
Lang Syne Plantation, 28, 41, 42, 46
Lawrence, Una Roberts, 173, 175
League of Women Shoppers, 232
Lee, Russell, 117, 123, 139, 191, 311 n 56
Lerner, Gerda, 2, 297, 318 n 83
Library of Congress, 65
Life, 1, 126, 138, 240–48
Lifson, Ben, 142, 155
Lippmann, Walter, 299 n 2
Literary Guild of America, 21
Lomax, Alan, 274
Look, 104, 126, 138
Lorentz, Pare, 91. *See also* documentary films
Louisville Courier-Journal, 183
Luce, Henry, 209–10, 216, 217, 240, 321 n 28
Lynd, Robert and Helen, 175–76, 244. Works: *Middletown*, 175–76; *Middletown in Transition*, 175–76, 244

Macdonald, Dwight, 213
MacDowell Writers' Colony, 41
machine-age aesthetic, 52, 198–99, 207–8, 210–11, *212*, *215*, 218, 221–22, *223*, 225–27, 228–30, *229*, 248–49, 271, 283, 295, 318 n 2, 322 n 36
machine technology, 6, 30, 34, 175, 198–99, 205, 207–8, 218, 221, 222–23, 248–49, 263–64, 272, 295, 296
MacLeish, Archibald, 114, 213–14, 311 n 50
Marchand, Roland, 199, 318 n 1, 319 n 15, 320 n 19
Marsh, Reginald, 232

McAlmon, Robert, 256
McCausland, Elizabeth, 163, 272, 273, 291
McMahon, Audrey, 273, 277–78, 286
Meikle, Jeffrey, 319 n 14, 320 n 17, 320 n 24, 322 n 36
Meldrum and Fewsmith Advertising Agency, 216–17, 222–23
Melrose Plantation, 27
Mencken, H.L., 21, 35, 240
Miami, 151, 153–56
migrant labor, 103, *111, 122,* 144, 148–53, *152*
Millay, Edna St. Vincent, 21, 35
Mississippi Delta region, 114, *115,* 129–30, *131,* 170–73, 274
Model, Lisette, 272, 327 n 25
modern dance, 129, 130, 132
modernism: in photography, 6, 197–99, 204, 211–13, 222, 227–28, 237–38, 240, 248–49, 252, 257, 259–60, 262, 266–72, 281, 283, 285, 288–89, 294, 300 n 7; in the U.S., 7, 227, 249, 288–89
Modotti, Tina, 5
Monuments Historiques (Paris), 260–61
Morgan, Arthur, 233
Moscow Art Theatre, 134
Mountain Workers' Conference (1930), 61–62
muckrakers, 13
Mumford, Lewis, 208, 320 n 17
Museum of Modern Art, 244, 265, 283
Museum of the City of New York, 265, 272–73, 285
Mydans, Carl, 138, 313 n 20

NAACP *(National Association for the Advancement of Colored People),* 44
Natanson, Nicholas, 171
National Broadcasting Company (NBC), 227–30, 234, 322 n 42
National Committee for the Defense of Political Prisoners, 231–32
Native Americans, 46–47, 84, 88–90, 106, 121, 243

nativist movement (1920s), 35–36. *See also* Ku Klux Klan
Nazism, 126, 132–33, 293–94
Necarsulmer, Henry, 18, 301 n 13, 304 n 50
Nesbit, Molly, 262, 263
New Deal, 2, 67, 72, 91, 94, 100, 103, 105, 126, 127, 144–45, 232–33, 241, 251, 273, 277–78, 286, 287–88
New England, 15–16, *16,* 19, 27, 176–80, *179,* 265–66, 292
Newhall, Beaumont, 244, 285, 288
New Orleans, 48, 50, 162, 192
New School for Social Research, 129, 232, 273
New Workers School, 232
New York Herald Tribune, 274
New York Historical Society, 265
New York Training School for Teachers, 78
New York World's Fair, 175, 287
New York World-Telegram, 281
Niles, John Jacob, 18, 20, 27, 28, 31, 34, 35, 36, 37, 38, 50, 62–66, 70–71, 147, 183, 301 n 13, 303 nn 34, 42
1933 Century of Progress Exposition, 52, 230
1920s youth movement, 38–40

Oakes, James, 121
Odets, Clifford, 134, 312 n 14. Work: *Waiting for Lefty,* 134, 312 n 14
Odum, Howard, 164, 166, 313 n 20, 316 n 56
Office of War Information (OWI), 194
Ohio State University, 253
Oliver Chilled Plow Company, 211
O'Neal, Hank, 255
O'Neill, Eugene, 207, 253, 304 n 49. Works: *Dynamo,* 207; *Moon of the Caribbees,* 253
Orhn, Karin, 132
Orvell, Miles, 225, 300 n 7 (introduction), 318 n 2

Page, Thomas Nelson, 41

Paris: art and culture (1900–1930), 3, 254–62, 271–72, 296
Parker, Dorothy, 21
Partridge, Roi, 81, 84
Peeler, David, 7, 84–85, 98, 164, 234, 238, 299 n 2, 315 n 55
People of the Cumberland, 127, 138
Peterkin, Julia, 27, 28, 41–44, 46–48, 50, 101, 304 n 48. Works: *Bright Skin,* 47; *Roll, Jordan, Roll,* 44; *Scarlet Sister Mary,* 41–42
Philadelphia Evening Bulletin, 126, 137–38
photographers. *See* women photographers *and individual photographers' entries*
photographic equipment, 3, 33; used by Berenice Abbott, 264; used by Eugène Atget, 259–60; used by Margaret Bourke-White, 208, 213, 320 n 25; used by Dorothea Lange, 89–90; used by Marion Post, 133, 137, 139, 156, 162, 176, 183; used by Doris Ulmann, 10, 30–31, 73, 292
photographic methods and styles, 3, 5, 8, 56, 132–33, 142–43, 193, 213–14, 237–38, 259–60, 262, 263, 264, 292–97, 304 n 51; oil pigment printing, 17; Photo-Secession, 13, 15, 17, 300 n 1.7; pictorialism, 13, 15, 28, 51, 58, 205, 211, 213, 252, 272, 292, 300 n 1.7, 305 n 64, 325 n 2; Workers' Photography Movement, 132, 185
photographs: analysis of, 2–3, 6–8, 17, 29–30, 105, 262–63, 296–97, 330 n 2; as autobiographical narratives, 1, 64–65, 88–89, 94, 110, 125, 173, 236, 246; as cultural texts, 6, 8, 29, 58, 68, 71–73, 92, 102, 118–23, 127, 136, 139, 142–43, 153–55, 173, 180, 195, 205, 231, 237–38, 244–46, 252, 295; as educational tools, 71–72, 103–5, 127, 164–65, 206, 208, 231, 262,

278–79, 285, 288, 291, 292; as historical records, 58–60, 66–68, 72, 123–24, 127, 164, 218–21, 237–38, 240, 252, 259–61, 262, 266, 269–71, 272–74, 279, 281, 288, 291, 295, 296; as political and social messages, 1, 28–30, 52, 68, 91–92, 104, 105, 108, 112, 123, 130, 132–33, 136, 139, 140, 142–43, 153–55, 163–64, 171, 232, 233, 236–37, 238–49, 273–89, 292, 294; role in the 1920s and '30s, 2, 4, 7, 103, 105, 132–33, 156, 163–64, 185, 205–6, 213–14, 230–31, 236–38, 248–49, 251–52, 259–61, 262–63, 269, 271–72, 285, 291–97, 299 n 2
photography clubs and societies, 17, 81–82, 132, 135–36, 213
photojournalism, 4, 136–37, 217–21, 240–49
photomurals, 227–28, 231
Photo Notes, 163
Photo-Secession. *See* photographic methods and styles
Picasso, Pablo, 259, 260
pictorialism. *See* photographic methods and styles
Pictorial Photographers of America, 13, 17, 18, 213
picture magazines, 126, 138, 139, 163, 237–38
Pine Mountain Settlement School, 60
Pittsburgh Salon of Photography, 17
PM's Weekly, 180
Popular Photography, 1
Post, Emily, 163
Post, Helen, 128, 134
Post, Marion, 1, 4, 5, 6, 8, 69, 247, 249, 264, 291, 299 n 6; camera methods/styles of, 123, 127, 128, 142–44, 149, 150–57, 159, 165–71, 176–83, 185–87, 193, 240, 293–94, 296, 311 nn 56, 57; 314 n 26, 315 n 46; education of, 129–30, 132–36, 312 n 7; family background of, 127–29; and Doris Humphrey, 129–30, 153; personal vision of,

127, 130, 150–51, 155, 170–73, 176–81, 293–94; and Ralph Steiner, 135–38, 213; and Paul Strand, 135, 138; and Roy Stryker, 138–39, 140, 144–46, 148, 150, 151, 156–59, 161–63, 176, 183, 184, 191, 193, 222, 224, 294, 313 n 20; and Lee Wolcott, 192, 193–94, 299 n 6. Works: *Boy Scouts Examining Weapons*, 194; *Cashiers Paying Off Cotton Pickers*, 170–71; *Child of an FSA Client*, 181, 182; *Counting Ballots after a Town Meeting*, 178–80, 179; *Couple Stripping Tobacco*, 166, 167; *Dancing in a Juke Joint*, 173, 174; *Dinner on Corn Shucking Day*, 168–70, 169; *Family Dinner*, 159, 160; *Land Use Planning Committee (Locust Hill)*, 188–90, 189; *Land Use Planning Committee (Yanceyville)*, 188–90, 189; *Man Entering Theatre*, 129–30, 131; *Men Carrying a Coffin*, 185, 186; *Miner's Wife*, 140–42, 141; *People Sunning Themselves*, 153–55, 154; *Pie Supper*, 187; *Town Meeting*, 178–80, 179; *Unsatisfied Farmers with Cotton Buyer*, 171–73, 172; *Visiting After a Memorial Service*, 185–87, 186; *Wife and Child of Migrant Laborer*, 150–51, 152. *See also* Wolcott, Marion Post

Post, Nan, 127–29, 312 n 5
Progressive reformers, 10, 13, 60–61, 68, 187, 207, 302 n 29
public art, 227–28, 272–89, 304 n 54, 329 n 39
Public Works of Art Project (PWAP), 277
Puckett, John Rogers, 102, 110, 311 n 50

Raper, Arthur, 161, 162, 164. Works: *Preface to Peasantry*, 161; *Sharecroppers All*, 162
Ray, Man, 255–56, 259, 262

regionalism, 48, 50, 52, 123, 146, 175, 176–80, 228, 274–77, 288, 292, 304 n 54
religious/spiritual life in America, 30, 40–42, 48–50, 173, 175, 185–87; as photographed by Doris Ulmann, 14, 30, 39, 50, 292
Resettlement Administration (RA). *See* Farm Security Administration (FSA)
Richardson, H.H., 266
Riis, Jacob, 102, 271. Work: *How the Other Half Lives*, 271
Roberts, Elizabeth Madox, 23–25. Work: *My Heart and My Flesh*, 25
Rockefeller Plaza, 227–30, 266–67, 272
Roll, Jordan, Roll, 44
Roosevelt, Eleanor, 65
Roosevelt, Franklin D., 91, 103, 231, 243, 274, 287
Roosevelt, Mrs. Theodore, 244
"Roosevelt's Depression" (1937), 144–45, 287
Rosenblum, Naomi, 299 n 3
Ross, Malcolm, 183. Work: *Machine Age in the Hills*, 183
Rosskam, Edwin, 114, 180
Rothstein, Arthur, 139, 140, 158, 193, 314 n 37
rural life in America, 30, 32, 39, 42, 44–48, 45, 55, 58, 59, 60–61, 72, 73, 104, 105–6, 107, 108, 109, 111, 112, 113, 114, 115, 118, 136–37, 144–51, 158–62, 160, 166–73, 167, 169, 175–90, 182, 186, 189, 194–95, 237–40, 239, 292–93, 294, 315 n 50
Russell Sage Foundation, 51, 61, 184

Sacco and Vanzetti case, 35
Sanger, Margaret, 128
San Joaquin Valley, 118
Saturday Evening Post, 222, 231, 263
Saxon, Lyle, 27, 28, 48, 50. Work: *Children of Strangers*, 48
Schlereth, Thomas, 7
Schuster, Max, 219, 220, 233

Scopes trial (1925), 35
Scottsboro Boys case, 232
Sears Century of Progress Exposi-
    tion, 52, 230
Sedgwick, Ellery, 26
segregation (racial), 114, *115*, 129–
    30, *131*, 136, 145–46, 165, 166,
    168–73, *172*, 180–81, 188–90,
    237, 249, 292, 296, 316 n 62
Sekula, Allan, 8, 105
Selznick, David O., 230
sexuality, 64–65, 77–79, 80, 84–86,
    85, 129, *141*, 142, 143, 145–47,
    151, 164, 185, 194, 202–4, 211,
    213, 224, 241, 243, 246, 251,
    257
Shahn, Ben, 104, 138–39, 142
sharecropping, 106, 108, *109*, 118,
    139, 144, 237, 240, 296
Shipman, Louis Evan, 30, 302 n 32
Significance of the Frontier, The
    (Frederick Jackson Turner), 121
Singal, Daniel, 7, 316 n 56
Sklar, Kathryn Kish, 2
skyscrapers, 205, 213, 221–22, *223*,
    241, 266–71, *267*, *268*, *270*,
    282, 295, 319 n 14
small town America, 123, *131*, 175–
    80, *179*, *189*, 188–90
Smith, Al, 41
Snyder, Robert, 314 n 33
social science, 50–51, 98, 100–103,
    117–18, 128–29, 163–64, 170,
    175, 244–46, 262, 291, 293,
    299 n 2
Soil Conservation Service, 190
South (in the U.S.), 27–28, 34–35,
    41–50, 51–56, 60–61, 66, 69–
    71, 72, 106, 108, *109*, 112,
    113, 114, *115*, 117–18, 120–21,
    129–30, 144–56, 158–63, *160*,
    164–65, 166–73, *167*, 180–82,
    182, *189*, 190–91, 232, 233,
    236–40, *239*, 241, 247, 274,
    292, 295, 315 n 50, 318 n 84,
    324 n 56
Southern Baptist Convention, 173,
    175
Southern Highland Handicraft Guild
    (SHHG), 51, 61, 292

Southern Mountain Handicraft Guild
    (SMHG), 51
Southwest (in the U.S.), 84, 88–90,
    120–21
Soviet Union: Five Year Plan in,
    217–21; U.S. recognition of, 231
Stalin, Josef, 217
Standard Oil Company, 80
Stange, Maren, 105, 310 n 41, 319–
    20 n 15
St. Denis, Ruth. *See* modern dance
steel industry, 205, 208–10; Krupp
    Works, 217; Otis Steel Com-
    pany, 205–6, 208–9, 213;
    Republic Steel, 208, 222
Steichen, Edward, 96, 205
Stein, Gertrude, 20–21, 90
Stein, Sally, 79, 155, 187
Steiner, Ralph, 135–38, 213, 216,
    232, 246, 320 n 25
Stieglitz, Alfred, 13, 15, 51, 205,
    252, 285, 300 nn 1.7, 10; 325 n
    2. Works: *Pictorial Photography*,
    51
stock market crash (1929), 7, 214–
    16, 263, 321 n 27
Stott, William, 29, 92, 98, 105, 142,
    180, 315 n 55
Strand, Paul, 135–36, 138
Stryker, Roy, 65, 68, 103, 104, 114,
    116–17, 123–24, 126, 138–39,
    140, 144–46, 148, 150–51,
    156–60, 161–63, 164, 165,
    175–76, 184–85, 206, 280, 291,
    294, 310 n 41, 311 n 52, 313
    nn 20, 24
Sturtevant, Roger, 81, 82, 91, 94,
    308 n 12
Susman, Warren, 228

Tarbell, Ida, 80
Taylor, Paul S., 100–102, 114, 117–
    18, 146, 147, 293, 310 n 35.
    Work: *An American Exodus*,
    117–23
television, 228
Tennessee Valley Authority (TVA),
    233
Terrebonne Project (FSA), 180–81,
    191

Theatre Arts Monthly, 47
Theatre Guild Magazine, 207
Thornton, Gene, 58, 60
Time, 209
tobacco farming, 166–68, *167*
Tocqueville, Alexis de, 120–21
Trachtenberg, Alan, 8, 127, 237,
   300 n 1.7
Transylvania Project (FSA), 180
Trend, 273
Tucker, Anne, 98
Tugwell, Rexford, 103, 206
Turner, Frederick Jackson, 121
Turner, Marie, 183–84

Ulmann, Bernhard, 10
Ulmann, Doris, 1, 4, 6, 7, 8, 101,
   147, 183, 249, 264, 291, 307
   nn 81, 85; and Sherwood
   Anderson, 22–23; camera
   methods/styles of, 5, 9, 18–23,
   30–33, 36–37, 44, 52–53, 55–
   58, 60, 63, 73, 75, 83, 86–87,
   102, 112, 126, 187, 214, 240,
   252, 257, 292–93, 294, 296,
   303 n 34, 305 n 64; death of, 9,
   71; education of, 10–13, 79–80;
   family background of, 10;
   financial independence of, 4,
   69–71; groups photographed by,
   14–15, 19, 30, 41–42, 44, 46–
   47, 50, 307 n 85; and Lewis
   Hine, 11, 13–14; and Charles
   Jaeger, 12, 13, 18, 301 n 12;
   personal vision of, 9, 19–21,
   25–27, 30, 33, 37, 46–47, 58,
   68, 71–73, 105, 143, 195, 292–
   93, 294, 307 n 85; poor health
   of, 12, 27, 38, 63, 70; and
   Clarence H. White, 12, 13, 17,
   27, 56, 58, 211. Works: *Aunt
   Cord Ritchie, Basketmaker*, 53,
   55; *Aunt Lou Kitchen*, 53; *Aunt
   Tish Hays*, 56; *The Back Stairs
   (The Orphan)*, 13, *14*; *A Book
   of Portraits of the Medical
   Faculty of Johns Hopkins
   University*, 18, 29; *Christopher
   Lewis, Wooten, Kentucky*, 39,
   40–41; *Doris Ulmann and Julia*

Peterkin, 42, 43; *Elizabeth
   Madox Roberts portrait*, 23–25,
   24; *Emery MacIntosh*, 53; *Enos
   Hardin*, 53; *Handicrafts of the
   Southern Highlands* (posthu-
   mous), 72; *Hands of Lucy
   Lakes*, 54; *Hayden Hensley*, 53;
   Lydia Ramsey series, 37; *Man
   with Sacred Image*, 48–50, *49*;
   Melungeon series, 46–47; *Mr.
   and Mrs. William J. Hutchins*,
   62–63; *Mrs. C.O. Wood*, 56–58,
   57; *Nancy Greer, Trade,
   Tennessee*, 31, *32*; *Niles-Virginia
   Howard series*, 64–65; *Old
   Timers' Day*, 64; *Oscar Cantrell*,
   53; *A Portrait Gallery of
   American Editors*, 25–27;
   Richard Walsh portrait, 26;
   Roll, Jordan, Roll, 44; *Scene at
   Georgetown Island, Maine*, 15,
   16; *Sherwood Anderson
   portrait*, 22–23; *III Clouds over
   the Mountain*, 15, 18; *Twenty-
   Four Portraits of the Faculty of
   Physicians and Surgeons of
   Columbia University*, 17–18, 29;
   *Untitled (South Carolina
   fisherman)*, 44, *45*; *Wilma
   Creech*, 59, 60; *Winnie Felther
   series*, 40;
University of North Carolina
   (1930s), 164, 166, 188
University of North Carolina Press
   (1930s), 164–65, 316 n 57
urban America, 10–11, 13–15, *14*,
   35, 48–50, 75, 77–78, *95, 97*,
   134–35, 137, 151, 153–55,
   197–98, 204–6, 208–10, 221–
   24, 251, 252, 262, 264, 265–76,
   *267, 268, 270, 276, 278–79,
   280–83, 282, 284, 285, 288–89,
   295–96*
U.S. Centennial Exhibition (1876), 3
U.S. Congress, 123, 126, 144–45,
   288–89
U.S. Department of Agriculture, 116,
   145, 158, 190
U.S. Department of the Treasury,
   233

U.S. History as Women's History, 2–3, 297
U.S. Public Health Service, 127, 145, 162, 180, 181

VanBeuren Corporation, 231
Vance, Rupert, 164
Van Doren, Carl, 21, 26
Van Dyke, Willard, 90, 99, 100, 102, 309 n 16
Vanity Fair, 231, 263
Van Sweringen family, 204, 319 n 13
Van Vechten, Carl, 41
Vincentini, Noel, 279

Wall Street, 269–70
Walsh, Richard, 26
Warren, Dale, 21, 25–26
Wengler, Harold, 209
West (in the U.S.), 118–23, *119, 122,* 192–94, 241–43, *242*
Weston, Edward, 82, 285
We Take You Now to Caswell County, 190
Whisnant, David, 302 n 29, 305 n 66
White, Clarence H.: and Margaret Bourke-White, 201, 211, 213; and Dorothea Lange, 79–80, 211; and Doris Ulmann, 12, 13, 17, 27, 56, 58, 211
White, Joseph, 199–200, 201, 202
White, Minnie Bourke, 199–200, 201, 210, 213, 217, 221, 225
White, Minor, 271
Wiggins, Myra Albert, 3
Wilder, Thornton, 303 n 35
Williams, Aubrey, 277
Williams, Jonathan, 29, 34
Wolcott, Lee, 192, 194, 299 n 6
Wolcott, Marion Post, 192–95, 300 n 7, 311 n 1. *See also* Marion Post

women in America: "career girl" stereotype of, 69, 133–34, 216, 224, 264, 311 n 56, 319 n 11, 322 n 38; expanding opportunities for, 12, 76, 80, 128–29, 137, 188, 199, 224, 232, 308 n 3; expectations of, 3, 4, 12, 18, 76, 77, 79, 80–81, *85,* 86, 133–34, 137, 145–46, 150–51, 155, 163, 168–71, *169,* 178, 188, 190, 194, 199–200, 203, 208, 222–23, 241, 243, 244–46, *245,* 251, 264, 308 n 5, 309 n 17
women photographers, 2, 297, 299 n 3, 314 n 33; and documentary, 3–4, 180, 244, 252, 264, 291–97; and fieldwork, 68–69, 116–18, 128, 137, 139–40, 145–49, 155, 161–63, 171, 185, 191, 194, 222, 224, 251, 264; Man Ray's treatment of, 255–56; Roy Stryker's treatment of, 117, 139–40, 145–47, 150, 156–59, 162, 191, 206, 222, 224, 294. *See also individual photographers' entries*
Woodward, Ellen, 286
Wootten, Bayard, 73
Workers' Film and Photo League, 135–36, 232
Workers' Photography Movement (Europe), 132–33, 185
Works Progress Administration (WPA), 67, 68, 183, 237–38, 252, 273–75, 277–80, 286–88, 296
World War I, 20–21, 35, 199, 260
World War II, 112, 190, 192, 193–94, 246–48
Wright, Richard, 114–16. Work: *Twelve Million Black Voices,* 114–16